SPECTRAL LINE SHAPES

Previous Proceedings in the Series of ICSLS

Year	Volume	Held in	Publisher	ISBN
2000	11	Berlin, Germany	AIP Conf. Proceedings Vol. 559	1-56396-991-2
1998	10	State College, Pennsylvania	AIP Conf. Proceedings Vol. 467	1-56396-754-5
1996	9	Florence, Italy	AIP Conf. Proceedings Vol. 386	1-56396-656-5
1994	8	Toronto, Canada	AIP Conf. Proceedings Vol. 328	1-56396-326-4
1992	7	Carry le Rouet, France	Nova Science Publishers, Inc.	1-56072-130-8
1990	6	Austin, Texas	AIP Conf. Proceedings Vol. 216	0-88318-791-4

Related Titles from the AIP Conference Proceedings Subseries on Atomic, Molecular, and Chemical Physics

637 Classical Nova Explosions: International Conference on Classical Nova Explosions
Edited by Margarita Hernanz and Jordi José, November 2002, 0-7354-0092-X

636 Atomic and Molecular Data and Their Applications: 3rd International Conference on Atomic and Molecular Data and Their Applications - ICAMDATA
Edited by David R. Schultz, Predrag S. Krstić, and Fay Ownby, October 2002, 0-7354-0091-1

635 Atomic Processes in Plasmas: 13th APS Topical Conference on Atomic Processes in Plasmas
Edited by David R. Schultz, Fred W. Meyer, and Fay Ownby, October 2002, 0-7354-0090-3

584 Resonance Ionization Spectroscopy 2000: Laser Ionization and Applications Incorporating RIS; 10th International Symposium
Edited by James E. Parks and Jack P. Young, August 2001, 0-7354-0024-5

551 Atomic Physics 17: XVII International Conference on Atomic Physics; ICAP 2000
Edited by Ennio Arimondo, Paolo De Natale, and Massimo Inguscio, February 2001, 1-56396-982-3

500 The Physics of Electronic and Atomic Collisions: XXI International Conference
Edited by Yukikazu Itikawa, Kazuhiko Okuno, Hiroshi Tanaka, Akira Yagishita, and Michio Matsuzawa, February 2000, 1-56396-777-4

To learn more about these titles, or the AIP Conference Proceedings Series, please visit the webpage **http://proceedings.aip.org/proceedings**

SPECTRAL LINE SHAPES

16th International Conference on
Spectral Line Shapes
Volume 12

Berkeley, California 3-7 June 2002

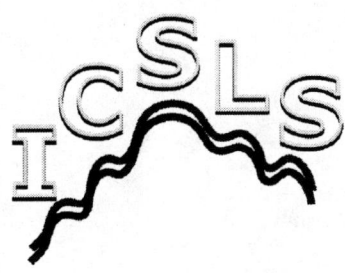

EDITOR
Christina A. Back
Lawrence Livermore National Laboratory
Livermore, California
The University of California, Davis
Davis, California

SPONSORING ORGANIZATIONS
The University of California, Davis
Lawrence Livermore National Laboratory

Melville, New York, 2002
AIP CONFERENCE PROCEEDINGS ■ VOLUME 645

Editor:

Christina A. Back
Lawrence Livermore National Laboratory
L-21
P.O. Box 808
Livermore, CA 94551
USA

E-mail: tinaback@llnl.gov

L.C. Catalog Card No. 2002114215
ISBN 0-7354-0100-4
ISSN 0094-243X
Printed in the United States of America

CONTENTS

LOW-DENSITY SOURCES

HIGH-DENSITY SOURCES

X-RAY LASERS

ULTRACOLD SYSTEMS

APPENDICES

Preface

The 16th International Spectral Line Shape Conference was held in Berkeley, CA, June 3-7, 2002, at the Clark Kerr Campus of the University of California. This conference is an international forum to discuss fundamental and applied research associated with the formation of spectral line profiles. It includes research involving line profiles observed in absorption, emission, and scattering by laboratory and astrophysical sources. Areas covered in the conference included: laser-produced plasmas, magnetically confined plasmas, stellar atmospheres, molecular and atomic systems, high-resolution spectroscopic applications and measurements, collision-induced effects, and ultracold regimes. Papers on experiments, theory, and applications were presented.

This conference is held every two years, alternating between Europe and the U.S. The last two conferences were held in Berlin, Germany (2000), and State College, PA, USA (1998).

Conference participants represented fourteen countries and presented sixty-eight papers. Unfortunately, due to the events of 9/11 some people were unable to get visas in time to come for the conference. However, some of those absent were able to submit their contributions for the proceedings. Invited talks were 30 minutes and contributed talks were 20 minutes. Other contributions were presented at two poster sessions. Sixty-one manuscripts are included in this conference proceedings.

The invited papers were chosen to pair theoretical and experimental papers as much as possible. The conference had no parallel sessions to facilitate more cross-fertilization between people of different disciplines. There was an emphasis on highlighting younger speakers in an effort to encourage those up and coming in their respective fields. The conference also tried to rebalance the program to recognize new efforts in astrophysics and ultracold physics.

The proceedings group the line shape themes into seven broad categories: Low-Density Sources, Astrophysical Plasmas, Atomic and Molecular Interactions, High-Density Sources, Innovative Theory and Experimental Techniques, X-ray Lasers, and Ultracold Systems. The Low-Density and High-Density source categories are imperfect descriptors and were generally chosen to include plasmas below and above densities of 10^{19} cm^{-3}, respectively.

We gratefully thank the sponsoring institution, the University of California, Davis for administrative and organizational support. In particular, we acknowledge the help of the conference administrator, Estelle Miller, who was the central contact, and Jane Keene, who provided assistance during the conference. Without them — both of the University of California, Davis — the conference would have been difficult to accomplish. The chair would also like to personally thank the local organizing committee for their help and advice. Finally, we also thank Lawrence Livermore National Laboratory for financial contributions.

The next conference will be chaired by Dr. Elisabeth Leboucher-Dalimier during the summer of 2004 in Paris, France. We offer our best wishes to her and her colleagues in the successful continuation of this conference.

Christina A. Back
Livermore, CA
10 September 2002

International Committee:

Roland Stamm	Université de Provence, Marseille roland.stamm@piimdgp.univ-mrs.fr	(France)
Chantal Stehlé	Observatoire de Paris, Meudon chantal.stehle@obspm.fr	(France)
William Stwalley	University of Connecticut stwalley@uconnvm.uconn.edu	(USA)
Jozef Szudy	Nicholas Copernicus University, Torun szudy@phys.uni.torun.pl	(Poland)
George Tabisz	University of Manitoba tabitsz@cc.umanitoba.ca	(Canada)
Richard Tipping	University of Alabama rtipping@bama.ua.edu	(USA)

Local Organizing Committee:

Mark Adams	Massachusetts Institute of Technology adams@mit.edu	(USA)
Kevin Fournier	Lawrence Livermore National Laboratory fournier2@LLNL.gov	(USA)
Carlos Iglesias	Lawrence Livermore National Laboratory iglesias@LLNL.gov	(USA)
Roberto Mancini	University of Nevada, Reno rman@physics.unr.edu	(USA)
Hyun-Kyung Chung	Lawrence Livermore National Laboratory hchung@LLNL.gov	(USA)
Howard Scott	Lawrence Livermore National Laboratory hascott@LLNL.gov	(USA)

LOW-DENSITY SOURCES

Characterization of Impurities in Tokamak Divertor Plasmas from Analysis of Spectral Profiles

R. C. Isler[*], N. H. Brooks[†], and B. Zaniol[‡]

[*]Oak Ridge National Laboratory, Oak Ridge, TN 37831
[†]General Atomics, San Diego, CA 92186-5608
[‡]Consorzio RFX, Assoociazone EURATOM-ENEA sulla Fusione, Padova, Italy,35127

Abstract. Studies of the production, transport, and radiative losses of impurities in present-day tokamak divertors provide input necessary for the design of future burning- plasma machines. Several types of information rely on detailed analysis of emission profiles. These include ion temperatures, ion flows along field lines, and impurity production mechanisms. Temperatures and flows are determined from Doppler broadening and shifts by comparing measured line shapes to theoretical profiles that include the nonlinear Zeeman/Paschen-Back effect. The two major production mechanisms for atomic carbon are physical and chemical sputtering. These processes can be distinguished by comparing atomic and molecular fluxes, which requires modeling the band emissions of CD and C_2. They can also be differentiated from measurements of effective temperatures of C I (best profile fits to thermal distributions). Careful inspection of profiles that give high effective temperatures reveals that they are not actually Gaussian but have asymmetries and shifts that can be correlated to energy distributions expected for physical sputtering. Examples of all these applications are discussed in this review.

I. INTRODUCTION

In most large present day tokamaks the particle and heat exhaust from core plasmas is channeled onto divertor targets rather than being allowed to dissipate more uniformly on the vacuum chamber walls. The targets, usually graphite tiles, are a major source of impurities that can produce deleterious effects if they migrate back into the core. Consequently, extensive efforts have been made to characterize the impurity behavior in divertors in as much detail as possible. The major tool for these studies is optical spectroscopy in both the visible and vacuum ultraviolet regions. The vacuum region is most useful for evaluating the radiated power from individual ions. The visible region is more useful for understanding the dynamics of the edge plasma, particularly through the analysis of emission profiles. Ion temperatures and flows along field lines are routinely measured for C II, C III, and B II from Doppler broadening and shifts. The line splitting caused by the magnetic field can be larger than the thermal broadening and must be taken into account accurately. During high recycling operation, flows of C V (He-like carbon) can also be evaluated using transitions between highly excited levels in the Li-like ions, which are populated by charge exchange from thermal deuterium. Visible spectroscopy is also employed to

CP645, *Spectral Line Shapes: Volume 12, 16th ICSLS*, edited by C. A. Back
© 2002 American Institute of Physics 0-7354-0100-4/02/$19.00

determine the release mechanisms for carbon (primarily physical and chemical sputtering) both by comparing influxes of CD and C_2 to the influx of C I, and by detailed analysis of the line profiles of C I.

II. INSTRUMENTATION

All the data presented here have been obtained from the DIII-D tokamak at General Atomics [1]. A schematic cross section of this toroidal machine is shown in Fig. 1a. Open and closed magnetic flux surfaces are indicated respectively by solid and dashed lines. Also shown are the quasi-vertical spectrometer views onto the divertor target. It is worth noting that the toroidal field strength varies as 1/R, where R is the major radius of the machine, so the closer the view to the major axis, the greater the spectral line splitting. An enlargement of the divertor region is shown in Fig. 1b, together with a portion of the separatrix that forms the boundary between open and close field lines. Also shown are five quasi-tangential views, which appear as hyperbolic segments when projected onto a poloidal plane. These views are used to measure the rapid ion flow velocities along the open field lines. Deuterium is employed as the operating gas, and carbon is the dominant impurity.

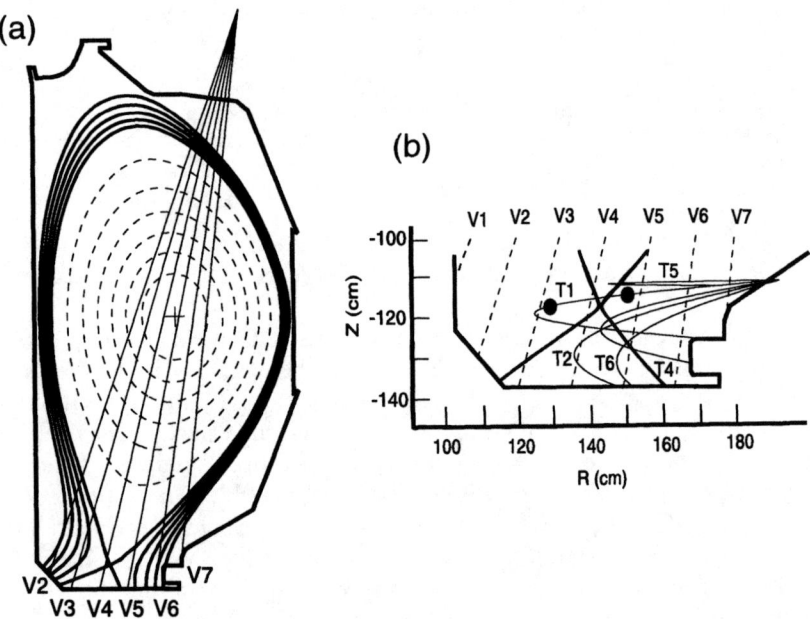

FIGURE 1. (a) Schematic diagram of a cross section of the DIII-D tokamak. Closed flux surfaces are shown by dashed lines, open surfaces by solid lines. The quasi-vertical spectrometer views are labeled V2-V7. (b) Enlargement of the divertor region showing typical quasi-tangential views which project as hyperbolic segments in a poloidal plane. R and Z are distances from the major axis and the midplane.

Spectra in the 4000 Å to 9000 Å range are acquired by means of a 1.3 m Czerny-

4

Turner spectrometer equipped with a 1200 l/mm grating. A two-dimensional charge-coupled device (CCD) simultaneously records the input signals transmitted from the vacuum vessel to the spectrometer through twelve 50 m optical fibers. The integration time is typically 125 ms. Pixel widths are 22μ, and the dispersion varies from 0.14 Å/pixel to 0.11 Å/pixel over the accessible wavelength range. At 9100 Å, the vicinity of the strongest C I lines, the instrument width corresponds to a temperature of approximately 1 eV. The system is absolutely calibrated to measure intensities.

III. ION TEMPERATURES

Ion temperatures are deduced, as usual, from Doppler widths of spectral lines, but a straightforward fit to a single Gaussian function is seldom possible. In magnetic fields of 1 to 5 T, which are typical of most tokamaks, splitting of the Zeeman components is insufficient to resolve individual transitions. Nevertheless, the magnitude of these fields is great enough to cause observable mixing within some multiplets of atomic states having the same value of magnetic quantum number m and differing in total angular momentum quantum number J by ±1. Temperatures are ordinarily determined by fitting measurements of the entire multiplet structure to theoretical calculations that employ the nonlinear Zeeman/Paschen-Back effect, i.e., by diagonalizing the matrix representation of the spin-orbit and magnetic interactions [2],

$$M_{mm'}^{JJ'} = \langle J'm' \mid \xi \vec{L} \cdot \vec{S} + \frac{\beta}{\hbar}\vec{B} \cdot (\vec{L} + g_s\vec{S}) \mid J\,m \rangle. \tag{1}$$

The fine structure increments are known very accurately, and can be obtained from standard compilations [3].

Figure 2 shows measured and calculated line shapes for the ^{3}P – ^{3}S transitions of B II near 7035 Å. Boron remains on the surface of the carbon tiles following glow discharges in a mixture of helium and diborane, which is employed to reduce

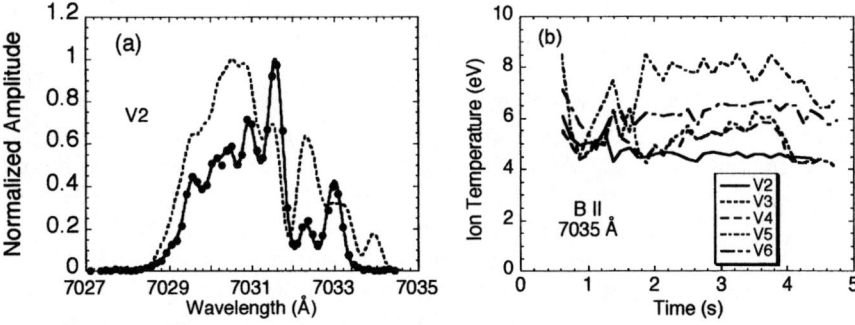

FIGURE 2. (a) Solid line: fit to measured B II data (filled circles) along view V2 using the nonlinear Zeeman/Paschen-Back effect. Dashed line: calculated profile using the low-field, linear approximation (Zeeman effect). (b) Time histories of B II ion temperatures along several views.

deuterium recycling. The solid line is the profile obtained from the exact calculation of magnetic field effects and fit the measured data quite well. In contrast, the dashed line is a low-field (Zeeman effect) approximation obtained by ignoring the off-diagonal matrix elements of Eq. (1), and the fit to the data is very poor. This discrepancy is more severe for B II than for the other multiplets usually monitored in DIII-D owing to the relatively small fine-structure intervals.

Ion temperatures in the inner divertor leg are between 4 and 6 eV, a range that is also typical for C II. They attain values as high as 8 eV along V5, which comes close to the outer strike point where the separatrix touches the carbon tile (see Fig. 1a). It is not certain that singly charged ions equilibrate with the background deuterons before being ionized to the next stage. Some studies suggest that the lower temperature range is an indicator of conditions in which chemical sputtering is the main production mechanism, whereas temperatures above 8 or 10 eV point to physical sputtering becoming an important process (see Sec. V). Similarly, C III temperatures tend to fall into two groups depending on operating conditions, 10 to 15 eV and 20 to 30 eV.

IV. PARALLEL FLOWS

Knowledge of carbon transport in the divertor and in the rest of the open-field line, scrape-off layer is the key to understanding how impurities can contaminate the core plasma. In order to develop a comprehensive picture of the transport it is necessary to construct codes that model the complicated plasma behavior because the amount of information that can be gained directly from experiment is limited. Data from Langmuir probes and from spectroscopy is used to benchmark the codes. Measurements of flow velocities parallel to the magnetic field are particularly useful for this purpose. They are determined from Doppler shifts of spectral multiplets observed along quasi-tangential views such as shown in Fig. 1b [4]. The same multiplet in a vertical view serves as the fiducial against which the shift is measured. This approach is feasible because the only shifts that might occur in the vertical views come from relatively slow cross-field convection.

Low Ionization Stages

Theory predicts that flow velocities along the open field lines may be directed either toward or away from the divertor target depending on location, and experiments support this conclusion [5]. Figure 3 illustrates typical data. A profile of the 6579 Å, $^2P - ^2S$ transition of C II, as seen from a vertical view, is shown in Fig. 3a. The centroid of the 3/2 – 1/2 component is plotted as a reference. A profile from view T6 is illustrated in Fig. 3b. Since the tangential views are nearly parallel to the field lines, the π-components are suppressed. The separation of the σ-components indicates the radial location of the emitting volume, owing to the 1/R dependence of the toroidal magnetic field. The pattern is blue shifted as observed by the fact that the minimum of the 3/2 – 1/2 feature does not coincide with the centroid calculated from the vertical view. The observed shift only provides the velocity component along the observer's line of sight. It is necessary to convert this number to a parallel flow velocity by using magnetic equilibrium calculations (EFIT's), which can provide the total field strength and direction cosines at all points. The data of Fig. 3b are fit quite well by assuming a single, average flow speed of 6×10^5 cm/s. However, the profiles of Fig 3c, which are recorded along view T3, can only be explained by the presence of two separate groups of radiating ions. One of these gives rise to a red shift (reversed flow) near the separatrix and a more intense group gives rise to a blue shift (normal flow) several cm farther out in the scrape-off layer. Figure 4 illustrates the changes in the observed line profiles as the separatrix is swept to larger radii during the discharge.

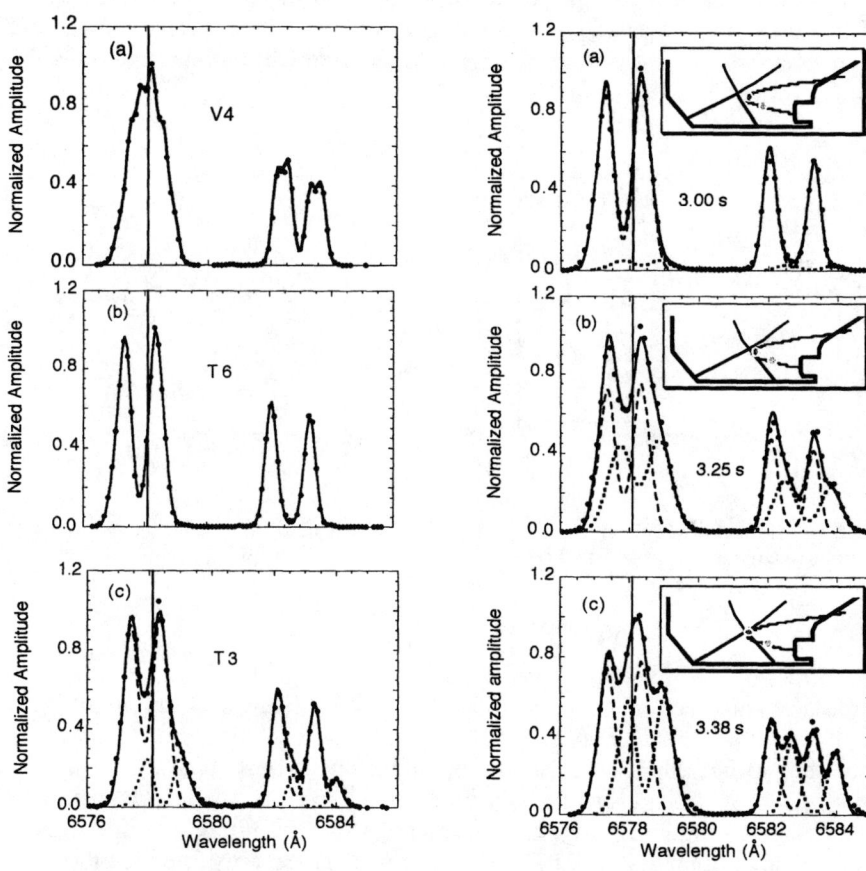

FIGURE 3. (a) Profile of 6579 Å doublet of C II. The vertical line shows the centroid of the low–wavelength feature. (b) a single blue shifted component along T6. (c) Blue- (dashed) and red- (dotted) shifted components observed along T3.

FIGURE 4. Normal (dashed line) and reversed (dotted line) components of C II as a function of separatrix position. Locations of emitting sources are shown by circles on the tangential viewing chord T4.

The reversed-flowing group of ions is detected at the tangency point of the view so some details of its spatial characteristics are probed during the sweep. It is obvious that the C II density grows and the flow speed increases close to the separatrix.

During operation at high densities the plasma near a divertor target may fall to 1 eV or less. At these low values, electrons and deuterons recombine and the ion current to the target disappears. The plasma is said to detach, or to partially detach if the condition occurs only over a segment of the divertor leg close to the separatrix [6]. Time histories of C II flows upstream of the recombining region are shown in Fig. 5 for a plasma which partially detaches at 2.5 s. Flows toward the target (Fig. 5a) indicate a clear rise at this time. It is believed that this effect results from a decrease of ion temperature and density gradients along the field lines and a concurrent reduction of neoclassical forces, which tend to force the particles in the divertor toward the midplane. However, corresponding decreases of the reversed flow (Fig. 5b) are not so broadly apparent, and more detailed modeling will be required to obtain an accurate interpretation of the results.

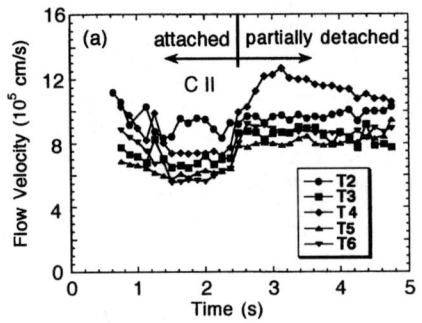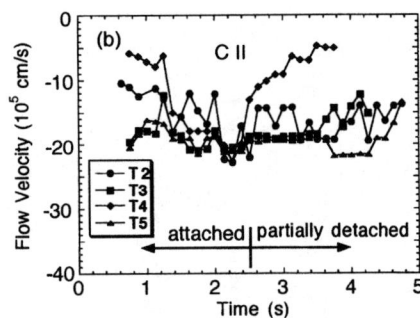

FIGURE 5. (a) Normal flow velocities in a discharge that partially detaches at 2.5 s. (b) reversed flows near the separatrix in the same discharge.

Flows of He-like Carbon

In order to obtain a complete picture of carbon in the divertor or scrape-off layer of DIII-D, it would be desirable to document all ionization stages for different types of operation, but so far, this has not proven possible. The lowest stages, C I – C III, are excited by electron collisions and are easily observed, but it is difficult to characterize the higher stages. Large transition probabilities for vacuum ultraviolet lines leave the accessible, visible branches of C IV relatively weak. The energy levels of the He-like and H-like ions, C V and C VI, lie too high above the ground state to be appreciably excited by the low temperature electrons of the divertor. But certain transitions between high n-levels of C IV do appear anomalously bright with complex line shapes in high recycling discharges [7]. It has been shown that these arise from charge exchange of deuterium in the $n = 2$ level with C V, and lead to preferential populations of high angular momentum states of C IV in the $n = 5 - 7$ levels [8]. In principle, these transitions can be analyzed to obtain temperatures and flow velocities of the He-like stages.

An example of data from a vertical view is shown in Fig. 6a, together with a spectrum of the underlying transitions that has been calculated without assuming any broadening mechanisms. Each of the three transitions exhibits the usual Paschen-Back pattern because the fine structure intervals are much smaller than the line splitting in the magnetic field. They are weighted to get the best fit to the data, but these weightings agree very well with relative intensities derived from Classical Trajectory Monte Carlo (CTMC) computations [8]. The assumption that thermal motion is the only broadening mechanism does not give adequate fits. A convolution of Gaussian and modified Lorentzian profiles has been employed. Although the angular momentum states are not degenerate, it appears that they may be close enough in energy that Stark broadening could affect the line wings. Unfortunately, the most appropriate mixture of the two profiles to get a good fit is not uniquely determined, so ion temperatures cannot be extracted.

This problem does not influence the shift measurements. Data from tangential chord T1 (Fig. 1b) which penetrates into the inner leg of the divertor is illustrated in Fig. 6b. Both blue- and red-shifted components contribute to the total signal. As in previous examples, the blue-shifted component indicates normal flows toward the target at 1.8×10^6 cm/s in the outer leg. But here, the red-shifted component in the inner leg also indicates normal flow at 1.6×10^6 cm/s, since the reversed direction of the poloidal field from the tokamak current changes the pitch of the total magnetic field relative to the observation direction. So although calculation of the theoretical line shapes from C IV can be somewhat involved, it is feasible to measure the flows of the He-like carbon ions, which are predicted to be the dominant impurity species in the scrape-off layer.

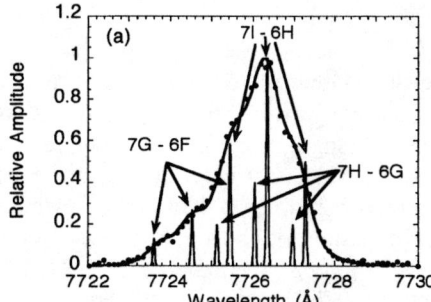 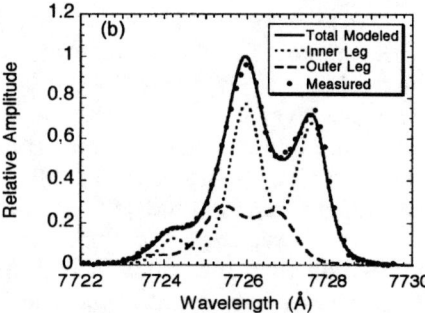

FIGURE 6. (a) n = 7 – n' = 6 transitions of C IV excited by charge exchange with deuterium in the n = 2 level. Data is from vertical view V5. (b) Profile of the same transition from a tangential view showing the difference of component shifts between the inner and outer legs.

V. CARBON PRODUCTION MECHANISMS

A great deal of research has been directed toward studying plasma-graphite interactions that relate to carbon release in fusion devices. The most important of these mechanisms are physical sputtering, chemical sputtering, radiation-enhanced sublimation, and thermal sublimation. Conditions favoring the latter two processes are not usually realized in the DIII-D tokamak, so efforts to identify the active processes on the carbon tiles of the divertor targets have been aimed at distinguishing physical and chemical sputtering. The procedure relies on analyzing spectroscopic data to relate influxes of CD, and C_2 to the influx of C I and on correlating effective temperatures and detailed profiles of C I to those expected for the two different production mechanisms [9].

Fluxes

The C I influx provides a measure of the total amount of carbon released from the divertor targets. It can be produced directly by physical sputtering or by dissociation of molecules generated by chemical sputtering such as CD_4 or C_xD_y where x is 2 or 3 and y ranges from 2 to 6. Deciding whether or not chemical sputtering is the dominant process for a given type of operation depends on using the measured CD and C_2 fluxes to infer what fraction of the observed C I results from production of molecules. The procedure is described in detail in Ref. 9. It relies on a mixture of theoretical [10,11] and experimental [12] results to obtain the expression

$$\Gamma_{CI}^{mol} = 52 \times \Gamma_{C_2} + (\Gamma_{CD} - 8 \times \Gamma_{C_2}). \qquad (2)$$

The first term on the right-hand side of Eq. (2) accounts for production of C_2D_y and C_3D_y and the second for production of CD_4 on the carbon tiles.

Because of spectral interferences, it is not feasible to make direct measurements of the total intensity of the molecular bands. In order to determine fluxes, measurements are made over a limited wavelength range, which permits the entire band to be modeled. Figure 7a shows the emission in the wavelength range that encompasses the (0,0) and (1,1) Q-heads of the 4300 Å system of CD, and Fig. 7b shows the best calculated simulation the data. They are very similar, but some details do not match. Measured wavelengths are employed for the calculation; they give a much better correlation than those obtained from tabulated molecular constants. Some of the discrepancies may result from insufficient accuracy in the measurements, but weak lines of D_2 may also be present. Typical rotational temperatures of CD are in the range of 0.2 – 0.3 eV. The (0,0) Swan band of C_2 (Fig. 7c) is well reproduced by modeling that makes use of the accurately known molecular constants.

Two sets of flux measurements are shown in Fig. 8. In the first, the estimates of C I produced from chemical sputtering are basically equal to the total measured influx of C I, an implication that all the carbon release results from chemical sputtering. In the second example, chemical sputtering seems important only in the very early stages of

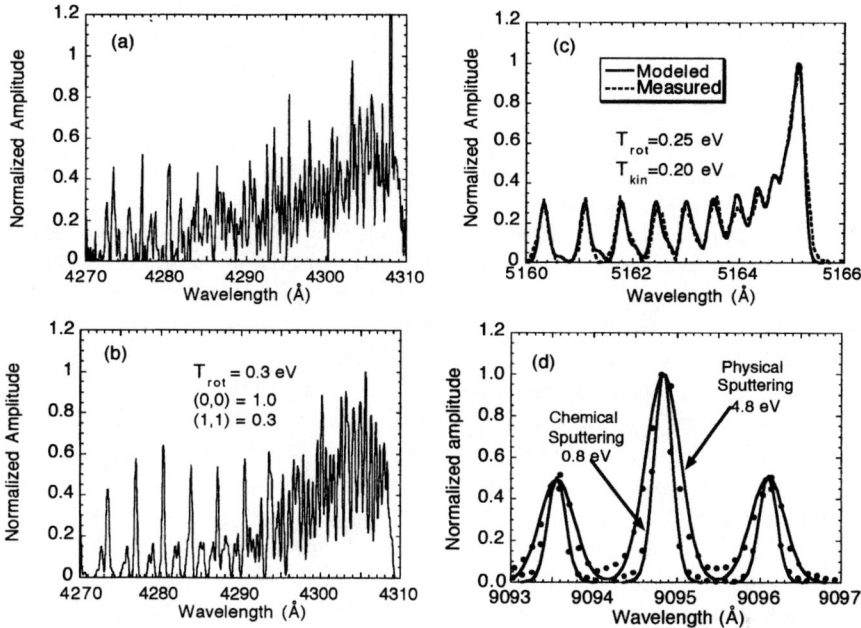

FIGURE 7. (a) Measured spectra around the Q-heads of the (0,0) and (1,1) bands of the CD 4300Å system. (b) Modeled spectrum that gives the best agreement with (a). (c) Measured and modeled spectral region around the (0,0) band of the C_2 Swan system. (d) Best Gaussian fits to C I lines that appear to result from either physical or chemical sputtering.

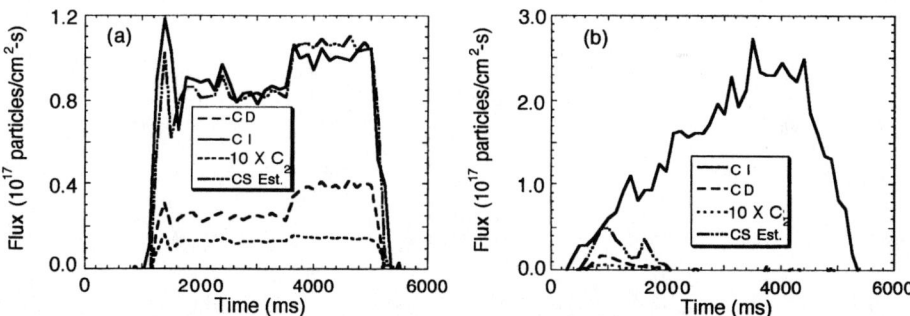

FIGURE 8. (a) Measured fluxes of C I, CD, and C_2 with an estimate of C I from chemical sputtering using Eq, (2). The total C I influx appears to come from chemical sputtering. (b) Same as (a) but the total C I influx appears to come from physical sputtering after 2000 ms.

the discharge; physical sputtering dominates through most of the shot. Effective C I temperatures, shown in Fig. 7d, are obtained as best fits to Gaussian line shapes and tend to support the identification of the carbon production processes. Whenever the flux analysis indicates the principle mechanism is chemical sputtering, the C I temperatures are in the range of 1.0 ± 0.2 eV as expected from a molecular-dissociation source. When physical sputtering is inferred, the effective temperatures are $5 - 8$ eV, a range that is more in line with expectations for ejection of carbon by fast particle impact [13].

C I Profiles

On close examination, the C I line profiles yielding high effective temperatures are seen to be asymmetric and to have peaks shifted from the tabulated line center ($\Delta\lambda=0$). Asymmetries are expected for physically sputtered particles if the thermal relaxation times with the background gas are longer than the ionization times. For views perpendicular to a flat surface, all the atoms move toward an observer, so the profile should have no red wing but still peak at the nominal line center. Since the views into the DIII-D divertor are not strictly perpendicular, the profiles should exhibit a weak red wing and a slight shift of the peak position. These characteristics result from the fact that, in a non-normal view, a small fraction of the particles traveling nearly parallel to the surface is eliminated from the flux moving toward the observer and appears as flux moving away from the observer. These are particles that contribute to the profile near the nominal line center, so a shift in the observed peak is expected. Additional shifts in the peak position may arise if the angular distribution is not isotropic above the surface. For example, a Thompson sputtering distribution[13] for normal-incidence impact is supposed to vary as $\cos\theta$, so even for perpendicular viewing, the line shape should have zero-intensity at $\Delta\lambda=0$.

In order to determine if the measured C I profiles were consistent with physical sputtering, they have been compared to the profiles expected from a Thompson distribution with a "soft" cutoff factor [14],

$$f(E)dE = \frac{E}{\left(E + U_0\right)^3} \, g(\theta) \left[1 - \left(\frac{E + U_0}{\gamma\left(\gamma - 1\right)E_{imp}}\right)^{1/2}\right] dE \qquad (3)$$

$$\gamma = \frac{4 m_D m_C}{\left(m_D + m_C\right)^2}. \qquad (4)$$

U_0 is the binding energy of carbon in the surface, 7.4 eV, and m_D and m_C are the atomic masses of deuterium and carbon.

The measured profile is first deconvoluted using a maximum entropy routine to obtain the most likely emission line shape [15]. An example is shown in Fig. 9a. for the most intense component of the $^3P - {}^3P^o$ multiplet. The line is clearly asymmetric around the peak, which is shifted -0.051 Å from the nominal center or about 1/2 the width of a pixel on the CCD camera. The shift measurements are believed to be accurate to ± 0.01 Å. Equation (3) can be converted into a distribution in $\Delta\lambda$ by using the Doppler shift relationship, $\Delta\lambda/\lambda_0 = v \cdot \cos\theta/c$, and integrating over θ.

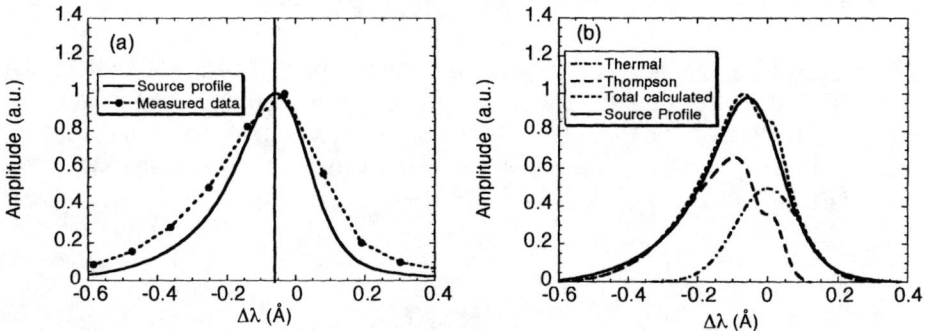

FIGURE 9. (a) Measured and deconvoluted profile of the brightest line of the $^3P - ^3P^o$ multiplet of C I. (b) Comparison of deconvoluted profile with the sum of Thompson and thermal distributions.

The deconvoluted profile and the constituents used to model it are shown in Fig. 9b. The blue wing is well described by the Thompson distribution with an impact energy of 140 eV and with $g(\theta) = 1$, but the red wing is far too prominent to come solely from physical sputtering [16]. If the observed profile is assumed to be a superposition of the Thompson profile and a thermal profile with a temperature of 1.2 eV, the overall fit is quite good. The implications of these results are not completely clear. The sheath potential at the divertor targets is estimated to be 110 to 120 eV based on Thomson scattering measurements, which is reasonably consistent with the inferred impact energy. The temperature of the thermal distribution is similar to that expected from molecular dissociation, but molecular bands are not observed in this experiment. So while it is tempting to associate the line fitting results with a combination of physical and chemical sputtering, more investigations are needed. In particular, a calculation of the thermal relaxation times of physically sputtered carbon ions could prove quite useful for sorting out the uncertainties of interpretation.

VI. SUMMARY

Analysis of spectral emission profiles has provided a wide range of information concerning impurities in tokamak divertors. By using Doppler widths and shifts, ion temperatures can be determined to an accuracy of 1 eV and parallel flows to an accuracy of 10^5 cm/s. The existence of theoretically predicted reversed-flow regions (away from the target plates) have been verified under a wide variety of operating conditions. Correlations of CD, and C_2 fluxes with C I fluxes are used to distinguish between physical and chemical sputtering as release mechanisms. Effective temperature measurements combined with detailed analysis of non-thermal profiles of atomic carbon support the inferences obtained from flux studies and can be exploited to gain insight into carbon production processes.

ACKNOWLEDGMENTS

This work has been supported by US DOE Fusion Contract DE-AC05-00OR22725, DE-AC03-99ER54463. Figure 2 has been reprinted from the Journal of Nuclear Materials, **266-269**, R. C. Isler *et al.*, "Spectroscopic measurements of impurity temperatures and parallel ion flows in the DIII-D divertor", p. 376, Copyright 1999, with permission of Elsevier Science.

REFERENCES

1. Luxon, J. , Anderson, P., Batty, F., et al., *Plasma Physics Controlled Nuclear Fusion Research*, Proceedings of the 11th International Conference, Tokyo, 1986 (International Atomic Energy Agency, Vienna, 1987) Vol. I, p. 159.
2. Isler, R. C., Wood, R. D., Klepper, C. C., Fenstermacher, M. E., and Leonard, A. W., *Phys. Plasmas* **4**, 355 (1997).
3. Bashkin, S and Stoner, J. O. Jr., *Atomic Energy Levels and Grotriann Diagrams 1* (North Holland, Amsterdam, 1975).
4. Isler, R. C., Brooks, N. H., West, W. P., Leonard, A. W., McKee, G. R., and Porter, G. D., *Phys. Plasmas* **6**, 541 (1999).
5. Porter, G. D, Isler, R. C., Boedo, J., and Rognlein, T. D., *Phys. Plasmas* **7**, 3663 (2000).
6. Petrie, T. W., Maingi, R., Allen, S. L., Buchenauer, D. A., Hill, D. N., and Lasnier, C. J., *Nucl. Fusion* **37**, 643 (1997).
7. Tunklev, M., Engström, I., Jupén C., and Kink, I., *Physica Scripta* **55**, 707 (1997).
8. Zaniol, B., Isler, R. C., Brooks, N. H., West, W. P., and Olson, R. E., *Phys. Plasmas* 8, 4386 (2001).
9. Isler. R. C., Colchin, R. J., Brooks, N. H., Evans, T. E., West, W. P., and Whyte, D. G., *Phys. Plasmas* **10**, 4470 (2001).
10. Erhardt, A. B., and Langer, W. D., "Collisional Processes of Hydrocarbons in Hydrogen Plasmas", Princeton Plasma Physics Laboratory report PPPL-2477, Sept. 1987.
11. Alman, D. A., Ruzic, D. N., and Brooks J. N., *Phys. Plasmas* 7, 1421 (2000).
12. Pospieszczyk, A., Ra, Y., Hirooka, Y., Conn, R. W., Goebel, D. M., LaBombard, B., and Ngyren, R. E., "Spectroscopic studies of carbon containing molecules and their breakup in PISCES-A", University of California at Los Angeles report UCLA-PPG-1251, Dec. 1989.
13. Thompson, M. W., *Phil. Mag.* **18** , 377 (1968).
14. McCracken, G. and Stott, P., *Nucl. Fusion* **19** , 89 (1979).
15. Frieden, B. Roy, *J. Opt. Soc.* **62**, 511 (1972).
16. Isler, R. C., Colchin, R. J., Brooks, N. H., Evans, T. E., West, W. P., Whyte, D. G., submitted for publication in the *Proceedings of the 15th Conference on Plasma Surface Interactions*, Gifu, Japan, May 27 – 31, 2002.

Line Shape Analysis of Particle Dynamics in Tokamak Edge Plasmas

L. Godbert-Mouret[1], R. Guirlet[2], A. Escarguel[1], M. Koubiti[1],

Y. Marandet[1], R. Stamm[1], H. Capes[2].

[1] *PIIM, UMR 6633 CNRS-Université de Provence, centre de St Jérôme,*
F-13397 Marseille cedex 20, France
[2] *Département de Recherche sur la Fusion Contrôlée, Association Euratom-CEA, CEA Cadarache,*
13108 Saint-Paul lez Durance cedex, France

Abstract. Characterizing edge plasmas is an important issue for a better control of the plasma wall interaction, and the recycling of thermonuclear fuel. We use line profiles emitted by hydrogen isotopes, helium and impurity ions, as a non intrusive probe for diagnosing the density and temperature of edge particles. An analysis of line wings of Dα has been performed with the aim of investigating the dynamics of edge particles.

1. INTRODUCTION

Plasma conditions in the edge strongly affect the operation of magnetic fusion devices. This interface region is expected simultaneously to recycle the particles flowing from the core plasma, to protect the wall from erosion, and to control the impurity contamination. These requirements are often achieved in Tokamaks by the existence of a divertor, a device creating a volume containing a partially ionized, strongly radiative plasma. Observing and modeling this radiative plasma is essential for improving the operation of such magnetic fusion devices, and many works in plasma spectroscopy [1,2,3,4] have been devoted to the characterization of edge plasmas in recent years. Spectroscopy is indeed a powerful and non intrusive tool to diagnose temperatures, densities, and particle fluxes.

Spectral line shapes emitted in the edge plasma carry a wealth of information on the plasma status and parameters. We have analyzed the line shapes for low principal quantum number transitions of hydrogen isotopes, helium and carbon neutrals and ions, with the aim of understanding the shaping of the line due to the presence of several populations of emitters, and of the perturbations by electric and magnetic fields. For recombining plasma conditions (with electron temperature lower than 2 eV), the analysis of lines with high principal quantum number near the limit of the deuterium Balmer series, significantly improves the temperature and density diagnostic.

Edge plasmas are interesting objects of study from the physics point of view since they are generally out of equilibrium due to the presence of sources of particles, and of

CP645, *Spectral Line Shapes: Volume 12, 16th ICSLS*, edited by C. A. Back
© 2002 American Institute of Physics 0-7354-0100-4/02/$19.00

strong density and temperature gradients. We investigate possible signatures of non thermal effects on the line shape, and discuss several plasma spectroscopy models which may be applied to these plasmas.

2. LINE SHAPE MODEL

To calculate line shapes, we use a model able to take into account fine structure, Zeeman, Stark and Doppler effects. The line shape results from different effects due to surrounding particles and external fields on the emitter. In presence of an external magnetic field, the light emitted by atoms or ions is polarized, and the profile depends on the angle of observation with respect to the magnetic field.

The line shape $I_{\hat{e}}(\omega)$ for radiation polarized along the unit vector $\hat{e}$, is given by a one sided Fourier transform of the dipole autocorrelation function :

$$I_{\hat{e}}(\omega, \vec{B}) = \frac{1}{\pi} \mathrm{Re} \int_0^\infty C_{\hat{e}}(t, \vec{B}) e^{i\omega t} dt , \qquad (1)$$

where $C_{\hat{e}}(t, \vec{B})$ is defined with the relation :

$$C_{\hat{e}}(t, \vec{B}) = \left\langle \left\langle \hat{e}.\vec{d} \left| \left\{ U_E(t, \vec{B}) \right\}_E \right| \hat{e}.\vec{d} \right\rangle \right\rangle . \qquad (2)$$

In this expression, written in the Liouville space formalism, $\vec{d}$ is the dipole operator, $U_E(t, \vec{B})$ is the time evolution operator of the emitter in fixed electric and magnetic fields $\vec{E}$ and $\vec{B}$.

The symbol $\{....\}$ stands for an average over the norm of electric field and its angle with the magnetic field. The time evolution operator can be formally written as :

$$U_E(t, \vec{B}) = e^{-i(L_E(\vec{B}) - i\Phi)t} , \qquad (3)$$

where $L_E(\vec{B})$ is the Liouville operator related to the Hamiltonian H in Hilbert space by :

$$L_E(\vec{B}) = \frac{1}{\hbar} \left\{ H \otimes I^d - I^d \otimes H \right\}. \qquad (4)$$

In this equation, superscript d means dual space operator. The purpose for the use of Liouville space formalism is to treat radiating atomic systems whose upper and lower levels are simultaneously perturbed.

The total Hamiltonian H describing the emitter can be written as :

$$H = H_0 + H_{FS} - \vec{d}.\vec{E} - \mu_B (\vec{L} + 2\vec{S}).\vec{B} . \qquad (5)$$

H_0 is the Hamiltonian of the isolated emitter, H_{FS} is the fine structure term, and $\vec{d}.\vec{E}$ is the Stark interaction potential due to electrons and ions. The last term represents the Zeeman effect, and contains the Bohr magneton μ_B, the orbital angular momentum $\vec{L}$ and the spin $\vec{S}$.

The linear Zeeman effect is taken into account without any approximation in our code. Using the quantum numbers L, S, J, M_J, the elements of the magnetic term can be written :

$$\frac{1}{B\mu_0}\langle \alpha LSJM_J | H_{mag} | \alpha LSJM_J \rangle = M_J \left\{ 1 + (g_s - 1)\frac{J(J+1) + S(S+1) - L(L+1)}{2J(J+1)} \right\} \quad (6)$$

for diagonal terms ; for non-diagonal terms it reads :

$$\frac{1}{B\mu_0}\langle \alpha LSJM_J | H_{mag} | \alpha LSJ\pm 1 M_J \rangle = (1-g_s)\left\{ \frac{(J^2 - M_J^2)(J+L+S+1)(J+L-S)(J+S-L)(L+S-J+1)}{4J^2(2J-1)(2J+1)} \right\}^{1/2} \quad (7)$$

In both equations, g_s stands for the Lande factor.

We choose a reference frame with the z axis along the magnetic field, the angles between the z axis and the direction of observation (resp. the direction of electric field) being denoted by θ_0 (resp. by θ) [5,6]. It is possible to define two intensities observed in directions parallel and perpendicular to the magnetic field, and depending on the modulus of the electric field E, and on the angle θ :

$$I_{\shortparallel}(\omega, E, \theta) = \frac{1}{2}\left[I_{\hat{x}}(\omega, E, \theta) + I_{\hat{y}}(\omega, E, \theta) \right], \quad (8)$$

$$I_{\perp}(\omega, E, \theta) = \frac{1}{4}\left[I_{\hat{x}}(\omega, E, \theta) + I_{\hat{y}}(\omega, E, \theta) + 2I_{\hat{z}}(\omega, E, \theta) \right]. \quad (9)$$

In the case of an observation with an angle θ_0, the total intensity may be written :

$$I(\omega) = \cos^2\theta_0 I_{\shortparallel}(\omega) + \sin^2\theta_0 I_{\perp}(\omega). \quad (10)$$

In the following, we apply this model to various edge plasma emitters and conditions.

3. APPLICATION TO THE DIAGNOSTIC OF EDGE PLASMAS

In a future fusion reactor, nuclear reactions will produce energetic helium nuclei which can then recombine with electrons to form singly ionized atoms or neutrals.

One of the challenges of fusion devices is to extract helium ashes from the core plasma or at least keep their concentrations at low values. Therefore, it is a fundamental point to understand the dynamics of helium recycling in nowadays Tokamak plasmas where helium is injected in some discharges. In that frame, one can get some information by analyzing helium emission spectra. Here we analyze three helium lines measured in the edge of the Tore-Supra Tokamak : the two singlet transitions 3d (^{1}D) → 2p (^{1}P) and 3s (^{1}S) → 2p (^{1}P) with respective wavelengths λ=6678 Å and λ=7280 Å, and the triplet transition 3s (^{3}S) → 2p (^{3}P) at λ=7065 Å.

The analyzed spectra are measured along a line of sight which is perpendicular to the magnetic field lines, but since the mirrors used to collect light have polarization dependent reflection coefficients, the intensity ratio of the different components are unusual.

The width of the main feature of the observed spectra (see figure 1) being equal to that of the apparatus function, a temperature upper limit of 1000 K can be determined. This temperature limit is consistent with thermal desorption of He atoms from the wall, whose temperature is between 500-1000 K. It should be noted that a second population is needed to explain the blue shoulder, as well as the enhanced intensity of the blue sigma component.

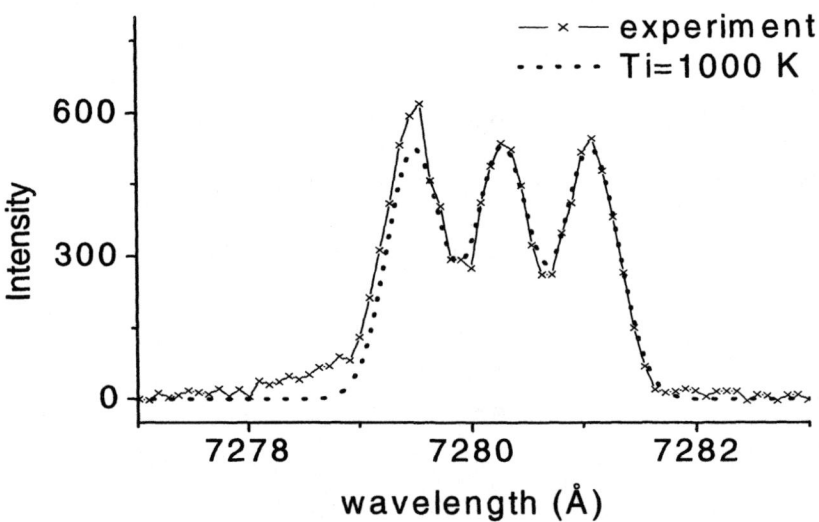

FIGURE 1. Comparison of the calculated and experimental profiles of the helium 3s (^{1}S)→2p (^{1}P) line. Dotted line : model, cross line : experiment.

In addition to helium ashes, impurities have to be controlled too and their concentrations minimized in the confined plasma of fusion reactors. Hence it is necessary to characterize impurity production in present Tokamaks to allow extrapolation of our knowledge to future fusion reactors. In Tore Supra the main impurity is the carbon released from the walls. We have studied the radiation emitted by various carbon ionization stages. The first analyzed experimental spectrum concerns a neutral carbon or C I line observed in the vicinity of a Tore Supra ergodic divertor neutraliser plate [7], and is presented in figure 2. The line is composed of six transitions (components) resulting from the fine structure of the energy levels 3s (^{3}S°) and 3p (^{3}P). Each one is composed of a triplet because of the degeneracy removal due to the magnetic field, the central feature (polarized parallel to the magnetic field) being suppressed due to the line of sight direction parallel to the magnetic field.

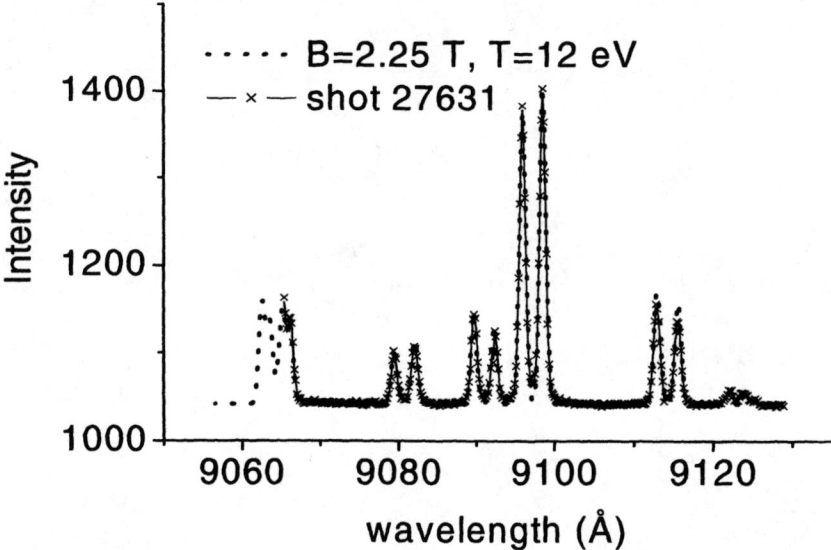

FIGURE 2. Fit of an experimental spectrum of neutral carbon measured near the neutralizer plates of the ergodic divertor. Observation is parallel to the magnetic field.

To obtain the correct intensity ratios, the line profile is calculated considering simultaneously all transitions. From the fit of the experimental spectrum, we deduce an emitter temperature of 12 eV in case of figure 2. This value (as well as its increase with the electron temperature, not presented here) is consistent with the expected energy of carbon atoms produced by physical sputtering, which indicates that chemical sputtering is less important in this case.

Illustrations of the capabilities of our line shape code are shown in figures 3 and 4 where the C II 2s^2(^{1}S)3s → 2s^2(^{1}S)3p and C III 2s(^{2}S)3s → 2s(^{2}S)3p lines have been respectively considered, both calculated in parallel observation. When a structure is visible only one couple of temperature-magnetic field values fit the measured spectrum. The code can thus be used with a fitting procedure to determine the location of emission and the temperature of the different emitting species of carbon.

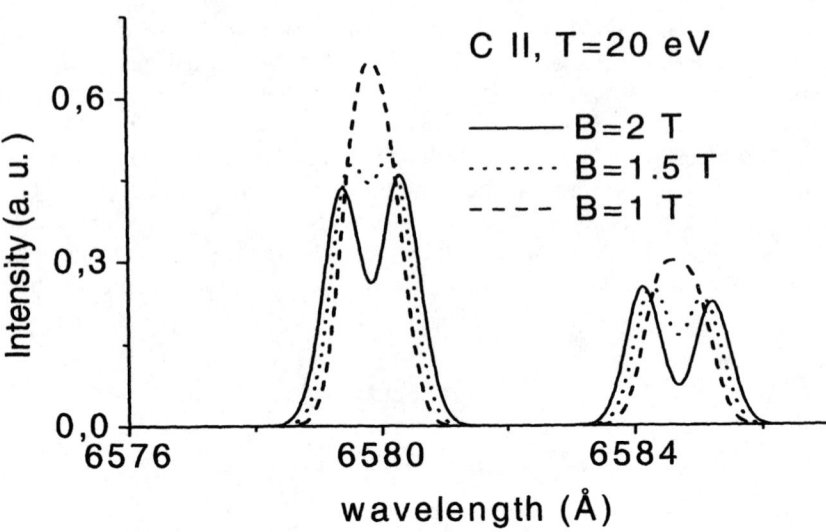

FIGURE 3. Theoretical profiles of the $2s^2(^1S)3s \rightarrow 2s^2(^1S)3p$ line of C II for different values of the magnetic field and the same value of the ion temperature of 20 eV. Observation is parallel to the magnetic field.

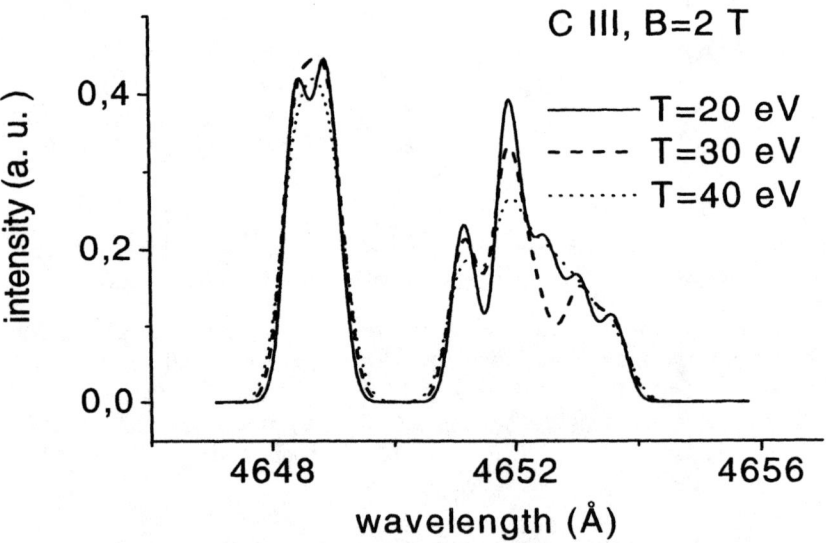

FIGURE 4. Calculated profiles of the $2s(^2S)3s \rightarrow 2s(^2S)3p$ line of C III for different values of the temperature and a same magnetic field value of 2 T.

4. Dα LINE AND MOLECULAR DENSITY DETERMINATION

A fit of the center of the Dα line requires at least two Maxwellian populations [8,9], having different temperatures. The colder one ($T_c \sim 2$ eV) corresponds to atoms created by the dissociation of molecules released from the wall. The temperature of this population can be explained considering the dominant dissociation path, together with the heating mechanism proposed by J. Hey [1]. The other population, whose temperature T_w is larger than 10 eV, is mainly attributed to fast reflected atoms and, probably to a lesser extent, to charge exchange reactions with deuterium ions. A calculation of the D_2 density at the edge was performed using this analysis [8] from the results of the fit, the brightness $B_{c/w}$ corresponding to each population can be obtained ; the density of cold $n_c(p=3)$ and warm $n_w(p=3)$ atoms in the 3p level is then deduced from equation (11), where A_{32} is the Einstein coefficient for the Dα spontaneous emission :

$$B_{c/w} = \frac{1}{4\pi} \int_{line\ of\ sight} A_{32} n_{c/w}(p=3)dl \qquad (11)$$

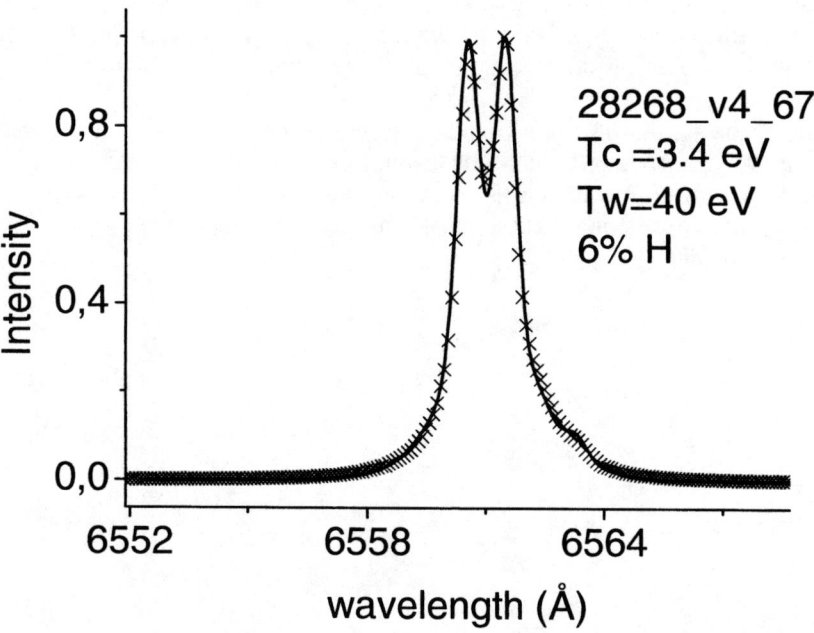

FIGURE 5. Fit of a Dα line with cold and warm populations and Hα line in red wing

The link between these densities and the molecular density n_{D_2} is provided by a collisionnal-radiative model developed by Fujimoto et al.[10]. The result was shown to agree with that deduced from the analysis of the D_2 molecular emission [11]. This method of analysis leads to high densities of molecules, which are comparable to the atomic density, for the whole range of edge electron temperatures (10 eV $<T_e<$40 eV):

$$\frac{n_D}{n_{D_2}} \approx 1.4 \qquad (12)$$

This confirms that the molecular contribution to the edge physics must be considered even for neutral temperatures larger than a few eV.

5. FAR WING OF THE Dα LINE

The strong brightness of the Dα line emitted near the neutralizer plates of the ergodic divertor of Tore Supra permits the observation of far line wings, which we have studied on the blue side, since the Hα line lies in the red wing. We have analysed the blue wing of Dα by using a log-log plot of the line intensity as a function of the wavelength separation from the peak of the blue σ component (figure 6). This method reveals a third Gaussian population of neutral emitters with a temperature around 100 eV, followed 5 Å further in the wing, by a feature which exhibits a linear dependence in the log-log plot. The power law $\Delta\lambda^{\alpha}$ dependence, with $1.5 < \alpha < 2.5$ suggests that this wing could be due to Stark broadening. However an estimate of thermal Stark broadening for the observed conditions cannot explain this wing [12]. An evaluation of possible broadening mechanisms involving non thermal effects has been made in our group, and involves non thermal Stark effects, or non diffusive transport effect on the Doppler profile.

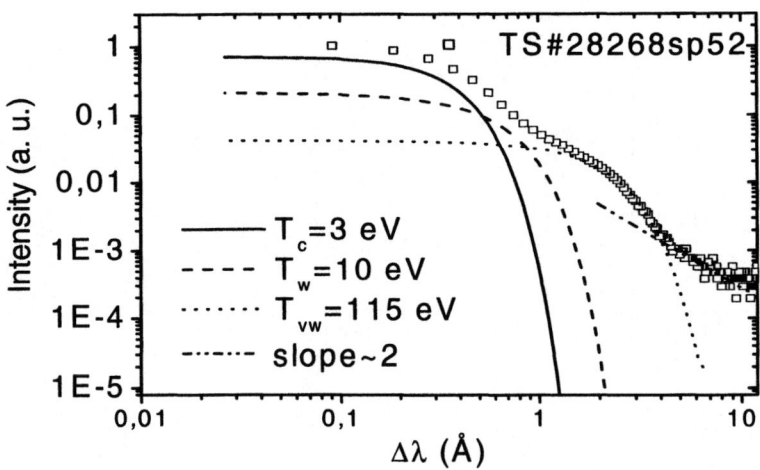

FIGURE 6. Dα line in a log-log scale. It is composed by three gaussian populations and a far wing with a slope of about 2.

6. HIGH *n* LINES IN RECOMBINING PLASMAS

For Tokamak with axisymmetric divertors, dense and cold (Te ≤ 2 eV) plasmas ("detached' plasmas) are obtained in the vicinity of the neutralizer plates. A spectroscopic diagnostic is then especially useful for detached plasmas, since the usual diagnostics (such as Langmuir probes) no longer provide reliable density and temperature measurements.

For conditions like those found in the divertor of JET, with electron densities between 10^{19} and $3x10^{20}$ cm^{-3}, and an electronic temperature of the order of the eV, the plasma is in a recombining regime. This means that recombination rates are larger than ionization rates, allowing a significant population of levels with high principal quantum number n, and thus the observation of lines in the JET divertor up to n=15 in the Balmer serie, and to n=12 for the triplet lines of helium (1s2p-1snl), with l=0,2[13]. A procedure intended to improve the simultaneous density and temperature diagnostics in the divertor has recently been developed. It combines the use of a collisional radiative model for the level populations, to our line shape code for a description of Stark broadening, and a contribution of the bound-free and bound-bound transitions modeled by lorentzians at continuum wavelengths. The latter contribution retains the electronically broadened transitions up to n=100. Our procedure provides a robust extraction of the plasma diagnostic information since it uses several features of the spectrum, such as the line shape, the line ratios and the slope of the continuum.

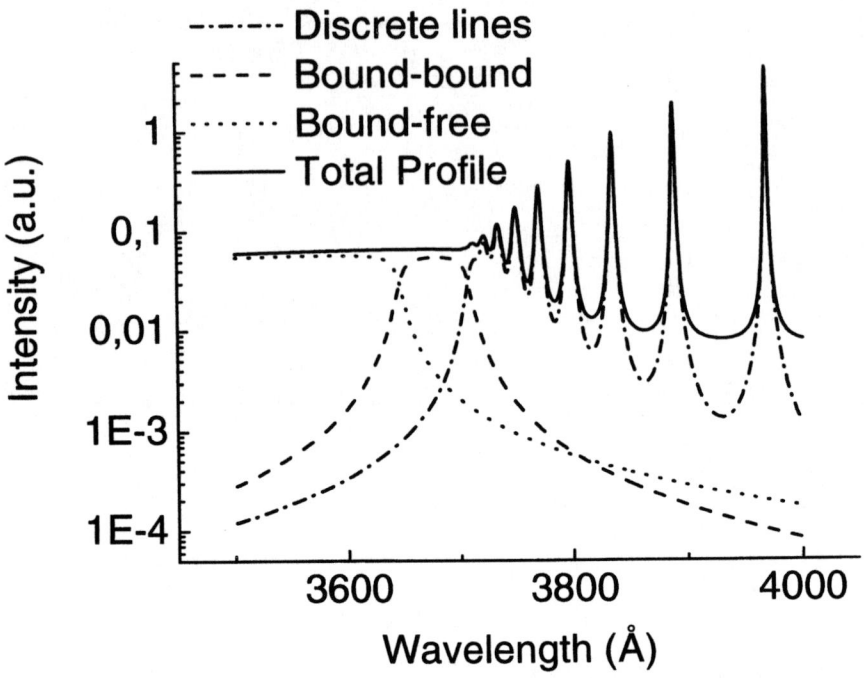

FIGURE 7. The total theoretical profile is the sum of the discrete lines calculated with our code, of a sum of lorentzians modeling the very high n (more than 16 up to 100) Balmer lines of deuterium, as well as the contribution of the bound-free transitions.

7. CONCLUSION

According to the edge plasma conditions, a line shape analysis can provide an access to the atom and ion temperature and to the electronic density. This requires a line shape model including Doppler, Zeeman and Stark effects. For the cases where the line shape of Dα is dominated by Doppler broadening, it is possible to determine several populations of emitters having different origins. A collisionnal-radiative model then allows to obtain the density of molecular deuterium in the line of sight. A possible spectroscopic signature of the turbulence in the edge region has been observed on the wings of Dα and discussed. In case of recombining plasmas, line shape analysis permits to determine both density and temperature.

REFERENCES

1. Hey, J. D, et al., *Contrib. Plasma Phys*. **34**, 725 (1994).
2. Welch, B., et al., Spectral Line Shapes IX, Editors M. Zoppi, L. Ulivi (AIP 386, New York 1996), p 113.
3. Isler, R. C., et al., *Phys. Plasmas* **4**, 355 (1997).
4. Terry, J. L., et al*., Phys. Plasmas* **5**, 1759 (1998).
5. Nguyen-Hoe, et al., JQSRT **7**,429 (1967).
6. Godbert-Mouret, L. et al., JQSRT **71**, 365 (2001)
7 Meslin, B. et al. J. Nucl. Mater. **266-269**, 318 (1999).
8. Escarguel, A., et al., *Plasma Phys. Control. Fusion* **43**, 1733 (2001).
9. Koubiti, M. et al., *Plasma Phys. Control. Fusion* **44**, 261 (2002).
10. Fujimoto, T., Sawada, K., Takahata, K., *J. Appl. Phys*. **66**, 2315 (1989).
11 Guirlet, R. et al., 28th EPS Conference on Control. Fusion. and Plasma Phys., Madeira (2001).
12. Marandet, Y. et al, these proceedings.
13. Koubiti, M. et al., these proceedings.

ZEEMAN SPECTROSCOPY OF TOKAMAK EDGE PLASMAS

J. D. Hey*, C. C. Chu* and Ph. Mertens[†]

*School of Pure and Applied Physics, University of Natal, Durban 4041, South Africa
(hey@nu.ac.za, chu@nu.ac.za)
[†]Institut für Plasmaphysik, Forschungszentrum Jülich, EURATOM Assoc.,D-52425 Jülich,
Germany (ph.mertens@fz-juelich.de)

Abstract. Zeeman spectroscopy is a valuable tool both for diagnostic purposes, and for more fundamental studies of atomic and molecular processes in the boundary region of magnetically confined fusion plasmas ($B \simeq 1$ to 10 T). The method works well when the Zeeman (Paschen-Back) effect plays an important, or dominant, rôle in relation to other broadening mechanisms (Doppler, Stark, resonant excitation transfer) in determining the spectral line shape.

For impurity species identification and temperature determination, Zeeman spectroscopy has advantages over charge-exchange recombination spectroscopy from highly excited radiator states, since spectral features practically unique to the species under investigation are analysed. It also provides useful information on probable mechanisms of line production (e.g. sputtering mechanisms, electron impact-induced dissociative excitation from molecules in the edge plasma), and on the temperature evolution of lower charge states in the process of convection inwards or diffusion outwards from the hotter plasma interior. Where different physical processes are responsible for different sections of the line profile — especially in the case of hydrogen isotopes — Zeeman spectroscopy can provide a set of characteristic temperatures for each section.

The method is introduced in both passive and active spectroscopy, and general principles of the Zeeman effect are discussed with special reference to régimes of interest for the tokamak. Relevant physical processes (sputtering mechanisms, electron impact-induced dissociative excitation from molecules in the edge plasma, and ion-atom collisional heating mechanisms) are illustrated by sample spectra.

1. INTRODUCTION

Because Zeeman spectroscopy exploits the wealth of detail contained in the shapes of multiplet transitions which are appreciably influenced by the Paschen-Back effect, it offers unique advantages in the study of tokamak boundary plasmas [1]. Besides its rôle as a diagnostic tool, in the determination of impurity temperature (velocity) and concentration [1-7], it can be used for more fundamental studies of the particular atomic processes responsible for line formation and/or for appreciable broadening of the observed spectrum [8-15]. The advantages of using Zeeman spectroscopy on transitions from low-lying levels over charge-exchange recombination spectroscopy (CXRS) via highly excited radiator states, for the purposes of impurity identification, have been discussed in some detail [15]. These advantages apply to both passive [1, 5, 7, 11] and active [6] spectroscopic applications, in particular to laser-induced fluorescence measurements at a test limiter. The analysis of spectral features practically unique to the species under investigation also provides valuable clues on possible, or probable,

CP645, *Spectral Line Shapes: Volume 12, 16th ICSLS*, edited by C. A. Back
© 2002 American Institute of Physics 0-7354-0100-4/02/$19.00

mechanisms of line production, and the temperature evolution of lower charge states in the process of convection inwards or diffusion outwards from the hotter plasma interior, can be followed systematically [1, 15].

We start with a discussion of principles of the Zeeman effect relevant to magnetically confined fusion plasmas ($B \approx 1$ to 10 T), and proceed to atomic and molecular processes of particular interest to investigations conducted by the present authors.

2. PRINCIPLES

The régime of interest is well characterised in terms of the atomic unit of magnetic field-strength, which may be expressed (in SI) as:

$$B_0 = \frac{\alpha^3 m_e c}{e a_0} = 12.52 \text{ T}, \tag{1}$$

where α denotes the fine-structure constant, m_e the electron mass, and a_0 the Bohr radius [16]. This should not be confused with the huge field-strength unit commonly employed for astrophysical purposes [17]:

$$B_0' = 2\alpha^{-2} B_0 = 4.701 \times 10^5 \text{ T}. \tag{2}$$

We proceed below on the assumption that, besides the Zeeman and Doppler effects, other broadening mechanisms, such as resonant energy transfer in gas discharges [14, 18] and the Stark effect [18-20], are relatively unimportant. The Zeeman effect already produces an appreciable distortion of Doppler-broadened line profiles, when the radiator temperature falls below [4, 21]

$$kT(\text{eV}) \le M(\text{u}) \left(\frac{n^4 B}{40 \, Z^2} \right)^2 \tag{3}$$

where M denotes the radiator mass (amu), Z its effective nuclear charge (ionisation stage: 1 for neutrals, etc.) and n the (effective) principal quantum number pertaining to the upper level of an n-α line. For hydrogenic species, the rôle of electric fields (both macroscopic and microscopic) in the edge region of the fusion plasma is of minor importance in relation to the Zeeman effect, provided that the electric field-strength and electron concentration do not exceed [11, 19]:

$$E \le 5 \times 10^4 \, \frac{ZB}{n} \quad \text{V m}^{-1}$$

$$n_e \le 3 \times 10^{19} \left(\frac{ZB}{n} \right)^{3/2} \quad \text{m}^{-3}. \tag{4}$$

For observational purposes, while the importance of the Doppler effect does not depend upon the choice of spectral region, the Zeeman effect is enhanced towards longer wavelengths. This is easily seen by comparing the expression for the Doppler width

(FWHM) with the splitting from the zero-field position of the $\sigma_{\pm}$ components in a simple Lorentz triplet:

$$\frac{\Delta\lambda_{1/2}^D}{\lambda} = 2\sqrt{2\,ln2\,\frac{kT}{Mc^2}}$$

$$\frac{\Delta\lambda_{\pm}}{\lambda} = \mp\frac{\mu_B}{hc}\lambda B \tag{5}$$

where μ_B denotes the Bohr magneton. Interactions linear in the magnetic field play the major rôle until field-strengths (in T) in excess of [21, 22]

$$B \geq 5 \times 10^4 \frac{Z^2}{n^4}, \tag{6}$$

are attained, after which the 'diamagnetic' quadratic Zeeman effect should not be ignored.

This may be clarified with reference to a hydrogenic (one-electron) atom or ion. The perturbed atomic Hamiltonian may be written [22]:

$$\hat{H} = \hat{H}_0 + \hat{H}_{nucl} + \hat{H}_{mag}^I + \hat{H}_{mag}^{II} \tag{7}$$

where the various contributions apply, respectively, to the isolated atom, the nuclear spin interactions (hyperfine structure splitting), and two distinct contributions for the interaction between the atom and the field. The first (I, 'paramagnetic') arises from the 'permanent' moments of the bound electron(s), while the second (II, 'diamagnetic') arises from magnetic moments induced by the field itself:

$$\hat{H}_{mag}^I = \frac{e}{2m}\vec{B}\cdot\left(g_\ell\hat{\vec{\ell}}+g_s\hat{\vec{s}}\right) \tag{8}$$

with [16, 23]

$$g_\ell = 1 - \frac{m_e}{M}$$

$$g_s = 2\left[1+\frac{\alpha}{2\pi}-0.328\left(\frac{\alpha}{\pi}\right)^2...\right],$$

and

$$\hat{H}_{mag}^{II} = \frac{e^2\vec{A}^2}{2m_e} = \frac{e^2}{8m_e}\left(\vec{B}\times\vec{r}\right)^2. \tag{9}$$

In our régime of interest, the hyperfine contribution is negligible compared to the first of the atomic magnetic terms in (7), and will not be considered further. This term (I) is responsible for the very important Paschen-Back effect, which produces ΔJ-forbidden components in the spectrum [24]. The term (II) effectively makes its appearance when the Paschen-Back effect is 'complete'. Although it has hitherto been neglected in tokamak studies, it may in fact play a rôle in distorting the appearance of higher series members in certain tokamak plasmas, e.g. when the Inglis-Teller limit is being determined [18, 25].

3. THE PASCHEN-BACK EFFECT

The term (I) in the atomic Hamiltonian $\left(\text{Eq. (7)}\right)$ results in the loss of spherical symmetry through the interaction. While the following commutation relations still hold:

$$\left[\hat{H}^{I}_{mag}, \hat{L}^2\right] = \left[\hat{H}^{I}_{mag}, \hat{L}_z\right] = \left[\hat{H}^{I}_{mag}, \hat{S}^2\right] = \left[\hat{H}^{I}_{mag}, \hat{S}_z\right] = 0, \qquad (10)$$

the total angular momentum is no longer a constant of the motion, and consequently both diagonal and off-diagonal matrix elements of the interaction occur [26-28]:

$$\langle SLJM \,|\hat{H}^{I}_{mag}\,|\, SLJM \rangle = \mu_B \, g_J \, B \, M$$

$$\langle SLJM \,|\hat{H}^{I}_{mag}\,|\, SL, J \pm 1, M \rangle = \mu_B \, B \, (g_s - g_\ell)$$

$$\times \sqrt{\frac{(L+S+1+J_>)(L+S+1-J_>)(L+J_>-S)(S+J_>-L)}{4J_>^2\,(2J_>+1)(2J_>-1)}} (J_>^2 - M^2) \qquad (11)$$

with the Landé g-factor

$$g_J = \frac{g_\ell[J(J+1)+L(L+1)-S(S+1)] + g_e[J(J+1)+S(S+1)-L(L+1)]}{2\,J\,(J+1)}. \qquad (12)$$

The corresponding tri-diagonal matrix equation must be solved for the perturbed eigenenergies and eigenfunctions [15, 27], subject to the non-crossing theorem of von Neumann and Wigner [29], in order to determine the perturbed line positions and strengths.

When all transitions between a particular pair of principal quantum numbers are included in the spectroscopic 'window' of observation, a large number of Zeeman components may be recorded, especially in CXRS of highly charged ions (see Table 1). Reduction of the number observed, by use of a linear polariser, or a quarter-wave plate combined with a linear polariser, will probably not eliminate the need for a computer code for fitting purposes (temperature determination) [4, 11, 12]. For multiplet transitions in non-hydrogenic atoms, an interactive code is also extremely useful [1, 5-7, 9, 15], if not essential, unless a single Zeeman component can be isolated properly by an optical polariser [8, 10].

The ΔJ-forbidden components in the spectrum appear when the original fine structure of the atom is appreciably perturbed. They increase in intensity with the magnetic field, but then weaken and disappear as the decoupling of orbital and spin angular momentum

TABLE 1. The number of Zeeman components in a set of transitions between energy levels of principal quantum number n and n' in a hydrogenic atom or ion [4].

$n \rightarrow n'$	all components	allowed	ΔJ-forbidden
sum	$2\,(12n'^2 - 14n' + 7)$	$2\,(9n'^2 - 8n' + 4)$	$2\,(3n'^2 - 6n' + 3)$
$\Delta M = 0\,(\pi)$	$2\,(4n'^2 - 4n' + 2)$	$2\,(3n'^2 - 2n' + 1)$	$2\,(n'^2 - 2n' + 1)$
$|\Delta M| = 1\,(\sigma)$	$2\,(8n'^2 - 10n' + 5)$	$2\,(6n'^2 - 6n' + 3)$	$4(n'^2 - 2n' + 1)$

under the influence of the field becomes complete – the 'swan-song' of the original multiplet before its destruction by the magnetic field [24].

It is important to note that, while itself linear in B, the magnetic interaction (I) produces terms quadratic in the magnetic field-strength, in the Taylor expansion of both the perturbed wavelength and the perturbed line-strength [30, 31]. These quadratic terms do not, however, in practice compete with those arising from magnetic interaction (II), for particular values of the magnetic field-strength [22].

4. THE 'DIAMAGNETIC' QUADRATIC ZEEMAN EFFECT

With the loss of the quantum number J, we now use the representation $|n\ell m_\ell m_s >$ for single-electron atoms. The Hamiltonian (II) is diagonal in m_ℓ, m_s and parity, but *not* in n, ℓ [22]. If the mixing of states of different principal quantum number is ignored, we obtain for the non-vanishing matrix elements [22, 32]

$$< n\ell m_\ell m_s \,|\hat{H}_{mag}^{\text{II}}\,| \, n\ell m_\ell m_s > = \frac{e^2 a_0^2 B^2}{8 m_e Z^2} \, \frac{n^2[5n^2 + 1 - 3\ell(\ell+1)][\ell(\ell+1) + (m_\ell^2 - 1)]}{(2\ell + 3)(2\ell - 1)}$$

$$< n\ell m_\ell m_s \,|\hat{H}_{mag}^{\text{II}}\,| \, n, \ell \pm 2, m_\ell m_s > = -\frac{5}{16} \, \frac{e^2 a_0^2 B^2}{m_e Z^2} \, \frac{n^2 \sqrt{(n^2 - \ell_>^2)(n^2 - (\ell_> - 1)^2)}}{(2\ell_> - 1)\sqrt{(2\ell_> + 1)(2\ell_> - 3)}}$$

$$\times \sqrt{(\ell_> + m_\ell)(\ell_> - m_\ell)(\ell_> + m_\ell - 1)(\ell_> - m_\ell - 1)}. \tag{13}$$

ΔL-forbidden components now arise in the spectrum, as first observed by Jenkins and Segrè [33] in the absorption spectrum of Na I for principal quantum number n between $n = 24$ and $n = 30$, with $B = 2.7$ T. This field régime has been termed 'Region II' by Schiff and Snyder [32]. At still higher fields ('Region III'[32]), the familiar level structure is entirely destroyed by the magnetic field, whose influence on the bound electron is eventually dominant over the nuclear Coulomb interaction [17,22]. For tokamak spectroscopy, however, we may safely confine our attention to magnetic field-strengths up to and including, for particular principal quantum numbers, the 'Region II'.

5. SPECTRAL LINE INTENSITIES AND THE RÔLE OF $\vec{B}$ IN PARTICLE COLLISIONS

We may write as follows the intensity of a multiplet transition radiated from a unit volume of the plasma under investigation [18-20,26]:

$$I\,(SL \rightarrow S'L') = \frac{4}{3} \frac{\alpha^2 m_e a_0^3}{c} \, \omega_0^4 \sum_{JM,J'M'} F(\omega) \, \frac{n_{JM}}{g_{JM}} \, \frac{S(JM - J'M')}{a_0^2 \, e^2} \tag{14}$$

where $F(\omega)$ denotes the angular frequency normalised line shape, n_{JM} the population density of an individual fine-structure sublevel (JM), with statistical weight g_{JM}, and S a line strength. The frequency throughout the multiplet is usually treated as constant (ω_0), equal to the value in the absence of fine structure, an assumption which constitutes one of the 'trivial' sources of asymmetry in line profile (line shift) analysis [18]. A much more important assumption is the following: the individual intensities are usually treated as 'statistical' rather than 'dynamical' [16], i.e. *even if* the major populating mechanism of the initial sublevels favours particular fine-structure components, *ion-radiator collisions are sufficiently frequent* to impose a statistical distribution. Then,

$$I\,(SL \to S'L') = \text{const.} \times \sum_{JM,J'M'} F(\omega)\,\frac{S(JM - J'M')}{a_0^2\,e^2} \qquad (15)$$

where, as in the previous equation, any angular dependence is contained in the line strengths.

Indications have indeed been found in tokamak boundary plasmas, that the conditions leading from Eq. (14) to Eq. (15) are not always completely fulfilled [15]. Spectra with potentially significant contributions from charge-exchange recombination have shown a small tendency to favour fine-structure levels of higher J; for spectra arising principally from electron impact excitation from the ground state, the reverse has been found. The inference may therefore be drawn, for this particular plasma, that the frequency of ion collisions in the boundary is insufficient to maintain a complete statistical distribution among the fine-structure sublevels (see Fig. 1 below). At any rate, the assumption of statistical intensities should be checked by a numerical model in particular cases of interest [11, 34, 35].

In performing estimates of, e.g., elastic scattering rates or population transfer between fine-structure sublevels, the rôle of the magnetic field may be neglected in ion-atom collisions, provided that the ion gyro-radius exceeds the local plasma Debye length [18]. This in turn implies that the electron concentration (assuming overall charge neutrality) should exceed [13, 15, 36, 37]:

$$n_e > \frac{m_e}{8\pi a_0^3\,M_f}\,\alpha^4 \left(\frac{Q_f B}{B_0}\right)^2. \qquad (16)$$

Envisaged here is a collision between a 'field' ion (mass M_f, charge Q_f) and the 'test' radiator . This requirement is generally fulfilled for singly-charged ion perturbers in the plasma edge.

In atomic excitation transfer processes (electron-radiator collisions, relative speed u), the relevant spatial scale is the impact parameter for strong collisions [18]:

$$\rho = \frac{\hbar}{m_e u}\,\left|\langle i|\frac{\vec{r}}{a_0}|j\rangle\right|, \qquad (17)$$

which is so much smaller than the electron gyro-radius, that the magnetic field may be ignored.

6. ATOMIC RELEASE (PRODUCTION) MECHANISMS

Intrinsic impurities [38] such as carbon and oxygen are released into the plasma through two types of sputtering mechanisms: chemical and physical. While the release of oxygen atoms at surfaces takes place almost entirely by chemical sputtering [9], the predominant mechanism of carbon release is strongly dependent upon edge plasma conditions. It has been found to be chemical only in the case of cold, dense, 'detached' plasmas $\left(kT_e < 8 \text{ eV (locally)}, \bar{n}_e > 3.5 \times 10^{19} \text{ m}^{-3}\right)$[8-10], largely independently of the actual heating power deposited in the plasma [15]. Since the energy acquired by the released atom is strongly dependent upon this mechanism, the transition between these two types of sputtering is even observable through temperatures deduced by Zeeman spectroscopy on ionised species [15]. Once released by physical sputtering, the atom acquires a Thompson velocity distribution [39], which relaxes fully towards a Maxwellian only after ionisation by electron impact. Another common impurity species released via physical sputtering is silicon, which has been studied through laser-induced fluorescence [6].

An important mechanism (for the hydrogen isotopes, O I and Si I) is molecular dissociation through electron impact. In this case, the destruction of molecular vibrational states results in the transfer of so-called Franck-Condon energy [40] to the dissociation products, shared (in the centre of mass frame) in inverse proportion to the atomic masses. These energies (generally in the range of a fraction to several eV) play an important part in explaining the final temperatures attained even by ionised species, derived from the dissociation products by subsequent electron impact ionisation [15].

By means of Zeeman spectroscopy, surprisingly low values were discovered for the atomic temperatures ($kT_C < 0.5$ eV) of the hydrogen isotopes, or, more precisely, atoms in excited states contributing to Balmer line radiation [11-14]. These low energies, which did not appear to depend strongly on bulk plasma conditions [11], were independently, albeit indirectly, confirmed by laser-induced fluorescence measurements on $L\alpha$ [41, 42]. More recently, the existence of these cold hydrogen atoms was inferred in spectra of H_α recorded on the HT-6M tokamak [43]. It was clear from available molecular data [44, 45] and earlier studies of electron beam collisions with molecules [46-48], that special excitation channels are required to produce atoms with such low kinetic energies. It is indeed remarkable, that such 'cold' atoms can survive sufficiently long under these conditions for their radiation to be measured in the plasma edge [11-13]. The following mechanism (electron impact-induced molecular dissociation) was proposed to explain the observations, where X represents a hydrogen isotope [11-13]:

$$X_2 + e(v) \rightarrow X_2^{**} + e(v') \qquad X_2^{**} \rightarrow X^*(n = 3, 4, 5) + X^*(n'\ell'). \qquad (18)$$

The molecule is first excited by electron impact to an unstable intermediate state (i.e. to the repulsive inner wall of one of the electronically excited potential curves), from which dissociation at low energy occurs into a pair of atoms. At least one of these is produced in an excited state corresponding to a Balmer line [47, 48]. Of the spectrum of possible atomic energies, corresponding to various excitation thresholds, the Franck-Condon energy equivalent to $kT_C \approx 0.3$ eV for H and D [47, 48], with threshold energy 17 eV, and here $n'\ell' = 1s$, is of great interest for present purposes. Subsequent investigations

32

have clarified the need, on the grounds of the Franck-Condon principle, for the initial molecule to be in an excited vibrational state of $^1\Sigma_g^+$, before the dissociative excitation reaction leading to 'cold', excited atom formation can proceed [42, 49]. Indeed, while longer molecular dwell times in the plasma, and perhaps the presence of surfaces, appear to favour the population of such vibrational states, a high injection rate of hydrogen molecules into the plasma appears to lead only to the 'usual' Franck-Condon energies of about 2.2 eV, with the absence of a 'cold' component to the D_α spectrum [50].

Because of the complexity of the above physical picture, it is of interest to learn whether other excitation channels are available for the production of 'slow' atoms, and this is indeed the case [44, 51, 52]. Provided that the reaction proceeds via the excited electronic state $^2\Sigma_g^+$, dissociative ionisation of H_2 also produces cold protons and cold excited atoms:

$$X_2 + e(v) \;\rightarrow\; X_2^+ + e(v') + e(v'') \qquad\qquad X_2^+ \;\rightarrow\; X^+ + X^*(n\ell). \qquad (19)$$

Since the threshold energy for (19) is well over double that required for (18), dissociative excitation might be considered as the favoured process involving molecules. Recent modelling based on [53], however, suggests otherwise: where both (18) and (19) are energetically allowed, i.e. where the electron temperature in the plasma edge is sufficiently high, (19) has been predicted to contribute more strongly to the line radiation from 'cold' atoms in the vicinity of a limiter surface [50]. This preliminary modelling certainly requires independent verification, however, in the light of [43]. One should comment, moreover, that a very similar Franck-Condon energy to the value of 0.2-0.3 eV has been assumed here as well.

It was also clear from the first measurements and analysis [11], that the molecular concentration at the plasma edge is significantly higher than previously assumed. This conclusion has indeed been supported by molecular spectroscopy using the Fulcher band $(3p\,^3\Pi_u \rightarrow 2s\,^3\Sigma_g^+)$ [49, 50].

We now return to the interesting case of oxygen, which can conveniently be introduced in molecular form through a graphite test limiter [15]. From the temperatures deduced for O^+, one concluded that an important fraction of these ions arises from atoms produced through electron impact-induced dissociation of O_2, a process which yields a variety of Franck-Condon energies [54]. Besides leading to the production of atoms in the metastable quintet state $2p^3\,3s\,^5S^o$, others are formed in highly excited Rydberg states ($2p^3\,np\,^5P$, $n \geq 8$), from which cascade to the metastable state takes place:

$$O_2 + e(v) \;\rightarrow\; O_2^{**} + e(v')$$
$$O_2^{**} \;\rightarrow\; O(2p^4\,^3P) + O(2p^3 3s\,^5S^o)$$
$$O_2^{**} \;\rightarrow\; O(2p^4\,^3P) + O^*(2p^3 np\,^5P) \;\; (8 \leq n \leq 15)$$
$$O^*(2p^3 np\,^5P) \;\rightarrow\; O(2p^3 3s\,^5S^o) + h\nu.$$

The Franck-Condon energy acquired per atom is 0.6, 1.0 or 2.0 eV, or more, for electron energies (roughly kT_e) above 14.9, 20.8 or 24.3 eV, respectively [54]. These energies are important for explaining the relatively high spectroscopic temperatures observed for O^+ [15].

33

7. ATOMIC HEATING MECHANISMS

Once these products of molecular dissociation ('test' atoms, mass M, temperature T) have been released into the plasma, they are immediately subjected to heating by collisions with the hot protons and deuterons (the 'field' particles, mass M_f, temperature T_f, concentration n_f). The momentum transfer (diffusion) cross-section $q_s(u)$, as a function of the relative speed $u = |\vec{v} - \vec{v}_f|$, governs the rate of the heating process [12, 55]:

$$\frac{dT}{dt} = \frac{16}{3\sqrt{\pi}} n_f \frac{M_r}{\Sigma M} (T_f - T) \beta^{5/2} \int_0^\infty u^5 q_s(u) \exp(-\beta u^2) du, \qquad (20)$$

$$\beta = \frac{M M_f}{2k (MT_f + M_f T)}.$$

with M_r denoting the reduced mass of the colliding partners, and $\Sigma M = M + M_f$.

The corresponding interaction potential may be written [12], for a 'field' ion of charge $+Q_f e$ and a polarisable 'test' atom (multipole polarisability $\alpha_{pol}^{(s)}$):

$$V(r) = -\frac{Q_f^2}{2} \hbar c \alpha \frac{\alpha_{pol}^{(s)}}{r^s}. \qquad (21)$$

Two types of interaction are relevant here: 'electrostatic' and induction (polarisation) [56]. The lowest orders of each correspond, respectively, to $s = 2$ (for a 'permanent' dipole moment, i.e. the linear Stark effect in hydrogenic atoms) and to $s = 4$ (the charge-induced dipole interaction). Following in order of importance are $s = 3$ ('permanent' quadrupole) and $s = 6$ (charge-induced quadrupole). The case of lowest s, when applicable, would be overwhelmingly the most important, were it not for the strong forward peaking in the differential cross-section [57]. The angular average

$$q_s(u) = 2\pi \int_0^\pi \sigma_{el}(u, \theta)(1 - \cos\theta) \sin\theta \, d\theta \qquad (22)$$

yields a drastic reduction in the contribution from this interaction [13]. For the next case ($s = 4$), the scalar contribution to the polarisability increases so rapidly with principal quantum number, that huge values (in relation to that for $n = 1$) are obtained for the upper levels of the Balmer α, β, γ, etc. lines [12, 13]. Based on the use of the (scalar) charge-induced dipole polarisability, quite approximate models (for H and D, and for O) have been presented for estimating the heating of the Franck-Condon atoms by ion collisions [12, 13, 15, 58]. While open to refinement in a number of respects (e.g. by the inclusion of n-mixing [18-20]), these models have provided clear indications that the proposed mechanism is indeed an effective source of heating of 'cold' atoms at the plasma edge. The first of these (for H and D) has, moreover, been employed in a line shape fitting program for D_α spectra from the edge of the Tore-Supra plasma, as part of a relaxation model for the atomic temperature evolution [59].

The above process is terminated by collisional ionisation, at which point interparticle Coulomb interactions control the temperature evolution, and the temperature attained by each successive ionisation stage is determined by a competition between ion-ion collisions (heating/cooling) and electron impact ionisation [1, 15, 60].

8. ILLUSTRATIVE EXAMPLES

Figure 1 illustrates a relatively simple Zeeman spectrum (C III multiplet 1), recorded in third order with the aid of the Littrow spectrometer and optical arrangement described in [1, 4, 11, 15]. Because of the observation direction in the equatorial plane of the tokamak, roughly tangentially to the magnetic flux surfaces, only the σ components contribute significantly. The very characteristic shape of this multiplet not only provides an excellent form of species identification, but also permits monitoring of the presence of impurity ions (mainly O^+), through additional radiation in this spectroscopic 'window' [15].

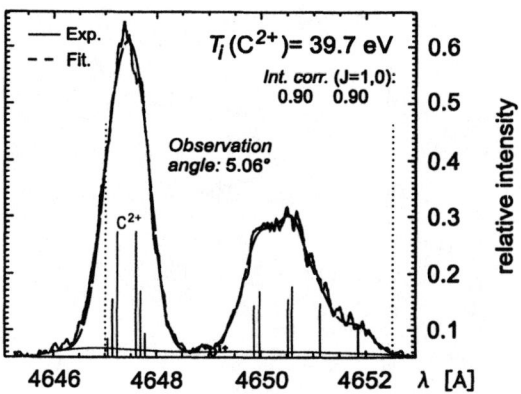

FIGURE 1. Zeeman spectrum of C III multiplet 1 from TEXTOR (B_{exp} =1.79 T, r= 44.6 cm, t = 2.53-3.03s, shot 59821) under longitudinal observation.

Figure 2 provides an example of 'active' Zeeman spectroscopy, in the form of laser-induced fluorescence via the Si I transition $3p^2\ ^3P_2 - 3p\ (^2P^o)4s\ ^3P^o_2$(multiplet 1 UV). The observation direction is this case was perpendicular to the toroidal magnetic field, while π-transitions were induced selectively with the aid of a pulsed dye laser pumped by an excimer laser, directed onto a test limiter surface. The plane of polarisation (**E** direction) of the laser could be varied with respect to the toroidal magnetic field direction, with the aid of a retardation ($\lambda/2$) plate, until the 'angle of no polarisation' [61], $\arccos \frac{1}{\sqrt{3}}$. At this point, π and σ components of equal intensity were seen in the fluores-

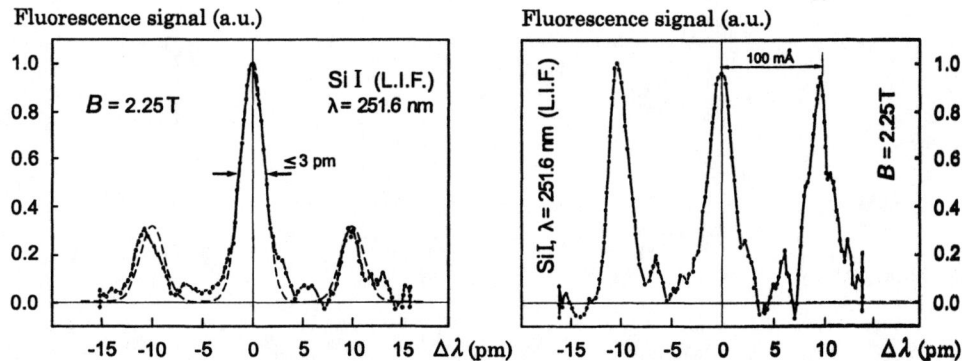

FIGURE 2. Zeeman patterns of atomic silicon, recorded *in situ* with laser-induced fluorescence. The UV-laser is linearly polarised; the pictures correspond to different angles of the plane of polarisation.

cence pattern. This method of selective tuning in fluorescence is clearly very powerful for investigating not only atomic velocity distributions, but also details of the magnetic field near the limiter position.

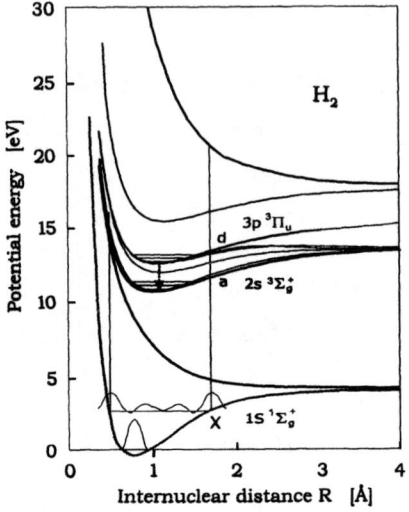

In Fig. 3, the left-hand vertical line shows how the Franck-Condon principle [40] may be used to explain the molecular dissociation process into 'slow' atoms. Vibrationally excited states, within the electronic ground state configuration, have a higher probability of being excited to the repulsive inner wall of an excited potential curve, i.e. to a vibrationally unstable intermediate state. Within half of the corresponding vibrational period, two unbound atoms are produced, each with a share $E = \frac{1}{2}[V(R) - V(\infty)]$ of the available vibrational energy.

The right-hand vertical line illustrates that both dissociative excitation and dissociative ionisation are possible mechanisms.

FIGURE 3. Electronic potential curves of H_2, with the states which correspond to the Fulcher-band [49, 50] labelled by "a" (lower) and "d" (upper). The vertical lines illustrate the Franck-Condon principle applied to the production of 'slow' atoms.

Figure 4 shows the results of a recent three-temperature fit [12, 13] to the first of the original D_β line profiles recorded in third order from the edge of TEXTOR [11], showing

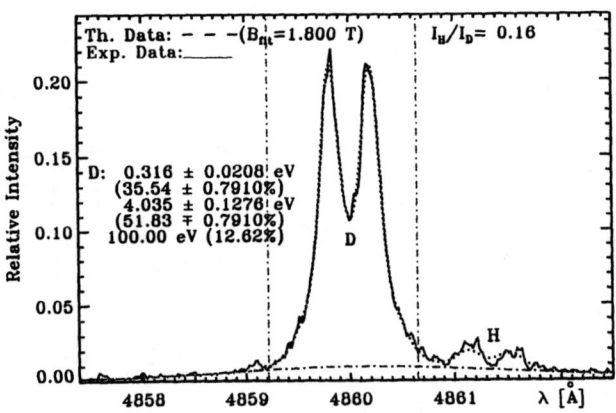

FIGURE 4. Zeeman spectrum (D_β, H_β) from the edge of TEXTOR (B_{exp} =1.76 T, r =48.3 cm, t = 2.0-4.0 s, shot no. 59408). The three temperatures classes deduced from the fit are shown, together with their percentage contributions.

a pronounced dip between the central σ components. After 'subtraction' of the hot (charge-exchange) temperature component, at least two temperature classes are revealed in the best-fit procedure between the vertical demarcation lines: a 'cold' component, $kT_C \approx 0.3$ eV, and a 'luke-warm' component $kT_w \approx 4.0$ eV. The first of these values is supported, in a remarkable way, by earlier experimental studies [44, 45, 62, 63] of the reaction represented by Eq. (18). The value of kT_w, on the other hand, provides evidence of the collisional heating mechanism discussed above, bearing in mind that the effective lifetime of the upper state of D_β is determined both by spontaneous emission and by collisional ionisation [13]. That both of these temperature components appear in the line of sight is an indication that the observation chord covers a region of some inhomogeneity in electron density [12].

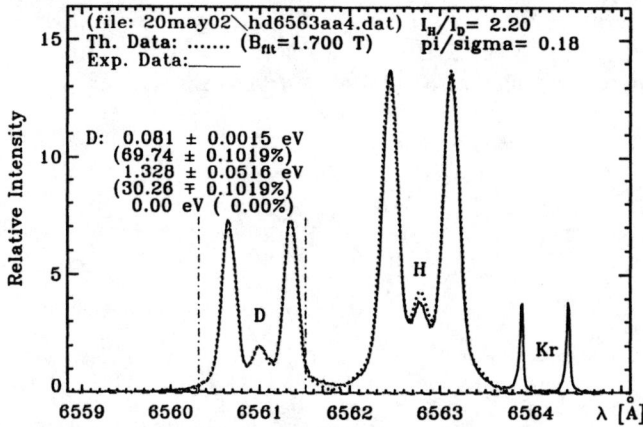

FIGURE 5. Spectrum of D_α and H_α from a Plücker tube discharge in Kr, measured transversely to the magnetic field. Owing to strong disturbance of the spectrum by atomic interactions, elimination of the π components by a linear polariser proved impossible ($B_{exp}= 1.60$ T).

Lastly, one may consider the possibility of elucidating these tokamak studies of Balmer spectra, with the aid of spectra taken from a Plücker discharge tube burning under the influence of a magnetic field of similar strength [12, 14]. For this purpose, the admixture of a heavy (noble) gas is essential, e.g. krypton. Figure 5 shows a Balmer-α spectrum recorded by photon counting in second order, together with the Kr I line $\lambda 4376.12$ Å in third order.

While the Zeeman splitting of the Kr I line was consistent with the B-field as measured by a Hall probe, the value derived from the fit to the Balmer-α lines exceeded this by 6%. This is strongly suggestive of additional broadening by atomic interactions.

Some details of the instrumentation are given in [12, 14]. The observation direction was perpendicular to $\vec{B}$. The linear polariser used to eliminate the π components passed a fraction of the π intensity, which was found to be depolarised, possibly by resonant excitation transfer collisions within the discharge column. In addition, the 'cold' temperature component was now in a range significantly below that predicted by Eq. (18), the electron energies being below the necessary threshold value of 17 eV. Such discharge tube studies therefore open up a new line of investigation of possible interest for the tokamak edge plasma. Here, a rich field for the study of resonant excitation transfer and polarisation rotation (depolarisation) is certainly available, as briefly discussed in [14].

Indeed, this final topic lies well within the domain of spectral line broadening by short-range interactions.

ACKNOWLEDGMENTS

Much of this work was performed as part of the project 'Spectroscopic Data for Fusion Plasmas' within the framework of a German (WTZ) - South African (NRF) scientific cooperation programme. Financial support for this project from this source, as well as from the University of Natal, is gratefully acknowledged. The authors also wish to thank Dr. A. Pospieszczyk and Dr. S. Brezinsek for valuable discussions, and appreciate the critical reading of their manuscript by Professor Hans R. Griem.

REFERENCES

1. Hey, J. D., Lie, Y. T., Rusbüldt, D., and Hintz, E., *Contrib. Plasma Phys.*, **34**, 725–747 (1994).
2. McCracken, G. M., Samm, U., Fielding, S. J., Matthews, G. F., et al., *J. Nucl. Mater.*, **176 & 177**, 191–196 (1990).
3. Schorn, R. P., Wolfrum, E., Aumayr, F., Hintz, E., Rusbüldt, D., and Winter, H., *Nucl. Fusion*, **32**, 351–359 (1992).
4. Hey, J. D., Lie, Y. T., Rusbüldt, D., and Hintz, E., *in* Proc. 20th Eur. Phys. Soc. Conf. on Controlled Fusion and Plasma Physics (Lisbon, Portugal), **17C** part III, 1111-1114 (1993).
5. Klepper, C., Isler, R. C., Tobin, S. J., and Hogan, J. T., *in* Proc. 21st Eur. Phys. Soc. Conf. on Controlled Fusion and Plasma Physics (Montpellier, France), **18B** part III, 1300-1303 (1994).
6. Mertens, Ph., and Silz, M., *in* Proc. 7th Int. Symposium on Laser-Aided Plasma Diagnostics (Fukuoka, Japan), **LAPD7**, 150-155 (1995).
7. Isler, R. C., Wood, R. W., Klepper, C. C., Brooks, N. H., Fenstermacher, M. E., and Leonard, A. W., *Phys. Plasmas*, **4**, 355–368 (1997).
8. Bogen, P., and Rusbüldt, D., *Nucl. Fusion*, **32**, 1057–1061 (1992).
9. Bogen, P., and Rusbüldt, D., *J. Nucl. Mater.*, **196-198**, 179–183 (1992).
10. Bogen, P., *Physica Scripta*, **T47**, 102–109 (1993).
11. Hey, J. D., Korten, M., Lie, Y. T., Pospieszczyk, A., Rusbüldt, D., Schweer, B., Unterberg, B., Wienbeck, J., and Hintz, E., *Contrib. Plasma Phys.*, **36**, 583–604 (1996).
12. Hey, J. D., Chu, C. C., and Hintz, E., *J.Phys. B: Atom. Mol. Opt. Phys.*, **32**, 3555–3573 (1999).
13. Hey, J. D., Chu, C. C., and Hintz, E., *Contrib. Plasma Phys.*, **40**, 9–22 (2000).
14. Chu, C. C., and Hey, J. D., *Contrib. Plasma Phys.*, **40**, 597–606 (2000).
15. Hey, J. D., Chu, C. C., Brezinsek, S., Mertens, Ph., and Unterberg, B., *J. Phys. B: Atom. Molec. Opt. Phys.*, **35**, 1525–1553 (2002).
16. Bethe, H. A., and Salpeter, E. E., *Quantum Mechanics of One- and Two-Electron Atoms*, Springer-Verlag, Berlin, 1957.
17. Fassbinder, P., and Schweizer, W., *Phys. Rev. A*, **53**, 2135–2139 (1996).
18. Griem, H. R., *Plasma Spectroscopy*, McGraw-Hill, New York, 1964.
19. Griem, H. R., *Spectral Line Broadening by Plasmas*, Academic Press, New York, 1974, ISBN 0-12-302850-7.
20. Griem, H. R., *Principles of Plasma Spectroscopy*, Cambridge University Press, 1997, ISBN 0-52-145504-9.

21. Hey, J. D., *Trans. Fusion Technol.*, **25**, 315–325 (1994).
22. Garstang, R. H., *Rep. Prog. Phys.*, **40**, 105–154 (1977).
23. Lamb, W. E. J., *Phys. Rev.*, **85**, 259–276 (1952).
24. Paschen, F., and Back, E., *Physica*, **1**, 261–273 (1921).
25. Welch, B. L., Griem, H. R., Terry, J., Kurz, C., et al., *Phys. Plasmas*, **2**, 4246–4251 (1995).
26. Condon, E. U., and Shortley, G. H., *The Theory of Atomic Spectra*, Cambridge Univ. Press, 1970, ISBN 52-109209-4.
27. Kiess, C. C., and Shortley, G., *J. Res. Nat. Bur. Stand.*, **42**, 183–207 (1949).
28. Semenov, R. I., and Tuchkin, V. I., *Opt. Spectrosc.*, **88**, 147–150 and **89**, 493–497 (2000).
29. von Neumann, J., and Wigner, E., *Phys. Z.*, **30**, 467–470 (1929).
30. Darwin, C. G., *Proc. Roy. Soc.* (London), **A 115**, 1–19 (1927).
31. Darwin, K., *Proc. Roy. Soc.* (London), **A 118**, 264–285 (1928).
32. Schiff, L. I., and Snyder, H., *Phys. Rev.*, **55**, 59–63 (1939).
33. Jenkins, F. A., and Segrè, E., *Phys. Rev.*, **55**, 52–58 (1939).
34. Purcell, E. M., *Astrophys. J.*, **116**, 457–462 (1952).
35. Pengelly, R. M., and Seaton, M. J., *Mon. Not. Roy. Astr. Soc.*, **127**, 165–175 (1964).
36. Mathys, G., *Astron. Astrophys.*, **139**, 196–210 (1984).
37. Günter, S., and Könies, A., *J. Quant. Spectr. Rad. Transfer*, **62**, 425–431 (1999).
38. Samm, U., *Plasma Phys. Control. Fusion*, **41**, B57–B76 (1999).
39. Thompson, M. W., *Phil. Mag.*, **18**, 377–414 (1968).
40. Condon, E. U., *Phys. Rev.*, **32**, 858–872 (1928).
41. Mertens, Ph., and Silz, M., *J. Nucl. Mater.*, **241-243**, 842–847 (1997).
42. Mertens, Ph., and Pospieszczyk, A., *J. Nucl. Mater.*, **266-269**, 884–889 (1999).
43. Wan, B., Li, J., Luo, J., Xie, J., Wu, Z., and Zhang, X., *Nucl. Fusion*, **39**, 1865–1869 (1999).
44. Tawara, H., Itikawa, Y., Nishimura, H., and Yoshino, M., *J. Phys. Chem. Ref. Data*, **19**, 617–636 (1990).
45. Kouchi, N., Ukai, M., and Hatano, Y., *J. Phys. B: Atom. Molec. Opt. Phys.*, **30**, 2319–2344 (1997).
46. Freund, R. S., Schiavone, J. A., and Brader, D. F., *J. Chem. Phys.*, **64**, 1122–1127 (1976).
47. Ito, K., Oda, N., Hatano, Y., and Tsuboi, T., *Chem. Phys.*, **17**, 35–43 (1976).
48. Ito, K., Oda, N., Hatano, Y., and Tsuboi, T., *Chem. Phys.*, **21**, 203–210 (1977).
49. Pospieszczyk, A., Mertens, Ph., Sergienko, G., Huber, A., et al., *J. Nucl. Mater.*, **266-269**, 138–145 (1999).
50. Brezinsek, S., *Untersuchung von atomarem und molekularem Wasserstoff vor einer Graphitober-fläche in einem Hochtemperatur-Randschichtplasma*, Forschungszentrum Jülich - Report No. **3962** (in German), 2002.
51. Dunn, G. H., and Kieffer, L. J., *Phys. Rev.*, **132**, 2109–2117 (1963).
52. Crowe, A., and McConkey, J. W., *J. Phys. B: Atom. Molec. Phys.*, **6**, 2088–2107 (1973).
53. Reiter, D., Bogen, P., and Samm, U., *J. Nucl. Mater.*, **196-198**, 1059–1064 (1992).
54. Freund, R. S., *J. Chem. Phys.*, **54**, 3125–3141 (1971).
55. Desloge, E. A., *Phys. Fluids*, **5**, 1223–1225 (1962).
56. Mason, E. A., and McDaniel, E. W., *Transport Properties of Ions In Gases*, Wiley, New York, 1988, ISBN 0-471-88385-9.
57. Landau, L. D., Lifshitz, E. M., and Pitaevskii, L. P., *Quantum Mechanics: Non-Relativistic Theory, 3rd edition* (translated by J. B. Sykes and J. S. Bell), Oxford: Pergamon Press, 1977, ISBN 0-08-020940-8.
58. Mertens, Ph., Brezinsek, S., Greenland, P. T., Hey, J. D., Pospieszczyk, A., Reiter, D., Samm, U., Schweer, B., Sergienko, G., and Vietzke, E., *Plasma Phys. Control. Fusion*, **43**, A349–A373 (2001).
59. Koubiti, M., Marandet, Y., Escarguel, A., Capes, H., Godbert-Mouret, L., Stamm, R., De Michelis, C., Guirlet, R., and Mattioli, M., *Plasma Phys. Control. Fusion*, **44**, 261–275 (2002).
60. McCracken, G. M., and Samm, U., *in* Proc. 8th Topical conference on Atomic Processes in Plasmas (Portland, ME), E.S. Marmar & J.L. Terry *edd.* (New York: AIP), 144-153 (1992).
61. Mitchell, A. C. G., and Zemansky, M. W., *Resonance Radiation and Excited Atoms*, Cambridge University Press, 1971, ISBN 0-521-08433-4.
62. Ogawa, T., and Higo, M., *Chem. Phys.*, **52**, 55–64 (1980).
63. Ogawa, T., and Higo, M., *Chem. Phys.*, **56**, 15–20 (1981).

Radiation Transport in Tokamak Edge Plasmas

H. A. Scott, M. L. Adams[*]

University of California, Lawrence Livermore National Laboratory
P.O. Box 808, Livermore, CA 94551, USA
[*]Department of Nuclear Engineering, Massachusetts Institute of Technology
NW16-230, 167 Albany St., Cambridge, MA 02139, USA

Abstract. Plasmas in edge regions of tokamaks can be very optically thick to hydrogen lines. Strong line radiation introduces a non-local coupling between different regions of the plasma and can significantly affect the ionization and energy balance. These effects can be very important, but they are not included in current edge plasma simulations. We report here on progress in self-consistently including the effects of a magnetic field, line radiation and plasma transport in modeling tokamak edge plasmas.

INTRODUCTION

High-density low-temperature plasmas with electron temperatures of order 1 eV and electron densities greater than 10^{15} cm^{-3} have been observed in the edge regions of the Alcator C-Mod [1] and D-IIID tokamaks [2]. At these temperatures and densities, the mean free path of a Lyman-α photon is less than 0.1 cm and the plasma can be very optically thick to line radiation. Optically thick Lyman lines have been confirmed experimentally [3]. However, simulations of these edge plasmas currently treat line radiation solely as an energy loss mechanism, assuming that the radiation can escape freely. Since the radiation field includes a significant amount of energy, a more realistic treatment of line radiation is needed for edge plasma simulations.

Wan, *et al* [4] addressed these issues for the first time in the context of a radiative divertor design for ITER. They used temperature and ion density profiles obtained with the UEDGE code [5] and recalculated the ionization balance and radiation flux to the divertor plate including line transfer effects. The results showed that the ionization balance changed significantly and the radiation flux decreased dramatically. Excitation by (primarily) Lyman-α line radiation allowed ionization to occur at much lower temperatures than otherwise predicted. These results were only suggestive, since the simulations were not self-consistent - changes in the ionization balance and radiative energy loss rate could not affect the ion and energy transport. Knoll saw similarly dramatic changes in self-consistent ITER-scale divertor simulations under the assumption that all Lyman-α radiation was trapped [6,7].

Our current research program is aimed at investigating the effects of radiation transport in these low temperature edge plasmas and providing computational tools for analyzing and simulating these plasmas. This paper is intended as both a brief description and a status report of the program to date.

CP645, *Spectral Line Shapes: Volume 12, 16th ICSLS*, edited by C. A. Back

For the plasma conditions considered here, which include magnetic fields of several Tesla, Stark and Doppler broadening and Zeeman splitting are of similar magnitudes for the lowest lines in the Lyman and Balmer series and must all be included in calculating line profiles. We calculate line profiles with the code TOTAL [8], which has been extended to include magnetic field effects. Line radiation transport is done with the code CRETIN [9], using profiles generated by TOTAL for the local conditions at each position in the edge plasma. This combined capability places an emphasis on robustness and computational efficiency in the line profile calculation. However, it allows for a consistent investigation of the interplay between the line profile, atomic kinetics and radiation transport, including an evaluation of the accuracy of less costly approximations. We demonstrate the use of the combined code with a sample calculation using conditions appropriate to Alcator C-Mod.

The following section discusses line profiles in magnetized plasmas, which are critical components in these investigations. The line profiles are important to the spatial and frequency distribution of the radiation field, which in turn influences the spatial variation of the excited-state populations and ionization balance, which ultimately affects the transport of (non-radiative) energy. Section 3 describes the use of magnetic line profiles in a line radiation transport code. While this is a useful and necessary step towards the desired simulation capability, it does not address the issue of adding radiation effects to an edge plasma simulation code. Section 4 discusses one possible avenue towards this goal and quantifies some optical depth effects. The paper concludes with a brief discussion of the next few steps in the research program.

MAGNETIC LINE PROFILES

In this section we describe the incorporation of magnetic effects into the line shape code TOTAL. The intent here is to highlight the salient points. More details are available in a companion paper [10] and a comprehensive discussion is given in Adams, *et al* [11]. We describe first the formalism used in TOTAL, then the modifications required in the plasma average and atomic Hamiltonian for the inclusion of an external magnetic field.

TOTAL Line Shape Formalism

TOTAL employs an electron collision operator formulated within the framework of a binary collision relaxation theory, plus a quasi-static electric ion microfield. The spectral line shape calculation considers only electric dipole transitions between bound states. Following standard line broadening theory, the line shape $\phi(\omega)$ is calculated as the Fourier transform of the time correlation function of the dipole operator, averaged over all atomic and plasma states:

$$\phi(\omega) = \int P(E)\phi(E,\omega)\,dE$$

$$= \int P(E)\frac{1}{\pi}\mathrm{Re}\int_0^\infty \Box\ \alpha\beta\,|\,\mathbf{d}e^{-\frac{i}{\hbar}\left(L_E - i\Gamma\right)t}\rho_i\mathbf{d}\,|\,\alpha'\beta'\,\Box\ e^{i\omega t}\,dt\,dE \tag{1}$$

where $P(E)$ is the quasi-static electric ion microfield distribution function, Γ is the electron collision operator, ρ_i is the initial density matrix of the atomic system and $|\alpha\beta\rangle$ represents the atomic state in Liouville space. L_E is the Liouville operator pertaining to the Hamiltonian H for a particular ionic field E

$$H = H_a - \mathbf{E} \cdot \mathbf{d} \tag{2}$$

where H_a is the atomic Hamiltonian, and $\mathbf{d}$ is the dipole operator. Matrix elements of the Liouville operator, i.e. energy levels and electric dipole matrix elements, are obtained from atomic structure calculations.

Following Calisti, *et al* [8], the computational approach centers around solving the complex eigenvalue problem for a fixed ionic field and then averaging over possible ionic fields. For a given ionic field E, the eigenvalues are easily obtained in the basis where $L_E - i\Gamma$ is diagonal, resulting in

$$\phi(E_j,\omega) = \sum_k \frac{c_{1k}\left(\omega - x_k(E_j)\right) + c_{2k}y_k(E_j)}{\left(\omega - x_k(E_j)\right)^2 + y_k(E_j)^2} \tag{3}$$

The line profile is then simply obtained as the sum of the components $\phi(E_j,\omega)$, weighted by the microfield distribution function. This approach has proved to be quite efficient, being limited primarily by the speed of the required eigenvalue calculations.

Magnetic Field Modifications

The plasma average consists of two parts: the electron collision operator, which is calculated within the framework of binary collision relaxation theory, and the quasi-static electric ion microfield. An external magnetic field introduces a preferential axis into the system that destroys the arbitrary orientation of the electric dipole operator. It can also potentially alter the dynamical properties of the plasma [11], but we do not address that issue in here.

The quasi-static ion microfield model assumes that the emitting atom is surrounded by a spherically symmetric plasma. The electric dipole moment can be chosen to point in an arbitrary direction, while the ion microfield distribution must be a function only of ionic field strength. With the introduction of a preferential axis, the spherical symmetry is broken and the integration over the microfield distribution becomes anisotropic:

$$\int dE\, P(E) dE \;\rightarrow\; \int dE_\parallel P_\parallel\left(E_\parallel\right) \int dE_\perp P_\perp\left(E_\perp\right) \tag{4}$$

where the subscripts refer to the direction of the electric field relative to the magnetic field. The plasma average must now include an average over electric field directions relative to the magnetic field. This modification increases computational time by introducing a double integration, as well as tripling the number of non-zero electric dipole matrix elements that enter the complex eigenvalue problem. We assume that the microfield itself remains unchanged by the introduction of a magnetic field. A discussion of this point is included in [11].

The external magnetic field also introduces additional terms into the Hamiltonian. We include here only the paramagnetic term, which produces the Zeeman effect:

$$H = H_a - \mathbf{E} \cdot \mathbf{d} - \mu_B \mathbf{B} \cdot \left[\mathbf{J} + \left(g_s - 1 \right) \mathbf{S} \right] \qquad (5)$$

for a state with quantum numbers $\mathbf{J}$ and $\mathbf{S}$, where μ_B is the Bohr magneton, and g_s is the electron gyromagnetic ratio. The additional (magnetic dipole) matrix element is evaluated by an atomic structure code in a similar manner to the electric dipole matrix element.

The positions of the Zeeman components in the resulting line profiles depend on the magnetic field strength, while the intensities of the Zeeman components vary with the direction of observation relative to the magnetic field. The line profile observed at an angle β relative to the magnetic field is given by

$$\phi(\omega, \beta) = \cos^2(\beta)\phi_{\parallel}(\omega) + \sin^2(\beta)\phi_{\perp}(\omega). \qquad (6)$$

where the subscripts here refer to the observation direction relative to the magnetic field.

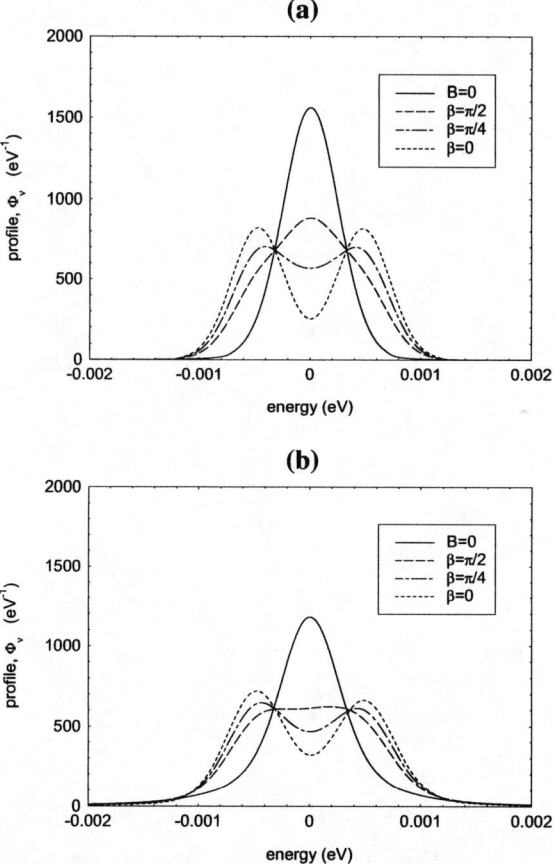

FIGURE 1. Line profiles for H Lyman-α at $T_e = T_i = 1$ eV with (a) $n_e = 10^{15}$ cm^{-3}, (b) $n_e = 10^{16}$ cm^{-3}. The energy scale here is relative to the position of the unbroadened transition. The solid curve is for B=0 while the other curves are for B=8 T for three different angles of observation β, measured relative to the magnetic field direction.

Hydrogen Lyman-α and Balmer-α Profiles

Figure 1 shows the emission profiles for hydrogen Lyman-α in the presence of an 8T magnetic field for plasma temperatures of 1 eV and electron densities of 10^{15} and 10^{16} cm^{-3}. For no magnetic field (B=0), the line shape is the usual Voigt profile. While the magnetic field significantly affects the spectral line shape, the characteristic Lorentz triplet is not apparent. This is due to thermal broadening, which at $T_i=1$ eV is comparable to the Zeeman splitting. Figure 2 shows the same profiles with $T_i=.001$ eV, effectively removing thermal broadening. The Lorentz triplet now becomes visible (for β≠0), as well as fine structure present in the individual peaks. Stark broadening remains evident, particularly at the higher electron density in Figure 2b. Note that Figure 2a uses a logarithmic scale.

Figure 3 shows the emission profiles for hydrogen Balmer-α for plasma temperatures of 1 eV and electron densities of 10^{14} and 10^{16} cm^{-3}. Doppler broadening is less important for this lower transition energy, so Zeeman splitting is the dominant feature at low densities, while Stark broadening dominates at higher densities.

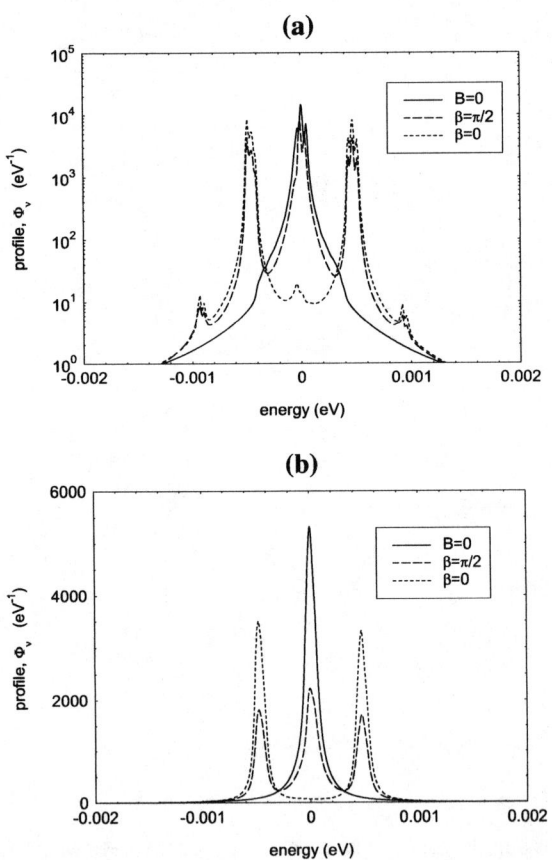

FIGURE 2. Same as Figure 1 with $T_i=.001$ eV, for two angles of observation.

MAGNETIC LINE RADIATION TRANSPORT

The modified TOTAL code has been fully integrated into the code CRETIN [9], which solves the equations of non-local thermodynamic equilibrium atomic kinetics with line radiation transport. As an example, we consider a one-dimensional plasma slab with conditions appropriate to Alcator C-Mod. The plasma has a thickness 5 cm and consists of hydrogen with atomic and electron number density both set to 10^{14} cm^{-3}, temperature of 1 eV, and magnetic field of 8T. Figures 4-6 show the results of a line transfer calculation using line profiles calculated for four different magnetic cases. The reference case (solid line) assumes no magnetic field. The next two cases assume that the magnetic field is in the plane of the slab (perpendicular to the direction of observation) or oriented normal to the slab, with line profiles varying with photon direction. The final case uses a direction-independent line profile, obtained by averaging the perpendicular and parallel line profiles.

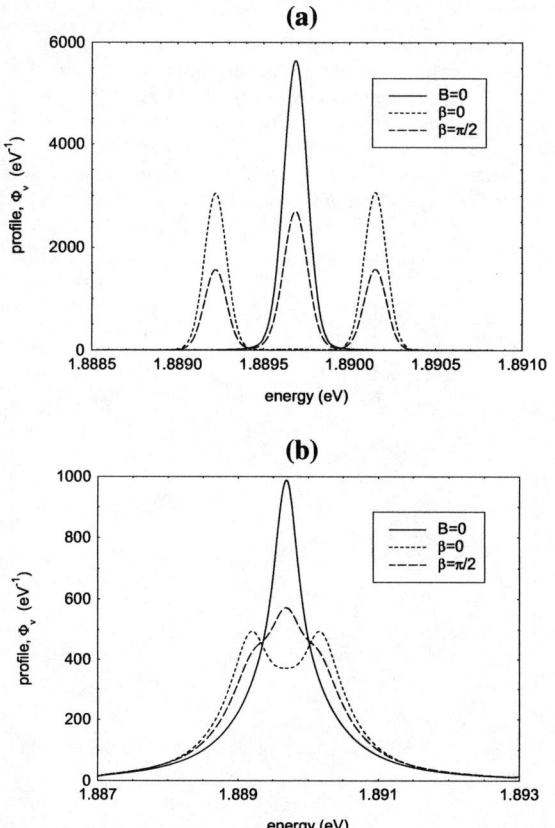

FIGURE 3. Line profiles for H Balmer-α with $T_e = T_i = 1$ eV and (a) $n_e = 10^{14}$ cm^{-3}, (b) $n_e = 10^{16}$ cm^{-3}. The solid curve is for B=0 while the other curves are for B=8 T for two different angles of observation β, measured relative to the magnetic field direction.

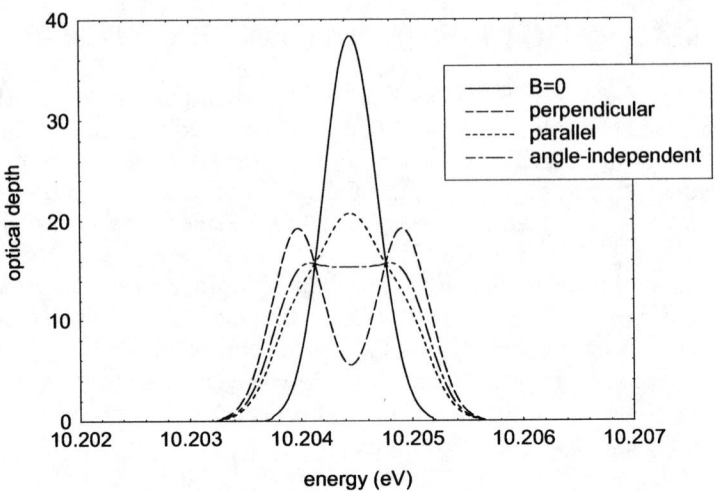

FIGURE 4. Optical depth for hydrogen Lyman-α as a function of energy for the 5 cm slab with $n_e=10^{14}$ cm^{-3} and $T_e=T_i=1$ eV, measured normal to the slab surface. The solid curve is for B=0 while the other curves are for B=8 T. The dashed curve was calculated using line profiles for a magnetic field in the plane of the slab (perpendicular to the final direction of observation, which is normal to the slab surface). The dotted curve was calculated using a line profile for a magnetic field normal to the slab surface. The dot-dashed curve was calculated using a line profile that was the average of the perpendicular and parallel profiles regardless of the actual angle relative to the magnetic field.

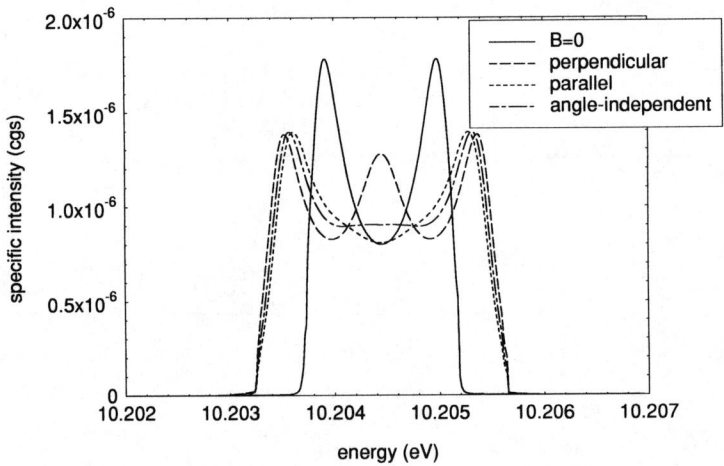

FIGURE 5. Specific intensity exiting normal to the slab surface, as a function of energy for the same cases as in Figure 4.

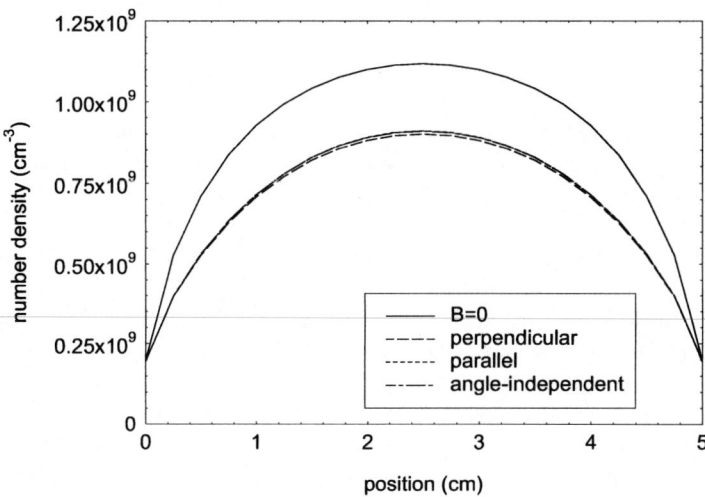

FIGURE 6. Upper state population for hydrogen Lyman-α as a function of position for the same cases as in Figure 4. The three curves with B=8 T very nearly overlap.

Figure 4 shows the optical depth across the slab in a direction normal to the slab surface. This quantity depends only on the line absorption profile in this single direction and the (essentially) uniform distribution of ground state populations throughout the slab. Although the optical depth varies with direction (and magnetic field orientation), the maximum of this quantity across the line profile is a commonly used measure for quantifying line radiation effects. Broadening due to Zeeman splitting is evident and the peak optical depth has decreased by approximately a factor of two.

Figure 5 shows the radiation spectrum emerging from the plasma normal to the slab surface. The B=0 case shows a strong inversion at line center, characteristic of an optically thick line. The parallel and perpendicular cases exhibit distinct spectral shapes, reflecting the magnetic field orientation. The total (integrated) radiation emitted increased by about 30% with the magnetic field present.

The effects of the radiation on the plasma are quite insensitive to the magnetic field orientation. This is demonstrated in Figure 6, which shows the upper state population density as a function of position. The magnitude of this quantity is much larger than the thermal value and is almost completely due to the effects of trapped line radiation. The value decreases towards the boundaries where radiation can escape more freely. The decreased trapping due to magnetic broadening results in a lower excited state population. Since this effect depends on the angle-integrated radiation intensity, it is insensitive to magnetic field orientation. This agrees with the results of Novikov, *et al* [12] in their investigation of magnetic field effects in radiation transfer through hydrogen plasmas.

INTEGRATION INTO EDGE PLASMA CODES

The result that an angle-independent treatment is sufficient for calculating plasma effects is significant, since that should simplify the process of integrating radiation effects into an edge plasma code. In this section we discuss one possible approach towards including radiation effects, based on a generalization of the collisional-radiative model of Stotler, Post and Reiter (SPR) [13].

SPR Collisional-Radiative Model

The plasma description generally used in an edge plasma code treats neutral hydrogen as a single species. Excited state distributions are not treated explicitly but become important at low temperatures and should be included in some manner. The approach of SPR uses a collisional-radiative model and assumes that the excited states are in equilibrium on plasma transport timescales. The excited states can then be expressed as functions of the ground state and ionized state. In this approach, the plasma transport deals explicitly with ground states and ionized states. The effects of excited state distributions enter through modified ionization and recombination coefficients. These coefficients are functions of temperature and electron density only, and are tabulated for use within the edge plasma code.

The excited state distributions affect the energy balance as well. Each ionization (recombination) results in an average gain (loss) of energy to the system of amount E_Z (E_R). These quantities differ slightly from the ionization energy of 13.6 eV because of the finite width of the electron distribution function as well as excited state effects, and the differences between these values and 13.6 eV are the tabulated quantities, defined by

$$\delta E_Z = E_Z - 13.6\,\mathrm{eV}\;,\;\delta E_R = 13.6\,\mathrm{eV} - E_R. \tag{7}$$

Generalization of SPR Approach

The SPR coefficients depend only on temperature and electron density. Optically thick line radiation introduces a spatial dependence as well, through the variation of the radiation field. As a first approximation, we assume that we can parameterize the coefficients by optical depth. This is equivalent to an escape factor treatment of line transfer and should be valid for simple geometries, but will need to be checked by more detailed calculations for complex cases.

As a further simplification, we assume that the line profile does not change dramatically across the region of interest so that the neutral column density is proportional to the optical depth. An equivalent parameterization is then the effective distance to boundary R, defined by

$$R = \int_0^r n_g(r')dr' \bigg/ n_g(r) \tag{8}$$

where n_g is the ground state neutral density at position r, which is measured to the nearest boundary.

Figure 7 displays the generalized ionization and recombination rate coefficients for an electron density of 10^{14} cm^{-3} and a wide range of values of R. The solid lines (R=0) give the optically thin SPR values. For temperatures below 2 eV, radiation effects dramatically change the rates. This is more easily seen in Figure 8, which displays the ratios of the generalization rate coefficients to the optically thin values.

Figure 9 shows the generalization of the tabulated energy changes per transition. The ionization values decrease slightly with increasing optical depth as ionizations occur out of excited states. The recombination values differ dramatically at low temperatures, as recombinations to the ground state are largely balanced by radiative excitations. The limiting case where recombinations only occur to the first excited state (E_R =3.4 eV) is equivalent to that used by Knoll in his simulations [6,7].

These coefficients were calculated using the radiation field present in the center of a spatially uniform plasma slab of extent 2R, assuming thermally broadened line profiles. The use of thermal broadening at this electron density can be justified by noting that trapping the first few Lyman lines, which have negligible Stark broadening, produces almost all of the change. Zeeman splitting, however, can decrease the optical depth by up to a factor of two, producing a corresponding decrease in the appropriate value of R for given plasma profiles. Future work will be required to clarify this correspondence, but the expected changes are much smaller than the overall effects of radiation trapping.

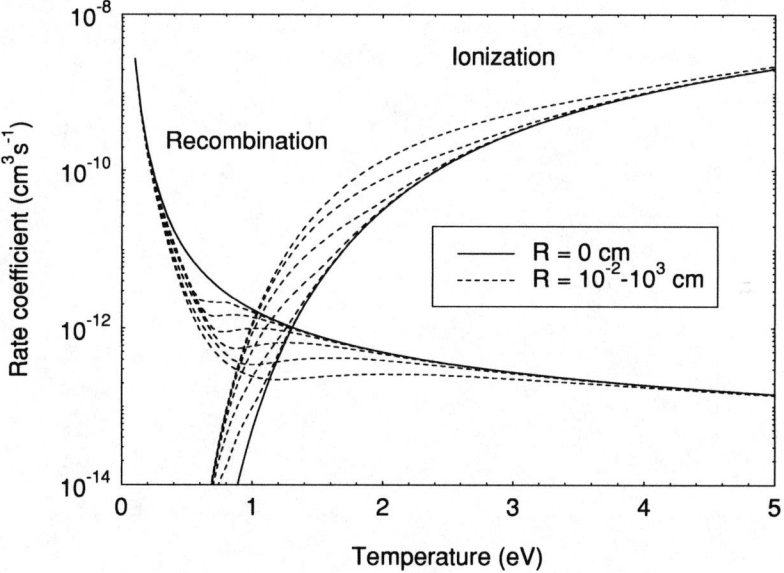

FIGURE 7. Effective ionization and recombination coefficients as a function of temperature for n_e=10^{14} cm^{-3}. The solid curves are for an optically thin plasma (R=0). The dotted curves are for six different optical depths, parameterized by the effective distance R with values ranging from R=10^{-2} cm to R=10^{3} cm. The effective width increases by a factor of 10 for each successive curve.

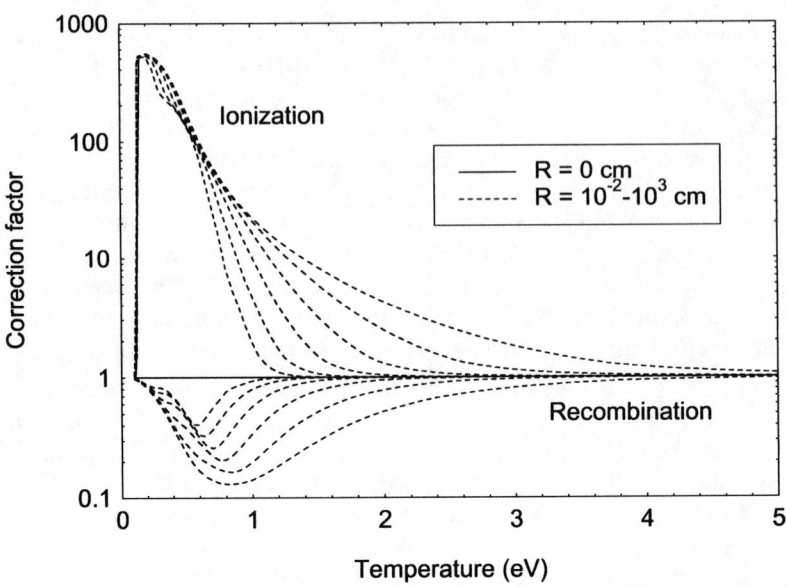

FIGURE 8. Correction factor for the optically-thin effective ionization and recombination rates for the same cases as in Figure 7. The correction factor is defined as the ratio of the rate to the optically-thin value.

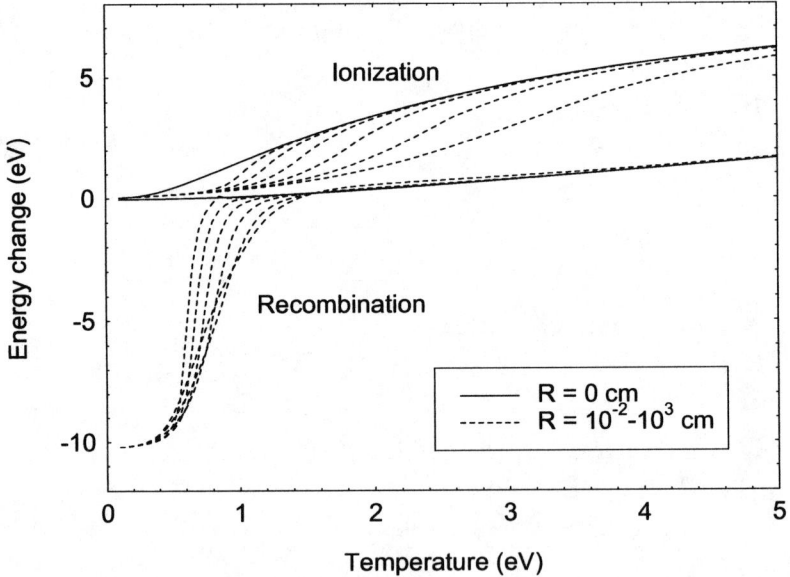

FIGURE 9. Average electron energy change per transition for ionization (δE_Z) and recombination (δE_R) for the same cases as Figure 7.

DISCUSSION

The first few steps towards incorporating radiation transport in edge plasma simulations are now complete. Line profiles in magnetized plasmas are available from the modified TOTAL code, for complex atoms as well as for hydrogen. The line profiles can be used for diagnostics in optically thin situations [10] and have been integrated into the CRETIN code for accurate calculations of radiation flow in optically thick plasmas.

As a step towards evaluating the effects of line radiation in edge plasmas, tables incorporating the modified SPR coefficients can be used by making minor modifications in existing edge plasma codes, and we have begun experimenting with these in the UEDGE code. We are also constructing a one-dimensional code that combines edge plasma transport processes and radiation transport. This will allow us to validate and improve the tabulated coefficients, as well as investigate other possible integration procedures. We expect these developments to substantially improve our understanding of radiation effects in tokamak edge plasmas.

ACKNOWLEDGMENTS

We would like to thank R.W. Lee and H.K. Chung (LLNL) for their valuable contributions.

This work was performed under the auspices of the U.S. Department of Energy by the University of California, Lawrence Livermore National Laboratory through contract number W-7405-Eng-48.

REFERENCES

1. Terry, J. L., Lipschultz, B., Bonnin, X., Boswell, C., Krasheninnikov, S. I., Pigarov, A. Yu., LaBombard, B., Pappas, D., and Scott, H. A., *J. Nucl. Mater.* **266-269**, 30 (1999).
2. Isler, R. C., McKee, G. R., Brooks, N. H., West, W. P., Fenstermacher, M. E., and Wood, R. D., *Phys. Plasma* **4**, 2989 (1997).
3. Terry, J. L., Lipschultz, B., Pigarov, A. Yu., Krasheninnikov, S. I., LaBombard, B., Lumma, D., Ohkawa, H., Pappas, D., and Umansky, M., *Phys. Plasma* **5**, 1759 (1998).
4. Wan A. S., Dalhed H. E., Scott H. A., Post D. E., and Rognlien T. D., *J. Nucl. Mater.* **220-222**, 1102 (1995).
5. Rognlien T. D., Milovich, J. L., Rensink, M. E., and Porter, G. D., *J. Nucl. Mater.* **196-198**, 347 (1992).
6. Knoll, D. A., P. R. McHugh, S. I. Krasheninnikov, and D. J. Sigmar, *Phys. Plasma* **3**, 293 (1996).
7. Knoll, D. A., *Nuclear Fusion* **38**, 133 (1998).
8. A. Calisti, Khelfaoui, F., Stamm, R., Talin, B., and Lee, R. W., *Phys. Rev A*, **42**, 5433 (1990).
9. Scott, H. A., *JQSRT*, **71**, 689 (2001).
10. Adams, M. L., Lee, R. W., Scott, H. A., Chung, H. K., and Klein, L., "Atomic Line Shapes in the Presence of an External Magnetic Field" in *Atomic Processes in Plasmas: 13th APS Topical Conference on Atomic Processes in Plasmas*, edited by D. R. Schultz and F. W. Meyer, AIP Conference Proceedings 635, New York: American Institute of Physics, 2002, pp. 261-272.
11. Adams, M. L., Lee, R. W., Scott, H. A., Chung, H. K., and Klein, L., *Phys. Rev E* (in press).
12. Novikov, V. G., Vorob'ev, V. S., D'yachkov, L. G., and Nikiforv, A. F., *JETP* **92**, 441 (2001).
13. Stotler, D. P., Post, D. E., and Reiter, D., *Bul. Am. Phys. Soc.* **38**, 1919 (1993).

Pulsed Doppler-Free Two-Photon Excitation of Hydrogen Lyman-α Fluorescence in Magnetically Confined Fusion Plasmas

D. Voslamber and J. Seidel

Physikalisch-Technische Bundesanstalt, Institut Berlin, Abbestr. 2-12, 10587 Berlin, Germany

Abstract. In previous theoretical investigations we found that two-photon Doppler-free laser-induced fluorescence (TPDF-LIF) in the Lyman-α line of the hydrogen isotopes may provide a diagnostic for magnetically confined fusion plasmas, in particular for the determination of the deuterium/tritium density ratio. The results obtained in these investigations were based on a quasi-stationary treatment of TPDF-LIF. So far, however, 243-nm laser radiation with sufficient intensity has only been generated as pulsed radiation with pulse lengths not much longer than the decay time of the fluorescence. Moreover, under suitable conditions short-pulse excitation may be more efficient than quasi-stationary excitation. For both reasons we have extended our treatment to include pulsed TPDF-LIF. Numerical calculations for representative plasma parameters show that up to about twice as many fluorescence photons as with quasi-stationary operation can be obtained with suitable laser pulse lengths and intensities.

INTRODUCTION

Magnetically confined fusion (MCF) plasma devices such as tokamaks are to exploit the deuterium-tritium nuclear fusion reaction

$$D + T = {}^4He + n + 17.6\,MeV$$

for energy production. Obviously, measurements of the D/T number-density ratio with spatial and temporal resolution are an important issue for the burn control and performance optimisation of such devices. In several previous publications [1–6] we proposed and investigated two-photon Doppler-free laser-induced fluorescence (TPDF-LIF), possibly combined with neutral beam injection, as a suitable means for this purpose (and perhaps for the determination of other plasma parameters as well). A schematic is shown in Figure 1. TPDF-LIF has the advantage of providing a truly direct measurement of the density ratio. Other methods suggested in the past, such as neutron flux measurements, the analysis of escaping fast neutral particles, or collective Thomson scattering rely on plasma modelling or are severely constrained by the magnetic field geometry in the scattering volume (see references quoted in Ref. [4]).

The essential step in TPDF-LIF is the Doppler-free excitation of the hydrogen isotopes from their ground states because this allows to resolve the isotope separation (227 GHz) of the D and T Lyman-α line frequencies. The process is made Doppler-

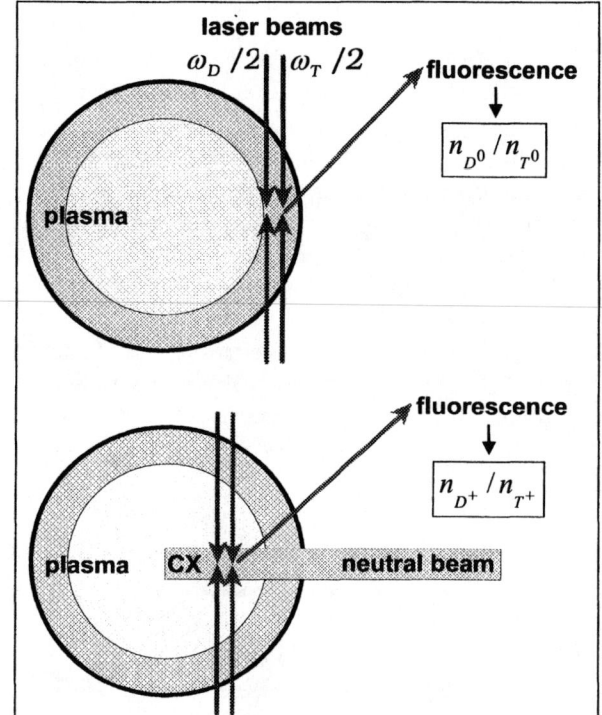

FIGURE 1. Schematic representation of TPDF-LIF for the determination of the D/T fuel mix in MCF plasmas. Probing the plasma edge yields the intrinsic neutral density ratio n_{D0}/n_{T0}. Probing the plasma core with the help of a powerful diagnostic neutral beam yields the density ratio of neutrals created by charge exchange (CX) between the beam neutrals and the plasma ions. This ratio therefore represents the ion density ratio n_{D+}/n_{T+}.

free by using a pair of two counterpropagating laser beams for each isotope, each of the beams having half the transition frequency and providing one photon to the excitation process. With opposite directions of the two beams of a pair, the two Doppler shifts have opposite signs and cancel each other in the sum of the frequencies. Thus, the excitation process is Doppler-free, but the fluorescence lines are not. However, the fluorescence signals can be separated temporally by introducing a short time delay between the laser shots for D and T.

In order to limit the expenditure of computation, the numerical results of our previous investigations of TPDF-LIF were based on a quasi-stationary treatment. However, 243 nm radiation with sufficient intensity to produce fluorescence distinctly exceeding the thermal Lyman-α background radiation has so far only been generated with laser pulses not much longer than the decay time of the fluorescence [7–9]. Moreover, the use of even shorter, more intense laser pulses may be advantageous on its own, as these may drive the two-photon transition closer to saturation [6]. Both aspects have motivated us to extend our theoretical treatment to include short-pulse operation, essentially with the goal of determining those laser radiation parameters for which the fluorescence outcome would be optimal. Results of first representative calculations are presented in the following.

HYDROGEN ATOMS IN MCF PLASMAS

A detailed account of the theory of TPDF-LIF for the hydrogen Lyman-α line in MCF plasmas has been given in Ref. [6], so that we can confine the discussion to an outline of basic facts here.

In an MCF plasma, hydrogen atoms move with high thermal velocities v in a strong magnetic field B. In the atomic rest frame, B and the electric Lorentz field $v \times B$ give rise to a combined Zeeman and motional Stark effect. By solving the nonrelativistic Schrödinger eigenvalue equation, the energy levels belonging to $n = 2$ are found to split as indicated in Figure 2. It is the one of the unshifted levels, with its eigenstate (denoted as |1> in the following) being a (velocity-dependent) mixture of angular momentum 2s and 2p states, that makes efficient TPDF 1s → 2s excitation possible at all, because the corresponding central transition frequency is essentially the same for all atomic velocities – apart from its second- order Doppler shift and some other small effects discussed below.

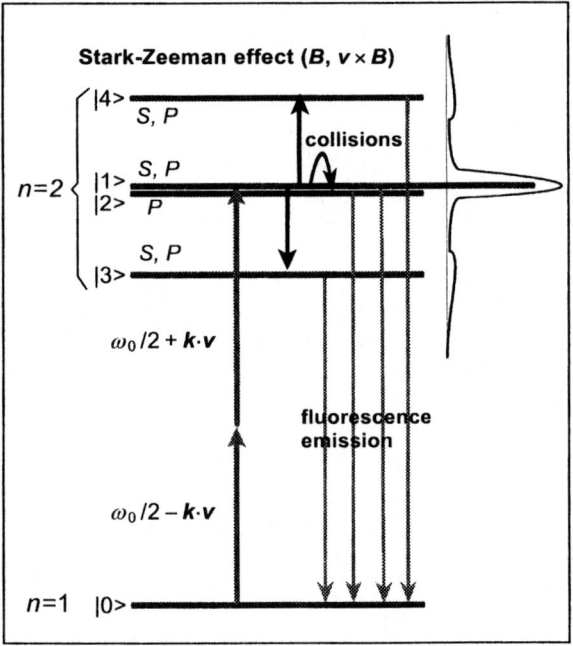

FIGURE 2. Lyman-α energy levels for a hydrogen atom moving with velocity v in a magnetic field B (nonrelativistic approximation without fine and hyperfine structure). Two of the upper levels are symmetrically shifted away from the unperturbed energy, but the other two levels remain unshifted (shown slightly separated in the figure). One of the two "central" eigenstates can be chosen to be a pure 2p state (with $m_l = 0$). The other eigenstate contains a 2s and a 2p component, as do the eigenstates with shifted energies. These are the states that can be reached by TPDF excitation from the 1s ground state. However, the large thermal velocity spread results in broad absorption profiles for the transitions to the shifted levels, as indicated in the figure. Therefore, efficient absorption can only be realised in the narrow central component. After two-photon excitation has created an upper-level population, this is redistributed among the other upper levels by collisions with plasma ions, and spontaneous one-photon decay (2p → 1s) of all these levels leads to the emission of fluorescence photons. The competing processes of two-photon induced emission and three-photon ionisation are not indicated.

For a quantitative treatment of TPDF-LIF, it is indispensable to go beyond the nonrelativistic Schrödinger equation. With the large linear Doppler broadening eliminated, the remaining homogeneous line widths are typically only some 100 MHz, so that the transition frequency for the excitation process should be known with an uncertainty of at most 10^8 Hz. This requires that relativistic effects and fine structure (including the Lamb shift) as well as hyperfine structure be taken into account at least to the lowest orders in the fine-structure constant α, and that various line-broadening mechanisms be included in the calculations.

The two-photon transition probability is affected by spectral broadening depending on both the laser characteristics (e. g., bandwidth and angular divergence giving rise to residual linear Doppler broadening) and the shape of the TPDF Lyman-α absorption line. Broadening effects from the laser can be minimised by technical means (except for the effect arising from the Fourier bandwidth limit in the case of short pulses), but this is not possible for the intrinsic broadening of the TPDF absorption profile. Apart from natural broadening resulting from the spontaneous decay of the 2p component of the TPDF excited level, the broadening is mainly due to four effects:

(i)   *ion-impact Stark broadening*, which outweighs electron and neutral broadening by far,
(ii) *three-photon-ionisation broadening* owing to the shortening of the lifetime of the two-photon-excited state as a result of ionisation by a third laser photon,
(iii) *motional Stark broadening* due to the thermal spread of the electric $v \times B$ field, which affects the fine structure components of the central upper level,
(iv) *quadratic Doppler broadening*, which remains after the elimination of linear Doppler broadening.

Of these effects, the motional Stark effect and the quadratic Doppler effect, both of them depending on v, are correlated and do not contribute independently to the line width. Also, these two effects are usually the most important ones in MCF plasmas (unless they are overtaken by ionisation broadening at high laser intensities or by ion-impact Stark broadening in the plasma edge). For TPDF-LIF, it is a fortunate circumstance that the corresponding line shifts for a given atomic velocity have opposite signs and cancel to a significant extent. This leads to a substantial reduction of the broadening and shift of the two-photon transition and an appreciable increase of the two-photon absorption probability.

RATE EQUATIONS FOR TPDF-LIF

With a given laser-pulse energy E_L, short-pulse operation may be expected to yield a higher fluorescence outcome than quasi-stationary operation since shorter pulse duration implies higher irradiation intensity and should thus lead to higher upper-level population. However, very short, intense pulses also give rise to effects going into the opposite direction: their strength leads to notable three-photon ionisation (causing the loss of fluorescing atoms and additional broadening of the TPDF absorption profile), and their shortness limits the overlap region for TPDF absorption and also increases the laser bandwidth. It seems likely therefore that the optimal fluorescence outcome is obtained for some intermediate pulse duration.

In a first attempt to determine those laser radiation parameters for which the fluorescence yield is highest, we took the laser irradiance to be exponentially decaying, $I_L(t) = E_L/(\pi r^2 \tau_L)\exp(-t/\tau_L)$, with the pulse energy E_L, the beam radius r, the decay time τ_L, and the frequency being the same for the two counterpropagating beams. We used Eqs. (3.6) and (3.9) of Ref. [5], the first to express the number of fluorescence photons in terms of the time-dependent atomic density matrix, and the second as the system of rate equations governing the evolution of the density matrix in the subspace relevant for TPDF-LIF. This subspace is spanned by the ground state |0> and the four excited Stark-Zeeman eigenstates |1>, |2> (unshifted) and |3>, 4> (shifted). The use of rate equations involves no approximation because "coherences" (off-diagonal density-matrix elements) can neither be created by the one- and two-photon transitions which go into the pure states |0> or |1>, nor by the collisions which are incoherent themselves. The inclusion of relativistic effects, fine and hyperfine structure and the Lamb shift does not increase the number of effective dimensions because all radiative and collisional transitions governing the rate equations are spin-conserving. Actually, owing to the symmetry between the two displaced upper levels the effective number of dimensions can be reduced to four.

The numerical procedure consists of first solving the rate equations for a single atom with velocity v and electron and nuclear spin magnetic quantum numbers m_s and m_I. The transition rates for the processes involved (two-photon absorption and stimulated emission, three-photon ionisation, one-photon spontaneous decay, and collisional redistribution) as well as the line profiles associated with these processes are taken from Reference [6]. Initially, until the laser irradiation starts at $t = 0$, the atom is taken to be in its ground state. During laser irradiation all the transition processes mentioned occur simultaneously. One of them, three-photon ionisation, causes the trace of the density matrix to decrease continuously from its initial value of one.

After the rate equations have been solved numerically for all combinations of m_s and m_I and sufficiently close-meshed values of v, the numerical integration over the observation time interval and the averages over the spin variables and the atomic velocity are performed according to Eqs. (3.7) and (3.6) of Reference [6]. Assuming the velocity distribution to be rotationally symmetric about the magnetic field direction (we took it to be Maxwellian), the integration over the azimuth angle of v can be carried out analytically, with the possibility of using Eqs. (5.2) of Ref. [6] if the extremely weak angular and polarisation dependence of the three-photon ionisation rate is neglected. Renouncing a polarisation analysis of the fluorescence radiation, the expressions in Eqs. (5.2) are summed over two orthogonal polarisations. Due to the finite effective length $c\tau_L$ of the laser light trains the length l of the observed cylindrical volume does not fully come into play. This is accounted for by a weight factor that reduces l to the time average of the length of the effective fluorescing overlap volume. The factor equals $1-l/(2c\tau_L)$ for $c\tau_L \geq l$ and $c\tau_L/(2l)$ for $c\tau_L \leq l$.

RESULTS AND CONCLUSION

We performed calculations for the hydrogen isotope tritium and three sets of plasma parameters which are about typical for the conditions of a large MCF device in the edge (Fig. 3), in the core (Fig. 5) and in the region in between (Fig. 4). The various

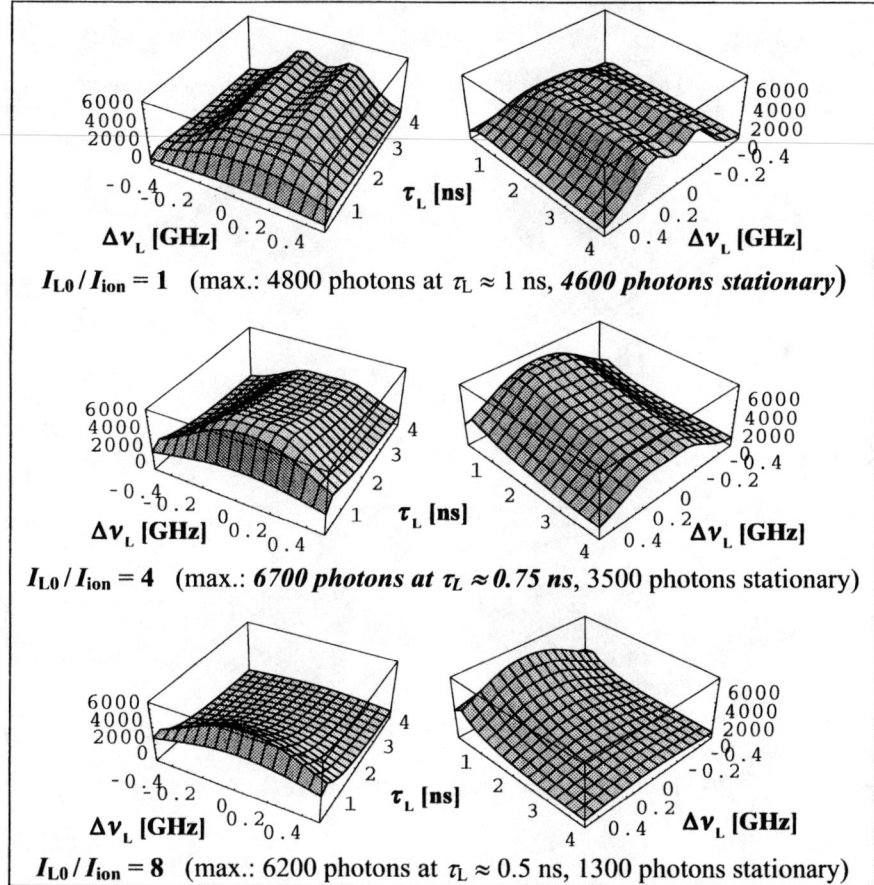

$I_{L0} / I_{ion} = 1$ (max.: 4800 photons at $\tau_L \approx 1$ ns, *4600 photons stationary*)

$I_{L0} / I_{ion} = 4$ (max.: *6700 photons at $\tau_L \approx 0.75$ ns*, 3500 photons stationary)

$I_{L0} / I_{ion} = 8$ (max.: 6200 photons at $\tau_L \approx 0.5$ ns, 1300 photons stationary)

FIGURE 3. Variation of the total number of observed Lyman-α fluorescence photons with laser-frequency detuning Δv_L and laser-pulse length τ_L for different values of the (initial) peak laser irradiance I_{L0} (in terms of the critical ionisation irradiance $I_{ion} = 436$ GW/m²). The photon numbers relate to a length $l = 100$ mm of the cylindrical observation volume, a solid angle of 1 mrad perpendicular to the counter-propagating laser beams, and an observation time of 8 ns starting with the onset of the laser pulses in the observation volume. The laser-beam radius is implicitly fixed by the laser-pulse energy $E_L = 100$ mJ, I_{L0}, and τ_L. The results shown in this figure correspond to plasma conditions typical for the *edge region* of a large MCF plasma device: magnetic field $B = 4$ T, temperature $T = 10^6$ K, electron density $n_e = 10^{19}$ m⁻³, atomic density $n_0 = 5 \times 10^{15}$ m⁻³. The results show that pulsed TPDF-LIF can produce fluorescence signals nearly 50% higher than the signal for (quasi-)stationary operation. It is also seen that the measurement does not benefit from increasing the laser irradiance beyond a few I_{L0}. The double-peak appearing for the lowest laser irradiance results from the hyperfine-structure splitting.

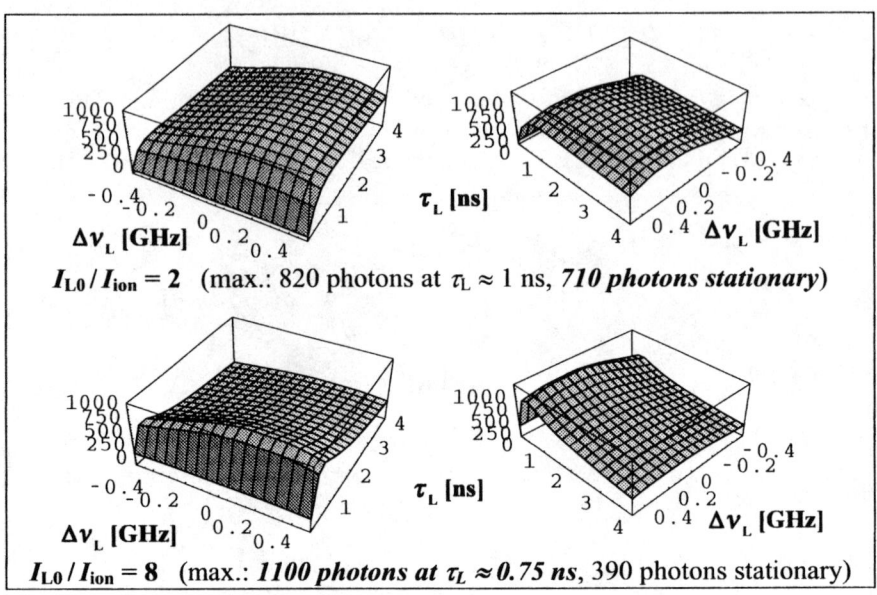

$I_{L0} / I_{ion} = 2$ (max.: 820 photons at $\tau_L \approx 1$ ns, *710 photons stationary*)

$I_{L0} / I_{ion} = 8$ (max.: *1100 photons at $\tau_L \approx 0.75$ ns*, 390 photons stationary)

FIGURE 4. Same as Fig. 3, but for plasma conditions *farther away from the edge*: magnetic field $B =$ 4 T, temperature $T = 10^7$ K, electron density $n_e = 10^{20}$ m^{-3}, atomic density $n_0 = 10^{15}$ m^{-3}.

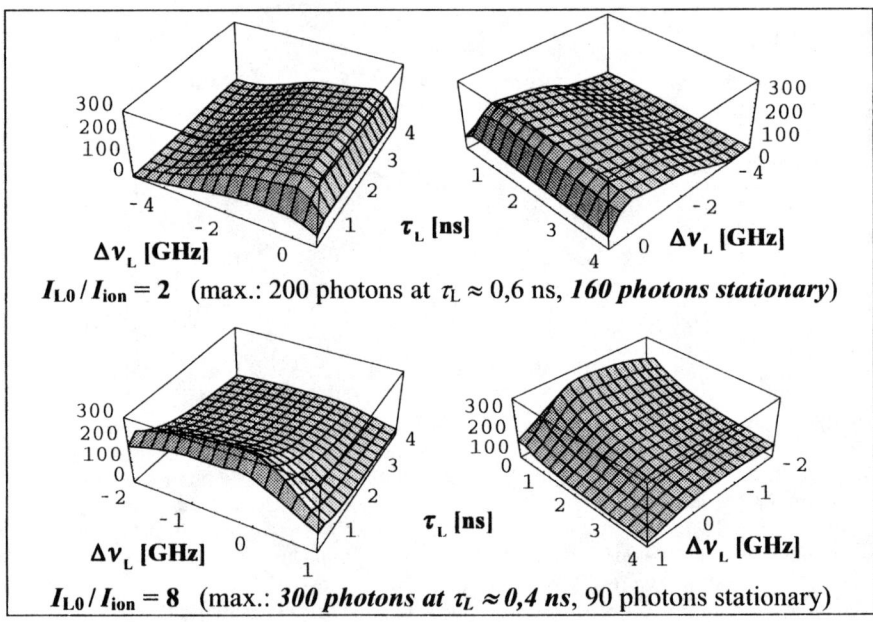

$I_{L0} / I_{ion} = 2$ (max.: 200 photons at $\tau_L \approx 0.6$ ns, *160 photons stationary*)

$I_{L0} / I_{ion} = 8$ (max.: *300 photons at $\tau_L \approx 0.4$ ns*, 90 photons stationary)

FIGURE 5. Same as Fig. 3, but for plasma conditions typical for the *core region* of a large MCF plasma device: magnetic field $B = 4$ T, temperature $T = 10^8$ K, electron density $n_e = 10^{20}$ m^{-3}, atomic density $n_0 = 10^{15}$ m^{-3} (requires injection of a neutral beam).

parameters chosen are indicated in the figure captions. The rather high neutral density assumed for the core ($n_0 = 10^{15}\,\text{m}^{-3}$) can only be reached by charge exchange with a powerful neutral beam [4]. For neutral densities n_0 below $10^{15}\,\text{m}^{-3}$, the photon numbers would be smaller by a factor of $n_0/(10^{15}\,\text{m}^{-3})$. The graphs shown in Figures 3–5 represent fluorescence photon numbers as functions of the laser-frequency detuning $\Delta \nu_L$ and the laser-pulse duration τ_L for given values of the initial irradiation I_{L0} of each laser beam. The graphs on the right-hand side correspond to those on the left-hand side "seen from behind".

In the figures, the initial irradiance I_{L0} is given in terms of the "critical ionisation irradiance" $I_{ion} = 436$ GW/m^2, which leads to an average ionisation rate equal to the average one-photon spontaneous decay rate. This is a convenient "natural" measure of the laser irradiance since three-photon ionisation will, in the end, make TPDF-LIF with very short laser pulses inefficient. The onset of this effect can be seen by comparing the graphs at the bottom of Fig. 3 (where I_{L0}/I_{ion} has exceeded its optimal value) with the other two pairs of graphs in the figure, and has also been confirmed for the conditions of Figures 4 and 5.

In Figs. 3–5, results are shown for laser-pulse lengths up to $\tau_L = 4$ ns, a value which clearly exceeds the mean inverse Einstein coefficient for spontaneous decay. Therefore, the results for the longest pulses in the figures are expected to give an approximate account of quasi-stationary operation, even if the observation time of 8 ns is longer than τ_L. The fluorescence outcome for these cases is highest for values of I_{L0}/I_{ion} varying between 1 and 2. The optimal values of I_{L0}/I_{ion} for short-pulse operation lie between 4 and 8 and are associated with laser-pulse lengths of about 0.5 ns to 1 ns. The maximum fluorescence-photon numbers indicated in the various graphs show that short-pulse operation may be more efficient than quasi-stationary operation: with suitably chosen parameters it is seen to provide a notable – though not spectacular – gain of up to a factor of two.

REFERENCES

1. D. Voslamber, *9th International Conference on Spectral Line Shapes*, edited by J. Szudy, Torun, Nicholas Copernicus University Press, 1988.
2. D. Voslamber, Report EUR-CEA-FC-1387 (1990).
3. D. Voslamber, in: *Spectral Line Shapes*, Vol. 8, edited by A. D. May, J. R. Drummond, and E. Oks (AIP Conf. Proc. 328), New York, AIP Press, 1995.
4. D. Voslamber and W. Mandl, *Rev. Sci. Instrum.* **69**, 1702–1715 (1998).
5. D. Voslamber and J. Seidel, *Rev. Sci. Instrum.* **70**, 928–931 (1999).
6. D. Voslamber, *Rev. Sci. Instrum.* **71**, 2334–2352 (2000).
7. K. Grützmacher, M. I. de la Rosa, J. Seidel, A. Steiger, G. Fußmann, W. Bohmeyer, and D. Voslamber, in: *Diagnostics for Experimental Thermonuclear Fusion Reactors*, edited by P. E. Stott, G. Gorini, and E. Sindoni, New York, Plenum, 1996.
8. K. Grützmacher, A. Steiger, Ch. Seiser, M. I. de la Rosa, and U. Johannsen, in: *Diagnostics for Eperimental Thermonuclear Fusion Reactors 2*, edited by P. E. Stott, G. Gorini, and E. Sindoni, New York, Plenum, 1998.
9. Ch. Seiser, *Ph.D. thesis*, Technische Universität Berlin, 1998.

Investigation of non thermal effects from the D_α line wings in edge plasmas

Y. Marandet*, H. Capes†, L. Godbert-Mouret*, R. Guirlet†, M. Koubiti* and R. Stamm*

*PIIM, Université de Provence, Centre de St-Jérôme F13397 Marseille France
†DRFC, Association Euratom-CEA, CEA Cadarache, 13018 Saint Paul lez Durance Cedex, France

Abstract. The far wings of intense D_α lines measured at the edge of the Tore Supra Tokamak are found to exhibit a power-law behavior. The characteristic exponent is not far from two. Since the low density rules out thermal Stark broadening, we discuss non thermal effects which may arise from the edge plasma drift-wave turbulence. We suggest that both the Stark and the Doppler profile could be affected by the turbulence.

INTRODUCTION

The D_α line ($\lambda_0 = 6560.1$ Å) is routinely used for diagnosing edge plasmas of Tokamaks. These plasmas are often driven far from equilibrium, since they are affected by sources and sinks of particles due to the wall and the hot core plasma. Therefore, strong gradients occur, leading to instabilities, and sometimes to turbulence. This work is devoted to a possible effect of this turbulence on the D_α line wings. We will first present the experimental setup and the measurements revealing a power-law scaling of the line wings in terms of the wavelength. We will then show that the thermal Stark effect cannot be held responsible for this behavior, due to low density of the edge plasma. As a consequence, non thermal effects should probably be accounted for in the spectral lineshape calculation. We will discuss here how the Stark and the Doppler profiles might be affected by the edge plasma turbulence.

D_α MEASUREMENTS AT THE EDGE OF TORE SUPRA

The spectroscopic measurements use four in situ optical fibres equipped with small telescopes. The lines of sight are roughly parallel to the magnetic field lines. The fibres transfer the emitted radiation through a high-resolution Czerny-Turner spectrometer (1200 or 2000 groove mm^{-1} gratings) to a two-dimensional CCD camera operating in the range $4000 - 10000$ Å, allowing $40 - 70$ Å wide spectra to be recorded. The instrumental function is well approximated by a Gaussian. As the electron density is low ($10^{18} - 10^{19} m^{-3}$) and the electron temperature lies between $10 - 50$ eV, the D_α line broadening is dominated by the Zeeman and Doppler effects, at least in the center of the line. Moreover, since the magnetic field is larger than 1 T in the Tore Supra edge plasma,

CP645, *Spectral Line Shapes: Volume 12, 16th ICSLS*, edited by C. A. Back
© 2002 American Institute of Physics 0-7354-0100-4/02/$19.00

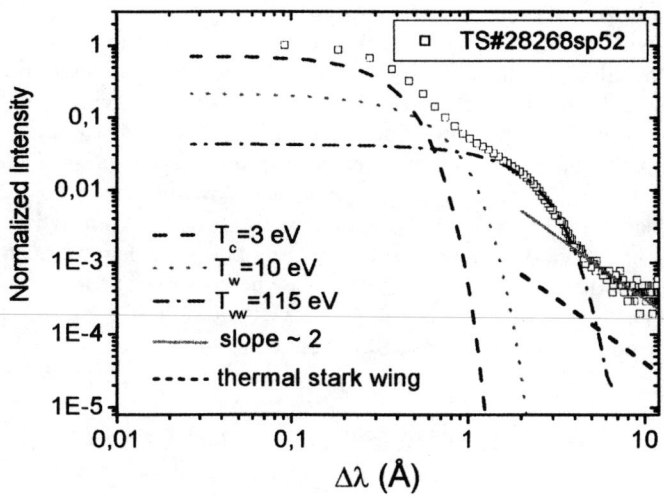

FIGURE 1. Plot of the D_α line wings in a logarithmic scale, shot TS28268. The squares correspond to the experimental data.

the strong magnetic field approximation is sufficient, i.e. fine structure can be neglected. Therefore, the D_α line splits into three Doppler-broadened components (one π and two σ). The lateral σ components are equally separated from the central π component. In our situation, under parallel observation with respect to the magnetic field, only the σ components are observable. The Zeeman profile has to be convolved with the Doppler profile I_D, related to the Velocity Distribution Function (VDF) $f(v)$ of the neutrals along the line of sight by :

$$I_D(\Delta\lambda)d(\Delta\lambda) = f(v)dv \tag{1}$$

If the emitters are in thermal equilibrium at temperature T, we have :

$$I_D(\Delta\lambda) = \frac{1}{\sqrt{\pi}\Delta\lambda_D} \exp\left(-\left(\frac{\Delta\lambda}{\Delta\lambda_D}\right)^2\right) \tag{2}$$

with the Doppler width $\Delta\lambda_D$ given by $\Delta\lambda_D = 4.63 \times 10^{-5}\lambda_0\sqrt{T/A}$, where A is the atomic weight of the emitter and T its temperature in eV.

In a previous work we showed [1, 2] that the line-shape cannot be fitted using only one Maxwellian population of neutrals. For the usual symmetric spectra (squares on Figure 1), two Maxwellian populations (dashed and dotted curves) with different temperatures were needed. These temperatures are related to the plasma-wall interaction and the edge plasma physics. However, in order to fit the line wings, a third population (dashed-dotted curve) of neutrals having a much larger temperature, of the order of 100 eV, has to be considered. Such a hot population may be related to charge exchange with core ions and is quite usual at the edge of Tokamaks [3, 4]. However, in the best signal to noise ratio cases, the line shows an additional feature. Indeed, a linear-like behavior in a logarithmic scale is then seen in the very far wing. This would imply a power law behavior for the profile. The signal to noise ratio, which can be improved by averaging over several successive spectra during the shot, is large enough to allow the observation of the line wing over four decades in intensity. The line-shape thus shows a power law-tail $\Delta\lambda^\alpha$, with a characteristic exponent α between 1.5 and 2.5. Although negligible at the center of the line, the Stark effect could lead to a power-law behavior with such an exponent in the far wing. The next section is thus devoted to the calculation of the D_α thermal Stark broadening for the conditions of the Tore Supra edge plasma.

ROLE OF THE THERMAL STARK EFFECT

For a density less than $10^{19}m^{-3}$ and a temperature of the order of 10 eV, a binary model is suitable both for electrons and ions. Moreover, in the case of the ions, the impact approximation is valid at the center of the line, whereas in the far wing the static approximation should be used. The electronic broadening can be treated using the impact theory, but in the wing incomplete collisions have to be accounted for. We are therefore in conditions where a unified theory [5, 6] is applicable. An accurate calculation of the D_α Stark profile should also include the coupling with the Zeeman effect. Neglecting this coupling leads to an over-estimation of the Stark width of a Zeeman component. Indeed, when the impact theory applies, the Zeeman splitting introduces an additional cutoff in the collision operator, which reduces the Stark width. In the far wings, where the static approximation is suitable for the ions, the Zeeman splitting becomes negligible. We have therefore calculated the Stark profile I_S using the following scalar model :

$$I_S(\Delta\omega) = \frac{\gamma(\Delta\omega)}{\Delta\omega^2 + \gamma(\Delta\omega)^2} \quad , \gamma = \gamma_i + \gamma_e \tag{3}$$

where the collision operator is the sum of an ionic and an electronic contributions. (Both $\gamma_i(0)$ and $\gamma_e(0)$ are computed using the GBK model). The electron density and temperature are measured from the probes, and we assume $T_i = T_e$. Since we are only interested in the far wing of the line, the convolution with the Doppler profile is not necessary to compare the resulting Stark profile to the experimental data. This comparison leads us to the conclusion that the thermal Stark effect cannot explain the observed power-law behavior in the wings, since the density is too low.

These results suggest that some non-equilibrium effect, which modifies either the Stark or the Doppler profile in the edge plasmas, is missing in the standard line broadening model. In the following, we will consider the effects the edge plasma turbulence may have on the D_α profile. Let us start with turbulence-enhanced Stark effect.

NON THERMAL STARK EFFECT

The edge plasma low-frequency drift-wave turbulence [7, 8] has been intensely studied since it is held responsible for the anomalous transport in Tokamaks. At the edge of Tokamaks, there are strong gradients of density and temperature which can lead to instabilities according to the direction of the magnetic field lines curvature. This mechanism is similar to the one underlying the Rayleigh-Taylor instability in fluid mechanics. The code RBM3D, developed in our lab, is designed to simulate this low frequency turbulence in the Tore Supra geometry. The main features that are observed are a characteristic time scale of the order of the microsecond and a typical length scale roughly given by the ion Larmor radius calculated using the electron temperature, ρ_s. The associated density or electrostatic potential fluctuations can be very large, up to 50% in some cases.

We shall now discuss the influence this turbulence may have on the Stark profile of the D_α line. A crude estimation of the turbulent electric field magnitude is given by :

$$E_{turb} = \frac{k_B T}{e \rho_s} \approx 100 kV.m^{-1},\qquad(4)$$

and is found to be at least of the order of the Holtsmark field, which is the typical thermal field. The Stark profile could therefore be significantly different than in the thermal case. Since the turbulence time scale is on the one hand much larger than the emitter lifetime, and on the other hand much smaller than the spectrometer integration time, a static modelling of turbulence is suitable. In order to introduce properly the turbulent electric field in the line-shape formalism, we use the following splitting of the total electric field created by the plasma :

$$\vec{E}(\vec{R},t) \simeq \underbrace{\sum_{j=1}^{N_D} q_j \frac{(\vec{R}-\vec{r}_j(t))}{|\vec{R}-\vec{r}_j(t)|^3}}_{E_{therm}} + \underbrace{\int \rho(\vec{r}',t) \frac{(\vec{R}-\vec{r}')}{|\vec{R}-\vec{r}'|^3} dr_M}_{E_{turb}} \qquad (5)$$

The particles in the neighborhood of the emitter, say within the Debye sphere (corresponding to the first term), lead to thermal-like Stark broadening, calculated using the local density and temperature. The VDF of both ions and electrons are assumed to be Maxwellian. This assumption is consistent with the use of fluid equations to describe the macroscopic behavior of the plasma, i.e. to calculate the local temperature and density. In addition to this first contribution, the second term in (5), corresponds to the turbulent electric field contribution, due to the large scale inhomogeneities of the plasma. Using an appropriate coarse-graining procedure, this field is easily shown to be the one which appears in the fluid equations. The hamiltonian of the emitter can thus be written as :

$$H = H_0 + H_B - \vec{d} \cdot \vec{E}_{therm} - \vec{d} \cdot \vec{E}_{turb} \qquad (6)$$

As explained before, the turbulent electric field can be considered as static. The Stark profile

$$I_S(\Delta\omega; E_{turb}, n(E_{turb}), T(E_{turb})) \qquad (7)$$

is then computed for a given E_{turb} and the corresponding local density and temperature. Indeed, it should be emphasized that all these plasma parameters fluctuate on the same time-scale. Finally, we have to average the profile over the integration time of the spectrometer. This average can be written as :

$$\langle I_S \rangle (\Delta \omega) = \int W(E_{turb}) I_S(\Delta \omega; E_{turb}, n(E_{turb}), T(E_{turb})) dE_{turb} \tag{8}$$

where W is the Probability Distribution Function (PDF) of the turbulent electric field. This distribution, as well as the correlations between n, T and E_{turb} can be computed using the fluid code. An application of this model to the D_α Stark profile is on progress.

Drift wave turbulence is often identified to plasma edge turbulence. However, high frequency Langmuir-like turbulence may also exist at the edge of Tokamaks. In the case of *broadband* high frequency turbulence, the electronic impact width may be increased [9, 10]. In our situation, an accurate calculation would require to take the magnetic field (which modifies the plasma wave dispersion relation) into account. However, since the use of additional heating does not modify the line wings, we assumed that Langmuir turbulence does not play the dominant role in our problem.

DOPPLER PROFILE AND STRANGE TRANSPORT

Numerical simulations of magnetized plasma turbulence have shown that *strange transport* (i.e. non-diffusive transport) may affect charged particles. In this section, we assume that strange transport may also affect neutral relaxation. We propose an approach based on a former work [2] in which we modelled the neutral relaxation using a Langevin equation :

$$\frac{dv}{dt} = -\nu v + \frac{F(t)}{M} \tag{9}$$

where the fluctuating force $F(t)$ is assumed to have Gaussian statistics and M is the mass of the neutral. The VDF can be calculated using $f(v,t) = \langle \delta(v - v(t)) \rangle$, where the brackets stand for an ensemble average. This VDF is solution of the following Fokker-Planck equation which describes the relaxation :

$$\frac{\partial f(v,t)}{\partial t} = v \frac{\partial}{\partial v}[vf] + \frac{k_B T_i}{M} \frac{\partial^2 f}{\partial v^2} \tag{10}$$

We now assume that due to the background turbulence, the statistic of the fluctuating force is no longer Gaussian and of a Lévy type. The PDF of this force is then given by :

$$L_\alpha(F) = \int_{-\infty}^{+\infty} dk e^{-ikF} e^{-B|k|^\alpha} \tag{11}$$

where the exponent α satisfies $0 < \alpha < 2$. This requirement ensures that $L_\alpha(F)$ is definite positive. The Lorentzian distribution corresponds to the case where $\alpha = 1$, whereas the Holtsmark distribution is found in the 3D case for $\alpha = 3/2$. The variance of these PDF is infinite :

$$\langle F^2 \rangle = \infty \tag{12}$$

This divergence means that the large values of the force are much more likely than in the Gaussian case. The Lévy distribution has indeed a power law tail, given by :

$$L_\alpha(F) \propto \frac{1}{|F|^{\alpha+1}}, \quad F \longrightarrow \pm\infty \tag{13}$$

Using the same approach than in Gaussian case, we are led to the following VDF in the Fourier space :

$$\tilde{f}(k,t) = \tilde{f}(ke^{-vt}) \exp\left(-\frac{D_\alpha}{\alpha v}(1 - e^{-vt})|k|^\alpha\right) \tag{14}$$

This VDF is the solution of a fractional Fokker-Planck equation :

$$\frac{\partial f(v,t)}{\partial t} = v\frac{\partial}{\partial v}[vf] + D_\alpha \frac{\partial^\alpha f}{\partial |v|^\alpha} \tag{15}$$

where the fractional derivative is a non-local operator defined in the sense of Riesz :

$$\frac{\partial^\alpha f}{\partial |v|^\alpha} = FT^{-1}\left[-|k|^\alpha \tilde{f}\right] \tag{16}$$

In the case where $\alpha = 2$, this equation reduces to the classical one. The physical difference between the classical and the fractional FP equation is the spatial non-locality of the latter. This non-locality arises because there is no characteristic length for the jumps in velocity space. The stationary solution of the equation is a Lévy distribution instead of a Maxwellian one :

$$f(v,t = \infty) = \int_{-\infty}^{+\infty} dk e^{ikv} \exp\left(-\frac{D_\alpha}{v}|k|^\alpha\right) \tag{17}$$

Such a stationary distribution is not necessarily inconsistent with statistical mechanics. Let us first recall that the background plasma is not in thermodynamic equilibrium. Moreover, the formalism of non-extensive thermodynamics, first introduced by C. Tsallis [11] is able to handle distributions having power-law tails. Let us now assume the initial distribution to be a Maxwellian, whose Fourier transform is :

$$f(v,0) = e^{-Ak^2} \tag{18}$$

The VDF at a time $t > 0$ is then found to be:

$$f(v,t) = \int_{-\infty}^{+\infty} dk e^{ikv} \exp\left(-Ae^{-2vt}k^2 - \frac{D_\alpha}{v}|k|^\alpha\right) \tag{19}$$

This distribution can be viewed at any given time as a generalized Voigt profile, since in the case $\alpha = 1$ we recover an usual Voigt profile. It can be shown that the tail of this distribution is again a power-law :

$$f(\mathrm{v},t) \propto \frac{C(t)}{|\mathrm{v}|^{\alpha+1}} \qquad (20)$$

The exponent of this power-law tail lies between 0 and 3. This result is consistent with the experimental data, for which the exponent was found between 1.5 and 2.5. The Doppler profile of the D_α line might consequently give an insight in some aspects of the transport of the neutrals at the Tokamak edges.

CONCLUSION

The observation of the D_α line wings in Tore Supra reveals a power-law behavior which can neither be explained by thermal Stark nor by thermal Doppler broadening. Therefore, we analyzed two non-thermal effects related to the edge plasma turbulence, which may affect the line-shape. First a Stark broadening model retaining the turbulent electric field and the related density and temperature fluctuations has been developed. A calculation of a D_α profile using the output of the RBM3D code is in progress. Moreover, as the emitter motion can also be affected by non-thermal electric fields, we have also presented a model assuming that the emitter experiences Lévy flights in the velocity space. The resulting Doppler profile could also explain the observations. It thus appears that relevant information on the turbulent fields distributions may be obtained through the analysis of the experimental spectral line-shapes. Such information could be of great interest to improve our understanding of plasma turbulence and anomalous transport in Tokamaks.

REFERENCES

1. Escarguel, A., and al., *J. Nucl. Matter*, **290-293**, 854 (2001).
2. Koubiti, M., and al., *Plasma Phys. Control. Fusion*, **44**, 261–275 (2002).
3. Hey, J., and al., *Contrib. Plasma Phys.*, **36**, 583 (1996).
4. Kubo, H., and al., *Plasma Phys. Control. Fus.*, **40**, 1115 (1998).
5. Voslamber, D., *Z. Naturforsch*, **24a**, 1458–1472 (1969).
6. Smith, E. W., Cooper, J., and Vidal, C. R., *Phys. Rev.*, **185**, 140–151 (1969).
7. Horton, W., *Rev. Mod. Phys.*, **71** (1999).
8. Krommes, J. A., *Physics Report*, **360**, 1–352 (2002).
9. Sholin, G. V., *Sov. Phys. Dokl.*, **15**, 1040 (1971).
10. Griem, H. R., *Spectral line broadening by plasmas*, Academic Press, New York-London, 1974, p153-162.
11. Tsallis, C., *J. Stat. Phys.*, **52**, 479 (1988).

Diagnostics Of Detached Plasmas Using High-n Lines And Continuum Spectra Of D and He

M. Koubiti[1], H. Capes[2], L. Godbert-Mouret[1], Y. Marandet[1], A. Meigs[3],
S. Loch[4,5], R. Stamm[1] and H. Summers[4]

1 PIIM, UMR 6633 CNRS-Université de Provence, centre St-Jérôme, 13397 Marseille cedex 20, France
2 Département de Recherche sur la Fusion Contrôlée, CEA Cadarache, St Paul Lez Durance, France
3 UKAEA/Euratom Fusion Association, Culham Science Center, Abingdon, OX14 3EA, UK
4 Department of Physics and Applied Physics, Strathclyde University, Glasgow, Scotland, UK
5 Allison Laboratory, Physics Department, Auburn University, Auburn, Alabama, USA

Abstract. A Stark line shape code has been coupled to a collisional-radiative model and to an analytical model for the merging into the continuum of the Balmer lines. The coupled codes have been used for temperature and density diagnostics of detached plasmas. In contrast with the occupation probability and the lowering of the continuum edge approaches, the analytical line merging model used here consists in the use of Lorentzian profiles for highly excited Balmer transitions of deuterium. In addition, high-n helium lines ($1s2p$-$1sn l$) up to $n=12$ observed in the JET divertor have been preliminary analyzed. The intensities of these experimental helium lines decrease more rapidly with the upper state nl than the intensities calculated with state populations at local thermodynamic equilibrium (LTE).

I. INTRODUCTION

In present magnetic fusion devices, the creation of a radiating cold dense plasma in the periphery (near the limiter or in the divertor chamber) is a common way to solve the problem of the wall erosion by strong particle and heat fluxes escaping from the core plasma. This cold layer has the important property of limiting both the impurity ionization to low stages and their penetration into the confined plasma. Through their interactions with the plasma particles, the main intrinsic impurities transform an important quantity of escaping energy into an isotropic radiation. Under such conditions, highly excited levels of deuterium are populated by recombination allowing high-n lines of the Lyman, Balmer, and Paschen series to be observed. Lines of the Lyman series are optically thick while those of the Paschen series are in the ultraviolet frequency domain. Lying in the visible domain, the lines of the Balmer series are easily measured in several machines. In JET [1-2] as well as in other Tokamaks like Alcator C-Mod [3], spectra extending beyond the Inglis-Teller limit and beyond the Balmer series limit have been measured. Being dominated by Stark broadening the full widths of the isolated lines of the spectra are usually fitted to deduce the electron density of the plasma. The electron temperature is deduced for example by using line intensity ratios and assuming a LTE for the level populations. To improve both the plasma electron and temperature diagnostics we fit the whole

Balmer spectrum of deuterium instead of using Stark widths of individual lines. The method is based on a Stark line-shape code, on a model for the line-merging into the continuum and on a collisional-radiative model for the level populations. These self-consistent codes to be described in the following are incorporated in a fitting procedure. The experimental spectra are fitted with this procedure in order to determine the electron density, the electron and ion temperatures, the recombination state of the plasma and even the concentration of impurities under some conditions.

II. THE SPECTRAL MODELLING CODES

The Stark profiles of isolated Balmer lines are calculated using a standard version of the PPP line shape code [4]. This version uses the impact and the quasi-static approximations [5] to treat the homogeneous and inhomogeneous broadenings due respectively to the plasma electrons and ions. Though one of the main characteristics of the PPP code is its capability to treat the ion dynamics effect, this latter has been ignored in the calculations presented here. For practical reasons we have limited the use of the PPP code to the profiles of Balmer lines with wavelengths $\lambda \geq 3712$ Å, i.e., the D_{15} line wavelength (transition $n=15 \rightarrow n'=2$). Examples of Stark profiles of some Balmer lines (lines D_{10} to D_{15}) calculated by PPP for three values of the electron density are shown in Figure 1. One can clearly see that as Stark effect increases with both the upper quantum number n and the electron density Ne, high n lines broaden and merge together.

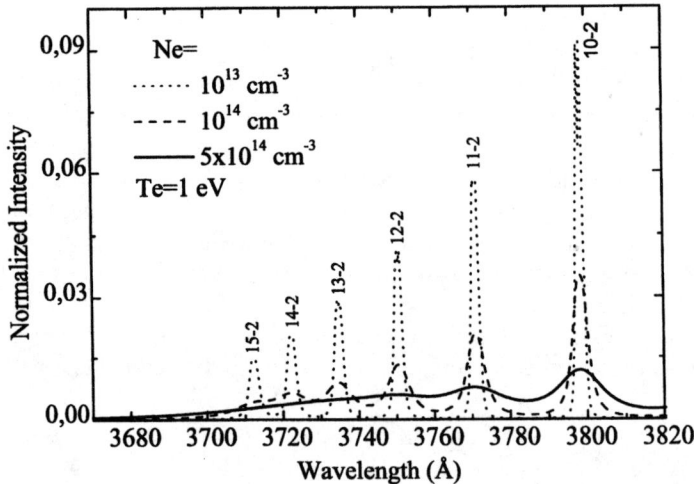

FIGURE 1. Stark profiles of some high-n deuterium Balmer lines calculated using the PPP code for an electron temperature Te=1 eV and three values of the electron density Ne: 10^{13} cm^{-3} (dot), 10^{14} cm^{-3} (dash), and 5×10^{14} cm^{-3} (solid). As Stark effect increases with the upper quantum number n and the electron density Ne, high-n lines broaden and merge together for the highest value of Ne.

It should be noted that the other part of the bound-bound emission, i.e., Balmer transitions with n greater than 15, is treated with an analytical continuity model of smooth line merging into the continuum.

II.1. An Analytical Continuity Model For The Line Merging Into The Continuum

Bound-bound emission is separated into two terms representing respectively the 'discrete' and the 'blended' lines. The former lines are calculated using the PPP code as mentioned above while the 'blended' part of the bound-bound emission spectrum is treated with a new approach. This approach of smooth merging of lines into the continuum [6-7], is completely different from the occupation probability [3,8] or the artificial lowering of the continuum edge [9] approaches. The basic idea of the line merging model used here is to introduce line profiles in the continuum part of the emission spectrum and to consider population elements when approaching the ionization threshold instead of the usual discrete quantum shells. The formalism applied for the isolated (or discrete) Balmer lines is extended to overlapping Balmer lines, i.e., transitions from highly excited levels (upper quantum number PQN n=16 to n=100) to the n'=2 lower level. The populations of these levels are calculated using the ADAS [10] collisional-radiative model. As the series limit is approached the lines broaden and merge together with the merged lines smoothly extending through the continuum. We have used Lorentzian profiles to model the emission of these highly excited levels. For simplicity all the Lorentzian widths γ have been taken equal to the width γ_{16} of the D_{16} line. This latter, γ_{16}, is consistently calculated for each couple of electron density and temperature (Ne, Te), by extrapolating the widths of the corresponding discrete lower lines. In conclusion, the bound-bound emission is treated by the PPP code and the Lorentzian profiles. Bound-bound emission has now a contribution at continuum wavelengths as it can be seen on figure 2. Inversely, as it will be shown in the next subsection, the bound-free emission treated here in a standard way has also a contribution at bound-bound wavelengths. Figure 3 shows the bound-free contribution.

II.2. Bound-Free Emission Profile

The bound-free emission profile is calculated from the corresponding emission coefficient [11] defined as the number of photons emitted at a wavelength λ per unit time, per unit volume, per unit solid angle at position L:

$$\varepsilon_{bf} = L \frac{1}{4\pi} [N_i f(v) b_\kappa dv][N_e v \sigma_{\kappa n}], \tag{1}$$

where $f(v)$ and $\sigma_{\kappa n}$ are respectively the Maxwellian velocity distribution of free electrons and the target area for their capture, b_κ is the bound b-factor (deviation factor from a Maxwellian distribution) extended into the continuum, and Ne and Ni are the electron and ion densities.

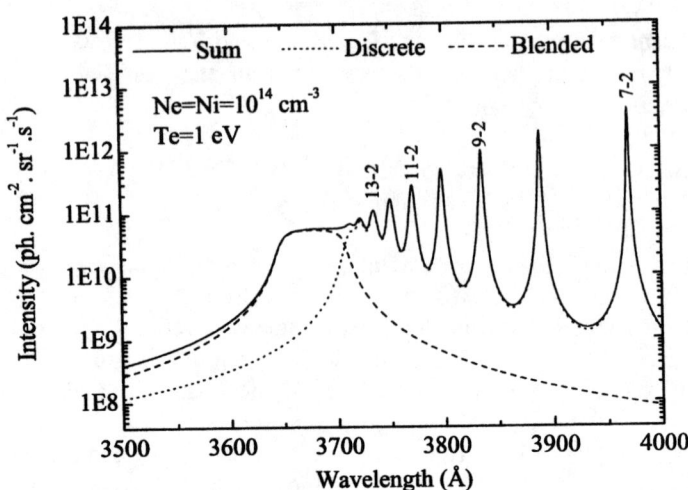

FIGURE 2. Bound-bound emission intensity. The bound-bound emission profile (solid) is the sum of the lower Balmer lines (with $n \leq 15$) labeled 'discrete' here (dot) and the 'Blended' part (dash). Note the semi-logarithmic scale and that the wavelength domain includes the D_7 (transition $n=7 \to n'=2$) line at ~3970 Å.

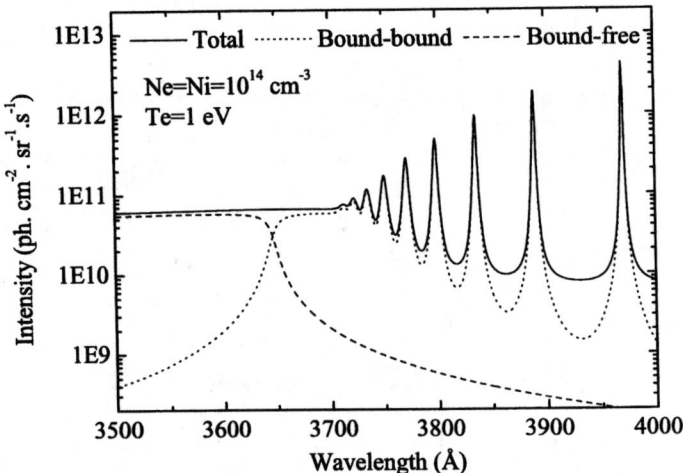

FIGURE 3. A synthetic profile of the deuterium Balmer emission for $Ne=Ni=10^{14}$ cm^{-3} and $Te=1$ eV. The total profile (solid) is the sum of the bound-bound (dot) and bound-free (dash) emission profiles. The wavelength domain is the same as in figure 2, i.e., lines D_n with $n \geq 7$ are included in the profile.

III. THE FITTING ROUTINE AND ITS APPLICATION

The codes mentioned above have been coupled together and incorporated in a powerful fitting procedure capable of extracting more accurate electron density and temperature. Other information can be extracted too as the recombination state, the ion temperature and the impurity concentration. To illustrate that, we show in figure 4 an example of a fitted JET spectrum and in figure 5 the residual.

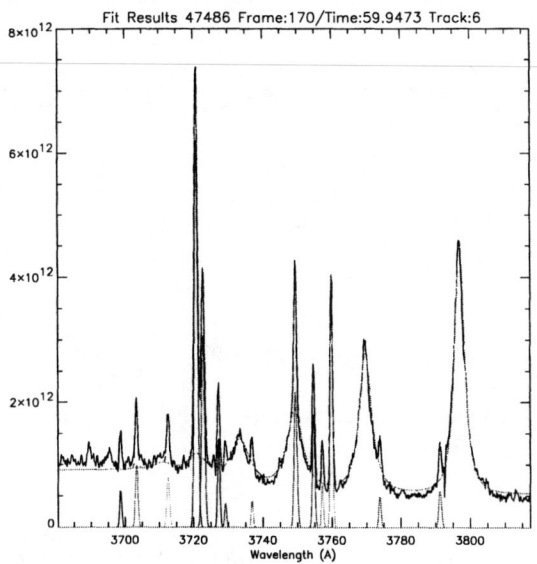

FIGURE 4. A fit of a Balmer spectrum emitted by deuterium in the axisymmetric divertor of JET. The experimental spectrum is in solid lines while the calculated one is in dotted lines. Note the presence of impurity lines fitted with gaussians. The plasma parameters inherent from the fit are $Ne=Ni=8x10^{13}$ cm^{-3} and $Te=0.75$ eV.

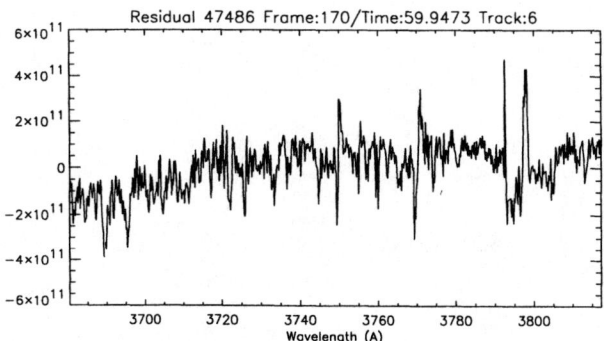

FIGURE 5. The residual of the fitting of the experimental spectrum presented in figure 4. It should be noted the likely presence of an extra impurity line at ~3797 Å.

On figure 4, one can see the good agreement between the experimental spectrum (solid) and the fitting feature (dot). The plasma parameters deduced from the fit are Te=0.75 eV and Ne=Ni=8x10^{13} cm^{-3}. Impurity lines are also fitted here with Gaussians. One can also see that the last resolved member of the Balmer series is the D$_{15}$ line, indicating an upper limit for the electron density of 10^{14} cm^{-3} which is the Inglis-Teller limit [12]. Note that the continuum is reached at ~3705 Å instead of ~3645 Å, which is the conventional value of the continuum edge of the hydrogen Balmer series.

IV. HIGH *n* LINE SPECTRA OF NEUTRAL HELIUM

In some JET helium discharges with a strong deuterium puffing spectra of high-n lines of neutral helium have been measured in the divertor region. A typical spectrum is shown in figure 6. One can identify one singlet line and five triplet lines. The triplet lines represent transitions 1s2p-1s*nl* with *n*=8,9,10,11 and 12, *l* being equal to 0 or 2 (s or d states). In figure 6 we have added to the experimental spectrum (solid) a calculated one using the PPP code with a hydrogen-like electron collision operator, and assuming a LTE for the state populations. It is clear that the experimental intensities of the lines decrease more quickly with the PQN *n* than the calculated one. This indicate that a collisional-radiative model allowing population transfer between the singlet and triplet state systems is needed. This is in progress.

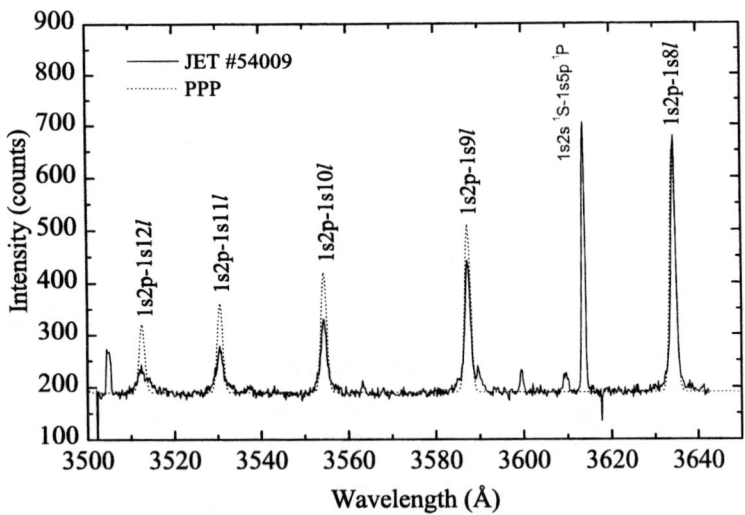

FIGURE 6. A comparison between an experimental spectrum of neutral helium (solid lines) with a synthetic profile (dashed). The latter was calculated using the PPP code with a hydrogen-like electron collision operator and assuming a LTE for the level populations. Note that experimentally the line intensities decrease more rapidly with the PQN n than theoretically.

V. CONCLUSION

We have presented a robust procedure to fit high-n lines and continuum spectra of deuterium in dense plasmas (detachment conditions or recombining plasma regimes). The procedure is based on the use of a Stark broadening lineshape code, a collisional-radiative model and an analytical model for the merging of high members of the Balmer series into the continuum. The latter requires the introduction of Lorentzian profiles which have contributions at bound-free wavelengths. An illustration of the capabilities of the procedure to determine plasma parameters has been presented. On the other hand, a spectrum of neutral helium including high-n triplet lines 1s2p-1snl with n between 8 and 12 has been compared to a calculation using a standard Stark lineshape code and assuming a LTE for the state populations. The rapid decrease of the intensities of the lines with n indicate the necessity to couple singlet and triplet state systems in any collisional-radiative model to fit the experimental spectrum.

REFERENCES

1. Meigs A et al, JET report JET-CP (98) 28 (1998)
2. Meigs A et al, Proc. 27th conf on Conrol. Fusion and Plasma Phys. Budapest 12-16 June 2000.
3. Pigarov A et al, *Plasma Phys. Control. Fusion* **40**, 2055 (1998)
4. Talin B et al, Phys. Rev. **A51**, 1918 (1995).
5. Griem H R, *Spectral Line Broadening by Plasmas*, New York and London: Academic Press, 1974
6. Loch D 2001, PhD thesis University of Strathclyde, Glasgow, UK.
7. Koubiti M et al, *Contrib. Plasma Physics* **42**, 206 (2002).
8. Hummer D and Mihalas D, *Astrophys. J.* **333**, 794 (1988).
9. Stewart J and Pyatt K, *Astrophys. J.* **144**, 1203 (1966).
10. Summers H P, ADAS Manual (1999); see also http://adas.phys.strath.ac.uk/
11. Menzel D H and Pekeris C L, *Mon Not R. Astr. Soc.* **96**, 77 (1935)
12. Inglis D R, Teller E, *Astrophys. J.* **90**, 439 (1939)

Lifetime of the $1s2p\ {}^1P_1$ excited level in Fe^{24+}

A. Graf*, P. Beiersdorfer†, C. L. Harris**, D. Q. Hwang* and P. A. Neill**

*University of California-Davis, Davis, CA 95616, USA
†University of California-LLNL, Livermore, CA 94551, USA
**University of Nevada-Reno, Reno, Nevada 89557, USA

Abstract. Measurements of the spectrum of Fe^{24+} in the 1.845 Å to 1.885 Å range obtained on the EBIT-I electron beam ion trap at Lawrence Livermore National Laboratory were used for determining the radiative lifetime of the $1s2p\ {}^1P_1$ excited state. The spectrum contains electric dipole forbidden transitions at 1.855Å ("x") and 1.868Å ("z") whose lineshape is well represented by a Gaussian line profile and is assumed to be due primarily to Doppler and instrumental broadening. The Gaussian contribution is assumed to be the same for all lines in the spectrum. This assumption simplifies the problem when considering a more complex combination of broadening mechanisms. For allowed transitions such as $1s2p\ {}^1P_1 \rightarrow 1s^2\ {}^1S_0$, "w", at 1.850 Å we assume a Voigt profile. In the simplest case this combines both natural (Lorentzian) and Doppler (Gaussian) broadening effects which contribute to the width of the spectral line. With the Gaussian contribution determined from lines "x" and "z", deconvolving the Gaussian from the Voigt profile gives the natural line width. This then is directly related to the radiative lifetime of the $1s2p\ {}^1P_1$ excited level.

INTRODUCTION

The natural line width demonstrates the uncertainty in the discrete energy levels of a bound electron. There is a corresponding time uncertainty given by Heisenbergs' uncertainty relation [1, 2].

$$\Delta E \Delta t = \frac{h}{2\pi} \tag{1}$$

For the $1s2p\ {}^1P_1 \rightarrow 1s^2\ {}^1S_0$ transition, the lower energy level is the ground state, which has no energy uncertainty. This implies that ΔE in Eq. (1) is the energy spread for the upper level in this transition and so Δt is the lifetime of the electron in the upper energy level. The energy uncertainty, ΔE, can also be written

$$\Delta E = h\frac{\Delta \omega}{2\pi} = hc\frac{\Delta k}{2\pi} = hc\Delta\left(\frac{1}{\lambda}\right) = hc\frac{\Delta\lambda}{\lambda^2} \tag{2}$$

where λ is the peak wavelength for the spectral line of interest. Putting this form into Eq.(1) we build an expression for the mean amount of time the electron spends in the excited state in terms of the measureable quantities λ and $\Delta\lambda$,

$$\Delta t = \frac{\lambda^2}{2\pi c \Delta\lambda} \tag{3}$$

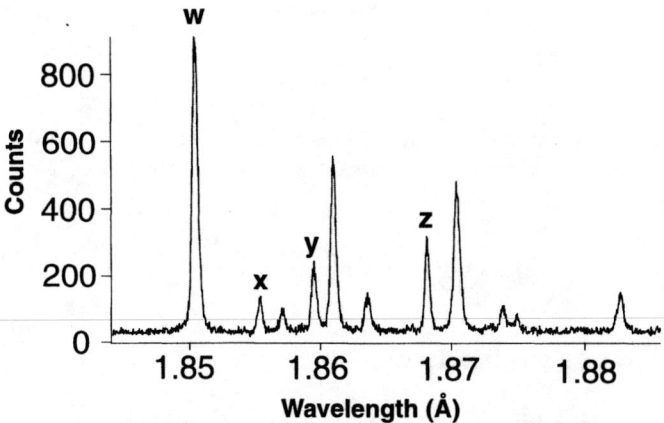

FIGURE 1. Spectrum of K-shell emission lines from heliumlike and lithiumlike iron ions. The transitions in heliumlike ions labeled w,x,y,z correspond to the transitions from $1s2p\ {}^1P_1$, $1s2p\ {}^3P_2$, $1s2p\ {}^3P_1$ and $1s2p\ {}^3S_1$ levels to the ${}^1S_0$ ground state, respectively.

A measurement of the natural linewidth thus can be exploited to derive the upper level lifetime. To our knowledge, this method had not been used previously to determine the lifetime of the $1s2p\ {}^1P_1$ excited level.

EXPERIMENT

The spectrum of Fe^{24+} and the strong lines of Fe^{23+} in the 1.845 Å to 1.885 Å range (Fig. 1) was obtained on the electron beam ion trap (EBIT-I) at Lawrence Livermore National Laboratory by a von Hámos bent crystal (Si(400)) X-ray spectrometer [3, 4]. For observations of extended light sources this spectrometer requires an entrance slit. Here we make use of the fact that the light source, the electron beam, is as narrow, $50\mu m$ diameter, as the required slit for the spectrometer. So no physical slit is needed. The spectrometer offers a throughput comparable to that of more common focusing devices better suited for larger light sources (e.g. Johann type spectrometer [5]). In addition, a simultaneous coverage of a large wavelength band is possible. Iron pentacarbonyl $Fe(CO)_5$ was introduced into EBIT-I with a gas injector. The electron beam energy was 8.028 keV, with a beam current range of 80 to 130 mA, appropriate for creating a majority of He-like iron ions in the trap ($I_p(Fe^{23+}) \approx 2.04$ keV and $I_p(Fe^{24+}) \approx 8.82$ keV). The ion temperature, obtained from the Doppler relation

$$T_i = m_i c^2 \left(\frac{(\Delta E/E)^2}{8\ln 2} \right) \tag{4}$$

was ≈ 464 eV.

ANALYSIS

Model Fit

Test Gaussian and Lorentzian distributions were created and convolved to verify the fitting method used. The Gaussian form used was as follows,

$$Gaussian_{MOD}(\lambda_D) = exp^{-\left(\frac{1}{2}\left(\frac{\lambda_D}{\sigma}\right)^2\right)} \tag{5}$$

with $G_{FWHM} \equiv \Delta\lambda_D \equiv 2(2\ln(2))^{\frac{1}{2}}\sigma$, the full width at half maximum for the Gaussian line shape and σ is the standard deviation defined by

$$Gaussian_{MOD}(\lambda_D = \pm\sigma) = exp^{-\frac{1}{2}} Gaussian_{MOD}(0) \tag{6}$$

The Lorentzian form was,

$$Lorentzian_{MOD}(\lambda_N) = \frac{1}{\left(\left(\frac{2\lambda_N}{\Delta\lambda_N}\right)^2 + 1\right)} \tag{7}$$

with $L_{FWHM} \equiv \Delta\lambda_N$, the full width at half maximum for the Lorentzian line shape. A Voigt fit was then performed on the convolution of the two, returning the Gaussian and Lorentzian contribution to V_{FWHM}, the Voigt full width at half maximum. The fitting program allowed the width due to Doppler broadening to be chosen and held fixed. This allowed for a verification of the Lorentzian width origionally put in.

Results

The Doppler broadening must be the same for all Fe^{24+} lines. This was tested using two forbidden transitions, "x" and "z", that appear in the spectrum. The natural width of a forbidden transition should be vanishingly small providing a good scenario for determining the other broadening mechanisms. In this case only the Doppler broadening (and the broadening from the instrumental response) is considered to be significant. The Gaussian fits to "x" and "z" are shown in (Fig. 2 b) and c)). The tables show the Doppler contribution to "x" to be within $\sim 0.37x10^{-4}$ Å that of "z", verifying our

TABLE 1. The Lorentzian (natural) width of "w" using the Gaussian (Doppler) width determined from the forbidden transition, "z" and the corresponding lifetime for "w".

Voigt shape	G_{FWHM}(Å)	L_{FWHM}(Å)	$\Delta t(s)$
$0.34^{+.02}_{-.03}$	$4.10x10^{-4+8E-6}_{-8E-6}$	$1.67x10^{-4+8E-6}_{-1E-5}$	$1.09x10^{-15+6E-16}_{-5E-17}$

TABLE 2. The Lorentzian (natural) width of "w" using the Gaussian (Doppler) width determined from the forbidden transition, "x" and the corresponding lifetime for "w".

Voigt shape	$G_{FWHM}(\text{Å})$	$L_{FWHM}(\text{Å})$	$\Delta t(s)$
$0.25^{+.05}_{-.06}$	$4.37x10^{-4+2E-5}_{-2E-5}$	$1.30x10^{-4+2E-5}_{-3E-5}$	$1.40x10^{-15+4E-16}_{-2E-16}$

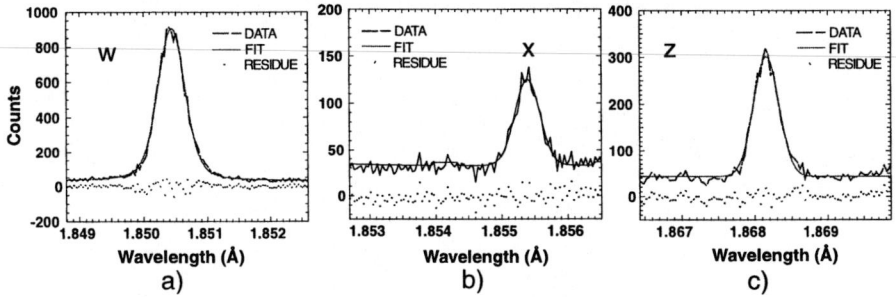

FIGURE 2. a) Voigt fit to "w" , b) Gaussian fit to "x", and c) Gaussian fit to "z".

contention that Doppler (and instrumental) broadening are the same for all Fe^{24+} lines. This allows some confidence when adopting this measured Doppler width for "w". The corresponding Lorentzian widths for "w" obtained by using the Doppler widths from both "x" and "z" separately were then used to determine two values for the lifetime of the $1s2p\ ^1P_1$ excited state in Fe^{24+}. The average of the two rather similar lifetimes gives a value of $1.24x10^{-15}$ s. This disagrees with the theoretical values of $2.18x10^{-15}$ [6] or $2.19x10^{-15}$ [7] by roughly a factor of two.

Neglecting all but the natural and Doppler broadening may be too much of a simplification. Any additional contributions to the width of the $1s2p\ ^1P_1$ excited state would decrease the natural width and thus increase the lifetime. Also, the data are a sum of many sets which could create a degradation in the resolution resulting in a slightly larger width in the final analyzed spectrum. One final approach to a more accurate result would be to decrease the ion temperature until the Doppler broadening contribution becomes negligible. This idea was successfully applied by *Beiersdorfer et al.* to Cs^{45+} to determine the lifetime of the 2p-3d transition [8].

ACKNOWLEDGMENTS

This work was performed at Lawrence Livermore National Laboratory under the auspices of the Department of Energy under contract No. w-7405-ENG-48.

REFERENCES

1. Cohen-Tannoudji, C., Diu, B., and Laloe, F., *Quantum Mechanics*, John Wiley and Sons and Hermann, 1977.
2. Weisskopf, V., and Wigner, E., *Z. Physik*, **63**, 54 (1933).
3. von Hámos, L., *Ann. Phys.*, **17**, 716 (1933).
4. Beiersdorfer, P., et al., *Review of Scientific Instruments*, **61**, 2338–2342 (1990).
5. Johann, H., *Z. Physik*, **69**, 185 (1931).
6. Lin, C. D., et al., *Phys Rev A*, **15**, 154–161 (1977).
7. Johnson, W. R., et al., *Advances in Atomic, Molecular, and Optical Physics*, **35**, 255–329 (1995).
8. Beiersdorfer, P., et al., *Phys Rev Letters*, **77**, 5353–5356 (1996).

Modeling of Doppler-Broadened H_α and D_α Spectra Including Fine Structure for Moderate Static Electric Fields

Alexander Demura[*], Chantal Stehlé[**], Zekerija Altug[***], Daniel Esch[***], Torge Rieper[***], and Volkmar Helbig[***]

[*]Hydrogen Energy and Plasma Technology Institute,
Russian Research Center "Kurchatov Institute", Moscow 123182, Russia
[**]Laboratoire de l'Univers et de ses Theories", Observatoire de Paris, Place J. Janssen 5,
F-92195 Meudon, France
[***]Institut für Experimentelle und Angewandte Physik, Universität Kiel, Leibnizstr. 19,
D-24098 Kiel, Germany

Abstract. Opto-galvanic spectra of the Doppler-broadened line profiles of H_α have been obtained in the cathode fall of a low pressure gas discharge in order to determine the local electric field strength by comparison with calculated Stark-profiles. Small but significant differences between the experimental and the theoretical profiles caused us to i) repeat the calculations with a second computer code to check the consistency of the respective results and ii) to repeat the measurements with deuterium instead of hydrogen to look for a possible influence of the hyper- fine-structure.

INTRODUCTION

For the diagnostics of low temperature gas discharges that have a variety of applications in industry new sensitive and reliable methods are needed for the measurement of the local electric field strength. Optical techniques that do not require a complex experimental set-up and expensive and elaborate laser systems are desirable. We [1, 2, 3] have investigated the possibility to use the Stark-splitting of the first line of the Balmer series of hydrogen to obtain the magnitude and the direction of the electric field strength in the cathode fall of a low pressure, low current DC gas discharge. We used opto-galvanic detection to scan the line profile by means of a diode laser system developed in our laboratory. For details see [1, 3]. To deduce the electric field strength from the Doppler-broadened experimental profiles the theoretical Stark profiles including fine-structure were needed. In [1, 4] those profiles were calculated following the theory outlined in [5, 6]. The computed spectra were stored on a PC as input data for the fitting routine in the evaluation process. Though a very reasonable fit could be obtained for the spectra taken in $\pi-$ and $\sigma-$polarization, respectively, there still exist small but significant differences between the theoretical and the experimental profiles as demonstrated in Fig. 1. Here the experimental profile taken at a distance of d = 0.5 mm from the cathode in $\pi-$ polarization is fitted to the

CP645, *Spectral Line Shapes: Volume 12, 16ᵗʰ ICSLS,* edited by C. A. Back
© 2002 American Institute of Physics 0-7354-0100-4/02/$19.00

model function from the calculations. The electric fieldstrength obtained by the least squares fit was E = 6330 V/cm.

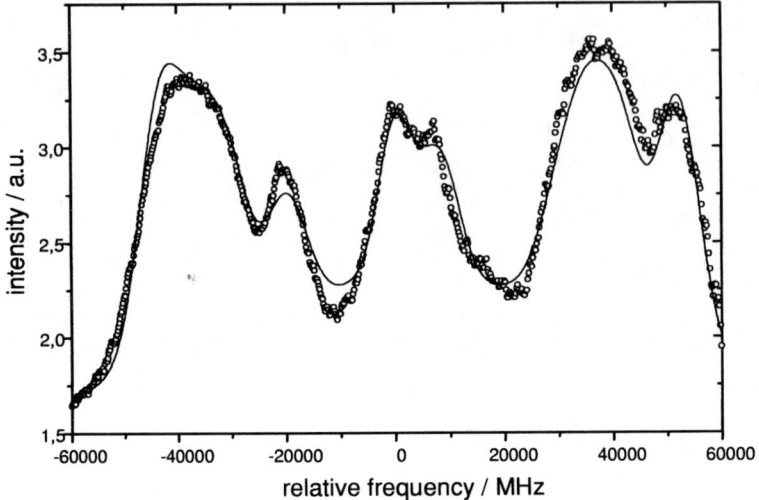

Figure 1. Experimental profile taken close to the cathode. The least squares fit to the calculated Stark-profile yielded an electric field strength of F = 6330 V/cm.

Two explanations for the remaining differences between the experimental and theoretical profiles seemed possible: i) the Stark-effect calculations for the dependence of the splitting and the intensities of the single fine-structure components on the electric field in [1, 4] were not correct and ii) the neglect of the hyperfine-structure causes the deviations.

CALCULATING THE STARK-EFFECT OF THE FINE-STRUCTURE

The calculations performed in [1, 4] were repeated with a computer routine developed in Meudon. Fig. 2 and 3 show a graphical comparison of the respective results for the isolated fine-structure components and their Stark-effect splitting for the case of zero field strength and a moderate field strength of F = 1600 V/cm. The calculated intensities are shown with a half-width corresponding to the natural line width. No differences were observed. The same is true for the corresponding profiles when convoluted with the Doppler-width corresponding to the experimental conditions as can be seen in Fig. 4 and 5.

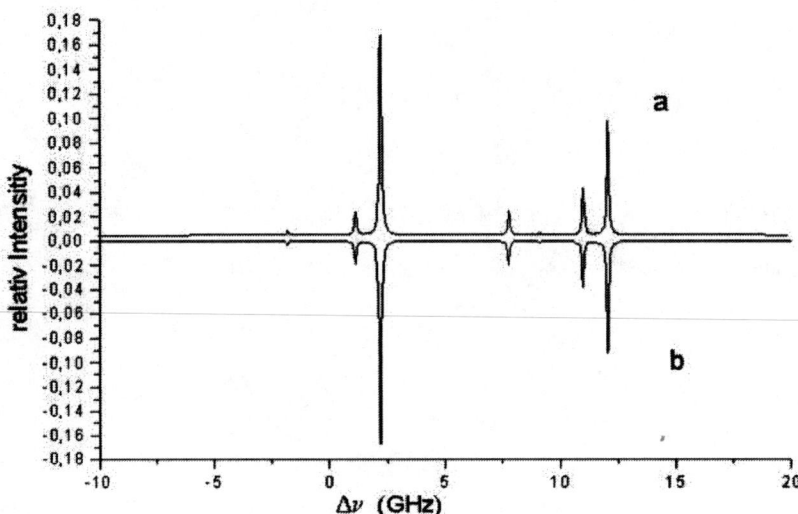

FIGURE 2. The fine-structure spectrum of H_α for zero electric field strength calculated a) with the computer code in Kiel [1, 4] and b) with the computer routine developed in Meudon.

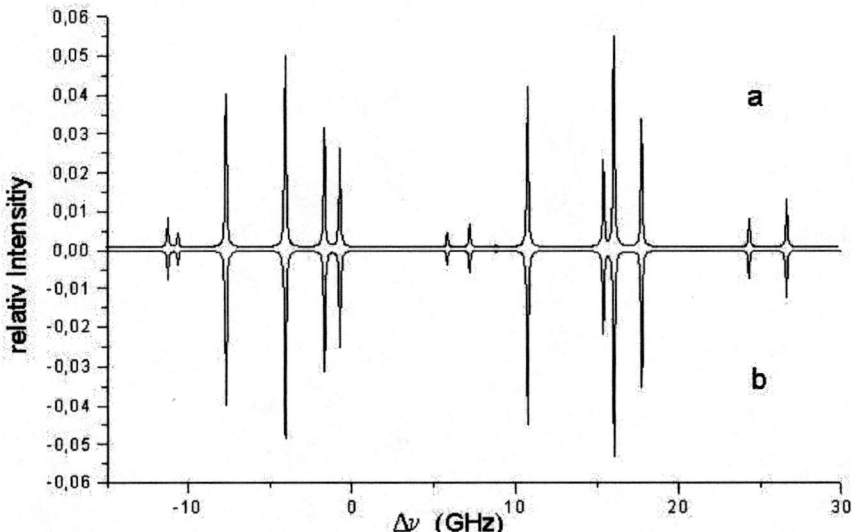

FIGURE 3. The fine-structure spectrum of H_α for an electric field strength of F = 1600 V/cm calculated a) with the computer code in Kiel [1, 4] and b) with the computer routine developed in Meudon.

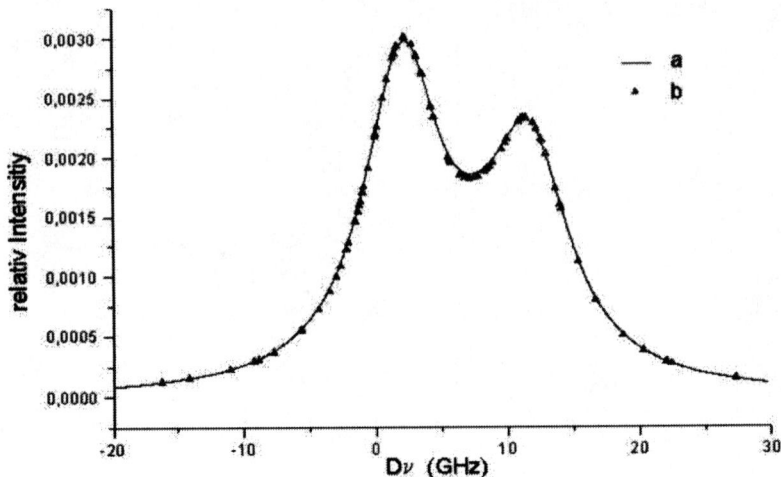

FIGURE 4. The Doppler profile of H_α for zero electric field strength calculated a) with the computer code in Kiel [1, 4] and b) with the computer routine developed in Meudon.

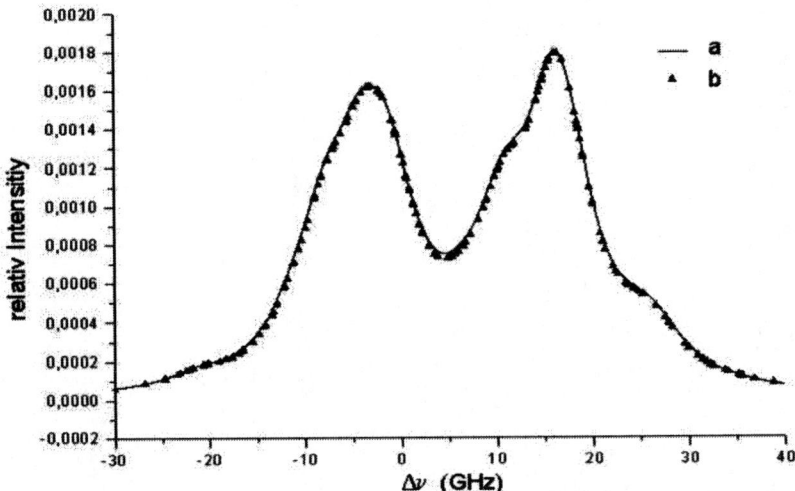

FIGURE 5. The Doppler profile of H_α for an electric field strength of F = 1600 V/cm calculated a) with the computer code in Kiel [1, 4] and b) with the computer routine developed in Meudon.

RUNNING EXPERIMENTS WITH HYDROGEN AND DEUTERIUM

To repeat the calculations including the hyperfine-structure is quite cumbersome because of the increasing number of components. We therefore started out as a first step with experiments running the discharge under identical conditions in hydrogen and deuterium which can be done easily. The idea was to get information whether or not there is a detectable difference for the respective spectra that could give a hint whether the neglect of hyperfine-structure could be the reason for the above mentioned differences. We performed experiments in pure hydrogen and pure deuterium for same pressure, the same flow-rate, the same current and identical distances from the cathode surface. Because it could not be ruled out that the slope of the cathode fall is different for the two gases in addition spectra were taken in a one to one mixture of hydrogen and deuterium. These spectra suffer from some additional noise because the signals get weaker. But within the experimental accuracy they showed no deviations from the spectra in the pure gases. In Fig. 6 two σ−spectra and in Fig. 7 two π−spectra for the H_α and the D_α lines are shown for different distances from the cathode. Using the hydrogen calculations model spectra were fitted to the experimental ones using the electric field strength and the Doppler-width as free fitting parameters. Due to the smaller Doppler-broadening the deuterium spectra show more

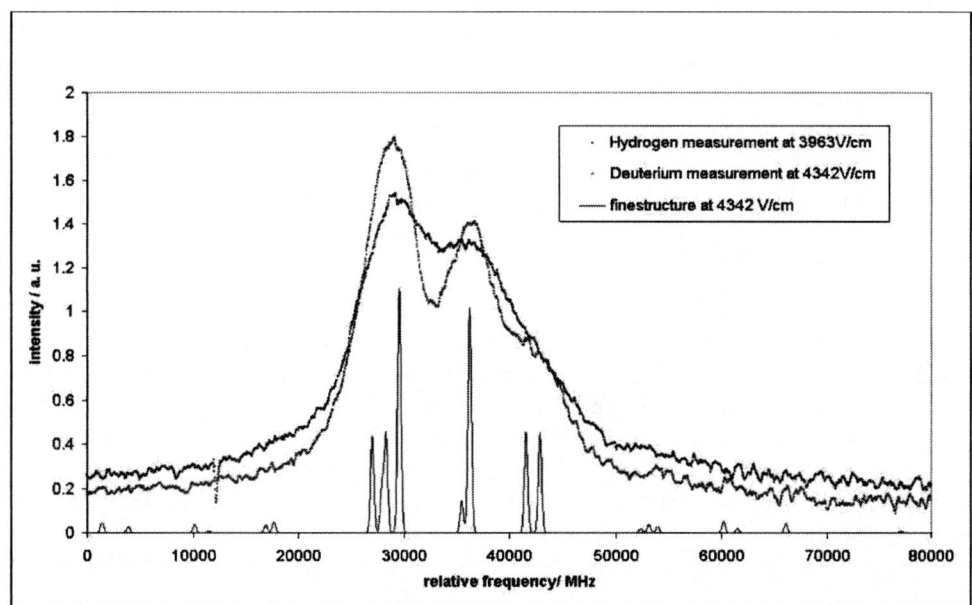

FIGURE 6. Experimental σ−profiles of the H_α and the D_A lines taken in the pure gases in the discharge. The field strengths given in the figure are the result of fitting the hydrogen calculations to both of the experimental spectra running the Doppler-width as a free fitting parameter. The position of the single fine-structure components are shown at the bottom without Doppler-broadening.

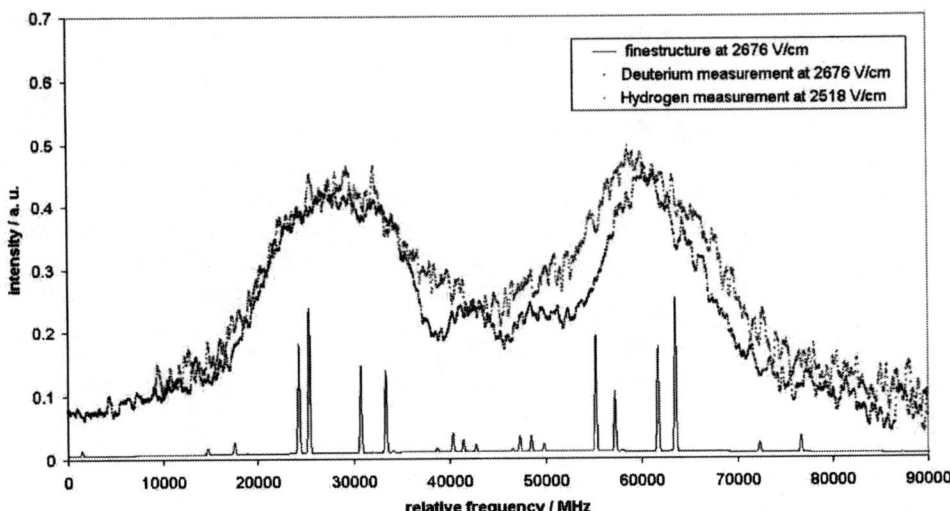

FIGURE 7. Experimental π–profiles of the H_α and the D_α lines taken in the pure gases in the discharge. The field strengths given in the figure are the result of fitting the hydrogen calculations to both of the experimental spectra running the Doppler-width as a free fitting parameter. The position of the single fine-structure components are shown at the bottom without Doppler-broadening.

structure than the hydrogen ones. But at least for the π–spectra the Doppler-broadening alone could not explain the observed differences between the two sets of spectra. For comparison the single Stark components of the fine-structure lines are shown without Doppler convolution in the bottom of the figures.

CONCLUSIONS

Trying to explain differences between experimental and theoretical Stark-profiles of the H_α - line of hydrogen we repeated the calculation of the Stark-effect of the fine-structure of H_α using a second computer routine. Both programs gave consistent results so that the origin of the mentioned differences have to be traced back to different reasons. Among the possible explanations the neglect of the hyperfine-structure in the calculations seems to be the most promising one. Experiments with hydrogen and deuterium for identical plasma conditions showed that the deuterium spectra could not be fitted to the hydrogen calculations with the same accuracy as the hydrogen ones. This suggest that it is worthwhile to include the hyperfine-structure in future calculations.

ACKNOWLEDGMENTS

This investigation was initiated during the stay of A.D. as a visiting professor at the university of Kiel. The financial support of this stay by the DAAD is greatfuly acknowledged.

REFERENCES

1. Rieper, T., PhD thesis, 2000, Kiel.
2. Rieper, T., Neumann, J., and Helbig, V., in *Spectral Line Shapes, Vol. 11,* edited by J. Seidel, AIP Conference proceedings 559, New York: American Institute of Physics, 2001, pp. 82 - 84.
3. Altug, Z., Neumann, J., Rieper, T., and Helbig, V., in *Proc. Xth LAPD,* edited by K. Muraoka, Fukuoka, 2001
4. Neumann, J., Diploma thesis, 2000, Kiel
5. Lüders, G., Z. Naturforschg., 5a, 608 (1950)
6. Lüders, G., Ann. d. Physik, 6, 301 (1951)

Measurements of Local Electric Fields in Low Density Plasmas via Stark-Splitting of Hydrogen Resonance Lines

A. Steiger[1], K. Grützmacher[1], C. Pérez[2], M. I. de la Rosa[2], and J. Seidel[1]

[1]Physikalisch-Technische Bundesanstalt, 10587 Berlin, Germany
[2]Dpto. de Óptica, Universidad de Valladolid, 47071 Valladolid, Spain

Abstract. The aim of this paper is to discuss in detail the schemes for electric field measurements in low density plasmas via Stark-splitting of the first two hydrogen resonance lines. Beyond it, we have proved by experiment that advanced UV-laser spectrometers, especially developed for two-photon spectroscopy of plasmas, are well suited for this purpose. As a first demonstration at low electric fields, simple opto-galvanic detection was chosen for measuring the Stark-splitting of the Doppler-free 1S-2S and 1S-3S/D two-photon transition of atomic hydrogen produced by thermal dissociation in a small reference cell. In addition, we have performed measurements in a hollow cathode discharge which provides higher electric fields in its cathode fall region and the 1S-2S spectrum was detected spatially resolved by means of an opto-galvanic signal and polarization spectroscopy as well. Finally, another detection scheme, namely two-photon induced Balmer-α fluorescence applied as sheet diagnostic will be described.

INTRODUCTION

Low pressure plasmas are widely used in many kind of surface modification processes. A famous example is the very large scale integrated circuit manufacturing in semiconductor industry. Typical applications are e.g. plasma etching or plasma enhanced chemical vapor deposition. In such low density discharges, electrons, ions and reactive species are generated mainly in the plasma bulk. However, the energy distribution of particles impinging the surface are determined by the characteristics of the plasma sheath. A key parameter for the understanding of the processes in this boundary region is the electric field strength.

Without causing any perturbation, this important quantity can be determined by Doppler-free two-photon spectroscopy [1] of the Stark-splitting of atomic hydrogen mostly present in such discharges. For this purpose, we take advantage of our advanced UV-laser spectrometers. They allow to obtain all at once: the determination of local electric fields beside their usual application of measuring atomic ground state densities, kinetic temperatures and electron densities. But all these kind of diagnostics require pulsed ultraviolet laser radiation of single longitudinal mode (SLM) quality.

The efficient generation of pulsed ultraviolet laser radiation is based on a special concept for nonlinear frequency conversion, especially developed in our lab for two-photon spectroscopy of plasmas more than eight years ago [2]. Starting with a

CP645, *Spectral Line Shapes: Volume 12, 16th ICSLS*, edited by C. A. Back

commercial injection-seeded pulsed Nd:YAG laser, we obtain tunability by frequency splitting of its second harmonic in opto-parametric processes of single mode quality. The narrow band output of the opto-parametric oscillator is converted into the ultraviolet range by sum-frequency generation with higher harmonics. Up to now, our results have not been achieved by any commercial laser system, e.g. 10 % efficiency for pulse peak intensity conversion from fixed frequency 1064 nm to tunable 243 nm radiation [2].

Our concept allows to cover the whole ultraviolet spectral range down to wave-lengths below 200 nm. The Doppler-free excitation of the 1S-2S and 1S-3S/D two-photon transition of atomic hydrogen requires radiation at 243 nm and 205 nm. Even at 205 nm, the laser spectrometer [3] provides sufficient pulse energy of the order of 5 mJ and a spectral bandwidth of about 300 MHz close to the Fourier-limit for the pulse duration of about 3 ns. The spectral bandwidth of about 300 MHz is sufficiently narrow to resolve e.g. hyperfine splitting of the 1S-2S transition of atomic hydrogen, and is therefore well suited to measure electric fields by laser spectroscopy.

SCHEMES FOR LOCAL ELECTRIC FIELD MEASUREMENTS

The principle of determine electric fields directly from the Stark-splitting of the resonance lines is based on the fact that the Lamb-shift provides a small but distinct energy gap between the $S^{1/2}$ and $P^{1/2}$ levels with the total angular momentum ½ of the fine-structure of the excited states of atomic hydrogen. If there is no external electric field interfering with the atom, two-photon transition from the 1S ground level is allowed only to an S but not to an P level. However, at non-vanishing field, the Stark effect causes line-mixing and line-splitting [4]. Therefore, the intensity ratio and the frequency separation of the forbidden to the allowed component is a direct measure of the perturbation by the external electric field.

Due to the Lamb-shift, the Stark splitting is not linear in the region of low electric fields as for large fields, where the Stark splitting scales as n(n-1) with the principal quantum number n. Reasonable n is restricted to n = 2 or 3, as long as direct access by two-photon absorption from the ground state is concerned. Even so, this allows to measure electric fields as small as 30 V/cm, which is no real limitation because for typical applications, interesting field strengths lie one order of magnitude above.

For simplicity, the following theoretical results are presented without hyperfine-structure which causes each fine-structure component splitting into two components for atomic hydrogen with an amplitude ratio of 2/3 to 1/3. Of course, any experimental result will exhibit this hyperfine-splitting. But this causes no problem because the Lamb-shifted forbidden component grows up red shifted and therefore can be easily distinguished from the blue shifted weak hyperfine-component.

Figure 1 shows the calculated two-photon transition probability for the Stark-splitting of the three fine-structure components of the n = 2 level of hydrogen as a function of their frequency relative to the unperturbed 1S-2S resonance. The curves result from varying the electric field and the dots corresponds to fixed field strengths i.e. every 100 V/cm from zero to 1 kV/cm and every kV/cm above. At zero field, only the $2S^{1/2}$ level is allowed, and at 200 V/cm the $2P^{1/2}$ has taken over 10 % of its intensity.

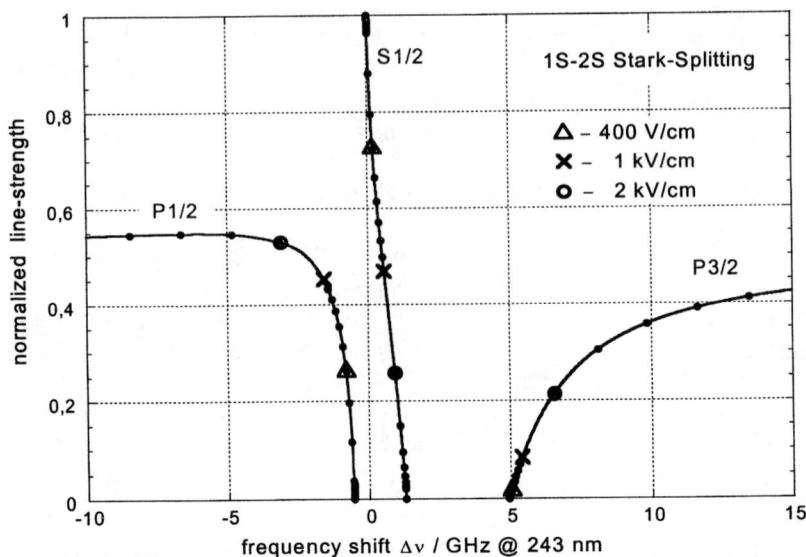

FIGURE 1. To illustrate the Stark-splitting of the 1S-2S transition of atomic hydrogen, line strength of all three fine-structure components is plotted as a function of their two-photon excitation frequency for different electric fields. At zero field, only the central component $S^{1/2}$ ($\Delta\nu = 0$) is allowed. For small electric fields, increasing part of its intensity is transferred to the $P^{1/2}$ component, but up to 400 V/cm (Δ), only tiny line shifts occur. At 1 kV/cm (**X**), both components reach equal intensity and the $P^{3/2}$ component has gained already 10 %. For fields > 2 kV/cm, the central component rapidly disappears and the linear Stark-effect shows up in equally separated frequency-positions of $P^{1/2}$ and $P^{3/2}$. However, the line intensity is still not symmetric.

Although Stark-shift is still negligible at this low electric field, the intensity ratio of 1:9 can be measured precisely with our laser because its bandwidth is only about half the intrinsic line separation by the 2S Lamb-shift. At about 500 V/cm, the Stark-effect increases the $2P^{1/2}$ - $2S^{1/2}$ line-splitting by a factor of two. And for about 3 kV/cm, the Stark-splitting enlarges the fine-structure splitting both P levels by a factor of two.

Compared to the 2S level, the splitting of the 3S/D level shown in figure 2 is rather complex. The additional D level leads to much more Stark-components and makes a $\Delta L = 2$ two-photon transition possible. As a consequence, three different polarization conditions for two-photon excitation have to be considered, namely linear polarization and two different cases ($\Delta L = 0$ or 2) for circular polarized laser beams. Besides the $\Delta L = 0$ condition which corresponds to the 1S-2S case, good conditions for low E-field measurements are found for the case of linear polarized laser beams, i.e. the smallest number of Stark components and large intensities. The bold bars given in the plot that show the frequency displacement in figure 2 indicate the components for the field free case. Please note, that the Lamb-shift of the $3S^{1/2}$ level is more than three times smaller than of the $2S^{1/2}$ level. Therefore, the $3P^{1/2}$ $3S^{1/2}$ line pair reacts much more sensitive to external electric fields. This can be seen in figure 2 where the interval of the electric field strength in the plots is up to 300 V/cm only and the favourable splitting indicated by circle allow field measurements down to 30 V/cm.

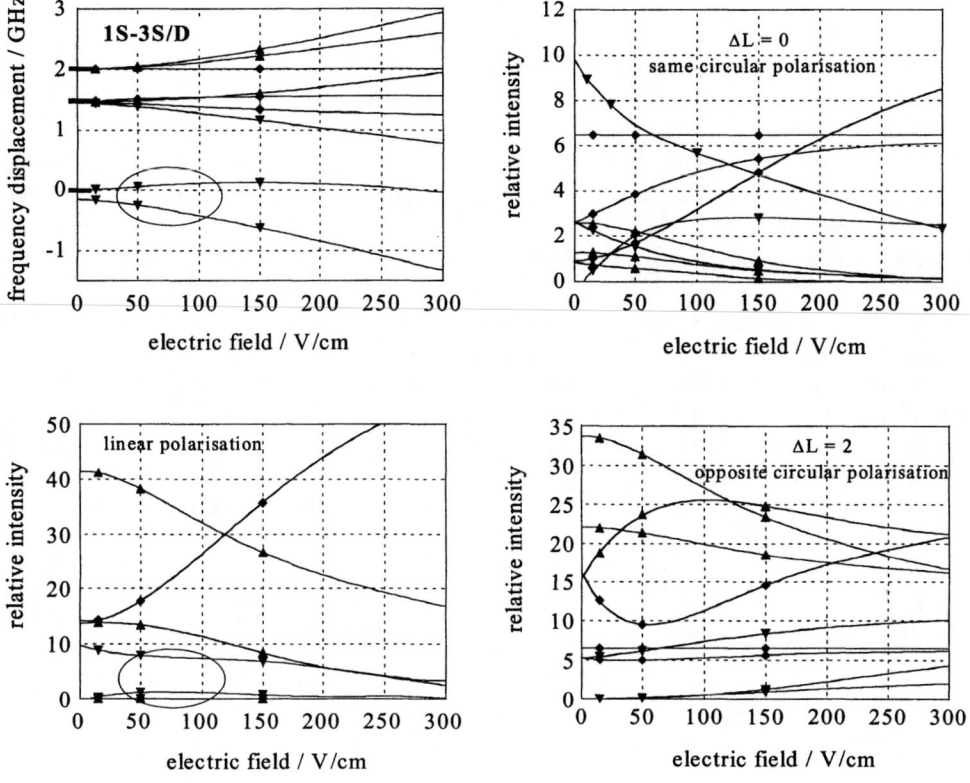

FIGURE 2. The plot at the upper left side shows the Stark splitting of the 3S/D level versus the electric field, and the frequency displacement is given for radiation at 205 nm as required for two-photon excitation. The selection rule for the angular momentum for the 1S-3S/D two-photon transition allows $\Delta L = 0$ and 2, hence their are three sets for the relative intensities of the Stark components corresponding to excitation based on linear polarized beams of same orientation (lower left side), circular polarized beams of identical (right upper side) and opposite orientation (right lower side).

However, the detection limit of 30 V/cm only holds, if all other broadening mechanisms are well below the laser bandwidth of 300 MHz, e.g.:

- power broadening and life-time reduction due to photo-ionization by the pulsed laser radiation become negligible at irradiances below 50 MW/cm^2,
- Stark broadening by electrons and ions is negligible at $n_e < 2 \cdot 10^{18}$ m^{-3}.

In addition, the electric field must not vary significantly in the measurement volume given by the overlap of the two laser beams, which can be focused to a diameter of 200 μm. The last limitation does not effect two-photon induced fluorescence spectroscopy, e.g. Balmer-α fluorescence excited by 205 nm applied as sheet diagnostic, because the spatial resolution is limited mainly by the detection system.

MEASUREMENTS OF LOW ELECTRIC FIELDS

The low field measurements via the 1S-2S and the 1S-3S/D transition of atomic hydrogen were conducted in a small opto-galvanic reference cell. The cell was filled with about 500 Pa hydrogen and atomic hydrogen is created by thermal dissociation at a hot tantalum wire of about 1500 K. The counter propagating laser beams are passing parallel to the hot wire. Resonant two-photon excitation and subsequent ionization due to absorption of a third photon from the laser beams is generating ions, which are attracted by a pickup wire, located at a distance of about 4 mm parallel to the hot one. Reasonable signals can be obtained with a pickup voltage of 10 V. This leads already to a field of the order of 30 V/cm between pickup and hot wire, and increasing the pickup voltage results in larger fields. In figure 3 two measurements of the 1S-2S and 1S-3S/D transition of hydrogen are shown and the Stark splitting reflects quite well the electric field, that corresponds to the pickup voltage. Please note, that each spectra represents a recording of single pulse signals while tuning the laser frequency, hence the spectra reveal the high pulse to pulse reproducibility of the laser spectrometers. In order to obtain spectra not affected by power broadening or depletion of the atomic ground state density, the pulse energy was attenuated well below 100 µJ.

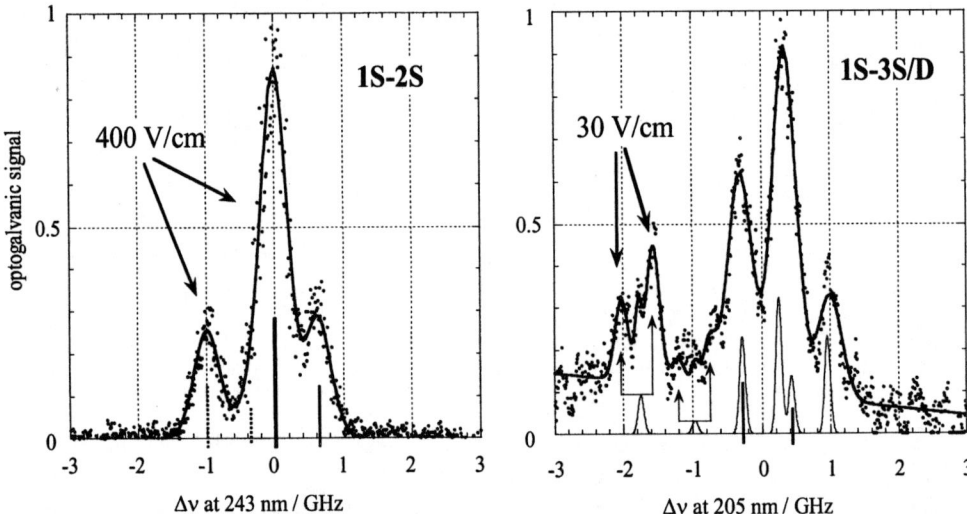

FIGURE 3. The Doppler-free spectrum on the left shows the Stark-splitting of the 1S-2S resonance. The opto-galvanic signal of individual laser pulses is plotted as a function of the laser frequency at 243 nm. The hyperfine-splitting with its weak component on the blue side is clearly resolved for the $2S^{1/2}$ level. However, the weak hyperfine-component of the red-shifted $2P^{1/2}$ level is hidden (small dotted vertical line). The splitting corresponds to an electric field of about 400 V/cm. On the right side, a similar measurement is shown for the 1S-3S/D resonance with the laser tuned at 205 nm. In this case, the splitting of the Stark-components indicated by arrows reveals an electric field of about 30 V/cm. In both measurements, the counter-propagating laser beams were linearly polarized.

MEASUREMENTS OF HIGH ELECTRIC FIELDS

As a further example, opto-galvanic spectroscopy was applied to measure the field distribution in the cathode fall region of a hollow cathode discharge: 15 mm inner diameter, 60 mm long, operated at 200 Pa of hydrogen and 100 mA of discharge current. Two counter-propagating laser beams of about 120 µJ pulse energy were overlapping about 50 mm out of focus of two cylindrical lenses (1 m focal length), providing a measurement volume of about 2 mm wide and 0.3 mm high in the upper central part of the cathode fall region, parallel to the cathode axis. Vertical displacement of the hollow cathode allows to measure the spectrum at various distances from the inner cathode surface. A set of measurements of the 1S-2S Stark-splitting is shown in figure 4. The spectra exhibit a high reproducibility and the spectrum at the bottom shows only the hyperfine structure splitting, i.e. an electric field smaller than 100 V/cm. The unexpected shoulders at the inner side of the red and blue shifted Stark component that appear for large splitting indicate, that the field is not quite homogeneous in measurement volume.

A simple data evaluation that relates the frequency displacement of the dominant part of the red and blue shifted component to the electric field strength is shown at the right side in Fig. 4. The resulting voltage drop in the cathode fall is about 340 V, which is about 120 V less than that of the entire voltage drop between anode and cathode.

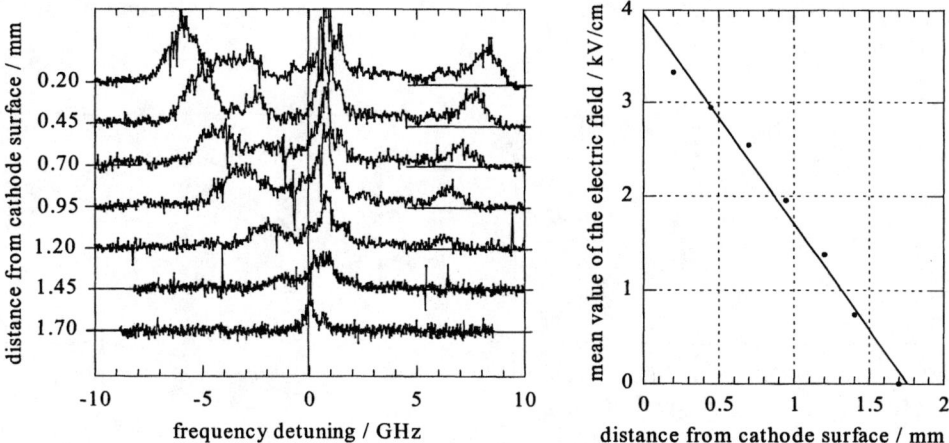

FIGURE 4. Set of measurements: each 1S-2S spectrum is recorded as a series of single pulse opto-galvanic signals while tuning the laser rapidly: 100 MHz at 243 nm. The base line of each spectrum is indicated together with the displacement from the inner cathode surface, and all signals are presented on the same linear scale. The plot at the right side shows the corresponding electric field strength versus distance from the cathode surface.

Comparison of Opto-Galvanic and Polarization Spectroscopy

The following comparison of measurements of electric fields in hollow cathode discharges performed with opto-galvanic detection and polarization spectroscopy [5] of the 1S-2S two-photon resonance of atomic hydrogen allows some interesting insight. In figure 5, the left side plot shows spectra obtained by polarization and opto-galvanic detection at similar electric field strength, i.e. both near to the inner surface of hollow cathode discharges. Although the discharges were somehow different, the spectra agree quite well, even in details like the shoulder at the left side of the red-shifted component. The difference in the height of the central component is due to the fact, that the signal of polarization spectroscopy corresponds to the sum of the squares of absorption and dispersion. As a consequence, the narrow central components appears much higher.

Another quite good agreement is obtained concerning the field distribution in the cathode fall: both measurements show a quite linear voltage drop, and the difference in the slopes is attributed to the difference of the discharges: 100 mA at a pressure of 200 Pa (PTB Berlin) result in a more extended cathode fall region of 1.7 mm than the discharge of 400 mA at a pressure of 600 Pa (Uni. Valladolid), which compresses the cathode fall region to 1 mm. Furthermore, in both discharges the cathode fall is only about 75% of the entire voltage drop. This does not confirm the common understanding, that the cathode fall represents nearly the entire voltage drop of the discharge. The large difference is probably related to the special characteristic of the extremely high electronic recombination due to molecules in a pure hydrogen discharge with a dominant molecular density.

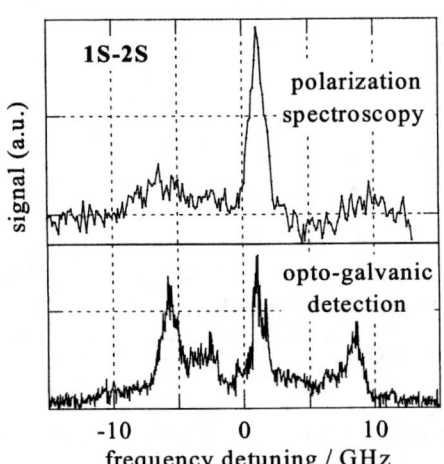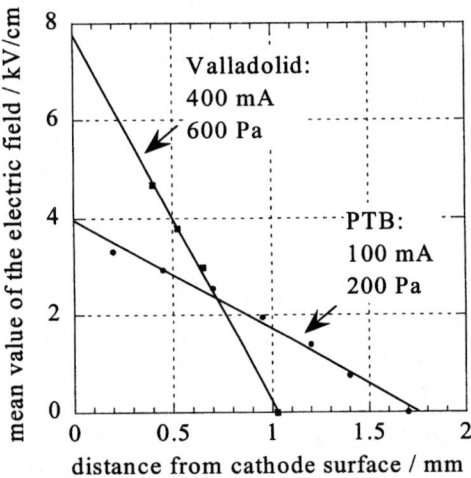

FIGURE 5. Spectra measured at similar mean electric field strength, about 3.7 kV/cm, obtained with two different detection techniques: two-photon polarization spectroscopy (deuterium discharge) and opto-galvanic spectroscopy (hydrogen discharge). On the left side the electric field versus distance from cathode surface in the two different experiments is shown.

Finally we like to compare the dynamic of the signals measured with opto-galvanic and polarization spectroscopy having in mind the polarization signal reflecting the ground state density. Therefore, the polarization measurements exhibit a strong decrease of atomic hydrogen density with decreasing distance from the cathode surface [5]. However, the opto-galvanic detection behaves opposite, i.e. the signal is increasing towards the cathode surface where larger electric fields allow the resonantly produced charged particles to change the resistance of the discharge more easily.

CONCLUSION AND OUTLOOK

Based on reliable quantum mechanical calculation of the Stark splitting of the 2S and 3S/D level of hydrogen low and high electric field measurements have been demonstrated by experiment with simple opto-galvanic detection. As an example for application, the field distribution in the cathode fall region of a hollow cathode discharge was measured. However, the more important message of this paper is, that advanced UV-laser spectrometers allow to generate high quality pulsed radiation in the entire UV-range: pulse energy of the order of 5 mJ, pulse duration of some ns and a spectral bandwidth of about 300 MHz. Therefore, they allow to extend remarkably the field of application of two-photon spectroscopic techniques, because local electric fields, ground state densities and kinetic temperatures can be determined with two-photon spectroscopy applied to a single transition of atomic hydrogen.

Actually we have changed to a different kind of plasma discharge, similar to those used in industry, namely a Gaseous Electronics Conference rf reference cell. This standard source defined several years ago is a stainless-steel reactor with parallel plate electrodes, which provides good radial access for spectroscopy. The rf voltage produces a time varying electric field between the electrodes which will be measured with high sensitivity by two-photon induced fluorescence in a sheet diagnostic scheme. The pulsed laser beams will be focused to a line by cylindrical lenses and shined in parallel to the electrode surface. Using a gated intensified CCD camera, the two-photon absorption at 205 nm will be detected by the Balmer-α fluorescence perpendicular to the sheet with good spatial resolution of less than 50 μm defined by the detection optics.

REFERENCES

1. Steiger A., *Physica Scripta* **86**, 68-71 (2000).
2. Steiger A., Grützmacher K., de la Rosa M. I., in *Laser in Research and Engineering*, Springer-Verlag Berlin, 1996, pp. 308-311.
3. Bustillo A., Grützmacher K., Steiger A., and Werner L., in *Proceedings 10th International Symposium on Laser-Aided Plasma Diagnostics*, Fukuoka, Japan 2001, pp. 247-252.
4. Lüders, G, *Ann. d. Phys.* [6] **8**, 301 (1951)
5. de la Rosa M. I., Pérez C., Gonzalo A.B., Mar S., and Grützmacher K., in *Proceedings 10th International Symposium on Laser-Aided Plasma Diagnostics*, Fukuoka, Japan 2001, pp. 294-298.

Spectral Line Profiles of the 5p-6d Transition in CIV

N.P. Kyrie[a], W. Freese[b], V.P. Gavrilenko[a,c], and H.-J. Kunze[b]

[a] General Physics Institute, Russian Academy of Sciences, Vavilov street 38, 119991, GSP-1, Moscow, Russia

[b] Institut für Experimentalphysik V, Ruhr-Universität, D-44780 Bochum, Germany

[c] Center for Surface and Vacuum Research, Russian State Committee for Standards, Vavilov street 38, 117334 Moscow, Russia

Abstract. Spectral line profiles of the *5p-6d* transition in the lithiumlike ions C IV have been measured in a theta pinch discharge. Theoretical calculations of the profiles of the *5p-6d* transition in C IV are based on the model microfield method. The electron density in a theta pinch discharge obtained by the theoretical analysis of the experimental profiles of the *5p-6d* transition in C IV is in the range $(1.2 - 1.8) \times 10^{17}$ cm^{-3} for different instants of time.

INTRODUCTION

The broadening of spectral lines of lithiumlike ions can be used for the diagnostics of hot and dense plasmas. In Refs. [1,2], measurements of the Stark broadening of a number of CIV, NV and OVI lines were reported. In the present paper, we study the Stark broadening of the transition *5p-6d* of the ion CIV ($\lambda \approx 444.1$ nm). The experiments were performed in a plasma of a 82 kJ theta pinch discharge in helium with an admixture of methane. Initial partial pressures of helium and methane were 150 mTorr and 40 mTorr, respectively. The details of the experimental apparatus are given in Refs. [3-5].

RESULTS

The plasma electric microfield mixes the wave functions of the closely spaced upper levels *6d*, *6f*, *6g*, and *6h* of CIV. The separation between these levels is as follows [6]: $\Delta(6f,6d) \approx 32.3$ cm^{-1}, $\Delta(6g,6f) \approx 2.96$ cm^{-1}, $\Delta(6h,6g) \approx 0.64$ cm^{-1}. This mixing of the wave functions of the levels *6d*, *6f*, *6g*, and *6h* leads to a considerable broadening of the CIV spectral line at 444.1 nm. Typical experimental profiles of this line at 444.1 nm recorded at different instants of time are shown by the thin solid line in Fig.1. An important feature of these profiles is that they are rather broad and asymmetrical with a more intensive "blue" part of the profiles. This "blue" part of the profiles corresponds to the dipole-forbidden transitions *5p-6f, 5p-6g*, and *5p-6h*.

Theoretical line profiles of the *5p-6d* transition in CIV were obtained by using the model microfield method [7]. This method takes into account the effects of ionic dynamics, and at the same time it enables one to make a transition to the static Stark profiles when such effects can be neglected. Figure 2 shows two theoretical Stark profiles of the CIV line at 444.1 nm calculated at electron densities $N_e = 1.2 \times 10^{17}$ cm^{-3} (curve 1) and $N_e = 2.0 \times 10^{17}$ cm^{-3} (curve 2). The perturbing particles were assumed to be hydrogen-like ions of helium. The electron temperature T_e was assumed to be 15 eV. For estimating this value of the temperature T_e, we used, first, the data from Ref. [5] where the electron temperature in the theta pinch discharge was found to be 10-20 eV. This estimate was extracted from the intensity ratio of the HeII P$_\alpha$ and HeI *2p-3d* transitions by using an approach presented in Ref. [8]. Second, the estimate of $T_e \approx 15$ eV was derived from the experimental data on the typical time τ_g of the growth of the intensity of the CIV line at 444.1 nm ($\tau_g \approx 0.5$ µs) for the electron density $N_e = 2.0 \times 10^{17}$ cm^{-3}. The choice of the electron density $N_e = 2.0 \times 10^{17}$ cm^{-3} followed from the comparison of the preliminarily calculated Stark profiles of the CIV line at 444.1 nm with the experimental profiles of this line. In this preliminary calculation we have neglected the ion motion effects. Our estimations also show that for the experimental conditions the ion temperature T_i is close to the electron temperature T_e. The theoretical profiles presented in Fig. 2 demonstrate the asymmetry similar to that of the experimental profiles. In our calculations of Stark broadening of the CIV line at 444.1 nm, we also took into account the relatively far lying upper level *6p* [$\Delta(6d,6p) = 594$ cm^{-1}], although its contribution to the Stark profile of the CIV line at 444.1 nm is small as compared with that of the levels *6d, 6f, 6g*, and *6h*. Although there is no direct Stark coupling of the levels *6g* and *6h* with the level *6d*, the contribution of the levels *6g* and *6h* to the Stark profiles of the CIV line at 444.1 nm is significant when the electron density is of the order of 10^{17} cm^{-3}. This can be seen from the comparison of two profiles (profiles 1 and 3) presented in Fig. 2 where profile 3 was calculated without taking into account levels *6g* and *6h*.

The thick solid curves in Fig. 1 show the theoretical profiles of the CIV line at 444.1 nm that provide the best fit to the experimental profiles of this line. While calculating these theoretical profiles we also took into account the Doppler effect and the instrumental broadening. The instrumental profile was a Gaussian with the full width at half maximum equal to 4.53 cm^{-1}.

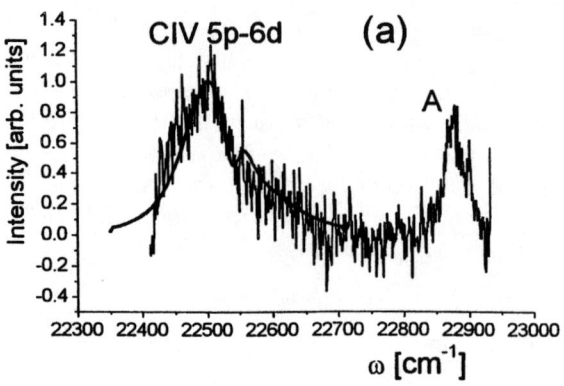

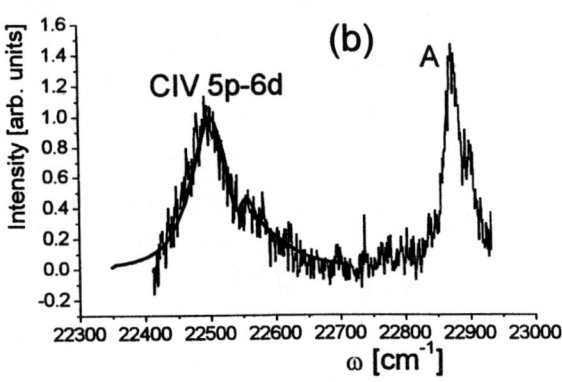

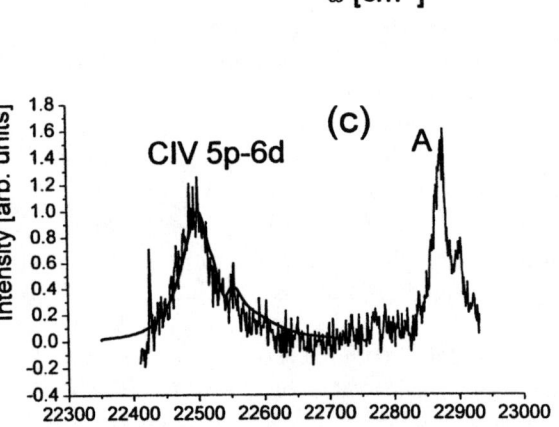

FIGURE 1. Experimental profiles of the CIV line at 444.1 nm (the *5p-6d* transition) recorded from a theta pinch discharge at three instants of time (thin curves). The line labeled 'A' corresponds to the transitions $3d'\,^4P^\circ - 4f'\,^4D$ in CII and $4f'\,^3F^\circ - 5g'\,^3G$ in CIII. The thick curves present the theoretical profiles on the CIV line at 444.1 nm calculated for $N_e = 1.8 \times 10^{17}$ cm^{-3} (Fig. 1,a), $N_e = 1.4 \times 10^{17}$ cm^{-3} (Fig. 1,b), and $N_e = 1.2 \times 10^{17}$ cm^{-3} (Fig. 1,c).

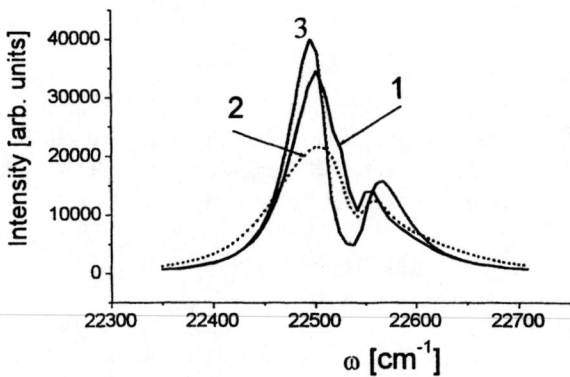

FIGURE 2. Theoretical Stark profiles of the CIV line at 444.1 nm calculated for $N_e = 1.2 \times 10^{17}$ cm^{-3} (curve 1) and $N_e = 2.0 \times 10^{17}$ cm^{-3} (curve 2). Curve 3 shows the theoretical Stark profile of the CIV line at 444.1 nm calculated for $N_e = 1.2 \times 10^{17}$ cm^{-3}, but without taking into account the levels *6g* and *6h*. All three profiles are calculated for $T_e = T_i = 15$ eV.

CONCLUSIONS

We have performed measurements of the CIV line at 444.1 nm in a theta pinch discharge. The profiles turn out to be rather broad and asymmetrical with respect to the vertical axis drawn through their maximum. This is due to the fact that the CIV line at 444.1 nm involves both allowed (*5p-6d*) and a few forbidden transitions (*5p-6f*, *5p-6g*, and *5p-6h*). Our calculations of the profiles of the CIV line at 444.1 nm are based on the model microfield method. The electron density in the theta pinch discharge obtained by the theoretical analysis of the experimental profiles of the CIV line at 444.1 nm is in the range $(1.2 - 1.8) \times 10^{17}$ cm^{-3} for different instants of time. We also performed measurements of the HeII P$_\alpha$ line in the theta pinch discharge at few instants of time. These measurements also enable us to determine the electron density by using the relation proposed in Ref. [9]. We obtained the following values of the electron density N_e for the same two instants of time for which the profiles of the CIV line at 444.1 nm are presented in Fig. 1(b,c): $N_e^{(b)} = 1.5 \times 10^{17}$ cm^{-3} and $N_e^{(c)} = 1.25 \times 10^{17}$ cm^{-3} which are in a good agreement with the values of N_e deduced from the profiles of the CIV line at 444.1 nm.

ACKNOWLEDGEMENTS

This work was partially supported (for N.P.K. and V.P.G.) by the Russian Foundation for Basic Research (project No 01-02-17810). For N.P.K. it was also supported by the "Deutscher Akademischer Austauschdienst" (Referat: 325, Kennziffer: A/00/06417). One of us (N.P.K.) greatly appreciates the hospitality of the Institute for Experimental Physics V at the Ruhr-University-Bochum.

REFERENCES

1. Böttcher, F., Musielok, J., and Kunze, H.-J., *Phys. Rev. A* **36**, 2265-2271 (1987).
2. Glenzer, S., Wrubel, Th., Büscher, S., Kunze, H.-J., Godbert, L., Calisti, A., Stamm, R., Talin, B., Nash, J., Lee, R.W., and Klein, L., *J. Phys. B: At. Mol. Opt. Phys.* **27**, 5507-5515 (1994).
3. Greve, P., Haumann, J., Kunze, H.-J., and Ullrich, L.K., *Phys. Fluids* **25**, 452-456 (1982).
4. König, R., Kolk, K.-H., and Kunze, H.-J., *Phys. Fluids* **30**, 3579 (1987).
5. Freese, W., and Kunze, H.-J., *Nukleonika* **45**, 173-177 (2000).
6. Bashkin, S., and Stoner, J.O., Jr., *Atomic Energy Levels and Grotrian Diagrams. Vol. I.*, North-Holland, Amsterdam, 1975.
7. Brissaud, A., and Frisch, U., *J. Quant. Spectrosc. Radiat. Transfer* **11**, 1767-1783 (1971).
8. Griem, H.R., *Principles of Plasma Spectroscopy*, Cambridge University Press, New York, 1997.
9. Musielok, J., "New Empirical Formula for N_e - Determination from the Broadening of the HeII 4686 Å Spectral Line" in *Proc. 19th Int. Conf. On Phenomena in Ionized Gases (Belgrade, Yugoslavia, 1989)*, Vol. 2, edited by J.M. Labat, Publisher: Faculty of Physics, University of Belgrade, Belgrade, 1989, pp. 326-327.

Stark broadening of Kr II spectral lines in dense plasmas produced in flashtubes

S.A. Flih and Y. Vitel

Laboratoire des Plasmas Denses, Université Pierre et Marie Curie, T12-E5, P.O.Box 90, 4 Place Jussieu, 75252 Paris, France

Abstract. We report measurements of the Stark parameters of three Kr II spectral lines belonging to the 5s-5p transitions, emitted by dense plasmas with electron densities in the range $1.1 - 4.4 \ 10^{18}$ cm^{-3} and temperatures from 13400 to 19600 K.

INTRODUCTION

The Stark effect is the main broadening mechanism in moderately dense plasmas and is widely used for diagnostic purpose. A knowledge of the Stark broadening parameters of Kr II spectral lines has become of interest because of the increasing role of krypton in laser and light sources. In this work, high density plasmas are produced in special flashtubes allowing variable filling up of an krypton-hydrogen mixture. Plasmas obtained have a good cylindrical symmetry and reproducible from shot to shot. They are in local thermodynamic equilibrium and their on axis electron densities vary from 1.1 to 4.4 10^{18} cm^{-3} and temperatures from 13400 – 19600 K. Stark parameters (FWHM and d) of three Kr II spectral lines belonging to the 5s-5p transitions are measured as a function of electron density N_e, and compared with other experimental data [1,4] and calculations [5].

EXPERIMENTAL SET-UP AND PLASMA DIAGNOSTICS

The plasma sources are created in linear flashtubes (inner diameter 3 mm or 5 mm, interelectrode length 100 mm) filled with a mixture of Krypton-Hydrogen at low H$_2$ concentrations (less than 4%), under initial pressures ranging from 200 to 600 torr. The gas breakdown is performed applying a high voltage pulse (30 kV, 1μs) on an external auxiliary electrode. Then, a low current (less than 3 A) simmer is maintained during a few ms before triggering the main discharge to ensure the discharge centering on the tube axis. This discharge is produced by means of a LC cell with variable inductance which permits to adapt the source impedance to the plasma impedance. The maximum current intensity is set in the range 0.2 kA - 1.1 kA and its pulse duration at half height is around 100 μs.

In figure 1, a schematic view of the optical set-up is displayed. The emitted spectra are recorded by using two detection systems: one with low and medium dispersion

CP645, *Spectral Line Shapes: Volume 12, 16th ICSLS*, edited by C. A. Back
© 2002 American Institute of Physics 0-7354-0100-4/02/$19.00

using a Czerny-Turner spectrograph (Spectro 1: focal length 300 mm, three grating turret 600-1200-2400 lines/mm) coupled with a two-dimensional intensified detector (ICCD), the other using a high dispersion Czerny-Turner spectrograph (Spectro 2: focal length 1500 mm, grating 1800 lines/mm) coupled with a one-dimensional intensified detector (OMA). For standardization purposes, we use a deuterium lamp for $\lambda < 380$ nm and a tungsten ribbon lamp for $\lambda > 380$ nm. The wavelength scale is calibrated using spectral lamps.

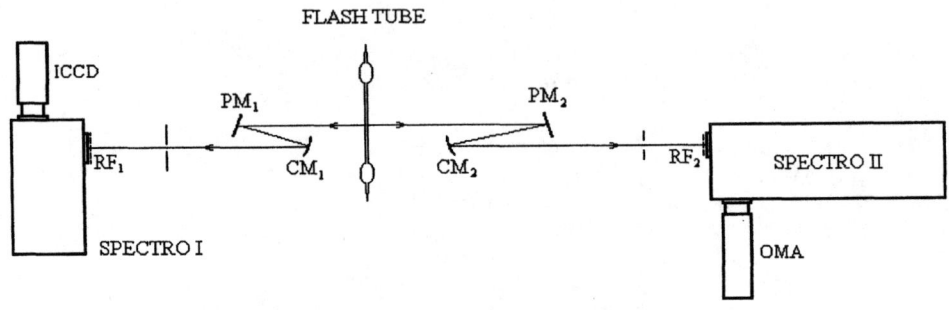

FIGURE 1. Schematic diagram of the optical set-up : CM_1, CM_2 : concave mirrors, PM_1, PM_2 : plan mirrors, RF_1, RF_2 : rejection filters.

All spectral measurements are performed during a short period of time (≈ 1 μs) at the current maximum, when the best filling of the plasma is obtained inside the flashtube. Radial profiles of temperature and particle densities are deduced from :
- measurements of optically thick Kr I line intensities in the near infra-red (810.4 and 811.2 nm)
- determination of the radial emissivity $\varepsilon(r)$ by measuring the transverse distribution of the continuum at $\lambda = 320$ nm where the plasma is optically thin, and applying the method of the Abel integral
- thermodynamic equilibrium equations.
The radial profiles so deduced, are practically flat over one half of the inner diameter as shown in figure 2.
To check the values found for the electron density profiles, we have determined those profiles by He-Ne laser interferometry at the wavelength of 3.39μm where the refractive index of the plasma is only due to the electron density. A good agreement is found from both methods.
The accuracy is estimated at 8% for the electron density and 5% for the temperature.

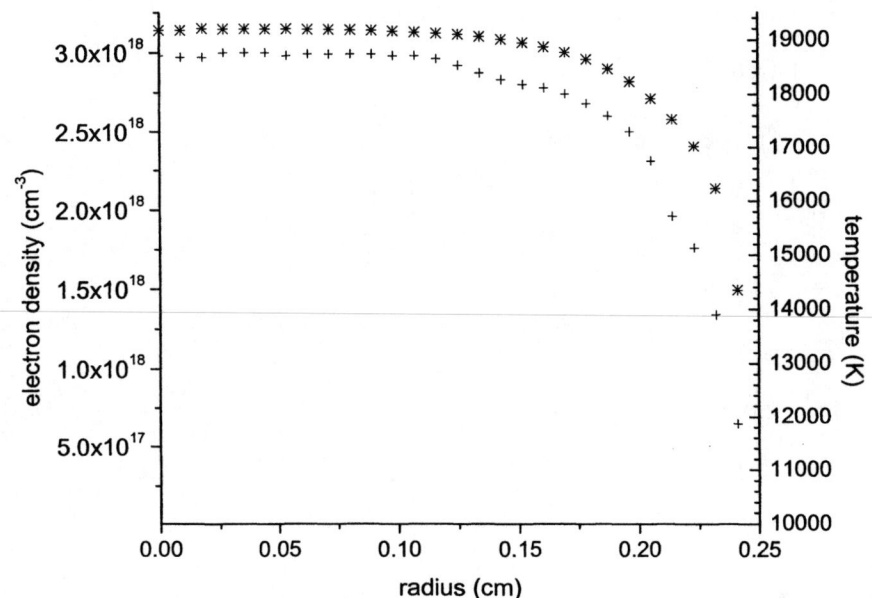

FIGURE 2: Example of electron density and temperature profiles: + electron density, ⊠ temperature.

EXPERIMENTAL RESULTS AND COMPARISONS

The different Kr II lines are registered by means of the high dispersion spectrograph coupled with the optical multichannel analyzer, giving a precision on the wavelength scale of 0.01nm. Each spectral line profile is obtained after correction for the optical thickness (measured with a concave mirror placed behind the flashtube on the optical axis of the detection system) and subtraction of the continuum background.

Then, experimental profiles are fitted to a Lorentzian shape taking into account the radial profiles of particles and temperature, on the basis of a $N_e/T^{1/2}$ dependency for the Stark parameters. In our conditions, the other broadening mechanisms like the Doppler effect or the Van der Waals effect are negligible in comparison to the Stark effect.

These fits are very good as shown in figures 3 and 4 for the three lines. From this processing we can conclude that the measured widths and shifts are mainly due to the hot region of the plasma where the radial profiles are practically flat.

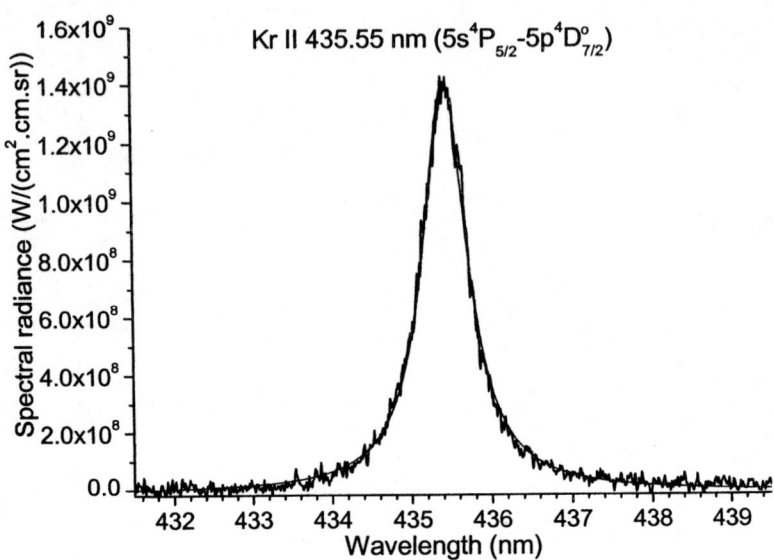

FIGURE 3. Experimental Profile and Lorentzian fit for $N_e(0) = 3.0\ 10^{18}$ cm^{-3} and $T_e(0) = 19200$ K.

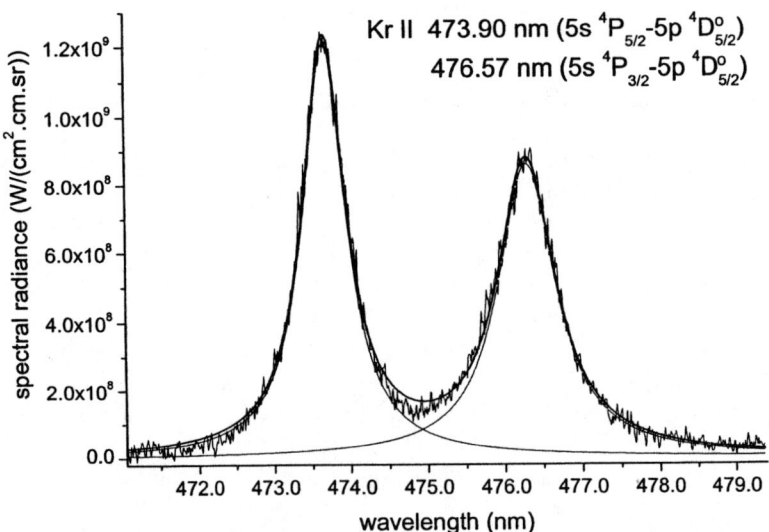

FIGURE 4. Experimental Profiles and lorentzian fits for $N_e(0) = 3.0\ 10^{18}$ cm^{-3} and $T_e(0) = 19200$ K.

In figure 5 are displayed for the Kr II line at 435.548 nm, our measured values of the full width at half maximum (FWHM) versus N_e, which are compared with other experimental results [1,4] and calculations based on a modified semi-empirical approach [5].

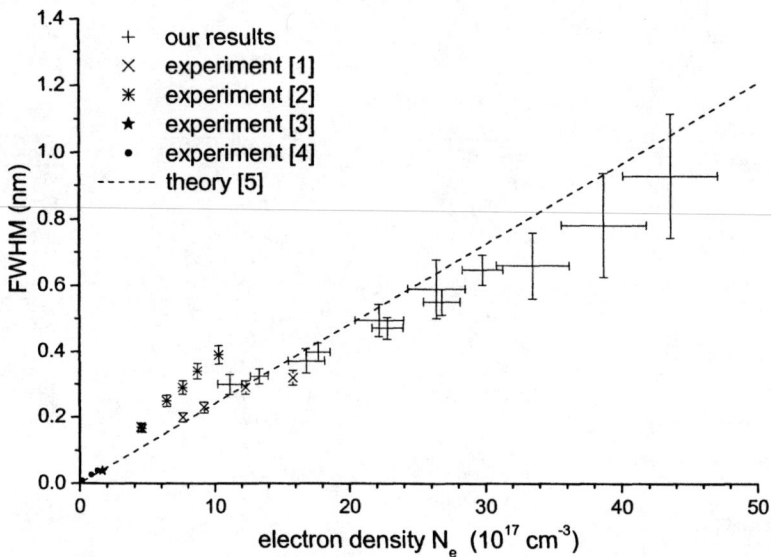

FIGURE 5. Experimental and theoretical FWHM versus electron density for the Kr II line at 435.55 nm

In figure 6 the comparisons are done for the shift (d) of this line.

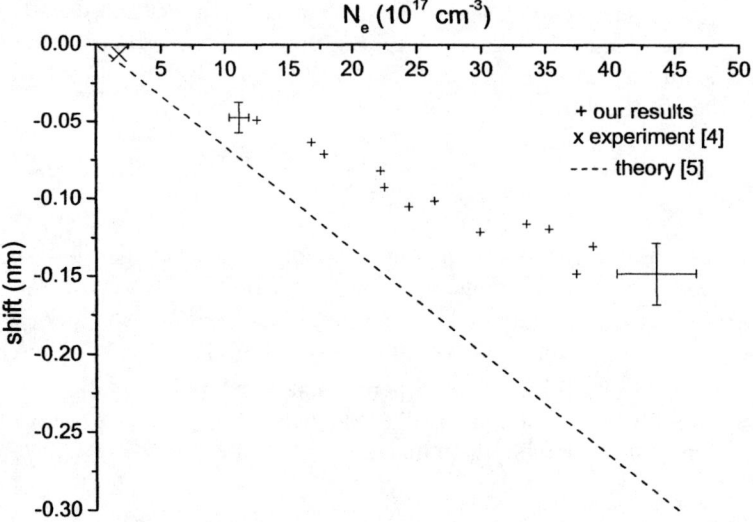

FIGURE 6. Experimental and theoretical shift versus electron density for the Kr II line at 435.55 nm

Our results show a rather good linear dependency between the Stark parameters (FWHM and d) and the electron density, making possible their use for determining electron density in dense krypton plasmas.

In the table 1, we compare, for the three lines, our measured FWHM and d normalized at $T = 17000$ K and $N_e = 1.10^{17}$ cm^{-3}, with other experimental and theoretical values.

TABLE 1. Comparisons of the normalized FWHM and d.

λ (nm)	FWHM (nm)	d (nm)	reference
	0.024 ± 0.002	-0.0037 ± 0.0004	[1]
	0.023 ± 0.002	-0.001 ± 0.0007	[2]
435.548	0.032 ± 0.002	-0.0041 ± 0.003	[3]
	0.023	-0.0066	[4]
	0.029 ± 0.0020.024		[5]
	0.024 ± 0.003	-0.00480 ± 0.0004	[1]
	0.026 ± 0.003	-0.0052 ± 0.001	[2]
473.900	0.033 ± 0.002	-0.0008 ± 0.0001	[3]
	0.021 ± 0.002	-0.0054 ± 0.0005	[4]
	0.034 ± 0.003	-0.0049	[5]
	0.028		
	0.036 ± 0.005	-0.0057 ± 0.001	[3]
	0.028 ± 0.003	-0.0039 ± 0.001	[4]
476.574	0.041 ± 0.003	-0.0071	[5]
	0.030		

On the basis of this table, one can conclude that the comparisons between our normalized FWHM with the other experimental values and theoretical predictions, lead to a satisfactory agreement taking into account the error bars. In the case of the shift, the agreement with the other experimental values is less good due to the fact that the blue shift of these lines is weak and difficult to measure at electron densities less than 10^{17} cm^{-3}. Concerning the theoretical values, they have the same negative sign and are in satisfactory agreement with ours for the lines at 473.900 and 476.574 nm, but under evaluated for the line at 435.548 nm.

CONCLUSIONS

We have measured the Stark broadening parameters of three Kr II lines, as a function of electron density in the range of 1- 4.4 10^{18} cm^{-3}. Our results show a good linear dependency of these parameters versus N_e, allowing their use for the determination of N_e in krypton plasmas. On the whole, there is satisfactory agreement (to within the error bars) with other experimental results. Calculations using a modified semi-empirical approach give a good evaluation of our measured Stark widths and a rough estimation of shifts in this range of plasma parameters.

REFERENCES

1. Vitel Y., Skowronek M. *J, Phys. B* **20**, 6493- 6506 (1987).
2. Uzelac N.I., Konjevic N., *J. Phys. B* **22**, 2517- 2525 (1989).
3. Milosavljevic V., Djenize S., Dimitrijevic M.S., Popovic L.C., *Phys Rev E* **62**, 4137- 4145 (2000).
4. De Castro A., Aparicio J.A., Del Val J.A., et al,. *J. Phys. B* **34**, 3275- 3286 (2001).
5. Popovic L.C., Dimitrijevic M.S., *Astron. Astrophys. Suppl. Ser.* **127**, 295-297 (1998).

Experimental Stark Widths and Shifts for Spectral Lines of Neutral and Ionized Atoms
(A Critical Review of Selected Data for the Period
1989 through 2000)

Wolfgang L. Wiese[*], Jeffrey R. Fuhr[*], Alain Lesage[†] and Nikola Konjević[‡]

[*]National Institute of Standards and Technology, Gaithersburg, MD 20899 USA
[†]Observatoire de Paris-Meudon, 92195 Meudon Cedex, France
[‡]Institute of Physics, 11081 Belgrade, P.O. Box 68, Yugoslavia

Abstract. We have extended our critical review of the available experimental data on Stark widths and shifts for spectral lines of non-hydrogenic neutral atoms and positive ions by one year, including the year 2000, so that we now cover the period from 1989 through the end of 2000. Our recently completed new review represents a continuation of earlier critical reviews up to 1988. Data tables containing the selected experimental Stark broadening parameters are presented with estimated accuracies. Guidelines for the accuracy estimates, developed during the previous reviews, are summarized. The data are arranged according to elements and spectra, and these are presented in alphabetical and numerical order, respectively. A total of 77 spectra are covered, and the material on multiply charged ions has significantly increased. Comparisons with comprehensive calculations based on semi-classical theory are made whenever possible, since the comparison with theory has often been a principal motivation for the experiments.

INTRODUCTION

Our new tabulation is a continuation of a series of critical reviews and tables on experimental Stark broadening data for spectral lines of non-hydrogenic atoms and ions which we started in 1976 [1,2] and continued in 1984 [3,4] and 1990 [5]. In this new installment, we cover the period from 1989 to the end of 2000, and–as in the last review–we have tabulated the data on atoms and ions in a single set of tables. We presented this new installment as work in progress at the last Spectral Line Shape Conference, [6] but have then extended our review by one year through the end of 2000 and have it now completed [7]. Generally, we have adhered to the format of our previous reviews, and we have subjected the data again to the same evaluation criteria as established earlier.

Our main source of literature references has been the master file of the Data Center on Atomic Line Shapes and Shifts at the National Institute of Standards and Technology (formerly the National Bureau of Standards) [8,9]. Also, two of us (N. K. and A. L.) have maintained independent literature searches during the entire period. A principal reason for many Stark-broadening experiments is to provide comparisons for theoretical Stark width and shift data. We have therefore presented comparisons with the generally successful and widely applied semiclassical calculations, similar to our previous reviews. Another reason for such experiments is (especially for heavier and higher ionized elements) the importance of Stark broadening parameters in stellar atmosphere opacity calculations, [10] and in the analysis of dense laboratory plasmas.

Our Evaluation Procedure

We have evaluated and tabulated the two principal Stark broadening parameters obtained from the experiments: the full width of a spectral line at half maximum intensity (FWHM) and the shift of a line, usually determined at the position of peak intensity. But in several cases the shifts are reported for the position of the halfwidth, and this is noted in the tables for these spectra. The shifts are listed as positive when they occur toward longer wavelengths (red shift), and as negative when they occur toward shorter wavelengths (blue shift).

We have provided detailed discussions of our evaluation procedure and of the adopted criteria in our earlier reviews [1-5, 11] and have given extensive literature references there. However, we summarize again in Ref. [7] our principal criteria for the selection of the experimental results and for the estimates of the uncertainties:

(a) The plasma source must be well characterized, i.e., it must be homogeneous in the observation region, be quasi steady-state during the observation time and must be highly reproducible. The last requirement is especially important for shot-to-shot line scanning techniques with pulsed plasmas. We have given a detailed listing of the applied plasma sources, their properties, range of applications, power requirements, etc. in a special table (Appendix A of our general introduction).

(b) The electron density in the observed plasma region must have been determined accurately by an independent method (i.e., a method other than utilizing the Stark broadening of the investigated lines).

(c) A temperature determination with an accuracy similar to that for the electron density must have been carried out.

(d) Competing line broadening and shift mechanisms must have been considered, such as Doppler, Van der Waals, and instrumental broadening and their contributions must have been estimated and subtracted. Recent line shape deconvolution techniques, especially for the slightly asymmetric profiles of neutral atoms, are discussed in Appendix B of the general introduction.

(e) Optically thin conditions must prevail at the line centers, or appropriate corrections must be made. Also, inhomogeneous plasma boundary layers must be either eliminated by appropriate experimental arrangements or must be taken into account in the uncertainty estimates.

We have generally found that in the large majority of the experiments, the above listed critical factors (a) to (e) have been addressed. Occasionally, one of the factors–like the measurements of the plasma temperature or the effects of plasma end-layers–has not been discussed. We have either noted this in the tables which list "Key data on experiments," or have commented on this in the introduction to the pertinent spectrum, and we have increased the uncertainty estimate.

As in our earlier reviews, we have again selected a few specific cases with fairly large amounts of experimental data, as well as those with some semi-classical theoretical comparison data, which were used for graphical illustrations of the quality of the data, and we provide three figures in this extended abstract.

In Fig. 1 we present the measured Stark widths (in angular frequency units) of the lithium-like 3s ^{2}S–3p ^{2}P° transition for six successive ions, B III through Ne VIII. A nuclear charge (Z) dependence is seen, which is also predicted by the semiclassical theory. In Figs. 2 and 3 we illustrate ratios of measured Stark widths to semiclassically calculated ones as a function of temperature for the 3s ^{2}S–3p ^{2}P° transition of Li-like N V (Fig. 2) and of boron-like O IV. For Li-like N V, the experimentally established temperature dependence is noticeably different from the semiclassically calculated one, while for O IV the agreement is very close.

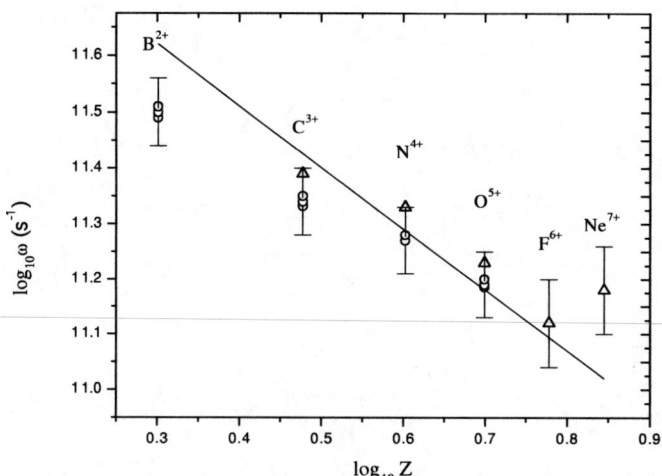

Fig. 1. Stark widths for the 3s ^{2}S–3p ^{2}P° transition of lithium-like spectral lines (in angular frequency units) as a function of $\log_{10}Z$. Experimental data are scaled linearly to a value of the electron density of 10^{17} cm^{-3} and to a value of the electron temperature of 87 000 K (7.5 eV) using the w_e (T_e) dependence from theoretical data [12]. Experimental results: Δ,Glenzer et al.[13]; and O, Blagojević et al.[12]. Error flags are calculated uncertainties including the error in the determination of the full-width at half-maximum and electron density measurements. The results of semi-classical calculations by Blagojević et al., [12] shown for comparison, are given by the solid line.

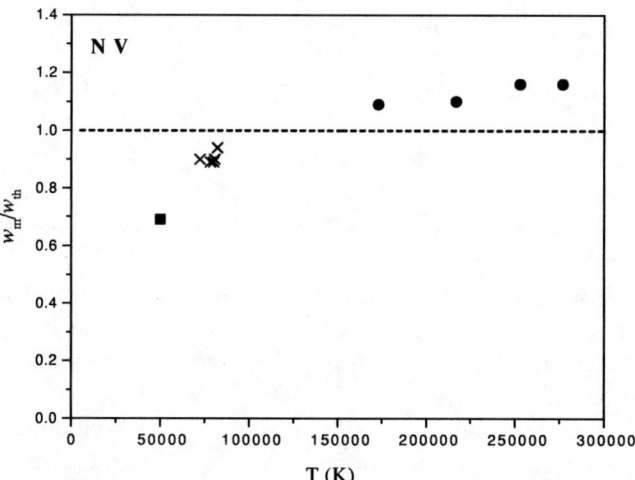

Fig. 2. Ratio of experimental Stark widths to semi-classical calculations as a function of temperature for the lithium-like 3s ^{2}S–3p ^{2}P° transition of N V. The experimental sources are: ■, Puric et al.[14]; ●, Glenzer et al.[13]; and ×, Blagojević et al.[12]. The semiclassical calculations were carried out by Blagojević et al. [12].

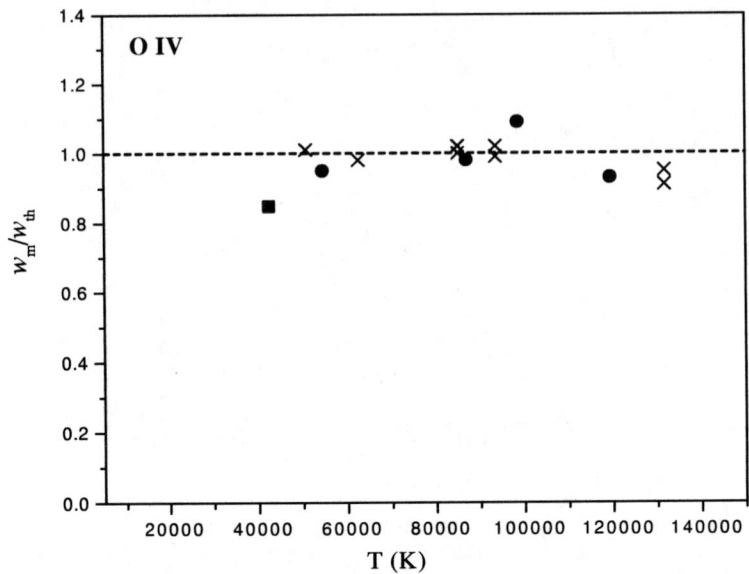

Fig. 3. Ratio of experimental Stark widths to semi-classical calculations as a function of temperature for the boron-like 3s ^{2}S–3p ^{2}P^o transition of O IV. The experimental data sources are: ■, Puric *et al.*[15]; ×, Blagojević *et al.*[16]; and ●, Glenzer *et al.*[17]. The measurements are compared with the semiclassical calculations of Blagojević *et al.*[16].

Comparisons with Theory

Whenever theoretical data are available, we have provided comparisons with the semiclassical theory developed by Griem [18] in 1962. For neutral atoms and singly-charged ions, the semiclassical results by Griem [19] are normally used. In a few cases, other calculations based on the semiclassical theory had to be applied and are clearly noted. For multiply-charged ions, the results of another semiclassical perturbation mechanism are used [20]. This approach was then modified and extended, especially by Sahal-Brechot, and Dimitrijević, and a large number of additional numerical results were provided by these authors (see the NBS/NIST bibliographies on spectral line broadening [8,9] for numerous references).

The ratios of measured-to-calculated widths and shifts tabulated by us for the various spectra provide guidance on the degree of agreement between the experiments and theory, and thus provide a valuable indication of the quality of the calculated as well as measured data. For most transitions the semi-classical calculations compare well with the experimental data, which was also observed in our earlier reviews.

We should note that the theoretical width data for ionic radiators are for the electron width only, and the usually very small additional width caused by ion broadening is neglected. Also, small shifts contributed by ions are neglected in the theoretical shift data for ionic radiators.

Arrangement of the Tables

The data are presented in separate tables for each spectrum (or stage of ionization) and the spectra are arranged according to chemical elements, which are given in alphabetical order.

Each data table is preceded by short comments providing important information on the selected literature and by a short tabular overview providing some key points on each selected experiment, such as the type of plasma source employed.

The data tables are subdivided into four principal parts. In the first part, which comprises three columns, each spectral line is identified by transition array, multiplet designation (wherever available), and wavelength (given in Angstrom units).

In the second part of the table, the principal plasma data are listed. Normally, these are the ranges of temperatures and electron densities at which the width and shift data have been measured. We have always listed the electron densities in units of 10^{17} cm^{-3}.

In the third part of the table, we present the measured Stark broadening data, specifically the full width at half maximum intensity (FWHM), i.e., the "Stark width," w_m, and the Stark shift, d_m. We also present the ratios of measured to theoretical, i.e., semiclassically calculated, widths and shifts, w_m/w_{th} and d_m/d_{th}, when available.

In the fourth part of the table, we provide estimates of the uncertainties of the data and we also identify the literature references. When Stark widths as well as shifts are measured, we provide two estimates of the uncertainties, where the first one refers to the width, while the second pertains to the shift. For the estimates we use again code letters, which indicate the following:

A = uncertainties within 15%
B$^+$ = uncertainties within 23%
B = uncertainties within 30%
C$^+$ = uncertainties within 40%
C = uncertainties within 50%, and
D = uncertainties larger than 50%

Pb II

Ground State: $1s^2 2s^2 2p^6 3s^2 3p^6 3d^{10} 4s^2 4p^6 4d^{10} 4f^{14} 5s^2 5p^6 5d^{10} 6s^2 6p\ ^2P^o_{1/2}$

Ionization Energy: 15.03248 eV = 121 245.14 cm^{-1}

Key data on experiments

Ref.	Plasma source	Method of measurement		Remarks
		Electron density	Temperature	
21	Low-pressure pulsed arc	Laser interferometer at 6328 Å and known Stark widths of S III and S II lines	Intensity ratio of S III and S II lines	Photographic technique
22	Pulsed capillary discharge	H$_\alpha$ Stark width	Relative intensities of two Cu II lines and source function of Cu I 3247 Å line	

Numerical results for Pb II

No.	Transit ion array	Multiplet	Wavelength (Å)	Temperature (K)	Electron density (10^{17} cm^{-3})	w_m (Å)	w_m/w_{th}	d_m (Å)	d_m/d_{th}	Acc.	Ref.
1.	6p–7s	$^2P^o$–^2S	2203.53	28000	1.90	0.094				C+	21
2.	6d–5f	^2D–$^2F^o$	4244.92	24000	11.0	8.25				C+	22
			4386.46	24000	11.0	7.70				C+	22
				27000	1.62	0.510		0.151		C+,C+	21
3.	7s–7p	^2S–$^2P^o$	6660.20	24000	11.0	9.90				C+	22

Table I: Data from references of Djenize et al.[21] and Fishman et al.[22]. See text for discussion.

Results of category D are used very sparingly, i.e., for important transitions only when no other data are available, and for uncertainties estimated not to exceed factors of three.

A sample of our data tables, PbII, is shown as Table 1. In order to fit this table into the book format, the introductory text is not included, but is listed in this paragraph as follows: Djenize et al.[21] have observed Pb II lines in a pulsed discharge end-on on a shot-to-shot basis. The plasma conditions were reproducible within 6-8%. Plasma homogeneity was assumed, but not experimentally tested. Self-absorption was assumed to be negligible due to the manner in which the lead was introduced into the plasma. The line widths were corrected for Doppler and instrumental contributions. An earlier result by Miller et al.[23] for the Pb II 4386 Å line is 3.5 times larger than that of Djenize et al. There is no line in common with the measurements by Purić et al. [24] Comparisons with Griem's [25] semi-empirical formula and its modified version yield ratios of 0.88 and 0.96, respectively, between measured and calculated values for the 2203.5 Å resonance line. Fishman et al.[22] have photographically observed Pb II lines with a high-density pulsed capillary discharge side-on. Radial electron density measurements based on the Abel inversion technique confirmed the homogeneity of their plasma. Self-absorption for the lines under study was measured and found to be rather small, not exceeding 15%. At the high electron density of their capillary discharge, the Stark widths of the investigated Pb II lines were so large ($\approx$ 10Å) that all other types of broadening as well as the instrumental width (of the order 0.1 Å) could be neglected. The results of Fishman et al. [22] disagree strongly with earlier shock-tube data by Miller et al. [23] Fishman et al. speculate that significant self-absorption in the earlier work is the cause of this discrepancy. For the Pb II 4245 Å line, good agreement is obtained with the results of Purić et al. [24].

In this critical review, we have expressed the relative uncertainties as "combined standard uncertainties" according to the recommendations by Taylor and Kuyatt [26]. This means that the overall uncertainties of the measurements are given as the root of the sum of the squares (rss) of the individual contributions. All reported–or otherwise estimated–individual standard uncertainties that are contributing have been combined to the "rss" relative error. This new procedure differs from the preceding reviews[1-5] where the straight sum of individual uncertainties was taken as the overall uncertainty. For example, in Refs. 1-5 the total error in the Stark width from uncertainties of ± 10% in the electron density and ± 15% in the half-width measured was ± 25%, while the rss estimate used here yields ± 18%. In many experiments, the two dominant contributions are the uncertainties in the electron density and line width, or shift, measurement. Smaller contributions arise usually from the deconvolution process, from plasma inhomogeneities such as cooler, lower electron-density end-layers, from optical depth problems and corrections, and from plasma temperature measurements.

In arriving at our uncertainty estimates, we have also taken into account occasional disagreements between authors outside their mutual error estimates. Furthermore, whenever possible, we have tested the adherence of the data to regularities and similarities predicted on the basis of atomic structure considerations [27,28]. For example, Stark half-widths should be normally nearly the same for all lines within multiplets, and we commented on cases where significant departures from this rule occurred. Studies of isoelectronic ions of the Li-sequence have shown clear systematic trends of the Stark widths with nuclear charge.

Some Observations and Conclusions

The growth of experimental Stark broadening data in the last 12 years has been especially pronounced for multiply-charged ions due to the development and applications of new plasma sources (see also Ref. [11]).

We have found that the new experimental data are generally very consistent with the earlier material, as well as with the results of the semiclassical theory. But most of the new data are not significantly better than those in our last review, indicating that improvements in the experimental techniques have not advanced significantly in recent years.

For lines of neutral atoms we observe that for Stark widths, the agreement between different experiments, as well as experiment and theory, is usually within ± 20%, and often within 10%. Stark width data of especially good quality and mutual consistency are found for several lines, of Ar I, C I and He I, so that these Stark widths can compete with other techniques as a high-accuracy plasma diagnostic approach for the determination of electron densities.

For lines of singly-ionized atoms, the accuracy of the experimental data is, however, not as well established yet, and more measurements are highly desirable. A few exceptions are several multiplets of Ar II and the first multiplet of Si II. For multiply-charged ions, good progress has been achieved in the last decade. For several spectra these are the first measurements. Stark widths have been measured over a large electron density and temperature range, and the agreement between experiments and theory is usually good.

References

1. Konjević, N., and Roberts, J. R., J. Phys. Chem. Ref. Data **5**, 209 (1976).
2. Konjević, N., and Wiese, W. L., J. Phys. Chem. Ref. Data **5**, 259 (1976).
3. Konjević, N., Dimitrijević, M. S., and Wiese, W. L., J. Phys. Chem. Ref. Data **13**, 619 (1984).
4. Konjević, N., Dimitrijević, M. S., and Wiese, W. L., J. Phys. Chem. Ref. Data **13**, 649 (1984).
5. Konjević N., and Wiese, W. L., J. Phys. Chem. Ref. Data **19**, 1307 (1990).
6. Konjević, N., Lesage, A., Fuhr, J. R., and Wiese, W. L., AIP Conf. Proceed. **559**, 126 (2001).
7. Konjević, N., Lesage, A., Fuhr, J. R., and Wiese, W. L., J. Phys. Chem Ref. Data, **31** 819 (2002).
8. Fuhr, J. R., Wiese, W. L., and Roszman, L. J., *Bibliography on Atomic Line Shapes and Shifts*, Spec. Publ. 366 (1972); Suppl. 1 (1974); Suppl. 2 (1975) (with Martin, G. A., and Specht, B. J.); Suppl. 3 (1978) (with Miller, B. J., and Martin, G. A.); Suppl. 4 (1993) (with Lesage, A.).
9. Fuhr, J. R., and Felrice, H. R., *Atomic Spectral Line Broadening Bibliographic Database*, http://physics.nist.gov/linebrbib.
10. Lesage, A., in Astrophysical Applications of Powerful New Databases, Adelman, S. J., and Wiese, W. L., Editors, Astron. Soc. Pacific Conf. Ser., Vol. 78 (1994).
11. Konjević, N., Phys. Reports **316**, 339 (1999).
12. Blagojević, B., Popović, M.V., Konjević, N., and Dimitrijević, M. S., J. Quant. Spectrosc. Radiat. Transfer **61**, 361 (1999).
13. Glenzer, S., Uzelac, N., and Kunze, H.-J., Phys. Rev. A **45**, 8795 (1992).
14. Purić, J., Srećković, A., Djenize S., and Platisa, M., Phys. Rev. A **36**, 3957 (1987).
15. Purić, J., Djenize, S., Srećković, A., Platisa M., and Labat, J., Phys. Rev. A **37**, 498 (1988).
16. Blagojević, B., Popović, M.V., Konjević, N., and Dimitrijević, M.S., Phys. Rev. E **50**, 2986 (1994).
17. Glenzer, S., Hey, J. D., and Kunze, H.-J., J. Phys. B **27**, 413 (1994).
18. Griem, H.R., Phys. Rev. **128**, 515 (1962).
19. Griem, H. R., *Spectral Line Broadening by Plasmas* (Academic Press, New York, 1974).
20. Sahal-Brechot, S., Astron. Astrophys. **1**, 91 (1969); **2**, 322 (1969).
21. S. Djenize, A. Sreckovic, J. Labat, R. Konjević, and M. Brnović, Z. Phys. D **24,** 1 (1992).
22. I. S. Fishman, E. V. Sarandaev, and M. Kh. Salakhov, J. Quant. Spectrosc. Radiat. Transfer **52**, 887 (1994).
23. M. H. Miller, R. D. Bengtson, and J. M. Lindsay, Phys. Rev. A **20,** 1997 (1979).
24. J. Purić, M. Cuk, and I. S. Lakicevic, Phys. Rev. A **32**, 1106 (1985).
25. H. R. Griem, Phys. Rev. **165**, 258 (1968).
26. Taylor B. N., and Kuyatt, C. E., NIST Tech. Note 1297, U. S. Government Printing Office, Washington, D. C. (1994).
27. Wiese, W. L., and Konjević, N., J. Quant. Spectrosc. Radiat. Transfer **28**, 185 (1982).
28. Wiese, W. L., and Konjević, N., J. Quant. Spectrosc. Radiat. Transfer **47**, 185 (1992).

ASTROPHYSICAL PLASMAS

Grating X-ray Spectroscopy of High-Velocity Outflows from Active Galaxies

W.N. Brandt* and S. Kaspi[†]

*Department of Astronomy & Astrophysics, 525 Davey Laboratory, The Pennsylvania State University, University Park, Pennsylvania 16802, USA
[†]School of Physics and Astronomy and the Wise Observatory, The Raymond and Beverly Sackler Faculty of Exact Sciences, Tel-Aviv University, Tel-Aviv 69978, Israel

Abstract. X-ray absorption and emission lines now serve as powerful diagnostics of the outflows from active galaxies. Detailed X-ray line studies of outflows have recently been enabled for a significant number of active galaxies via the grating spectrometers on *Chandra* and *XMM-Newton*. We will review some of the recent X-ray findings on active galaxy outflows from an observational perspective. We also describe some future prospects.

X-ray absorption lines from H-like and He-like ions of C, N, O, Ne, Mg, Al, Si, and S are often seen. A wide range of ionization parameter appears to be present in the absorbing material, and inner-shell absorption lines from lower ionization ions, Fe L-shell lines, and Fe M-shell lines have also been seen. The X-ray absorption lines are typically blueshifted relative to the systemic velocity by a few hundred km s^{-1}, and they often appear kinematically consistent with UV absorption lines of C IV, N V, and H I. The X-ray absorption lines can have complex profiles with multiple kinematic components present as well as filling of the absorption lines by emission-line photons. A key remaining uncertainty is the characteristic radial location of the outflowing gas; only after this quantity is determined will it be possible to calculate reliably the amount of outflowing gas and the kinetic luminosity of the outflow.

INTRODUCTION

Outflows are observed to be ubiquitous in active galactic nuclei (AGN), being seen in objects spanning a range of $\sim 10,000$ in luminosity. They have been studied in the most detail via observations of ultraviolet (UV) resonance lines from moderately ionized gas. In luminous Broad Absorption Line quasars (BALQSOs), outflows are observed to reach velocities up to a few 10^4 km s^{-1}, and they subtend ≈ 10–30% of the sky as viewed from the central source. In lower luminosity Seyfert galaxies, outflows are observed $\gtrsim 50\%$ of the time although they have velocities up to only $\approx 10^3$ km s^{-1}. These outflows are a major component of the nuclear environment, and they may carry a significant fraction of the accretion power. They may also be important in regulating the growth of the black hole and its host galaxy (e.g., Silk & Rees 1998; Fabian 1999) as well as in injecting matter, energy, and magnetic fields into the intergalactic medium (e.g., Turnshek 1988; Wu, Fabian, & Nulsen 2000; Furlanetto & Loeb 2001; Elvis et al. 2002). The observed outflows are photoionized by the radiation from the central source, and they are probably driven by radiation pressure. Despite their ubiquity and importance, their physical location and origin in the AGN system remain unclear; outflows may arise from winds driven off the surface of an accretion disk (e.g., Murray et al. 1995; Proga,

CP645, *Spectral Line Shapes: Volume 12, 16th ICSLS*, edited by C. A. Back
© 2002 American Institute of Physics 0-7354-0100-4/02/$19.00

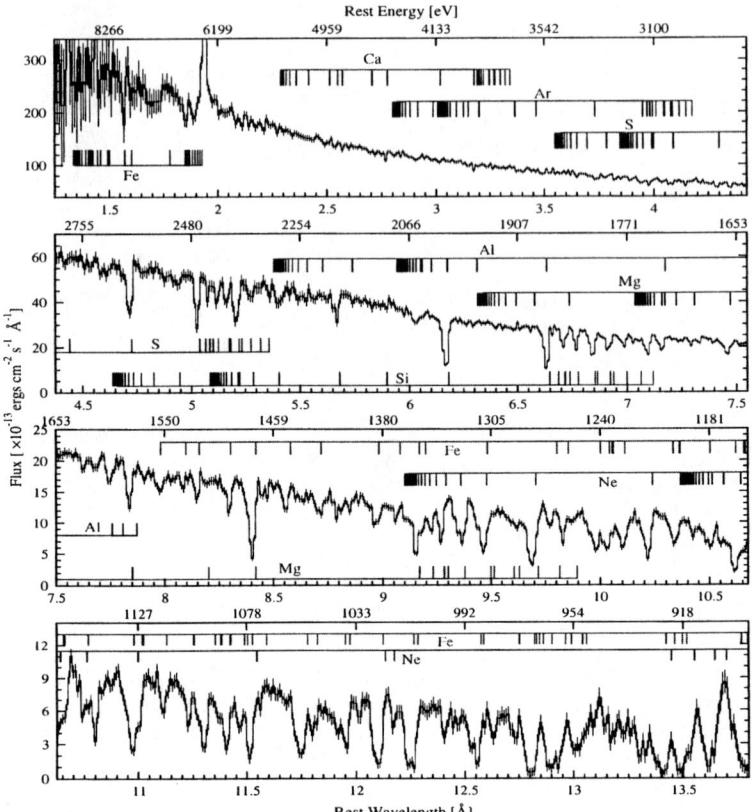

FIGURE 1. Part of the 10.4-day *Chandra* HETGS spectrum of the Seyfert galaxy NGC 3783. Marked are the large number of detected absorption lines as well as several emission lines. The lines are marked at their expected wavelengths in the rest frame of NGC 3783; the blueshifts of the absorption lines are noticeable. In total, more than 140 spectral features are detected in the X-ray spectrum of NGC 3783. Adapted from Kaspi et al. (2002).

Stone, & Kallman 2000; Elvis 2000), a dusty torus (e.g., Voit, Weymann, & Korista 1993; Krolik & Kriss 1995), or perhaps stars in the nucleus (e.g., Scoville & Norman 1995; Netzer 1996).

Ionized absorption in the X-ray band has been intensively studied in bright, low-redshift AGN for over a decade (e.g., the "warm absorbers" in Seyfert galaxies; Reynolds 1997; George et al. 1998). The luminous X-ray source in the nucleus acts as a "flashlight" allowing observers to "X-ray" material along the line of sight. However, prior to the launches of *Chandra* and *XMM-Newton*, such investigations were limited by a lack of spectral resolution. Over the past three years, the X-ray grating spectrometers on these

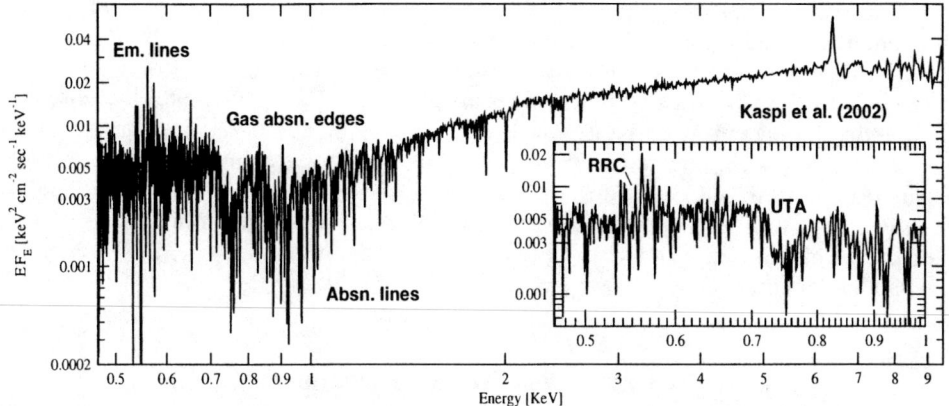

FIGURE 2. The 10.4-day *Chandra* HETGS spectrum of the Seyfert galaxy NGC 3783 shown in EF_E vs. energy format. The insert focuses on the spectrum below 1 keV. The different types of observed spectral features are labeled. Adapted from Kaspi et al. (2002).

two missions[1] have enlarged the number of spectral features available for study by a factor of ~ 50 (see Fig. 1; from 2–3 to more than 140). They have improved the velocity resolution available to observers from $\sim 15,000$ km s^{-1} to ~ 400 km s^{-1}. They have thereby provided qualitatively new information on the physical conditions, kinematics, and geometry/location of the absorbing material.

At present, efficient grating X-ray spectroscopy is possible only for ≈ 20 bright, low-redshift (mainly $z < 0.1$) AGN, mainly Seyfert galaxies. Even for these, the required observation lengths are typically $\gtrsim$ 1–2 days; the spectrum of NGC 3783 shown in Fig. 1 required a 10.4-day exposure with *Chandra*. Of necessity, the discussion below applies predominantly to this fairly small sample of objects. Highly luminous and distant quasars, for example, may have significantly different X-ray absorption properties. Furthermore, the discussion below will be closely tied to the X-ray observations, without detailed descriptions of theoretical models. For further information on theoretical models, the reader should consult one of the current reviews (e.g., Netzer 2001; Krolik 2002; and references therein).

SOME KEY RESULTS FROM X-RAY GRATINGS STUDIES

The spectral features seen in the current X-ray gratings studies include absorption lines [sometimes in unresolved transition arrays (UTAs)], emission lines, absorption edges

---

[1] The relevant instruments are the *Chandra* High-Energy Transmission Grating Spectrometer (HETGS; C.R. Canizares et al., in preparation), the *Chandra* Low-Energy Transmission Grating Spectrometer (LETGS; Brinkman et al. 1997), and the *XMM-Newton* Reflection Grating Spectrometer (RGS; den Herder et al. 2001).

(from gas and perhaps dust), and radiative recombination continua (RRCs).[2] The X-ray absorption lines are the most numerous features and are from H-like and He-like ions of C, N, O, Ne, Mg, Al, Si, and S. Inner-shell absorption lines from lower ionization ions, Fe L-shell lines, and Fe M-shell lines are also seen. Fig. 2 illustrates the observed features in the spectrum of NGC 3783. Taken together, these features are sensitive to a wide range of ionization parameter and column density, and they also provide useful temperature and density diagnostics. Furthermore, the ionizing continuum is directly visible (unlike the case for UV absorption lines, where the ionizing continuum is displaced in wavelength), and X-rays are relatively immune to dust extinction effects that can hinder studies at other wavelengths.

Physical Conditions in Outflows

Gratings observations have clearly established that a uniform, single-component model for the X-ray absorbing material in the outflow is too simple. Outflows contain gas with a wide range of ionization parameter. Fig. 3, for example, shows X-ray absorption lines from the wide range of ions observed in NGC 3783. It is not possible to fit both the low-ionization and high-ionization absorption lines simultaneously with a single "zone" of photoionized gas (e.g., Kaspi et al. 2002; Blustin et al. 2002). Similar results have been found for several other AGN (e.g., Sako et al. 2001; Kaastra et al. 2002). Current models for the outflow usually assume the absorbing gas is in photoionization equilibrium. This is a plausible assumption for some AGN, such as NGC 3783, that exhibit fairly slow and small-amplitude X-ray variability. However, AGN with more rapid and large-amplitude variability (e.g., NGC 4051 and other "Narrow-Line Seyfert 1" galaxies) may contain non-equilibrium X-ray absorbers (e.g., Nicastro et al. 1999; Collinge et al. 2001).

Early photoionization modeling of the ionized absorbers in Seyfert galaxies suggested that they have temperatures (T) of a few 10^5 K. The gratings data now confirm the expected temperatures. In NGC 3783, for example, modeling of the O VII and N VI RRCs indicate $T \gtrsim 6 \times 10^4$ K, and constraints from He-like triplet emission lines require $T \lesssim 10^6$ K (Kaspi et al. 2002).

Column densities for some of the stronger absorption lines detected from AGN outflows can be estimated via "curve of growth" analyses. However, in such analyses, it is essential to avoid lines that are saturated. The identification of saturated lines can be difficult, since they need not appear "black" (i.e., drop to zero intensity) due to the presence of nuclear X-ray scattering or multiple, unresolved line components (compare Hamann 1998 and Arav et al. 1999). Lines representing transitions to high atomic levels are the least likely to be affected by saturation (due to their lower oscillator strengths), and analyses of such lines indicate O VII and O VIII column densities of a few 10^{18} cm^{-2}. The corresponding total hydrogen column densities, assuming solar abundances and a reasonable ionization correction, range from a few 10^{21} cm^{-2} to $\approx 10^{22}$ cm^{-2}. Considering

---

[2] See the conference papers by E. Behar and T. Kallman for further discussion of some of these features.

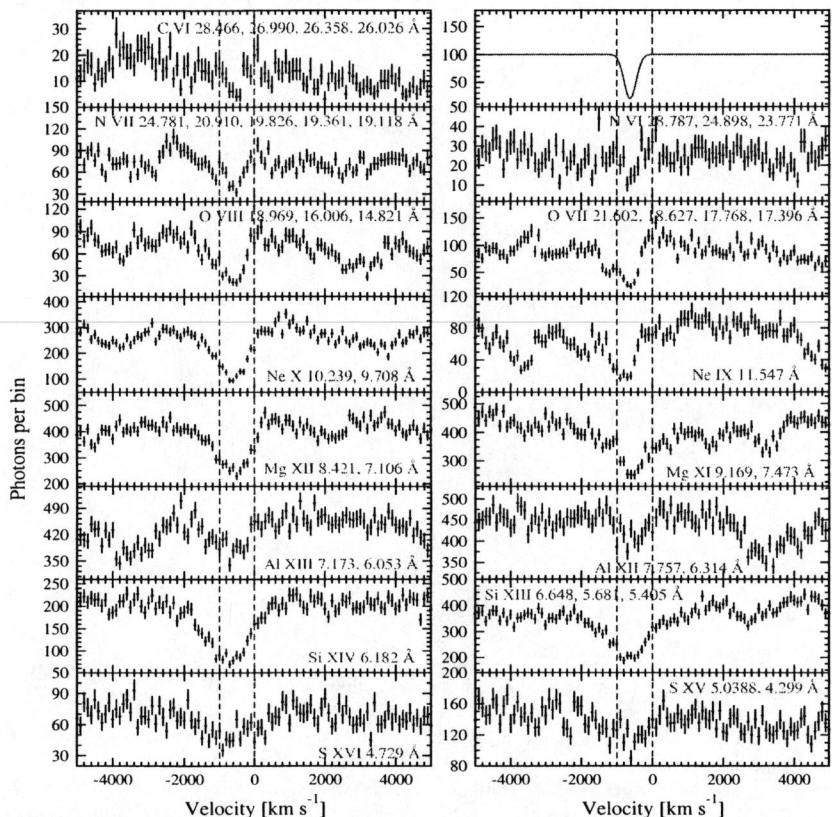

FIGURE 3. Velocity spectra showing co-added lines from different ions in the 10.4-day *Chandra* HETGS spectrum of the Seyfert galaxy NGC 3783. H-like ions are shown on the left, and He-like ions are shown on the right. The bin size is 100 km s^{-1}, and vertical dashed lines are given at velocities of 0 km s^{-1} and -1000 km s^{-1} to guide the eye. In the uppermost right panel we show a Gaussian absorption line representing the line response function of the instrument at 17.396 Å (the FWHM is 397 km s^{-1}); this is the poorest line response function applicable to the co-added velocity spectrum of O VII. Note the asymmetry of the O VII lines that is apparently from an additional absorption system. Adapted from Kaspi et al. (2002).

the statistical and systematic uncertainties currently present in curve of growth analyses, solar abundances usually appear consistent with the data. Some AGN outflows may also contain significant column densities of metals in the form of dust grains; X-ray gratings studies offer the exciting possibility of detecting these grains directly and measuring their chemical composition (e.g., Lee et al. 2001).

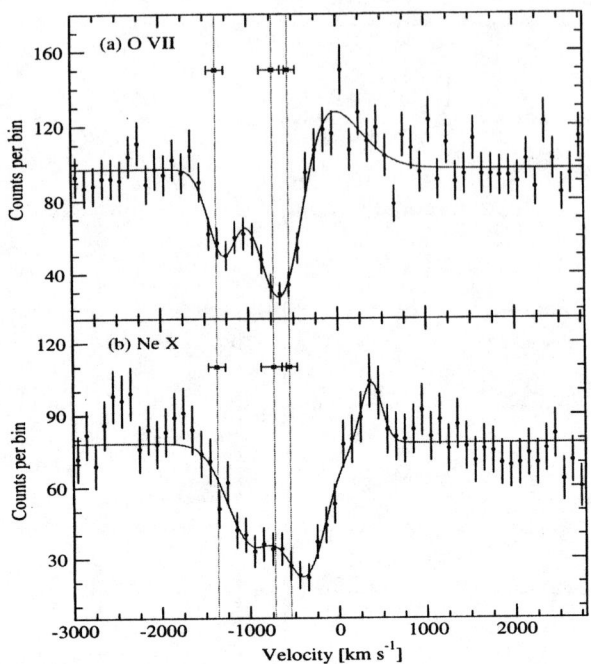

FIGURE 4. Velocity spectra, binned at 100 km s^{-1} resolution, of (a) four strong lines of O VII and (b) two strong lines of Ne X in the 10.4-day *Chandra* HETGS spectrum of the Seyfert galaxy NGC 3783. A three-Gaussian fit to each spectrum individually is overplotted on the data points (two Gaussians are in absorption, and one is in emission). Two absorption systems are clearly detected in O VII and probably exist in Ne X as well, although there are statistically significant differences between the O VII and Ne X absorption-line profiles. The vertical lines show the velocity shifts of the observed UV absorption systems; the squares with horizontal error bars show the FWHMs of these systems. Adapted from Kaspi et al. (2002).

Kinematics of Outflows

The most important, basic result on kinematics from X-ray gratings studies is that the X-ray absorbing gas is generally in a state of outflow with a bulk velocity of a few hundred km s^{-1} (as derived from the blueshifts of X-ray absorption lines). Prior to *Chandra* and *XMM-Newton*, it was not possible to speak reliably about the X-ray absorber as an outflow! In several cases, it has also now been possible to resolve the X-ray absorption lines. X-ray absorbing outflows can have velocity dispersions comparable to their bulk velocities, perhaps due to acceleration of the outflow or turbulence. Furthermore, in a few cases, multiple kinematic components in a single ion have been discovered (e.g., see Fig. 4). As illustrated in Fig. 4 and Fig. 5, different ions can have different kinematic properties; ionization level and kinematics are therefore connected. It is important to note that there may be systematic errors in velocity measurements derived from X-ray absorption lines. For example, some X-ray lines show "P Cygni" profiles where the emission line partially fills the red side of the absorption line. If this (geometry depen-

124

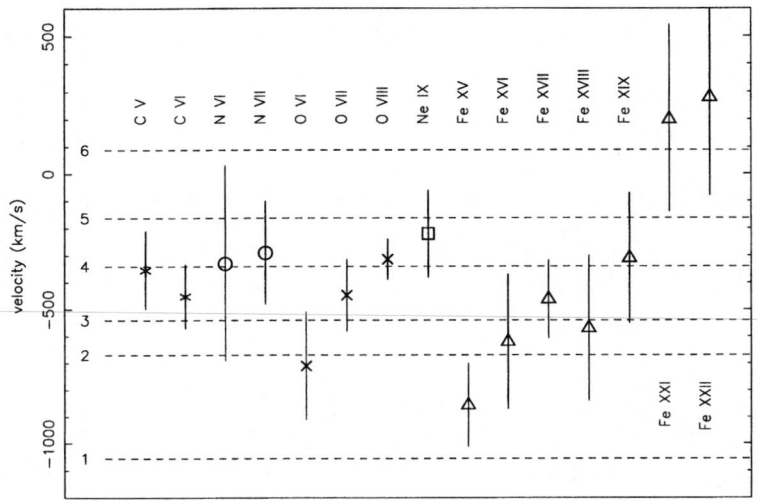

FIGURE 5. Average velocities (with respect to the AGN rest frame) for the X-ray absorption lines of different ions in the *Chandra* LETGS spectrum of the Seyfert galaxy NGC 5548. The dashed horizontal lines show the velocities of the six absorption components identified in the UV. Note the tendency for less ionized iron ions to have larger outflow velocities. Adapted from Kaastra et al. (2002).

dent) effect is not properly modeled, velocity measurements can be biased to be too large (because the effective centroid of the fitted absorption line is shifted blueward). This effect appears to be present in the 10.4-day *Chandra* HETGS spectrum of NGC 3783 (Kaspi et al. 2002; although alternative interpretations are possible), and it may well be present but difficult to correct in X-ray gratings spectra with lower signal-to-noise.

The outflow velocities measured from X-ray absorption lines often agree with those measured from UV absorption lines such as C IV, N V, and H I (e.g., see Fig. 4 and Fig. 5), supporting the general existence of a connection between these two types of absorption (e.g., Mathur, Elvis, & Wilkes 1995; Crenshaw et al. 1999). The exact nature of this connection, however, depends upon details of the modeling and remains debated (e.g., Crenshaw & Kraemer 1999; Arav, Korista, & de Kool 2002; and references therein).

Geometry and Radial Location of Outflows

From the relative strengths of the emission and absorption lines in the AGN with X-ray gratings spectra, the global covering factor of the X-ray absorbing outflow appears to be large.[3] The outflow covers $\gtrsim 50\%$ of the sky that is not already covered by the torus of AGN unification schemes. This direct constraint on the global covering factor agrees

---

[3] The global covering factor is the fraction of the sky, as seen from the central source, covered by the X-ray absorbing outflow.

well with the indirect constraint derived from counting the fraction of local Seyfert 1 galaxies with ionized X-ray absorbers. The line-of-sight covering factor also appears to be large in at least a few AGN; it can be constrained by measuring the extent to which saturated lines appear black.[4] In NGC 3783, for example, the Fe XX lines near 12.8 Å (see Fig. 1) limit any electron-scattered X-ray contribution to be $\lesssim 15\%$ (Kaspi et al. 2002).

The most important remaining uncertainty about X-ray absorbing outflows is their radial location; this key quantity is not directly constrained by a single observation of an AGN. Possible radial locations range from $\sim 10^{16}$ cm (e.g., an accretion disk wind) to $\sim 10^{18}$ cm (e.g., a torus wind), and there may be absorbing material across this entire range of radii. Knowledge of the radial location is essential for determining the mass outflow rate, the kinetic energy of the outflow, and the overall importance of the outflow in the AGN system. One physically appealing possibility is that the material in the X-ray absorbing outflow is the same material that scatters radiation to the observer in Seyfert 2 galaxies (e.g., Krolik & Kriss 1995; Krolik & Kriss 2001), but this connection cannot be firmly established until the radial location of the X-ray absorbing outflow is known. Variability studies combined with improved density diagnostics may allow the radial location to be determined (e.g., Krolik & Kriss 2001; Netzer et al. 2002), but this will require expensive *Chandra* and *XMM-Newton* observations at multiple epochs.

POSSIBLE FUTURE DIRECTIONS

The Need for Better X-ray Spectral Resolution

While the new data from *Chandra* and *XMM-Newton* represent an *enormous* advance, it is likely that X-ray spectroscopy of AGN outflows is still limited in some fundamental ways. One possible problem is illustrated in Fig. 6, where *Chandra*, *HST*, and *IUE* spectra of the Seyfert galaxy NGC 4051 are compared. Given the results from UV observations, it appears at least plausible that the ionized X-ray absorbers in AGN have significant velocity structure that cannot be resolved with current X-ray instruments. The velocity structure currently apparent in the X-ray spectra (e.g., Fig. 4 and Fig. 5) may just be the "tip of the iceberg." An optimist might argue that the high-temperature X-ray absorbing material will tend to be more volume filling than the lower temperature UV absorbing material, and therefore that it will be less clumpy. However, *FUSE* spectra of lines from the high-ionization ion O VI still show velocity structure finer than can be resolved with current X-ray instruments (e.g., Gabel et al. 2002).

If the X-ray absorption lines indeed possess significant unresolved structure, this will lead to systematic errors when attempting to derive column densities, ionization levels, and other physical parameters. Some line components may be saturated even though the unresolved average "line" does not appear black. A velocity resolution of ~ 100 km s^{-1}

---

[4] The line-of-sight covering factor is the fraction of the line-of-sight to the central source covered by the X-ray absorbing outflow. In reality, there may be multiple lines of sight to the central source if it has physical extent or if X-ray scattering is present.

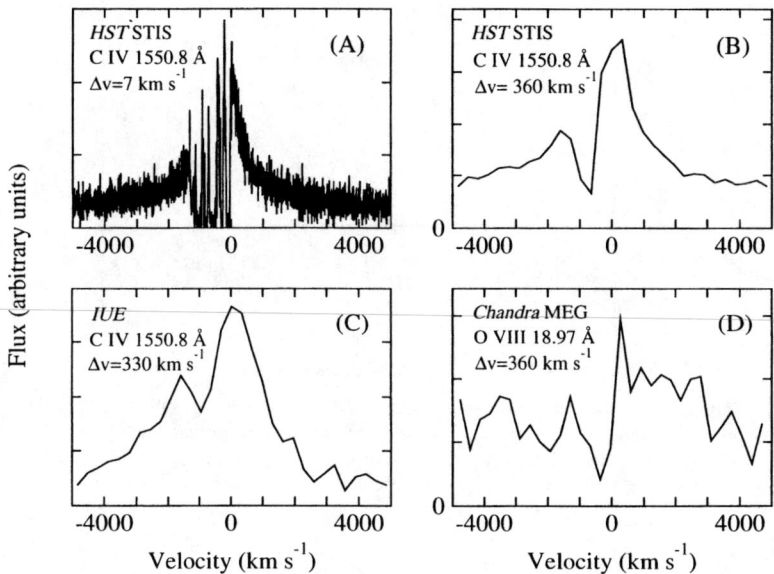

FIGURE 6. UV and X-ray absorption systems in the Seyfert galaxy NGC 4051 as seen by the *HST* Space Telescope Imaging Spectrograph (STIS; a and b), *IUE* (c; mean of 1978–1988 and 1994 epochs), and the *Chandra* HETGS (d; the MEG is part of the HETGS). The *Chandra* and *HST* observations were performed simultaneously in 2000. The approximate velocity resolution is listed for each panel. Panel (b) represents the STIS C IV line binned to the velocity resolution of the HETGS. Comparison of the panels stresses the point that the ionized X-ray absorber may be subdivided into further systems that cannot be resolved with current X-ray instruments. Note the similarity between panels (b) and (d), even though the high resolution in panel (a) reveals the C IV absorption to be extremely complex with at least eight distinct kinematic components. Adapted from Collinge et al. (2001).

or better will probably be required to resolve the X-ray lines. Such a resolution would also be valuable for addressing numerous other astrophysical issues (e.g., Elvis 2001). To our knowledge, no X-ray missions with the requisite resolution are currently scheduled for launch.

The Need for Higher Throughput

As mentioned above, efficient grating X-ray spectroscopy is presently possible only for ≈ 20 bright, low-redshift (mainly $z < 0.1$) AGN. Photon starvation has limited our ability to perform grating X-ray spectroscopy of more luminous but more distant AGN, such as BALQSOs. This is unfortunate, since the outflows in these objects are faster and probably more powerful than those in local AGN (e.g., see Laor & Brandt 2002 and references therein). The current X-ray spectra of BALQSOs, obtained with Charged Coupled Device (CCD) detectors, show that heavy X-ray absorption is often present (e.g., Green et al. 2001; Gallagher et al. 2002). In most cases, however, the dynamical state of the X-ray absorbing gas is unknown. One notable exception appears to be the

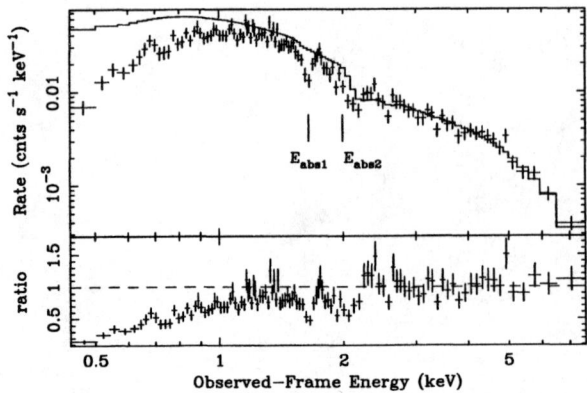

FIGURE 7. The observed-frame *Chandra* spectrum of the $z = 3.91$ BALQSO APM 08279+5255. The spectrum has been fit above 2.2 keV (10.8 keV in the rest frame) with a power-law plus Galactic absorption model; the model has then been extrapolated to lower energies to show the residuals. The lower panel shows the ratio of the data to the model. Note the absorption at low energies as well as the two apparent absorption lines from 1.5–2.1 keV. Adapted from Chartas et al. (2002).

gravitationally lensed BALQSO APM 08279+5255 at $z = 3.91$. Chartas et al. (2002) have recently claimed the detection of X-ray BALs from iron Kα that imply outflow velocities of $\approx (0.2$–$0.4)c$ (Fig. 7; also see Hasinger, Schartel, & Komossa 2002 for additional X-ray observations and an alternative interpretation).

High-throughput spectroscopy with enough resolution to determine the velocities of the X-ray absorbers in BALQSOs should provide a qualitative advance in our understanding of BALQSO outflows. Without basic kinematic information on the X-ray absorber, it is impossible to determine the total mass outflow rate and the kinetic energy of the outflow. Extremely high spectral resolution is probably not required for this science due to the high velocities observed in the UV; a velocity resolution of ~ 1000 km s^{-1} should be sufficient to allow major progress. Long *XMM-Newton* observations of 1–3 of the X-ray brightest BALQSOs can provide crucial information, but missions such as *Constellation-X* and *XEUS* will be needed to make high-resolution X-ray spectroscopy of BALQSOs routine.

ACKNOWLEDGMENTS

We thank all our collaborators, and we thank the conference organizers for a stimulating conference. We gratefully acknowledge support from NASA LTSA grant NAG5-8107 (WNB, SK), CXC grant GO1-2103B (WNB, SK), CXC grant GO0-1041X (WNB), and the Israel Science Foundation (SK).

REFERENCES

Arav N., Korista K.T., de Kool M., Junkkarinen V.T., Begelman M.C., 1999, 'Hubble Space Telescope observations of the broad absorption line quasar PG 0946+301', ApJ, 516, 27

Arav N., Korista K.T., de Kool M., 2002, 'On the column density of AGN outflows: The case of NGC 5548', ApJ, 566, 699

Blustin A.J., Branduardi-Raymont G., Behar E., Kaastra J.S., Kahn S.M., Page M.J., Sako M., Steenbrugge K.C., 2002, 'Multi-wavelength study of the Seyfert 1 galaxy NGC 3783 with *XMM-Newton*', A&A, in press (astro-ph/0206316)

Brinkman A.C., et al., 1997, 'Preliminary test results on spectral resolution of the Low-Energy Transmission Grating Spectrometer on board AXAF', Proc. SPIE, 3113, 181

Chartas G., Brandt W.N., Gallagher S.C., Garmire G.P., 2002, '*Chandra* detects relativistic broad absorption lines from APM 08279+5255', ApJ, in press (astro-ph/0207196)

Collinge M.J., Brandt W.N., Kaspi S., Crenshaw D.M., Elvis M., Kraemer S.B., Reynolds C.S., Sambruna R.M., Wills B.J., 2001, 'High-resolution X-ray and ultraviolet spectroscopy of the complex intrinsic absorption in NGC 4051 with *Chandra* and the Hubble Space Telescope', ApJ, 557, 2

Crenshaw D.M., Kraemer S.B., 1999, 'Intrinsic absorption lines in the Seyfert 1 galaxy NGC 5548: Ultraviolet echelle spectra from the Space Telescope Imaging Spectrograph', ApJ, 521, 572

Crenshaw D.M., Kraemer S.B., Boggess A., Maran S.P., Mushotzky R.F., Wu C., 1999, 'Intrinsic absorption lines in Seyfert 1 galaxies. I. Ultraviolet spectra from the Hubble Space Telescope', ApJ, 516, 750

den Herder J.W., et al., 2001, 'The Reflection Grating Spectrometer on board *XMM-Newton*', A&A, 365, L7

Elvis M., 2000, 'A structure for quasars', ApJ, 545, 63

Elvis M., 2001, 'Thermal limit spectroscopy as a goal for X-ray astronomy', in Inoue H., Kunieda H., eds, New Century of X-ray Astronomy. ASP Press, San Francisco, p. 180

Elvis M., Marengo M., Karovska M., 2002, 'Smoking quasars: A new source for cosmic dust', ApJ, 567, L107

Fabian A.C., 1999, 'The obscured growth of massive black holes', MNRAS, 308, L39

Furlanetto S.R., Loeb A., 2001, 'Intergalactic magnetic fields from quasar outflows', ApJ, 556, 619

Gabel J.R., Crenshaw D.M., Kraemer S.B., et al., 2002, 'The ionized gas and nuclear environment in NGC 3783: II. Averaged HST STIS and FUSE spectra', ApJ, submitted

Gallagher S.C., Brandt W.N., Chartas G., Garmire G.P., 2002, 'X-ray spectroscopy of quasi-stellar objects with broad ultraviolet absorption lines', ApJ, 567, 37

George I.M., Turner T.J., Netzer H., Nandra K., Mushotzky R.F., Yaqoob T., 1998, 'ASCA observations of Seyfert 1 galaxies. III. The evidence for absorption and emission due to photoionized gas', ApJS, 114, 73

Green P.J., Aldcroft T.L., Mathur S., Wilkes B.J., Elvis M., 2001, 'A *Chandra* survey of broad absorption line quasars', ApJ, 558, 109

Hamann F., 1998, 'Broad P V absorption in the QSO PG 1254+047: Column densities, ionizations, and metal abundances in broad absorption line winds', ApJ, 500, 798

Hasinger G., Schartel N., Komossa S., 2002, 'Discovery of an ionized Fe K edge in the $z = 3.91$ broad absorption line quasar APM 08279+5255 with *XMM-Newton*', ApJ, 573, L77

Kaastra J.S., Steenbrugge K.C., Raassen A.J.J., van der Meer R.L.J., Brinkman A.C., Liedahl D.A., Behar E., de Rosa A., 2002, 'X-ray spectroscopy of NGC 5548', A&A, 386, 427

Kaspi S., Brandt W.N., George I.M., Netzer H., et al., 2002, 'The ionized gas and nuclear environment in NGC 3783: I. Time-averaged 900 ks *Chandra* grating spectroscopy', ApJ, 574, 643

Krolik J.H., Kriss G.A., 1995, 'Observable properties of X-ray-heated winds in active galactic nuclei: Warm reflectors and warm absorbers', ApJ, 447, 512

Krolik J.H., Kriss G.A., 2001, 'Warm absorbers in active galactic nuclei: A multitemperature wind', ApJ, 561, 684

Krolik J.H., 2002, 'High-resolution X-ray spectroscopy and the nature of warm absorbers in AGN', in

Boller Th., Komossa S., Kahn S., Kunieda H., eds, X-ray spectroscopy of AGN with *Chandra* and *XMM-Newton*. MPE Press, Garching, in press (astro-ph/0204418)

Laor A., Brandt W.N., 2002, 'The luminosity dependence of ultraviolet absorption in active galactic nuclei', ApJ, 569, 641

Lee J.C., Ogle P.M., Canizares C.R., Marshall H.L., Schulz N.S., Morales R., Fabian A.C., Iwasawa K., 2001, 'Revealing the warm absorber in MCG–6–30–15 with the *Chandra* HETG', ApJ, 554, L13

Mathur S., Elvis M., Wilkes B.J., 1995, 'Testing unified X-ray/ultraviolet absorber models with NGC 5548', ApJ, 452, 230

Murray N., Chiang J., Grossman S.A., Voit G.M., 1995, 'Accretion disk winds from active galactic nuclei', ApJ, 451, 498

Netzer H., 1996, 'X-ray lines in active galactic nuclei and photoionized gases', ApJ, 473, 781

Netzer H., 2001, 'Physical processes in starburst and active galaxies', in Aretxaga I., Kunth D., Mújica R., eds, Advanced Lectures on the Starburst-AGN Connection. World Scientific, Singapore, p. 117

Netzer H., 2002, 'The density and location of the X-ray absorbing gas in NGC 3516', ApJ, 571, 256

Nicastro F., Fiore F., Perola G.C., Elvis M., 1999, 'Ionized absorbers in active galactic nuclei: The role of collisional ionization and time-evolving photoionization', ApJ, 512, 184

Proga D., Stone J.M., Kallman T.R., 2000, 'Dynamics of line-driven disk winds in active galactic nuclei', ApJ, 543, 686

Reynolds C.S., 1997, 'An X-ray spectral study of 24 type 1 active galactic nuclei', MNRAS, 286, 513

Sako M., et al., 2001, 'Complex resonance absorption structure in the X-ray spectrum of IRAS 13349+2438', A&A, 365, L168

Silk J., Rees M.J., 1998, 'Quasars and galaxy formation', A&A, 331, L1

Turnshek D.A., 1988, 'BALQSOs: Observations, models, and implications for narrow absorption line systems', in Blades J.C., Turnshek D.A., Norman C.A., eds, QSO absorption lines: Probing the Universe. Cambridge University Press, Cambridge, p. 17

Voit G.M., Weymann R.J., Korista K.T., 1993, 'Low-ionization broad absorption lines in quasars', ApJ, 413, 95

Wu K.K.S., Fabian A.C., Nulsen P.E.J., 2000, 'Non-gravitational heating in the hierarchical formation of X-ray clusters', MNRAS, 318, 889

Astrophysical X-ray Lines Observed by Chandra and XMM

T. R. Kallman

NASA Goddard Space Flight Center, LHEA, Code 665, Greenbelt, MD 20771

Abstract. This paper reviews the capabilities of Chandra and XMM and discusses some of the physical processes affecting line broadening in X-ray astronomy. Broadening due to relativistic effects is a process which is unique to this field, and which provides important insight into the nature of cosmic X-ray sources. The challenge of interpreting X-ray spectra is illustrated with an example of a well-known active galaxy, and the effect of line broadening on the K shell opacity of iron is discussed.

INTRODUCTION

The launches of Chandra and XMM in late 1999-2000 represented a major advance in the instrumentation available to X-ray astronomy, and in particular to spectroscopy. This has forced advances in the theory of X-ray spectral formation, and the use of spectra for diagnosing the conditions in cosmic sources. In this paper I will briefly introduce the capabilities of these two satellites, and then discuss some of the physics associated with line broadening and what we learn from cosmic X-ray sources from observation and modeling of line shapes.

CHANDRA AND XMM

Prior to the launch of Chandra and XMM, spectra were obtained by low resolution proportional counters, or medium resolution solid-state detectors. The most recent such instrument was the ASCA satellite. Chandra and XMM both combine the latter detector technology with diffraction gratings and grazing incidence optics to allow much higher resolution spectroscopy than previously was possible from many astronomical objects. The two satellites, although qualitatively similar, differ in details. These are summarized in the table, along with the corresponding quantities for ASCA. The pertinent distinction between Chandra and XMM is greater collecting area (XMM) vs. greater spectral (and spatial resolution) (Chandra). Each has its advantages and both have led to important new insights into the nature of cosmic X-ray sources, as will be shown in this paper and other papers in this volume. For more details on these two instruments, the reader is referred to the respective observatory web-sites.

CP645, *Spectral Line Shapes: Volume 12, 16th ICSLS*, edited by C. A. Back
2002 American Institute of Physics 0-7354-0100-4

THE CHALLENGE OF X-RAY ASTRONOMY

In addition to the intrinsic challenge of observational science, X-ray astronomy attempts to study some of the most exotic and extreme environments in the universe. It therefore must confront a shortage of terrestrial experiments with matter in such environments. X-ray astronomy is based on knowledge gained in other wavelength bands, such as stellar masses and compositions from optical spectroscopy, binary star periods from optical variability studies, or relativistic outflows detected from radio observations. The challenge is that we often don't know how much of this is directly applicable to the gas responsible for X-ray emission. And the theory for diagnosing cosmic X-ray plasma physical conditions (Temperature, density, radiation intensity), or emission mechanism based on observed spectra is in many cases not sufficiently developed to deal with the new data. This means that often more effort is devoted to modeling the plasma micro-physics than to addressing astrophysical issues, such as the sense and magnitudes of gas flows, elemental abundances, and evolution of the objects. In the rest of this paper we discuss some of what is known about diagnosing cosmic X-ray sources, and illustrate with examples.

LINE BROADENING MECHANISMS

Line broadening in X-ray astronomy is affected by a greater variety of processes than many other fields. Familiar mechanisms include natural broadening:

$$\left(\frac{\Delta\varepsilon}{\varepsilon}\right)_{natural} = \frac{8\pi^2 e^2 v^2 g_l f_{lu}}{m_e c^3 g_u} \tag{1}$$

$$\simeq 10^{-3}\varepsilon_{KeV}\frac{g_l f_{lu}}{g_u} \tag{2}$$

where ε_{KeV} is the line energy in keV, g_l and g_u are the statistical weights of the lower and upper levels and f_{lu} is the oscillator strength. The broadening due to thermal Doppler broadening:

$$\left(\frac{\Delta\varepsilon}{\varepsilon}\right)_{Thermal} = \sqrt{\frac{kT}{Mc^2}} \tag{3}$$

$$\simeq 4\times 10^{-5}T_4^{1/2}A^{-1/2} \tag{4}$$

where T_4 is the gas temperature in units of 10^4 K, and A is the atomic weight. The broadening due to bulk motion Doppler broadening:

$$\left(\frac{\Delta\varepsilon}{\varepsilon}\right)_{Bulk} = \frac{v}{c} \tag{5}$$

$$\simeq 3\times 10^{-3}v_{1000} \tag{6}$$

where v_{1000} is the velocity in units of 1000 km s^{-1}. The broadening due to pressure broadening :

$$\left(\frac{\Delta\varepsilon}{\varepsilon}\right)_{Pres} \simeq 1.5 \times 10^{-20} T_4^{1/6} n_e \qquad (7)$$

where n_e is the electron number density. Less familiar is broadening due to general relativistic effects, which can arise from emission near a black hole if light rays from a spatially extended source reach an observer by various trajectories:

$$\left(\frac{\Delta\varepsilon}{\varepsilon}\right)_{GR} = \frac{R_G}{R} \qquad (8)$$

where R is the distance from the black hole and $R_G = 2GM/c^2$ is the gravitational radius. The broadening due to special relativity for gas which is in a circular orbit around a point mass is:

$$\left(\frac{\Delta\varepsilon}{\varepsilon}\right)_{SR} = \sqrt{\frac{R_G}{R}} \qquad (9)$$

and broadening due to the recoil following scattering off of cold electrons (Compton scattering) leads to a broadening:

$$\left(\frac{\Delta\varepsilon}{\varepsilon}\right)_{Compton} = \frac{\varepsilon}{m_e c^2} \qquad (10)$$

$$\simeq 2 \times 10^{-3} \varepsilon_{KeV} \qquad (11)$$

These estimates show that the resolving power of Chandra and XMM detectors is not sufficient to detect natural broadening (although see below), and the density in most astrophysical environments is too low to cause appreciable pressure broadening. Thermal broadening is detectable for gas temperatures $\geq 10^7$K. Turbulent or bulk motion Doppler broadening are useful indicators of the conditions in many cosmic X-ray sources, and provide insight which is often not available from other wavelength bands. General relativistic broadening of the iron K line, at 6.4 – 6.7 keV has been detected from many of the brightest active galaxies and X-ray binaries using ASCA (Tanaka et al., 1996; Nandra, 2000). This provides evidence for strong gravitational fields, and hence for the existence of black holes in these objects.

THE O VII EDGE IN MCG-6-30-15

Prior to Chandra and XMM it appeared that the spectra of active galaxies below 3 keV was dominated by absorption from H-like, He-like medium-Z elements (N, O).

Detection of relativistically broadened iron K lines led to search for such features at lower energies. A notable target for such searches is the Seyfert galaxy MCG-6-30-15. This was the first object showing clear evidence for a relativistically broadened iron K line, and remains one of the best examples of such line broadening. The spectrum in the 15 – 20 A (0.6-0.9 keV) band provides evidence for such emission. Simple models for absorption by O VII and O VIII predict sharp edges at the energies of the bound-free continua of these ions, 16.8 A and 14.2 A (0.74 keV and 0.87 keV). Figure 1 panel b shows the predicted spectrum in the vicinity of the O VII edge for such a model. (solid curve) together with the count spectrum in this region as observed by Chandra (Lee et al., 2002). The absence of the predicted sharp edges led to the suggestion that the spectrum is dominated by relativistically broadened emission rather than absorption (Branduardi et al. 2000, Sako et al., 2001). Figure 1, panel a shows the predicted spectrum for such emission (solid histogram) and the fit to the Chandra data (error bars). The broad double-peaked line corresponds in energy to the Lα line of O VIII at 19 A (0.65 keV). The combined effects of special and general relativity in a rotating disk of gas near a black hole (less than ~10 gravitational radii) lead to the red-shifted, double-peaked feature.

Although this model fits the data adequately, there are a variety of other processes which can affect this spectrum, and which in combination can provide equally good fits. These include: the combined effects of absorption by the 1s-np line transitions near the edge in O VII. Panel c shows the effects of these on the predicted spectrum. The net effect is to decrease the sharpness of the edge. In addition, in the same spectral region, the resonance structure associated with the L edge of Fe I causes several sharp absorption features superimposed on a broader absorption. This is shown in panel d. As pointed out by Behar (2001) the complex of lines from n=2 to n=3 in various ions of iron lead to broad absorption feature near 15 and 17 A. For ions near neutral, the 17 A feature predominates. These UTAs (unresolved transition arrays) are displayed in panel e. The best fit to pure absorption is shown in panel f, and consists of a combination of O VII (including Rydberg series) and neutral iron L shell. An acceptable fit can be obtained from either broadened emission or pure absorption, although it is clear that the simplest absorption model cannot account for the detail in the spectrum in this energy band.

IRON K SHELL PHOTOIONIZATION AND THE EFFECT OF AUGER SPECTATOR RESONANCES

As an additional illustration of the effects of line broadening on astrophysical X-ray spectra, we consider the absorption from the K shell of iron. Traditional models for such absorption have considered only a sharp edge at the ionization threshold energy (eg. Verner and Yakovlev, 1995). However, photons at energies below the threshold energy can be absorbed by exciting 1s – np transitions. Such bound-bound resonance absorption is ubiquitous, but the case of inner shells is unique because the upper levels decay primarily by an alternate channel, KLL Auger transitions. The key point is that these 'spectator Auger resonances' have very large transition probabilities ($\geq 10^{14}$ s^{-1}), and so have large natural widths. And these widths are are constant, independent of the value of n in the initial 1s-np resonance absorption event. The spectator Auger resonance

TABLE 1.   Chandra, XMM and ASCA Comparison

Parameter	Chandra	XMM	ASCA
Launch	July 1999	December 1999	February 1993
Agency	NASA	ESA	Japan
Mirror Area, max (cm^2)	800	4000	1300
Spatial Resolution (arc-sec)	0.5	14	200
Spectral Resolving power at 1 keV	1000	500	12
Energy Band (keV)	0.1 – 10	0.3 – 3	0.5 – 12

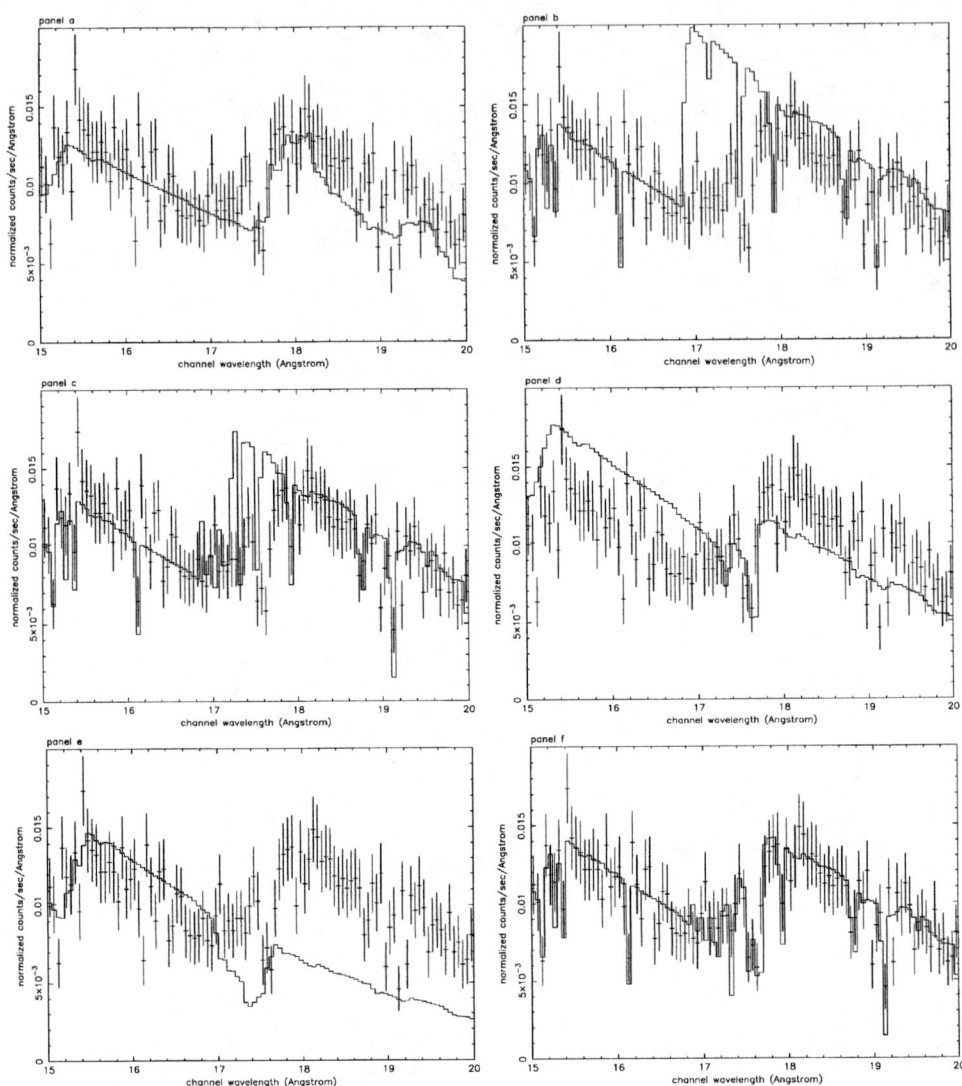

FIGURE 1. The spectrum of MCG -6-30-15 observed by Chandra in the 15 – 20 A band.

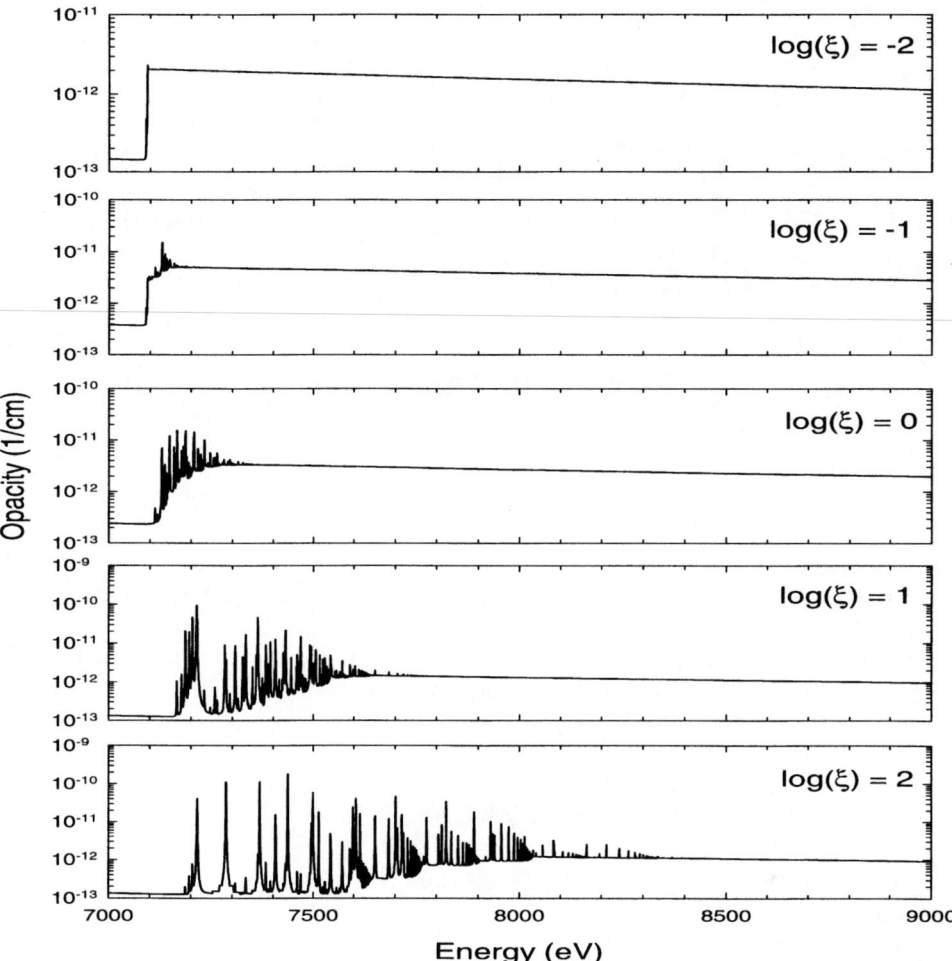

FIGURE 2. Opacity for a photoionized gas near the iron K edge showing the effects of spectator Auger resonances (Palmeri et al., 2002).

shapes are Lorentzian, and summation over the Rydberg series leads to significant pileup of the the resonances and net absorption at energies below the photoionization edge. This resonance structure is apparent in the opacity for medium-high ionization gas. Figure 2 shows a calculation of the opacity near the ionization threshold of iron, showing the resonance structure below the edges of the various ions of iron. The overall shape of the opacity thus depends on the ionization balance. In this figure, the ionization balance has been assumed to depend on the balance between photoionization and recombination. This can be conveniently parameterized in terms of the ionization parameter, ξ, which is the ratio of the flux of ionizing photons (in the 1-1000 Yr band) in energy units to the gas density (Kallman and Bautista, 2001). For intermediate values of this parameter the ionization balance favors second row ions, for which the edges are separated sufficiently to clearly reveal the resonances. At the highest ionization, the ions with less than 4 electrons do not have spectator Auger resonances, and at low ionization the edge energies pile up near 7.2 keV and the resonance structure is not resolved.

SUMMARY

Chandra and XMM have brought true spectroscopy to X-ray astronomy. However, the interpretion of astrophysical X-ray spectra in many cases must take into account exotic physical processes, such as general relativistic effects. Furthermore, the current instrumental resolution and data quality are such that blending of multiple features complicates modeling. Further progress requires progress in the atomic physics of X-ray absorption and emission, including inner shell process.

REFERENCES

Behar, E., et al., 2001, astro-ph/0109314, Ap. J. in press.B
Branduardi, G., et al., 2001, A and A, 365, 140
Kallman, T., and Bautista, M., 2001, Ap. J. Supp., 133, 221
Lee, J., et al., 2002 Ap. J. submitted
Nandra, P., 2000, astro-ph/0007356
Palmeri et al., 2002, astro-ph/0207324, Ap J. Lett. in press
Sako, M., et al., 2002, astro-ph/0112436
Tanaka, Y., et al., 1995 Nature, 375, 659
Verner, D., and Yakovlev, 1995 A and A S 109, 125

Autoionizing Line Shapes and Astrophysical Opacities

Balazs F. Rozsnyai

Lawrence Livermore National Laboratory. L-023,
P.O. Box 808, Livermore, California 94550

Abstract. Opacities are presented for the King IVa mixture for Cepheids in the temperature – density regions of interest. Two sets of opacities are presented, one with the inclusion of the autoionizing widths of the spectral lines, the other without it. The differences are analyzed and comparisons with other data are shown.

I. Introduction.

Photoabsorption by the large number of spectral lines in stellar atmospheres has a crucial role in determining stellar dynamics. In the case of Cepheid variable stars the pulsation period is determined by the Rosseland mean opacity along the stellar radius from the surface to the depth of the convective zone. Because of the interest in pulsation, Cepheid opacities have sustained an interest in the last few decades. Calculations by Simon [1] in 1982 were based on simple opacity models available at that time. Iglesias and Rogers [2] presented substantially improved Cepheid opacities with the inclusion of the detailed effects of the $\Delta n=0$ line transitions. In a 1989 work the author [3] made some estimates on the effect of detailed line accounting and using the method of "unresolved transition arrays" (UTA) for nl-n'l' line clusters. To the knowledge of the author, none of the opacity models made allowances for the inclusion of autoionizing line shapes. In the low density region of the outer envelope of a typical Cepheid the usual plasma broadenings are small, so the wings of the spectral line shapes are crucial in determining the Rosseland mean. In those regions the role of autoionizing widths of the weak but many satellite lines becomes important in calculating the Rosseland mean opacity. The purpose of this paper is to investigate the above effect by presenting computations for the King IVa mixture for Cepheids and showing the effect of the inclusion of the autoionizing line shapes.

II. Theoretical summary.

In this section we present the basic formulas for the line shapes. The details of the whole plasma model are given elsewhere [4,5,6,7]. For the electron impact broadening associated with a line connecting two quantum states labeled by m and n we assume a Lorentzian shape and for the widths we use Baranger's [8] approach

$$\gamma_{mn} = \frac{\hbar}{2}\left\{ n(v)v\left[\sigma_m^{in} + \sigma_n^{in} + \int |f_m(\vartheta\varphi) - f_n(\vartheta\varphi)|^2 d\Omega \right] \right\}_{av} \qquad (1).$$

CP645, *Spectral Line Shapes: Volume 12, 16th ICSLS*, edited by C. A. Back
2002 American Institute of Physics 0-7354-0100-4

In Eq. (1) σ_m^{in} stands for the sum of all the inelastic cross sections in the state m, likewise for n, and the fs stand for the elastic scattering amplitudes in the two states before and after phototransition. The free electron velocity distribution is given by n(v) and the symbol $\{\}_{av}$ means thermal averaging over the free electron velocities. Presently, we compute the inelastic cross sections and scattering amplitudes by using the self-consistent wave functions of the plasma model described in Refs. [4,7]. For the Stark broadening produced by heavy ions of the plasma we assume a static microfield distribution given by [6]

$$H^{\Gamma}(\beta) = \frac{2}{\pi} \beta \int_0^{\infty} x \sin \beta x \exp\left\{-\frac{x^{(3+q)/2}}{[\sqrt{x}+\Gamma]^q}\right\} dx. \tag{2a}$$

where

$$\beta = abs\frac{E-E_o}{E_o}; \quad \Gamma = \frac{Z_p Z_r e^2}{kT r_o} \quad ; q = 1.2876 \tag{2b}$$

In Eq. (2b) E stands for the perturbing Stark field, Z_p and Z_r for the effective charges of the perturber and radiator, respectively and kT for the temperature in energy units. Also, in Eq. (2b) r_o stands for the ion-sphere radius that is determined by the ion density. E_o is the 'normal field', which is the Stark field when the perturber – radiator distance is r_o. It should be noted that Eq. (2a) gives the Stark field due to all perturbing ions in contrast to the 'nearest neighbor' or binary approximation used in Ref [5], which considers the Stark field due to the nearest perturber only. In the case when $\Gamma=0$ Eq. (2a) yields the standard Holtzmark field. The microfield given by Eq. (2a) is used to calculate the Stark profiles for all lines as described in Ref. [6]. To account for the Doppler profile is a trivial matter, and in the case of statistical independence the effective line shape is obtained by convoluting the Lorentz, Stark and Doppler profiles. In addition to the above we consider the effect of autoionizing or Auger processes schematically represented by a Feynman diagram

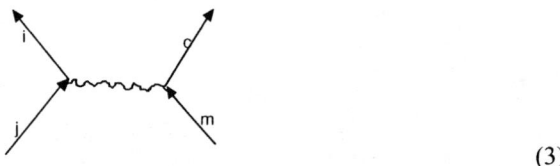

$$\tag{3}$$

indicating a process that an electron from a state j fills a hole state i while an other electron from a state m goes over to a continuum state c. It should be noted that interchanging the indices j and m describes the same physical process and the signs of the respective matrix elements are connected to the symmetric or antisymmetric (triplet/singlet) character of the two-electron wave function. In contrast, a simultaneous exchange of the indices i,j, and m,c describes the inverse process leading to dielectronic recombination. The autoionizing width associated with the Feynman diagram (3) is given by

$$\hbar \, A_{ji}^{mc} = 2\pi N_j N_m \sum_{l_c} (2l_i + 1)(2l_c + 1) a_i a_c (B_1 + B_2 + B_{12}) \qquad (4)$$

where

$$B_1 = \sum_k \frac{1}{2k+1} \begin{pmatrix} l_j & k & l_i \\ 0 & 0 & 0 \end{pmatrix}^2 \begin{pmatrix} l_m & k & l_c \\ 0 & 0 & 0 \end{pmatrix}^2 \left[R^k (n_j l_j n_m l_m ; n_i l_i c l_c) \right]^2 \quad (4a)$$

$$B_2 = \sum_p \frac{1}{2p+1} \begin{pmatrix} l_m & p & l_i \\ 0 & 0 & 0 \end{pmatrix}^2 \begin{pmatrix} l_j & p & l_c \\ 0 & 0 & 0 \end{pmatrix}^2 \left[R^p (n_m l_m n_j l_j ; n_i l_i c l_c) \right]^2 \quad 4b)$$

$$B_{12} = (-1)^\lambda \sum_k \frac{1}{2k+1} \begin{pmatrix} l_j & k & l_i \\ 0 & 0 & 0 \end{pmatrix} \begin{pmatrix} l_m & k & l_c \\ 0 & 0 & 0 \end{pmatrix} \text{x}$$

$$\sum_p \frac{1}{2p+1} \begin{pmatrix} l_m & p & l_i \\ 0 & 0 & 0 \end{pmatrix} \begin{pmatrix} l_j & p & l_c \\ 0 & 0 & 0 \end{pmatrix} \text{x}$$

$$\begin{Bmatrix} l_m & l_c & k \\ l_j & l_i & p \end{Bmatrix} R^k (n_j l_j n_m l_m ; n_i l_i c l_c) \, R^p (n_m l_m n_j l_j ; n_i l_i c l_c) \quad (4c)$$

and $\lambda = l_j + l_i + l_m + l_c + 1$, the R-s stand for the Slater integrals.

The a-s in Eq. (4) stand for the availability factor in the hole or in the continuum state and they are $a_i = 1 - \dfrac{N_i}{g_i}$ for a bound state and $a_c = \dfrac{\exp[(\varepsilon_c - \mu)/kT]}{\exp[(\varepsilon_c - \mu)/kT] + 1}$ for a continuum state.

The autoionizing widths given by Eq-s (4) are additive to all the other Lorentzian widths of a line when the final state after photoexcitation is an autoionizing state. This occurs for the weak but many satellite lines when there is a spectator electron in a high Rydberg level. In the next Section we present calculations for the King IVa mixture, which is widely used to calculate Cepheid properties.

III. Calculations.

The element composition of the King IVa mixture used in this report is given in Table 1. Presently we use only 12 elements and the largest Z is 26. The mixture contains traces of other elements in the vicinity of iron, like titanium, manganese and nickel in mole fractions one to two orders of magnitude less than iron. In the present work those mole fractions and lumped them into the iron component for computational convenience. Test calculations showed that the above simplification has negligible effect on the computed

141

results. In our example the mixture is characterized by z=0.02, which is the parameter of the mass fraction of all the heavier elements beyond helium.

Table 1. Element composition of the KingIVa mixture, typical for Cepheids.

Z	mole fraction	mass fraction*
1.0	9.0715(-1)	7.0000(-1)
2.0	9.1379(-2)	2.8000(-1)
6.0	2.8844(-4)	2.6181(-3)
7.0	8.0167(-5)	8.5959(-4)
8.0	6.3679(-4)	7.8038(-3)
10.0	3.5881(-4)	5.5437(-3)
11.0	1.4260(-6)	2.5095(-5)
12.0	1.7947(-5)	3.3401(-4)
13.0	1.1860(-6)	2.4496(-5)
14.0	2.2774(-5)	4.8860(-4)
18.0	2.3840(-5)	7.2904(-4)
26.0	3.6935(-5)	1.5790(-3)

*The numbers in parentheses are exponents of ten.

The temperature – density track from the surface or near surface toward the interior of a typical Cepheid was taken from Ref. [1] and a graph is shown in Fig. 1, the numbers in Table 2.

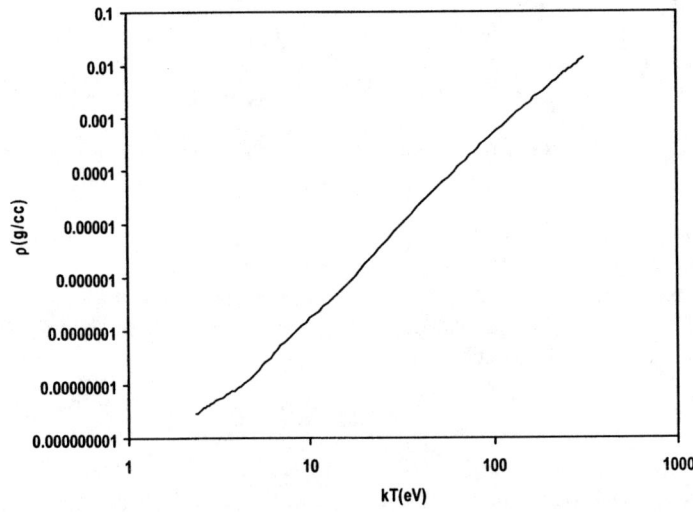

Fig. 1. Temperature – density trajectory of a typical Cepheid, taken from Ref [1]

142

Table 2. Temperature-density relations in Cepheid atmospheres.

kT(eV)	log T(K)	Density (g/cc)
2.3323	4.4324	2.8411(-9)
2.7185	4.4989	4.0440(-9)
3.2943	4.5820	5.9524(-9)
3.9963	4.6662	8.6672(-9)
4.8075	4.7466	1.4437(-8)
5.7381	4.8234	2.6855(-8)
6.9049	4.9037	5.2509(-8)
8.4822	4.9931	1.0425(-7)
1.0605(1)	5.0900	2.0562(-7)
1.3271(1)	5.1875	4.0918(-7)
1.6657(1)	5.2862	9.0296(-7)
2.0992(1)	5.3867	2.1871(-6)
2.6990(1)	5.4960	5.6523(-6)
3.5771(1)	5.6182	1.5551(-5)
4.9157(1)	5.7562	4.6704(-5)
7.0796(1)	5.9146	1.5947(-4)
1.0592(2)	6.0890	5.8811(-4)
1.5774(2)	6.2626	1.9808(-3)
2.4265(2)	6.4496	6.6857(-3)
3.1701(2)	6.5600	1.3623(-2)

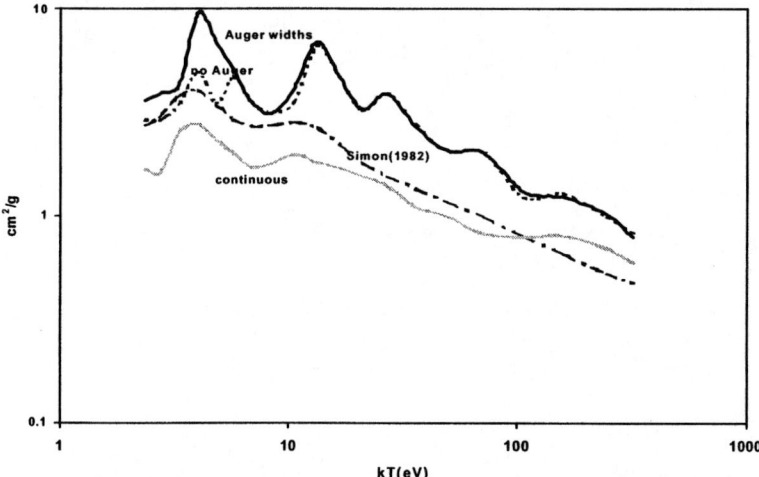

Fig.2. Rosseland mean opacities of the KingIV a mixture taken along the temperature-density track of Fig. 1. The dotted- dashed line is that of Simon (1982). The faint dashed line is obtained by considering continuous processes only.

In the calculations presented here we did not go below the 2.33 eV temperature to avoid the possibility of molecular formation, which is not included in the author's plasma model. The opacity calculations were performed along the temperature – density points of Table II with and without the inclusion of the autoionizing widths given by Eq-s (4). We show some of the important results in Figures 2 –4.

Figure. 2 shows the Rosseland mean opacities versus temperature along the temperature – density track of Table 2 or Fig. 1. The bumps in the opacities are very important in explaining pulsations. We also show in Fig.2 the data of Ref [1] and also the results obtained by neglecting all the lines, that is considering only photoionization and inverse bremsstrahlung. As is evident, the presence or absence of the autoionizing widths is significant in the low temperature and low density region. As we go more inward from the surface, the standard line broadening features tend to dominate over the autoionizing widths. We show the actual photoabsorption cross sections for the mixture in Fig. 3 for the case when the autoionizing widths seem to have the largest effect.

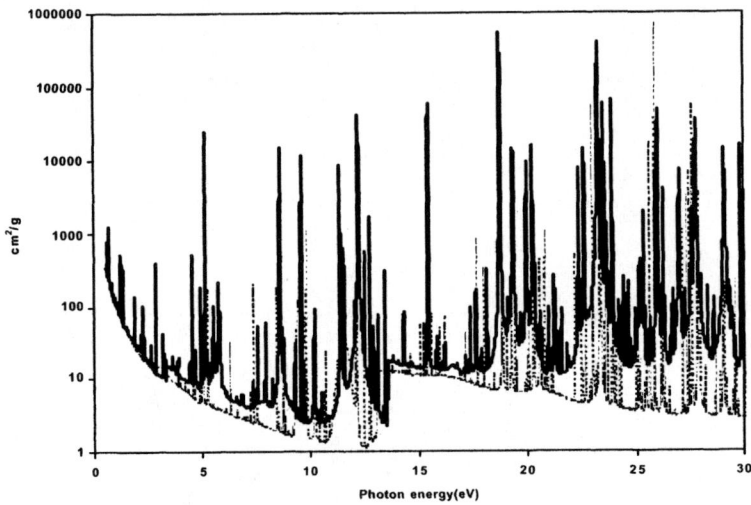

Fig. 3. Opacities of the KingIVa mixture at kT=3.996 eV and 8.667×10^{-9} g/cc density. The solid and dashed curves are with and without the autoionizing widths, respectively.

It is evident from Fig. 3 that the autoionizing widths tend to fill the gaps between line clusters thus yielding a larger Rosseland mean. We show the photoabsorption of the iron component in Fig. 4.

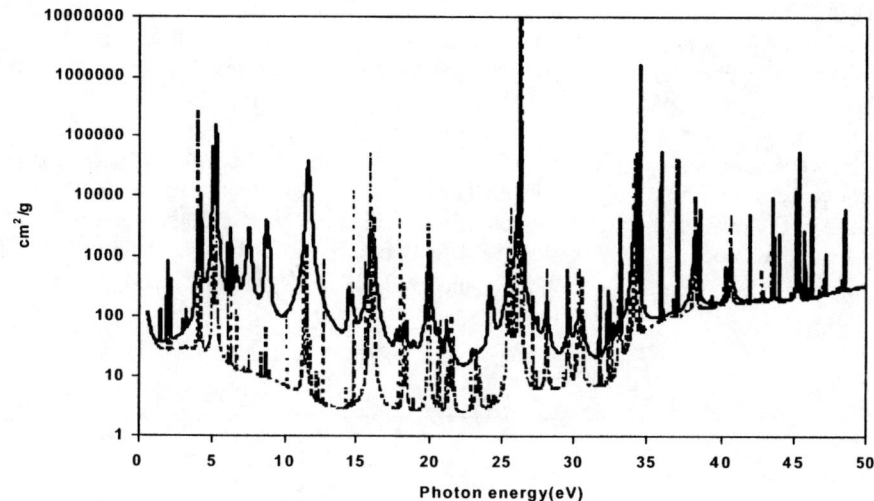

Fig 4. Opacity of the Iron component of the KingIVa mixture under the conditions of Fig.3.

In Table 3 we give a summary account of the electronic configurations of the iron component corresponding to Fig. 4. The many electron configurations over the valence levels are resolved up to 10^{-6} probability in the Boltzmann distribution. The actual number of J-J' lines of the nl-n'l' arrays are not computed, only their energy dispersion, the so-called 'unresolved transition array' (UTA) widths Ref [3]. The actual number of J-J' lines of the nl-n'l' arrays are mimicked by shifting one particular line evenly over the UTA width corresponding to the number of the actual J-J' lines. Table 3 is a typical representation of the computations that have to be performed in general for all Z components of the mixture.

Table 3. Summary of electronic configurations and spectral lines of the iron component at kT= 3.996 eV and 8.667x10⁻⁹g/cc density.

Core levels with 18 electrons.
       1s
       2s
       2p
       3s
       3p
Valence levels 3d to 6d.   25 Valence configurations involving 1 to 5 valence electrons.
Upper Rydberg levels are considered up to n=15 principal quantum numbers.

Number of 'average atom' (AA) lines....................................333
Number of 'UTA' clusters due to all valence configurations........8325
Number of 'DTA' lines, if all J-J' transitions are resolved.......671328.

In Table 3 the number of 'average atom' (AA) lines represents the simplest accounting due to the average occupancies of the one-electron levels. The presence or absence of the autoionizing widths is significant when the UTA groups are separated and are connected only by the wings of the line shapes. The autoionizing widths for inner shell excitations are always present, but at those photon energies they do not contribute to the Rosseland mean. However, autoionizing widths occur also at low photon energies when there is a spectator electron in a high Rydberg level. These are weak lines, but there are many.

Finally, in Fig.5 we show a comparison of Cepheid opacities computed by our model and by the OPAL code [2]. In this case the density points of an isotherm are given by the formula $\rho=Rx(T_6)^3$ where T_6 is the temperature in million Kelvin grades and ρ is the density in g/cc.

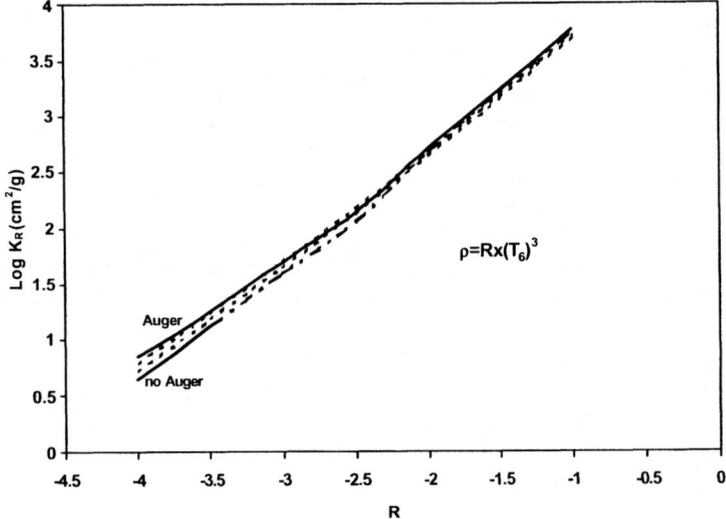

Fig. 5. Opacity of the T_6=0.045 (3.877 eV) isotherm of the KingIVa mixture.
The two inside dashed lines are 1991 and recent OPAL data.

Figure 5 shows the results of one isotherm at T_6= 0.045, which corresponds to 3.877 eV temperature in the energy scale. The top and bottom curves of Fig. 5 are the author's calculations with and without the inclusion of the autoionizing widths, respectively. In Fig. 5 we also show two OPAL results, one published in Ref [2], the other provided to the author by C.A. Iglesias [9]. The two sets of OPAL data are in between the author's two sets of calculations. The more recent OPAL data are next to the author's top curve, which includes the autoionizing widths. This is interesting because on the one hand OPAL does not include autoionizing widths, on the other hand the OPAL code does have a more detailed angular momentum algebra. As is evident again from Fig.5, the autoionizing widths are significant in the low density region.

IV. Conclusion.

The role of autoinizing widths in the Rosseland mean opacities of Cepheid variable stars is important in the low temperature and low density region of the stellar atmosphere. Accurate estimates of the Rosseland mean opacity in these regions are important for the calculation of pulsation periods. Presently, work is under way to compute opacity tables with the inclusion of the autoionizing widths.

Acknowledgment.

This work was performed under the auspices of the U.S. Department of Energy by University of California Lawrence Livermore National Laboratory under contract No. W-7405-Eng-48.

References

1. Simon. N.R. 1982, **Ap.J** (Letters) p260, **L87**
2. Iglesias, C. A. and Forrest R. J. 1991, **Ap.J** (Letters) 371, **L73**
3. Rozsnyai. B. F. 1989, **Ap.J, 341**, 414
4. Rozsnyai, B.F. 1972, **Phys. Rev. A5,** 1137
5. Rozsnyai, B.F. 1977, **J. Quant. Spectrosc. Rad. Transf. 17,** 77
6. Rozsnyai, B.F. and Einwohner, T. 1982, in *Spectral Line Shapes,* Vol. **2**, 315, ed. K. Burnett (Berlin: De Gruyter)
7. Goldberg, A, Rozsnyai, B.F. and Thompson, P. 1986, **Phys. Rev. A34,** 421
8. Baranger. M. in *Atomic and Molecular Processes.* Ed. D.R. Bates. Academic Press, (1962).
9. C.A. Iglesias, private communication.

ATOMIC AND
MOLECULAR INTERACTIONS

High-resolution studies on the influence of velocity-changing collisions on atomic and molecular line shapes.

R. Ciuryło*, A. Bielski*, J. R. Drummond†, D. Lisak*, A. D. May†,
A. S. Pine**, D. A. Shapiro‡, J. Szudy* and R. S. Trawiński*

*Institute of Physics, Nicholas Copernicus University, Grudziądzka 5/7, 87–100 Toruń, Poland.
†Department of Physics, University of Toronto, Toronto, M5S 1A7 Ontario, Canada.
**Alpine Technologies, 14401 Poplar Hill Road, Germantown, MD 20874, USA.
‡Institute of Automation and Electrometry, Siberian Branch, Russian Academy of Sciences,
Novosibirsk 630090, Russia.

Abstract. Starting with the transport/relaxation equation in the impact limit we have developed some new analytical and numerical approaches for modelling pressure-broadened spectral profiles of isolated atomic or molecular lines. Particular attention is paid to the influence of velocity-changing collisions on the line shape described by the speed-dependent broadening and shifting rates. These new models are compared to existing high-resolution experimental spectra.

LINE SHAPE AND TRANSPORT/RELAXATION EQUATION

The shape of a spectral line broadened due to atomic collisions and thermal motion effects can be written as real part of the velocity integral [1]–[5]

$$I(\omega) = \frac{1}{\pi}\text{Re} \int d^3\vec{v}\rho(\omega,\vec{v}). \tag{1}$$

where $\rho(\omega,\vec{v})$ describes the profile of the line corresponding to each velocity of absorber or emitter. For linear spectroscopy, the isolated line in the impact approximation $\rho(\omega,\vec{v})$ satisfies the following transport/relaxation equation

$$\rho_0(\vec{v}) = -i(\omega - \omega_0 - \vec{k}\cdot\vec{v})\rho(\omega,\vec{v}) - \widehat{S}\rho(\omega,\vec{v}), \tag{2}$$

where the ω_0 is the resonant frequency and $\rho_0(\vec{v})$ describes the initial velocity distribution of the absorbers or emitters in the initial state given by the Maxwellian velocity distribution $f_m(\vec{v}) = (\pi v_m^2)^{-3/2}\exp(-v^2/v_m^2)$. Here $v_m = \sqrt{2k_BT/m_A}$ is the most probable speed of the absorber or emitter of mass m_A, k_B is Boltzmann's constant and, T is the gas temperature. The collision operator $\widehat{S}$ describes the influence of the perturbers on the absorber or emitter interacting with the light. Equivalent approaches to this problem are also formulated in the framework of the correlation function formalism [1, 2, 6, 7].

The general expressions given above include all interesting effects such as velocity-changing collisions, speed-dependent broadening and shifting, and correlation between

CP645, Spectral Line Shapes: Volume 12, 16th ICSLS, edited by C. A. Back
© 2002 American Institute of Physics 0-7354-0100-4/02/$19.00

the velocity-changing and dephasing collisions. In the case where velocity-changing and dephasing collisions are uncorrelated the collision operator is expressed as the sum

$$\widehat{S} = \widehat{S}_D + \widehat{S}_{VC}, \tag{3}$$

of the dephasing collision operator $\widehat{S}_D$ and the velocity-changing collision operator $\widehat{S}_{VC}$. The operator $\widehat{S}_D$ is generally written as

$$\widehat{S}_D \rho(\omega, \vec{v}) = -\left[\Gamma(v) + i\Delta(v)\right] \rho(\omega, \vec{v}), \tag{4}$$

where $\Gamma(v)$ and $\Delta(v)$ are the speed-dependent collisional width (HWHM) and shift, respectively [6, 8]. The velocity-changing collisions are very often described using one of two models proposed a long time ago and known as the "soft collision" [9, 10] and "hard collision" [1, 11] models. In the soft collision model the velocity-changing collision operator $\widehat{S}_{VC}$ has the form [1]

$$\widehat{S}_{VC} \rho(\omega, \vec{v}) = v_{\text{diff}} \left(\frac{v_m^2}{2} \Delta_v + \vec{\nabla}_v \cdot \vec{v} \right) \rho(\omega, \vec{v}). \tag{5}$$

When velocity-changing collisions are described by the "hard" collision model the corresponding operator is written as follows [1]:

$$\widehat{S}_{VC} \rho(\omega, \vec{v}) = -v_{\text{diff}} \rho(\omega, \vec{v}) + v_{\text{diff}} f_m(\vec{v}) \int d^3 \vec{v}' \, \rho(\omega, \vec{v}'). \tag{6}$$

The "soft" collision model was explored by Galatry [10] to obtain the first analytical expression describing the shape of Dicke narrowed [12] lines assuming that the collisional width and shift are independent of the absorber or emitter speed. Unfortunately, when the collisional width and shift are dependent on the absorber or emiter speed there is no general analytical expression for the line shape and only some approximate formulas can be given [13]-[15]. However, when the speed-dependence of the collisional width and shift is assumed to be a quadratic function, the analytical expression can be derived [16]. On the other hand, the mathematical simplicity of the "hard" collision model leads to an analytical expression for the shape of the line for any speed-dependent width and shift [1, 17].

Very flexible analytical expressions for the line shape were obtained using an idea due to Rautian and Sobelman that velocity-changing collisions in one part can be described by the "soft" collision model and in another part by the "hard" collision model. In such a case an additional parameter (a "hardness" factor) is introduced. As was shown in Ref. [18] the Rautian-Sobelman model can lead to profiles which are very close to the line shapes [19, 20] obtained using the Kielson-Storer model [21]. It was also assumed that the speed derivatives of the collisional width and shift can be neglected. These analytical expressions were independently derived by Robert and Lance [22, 23] and by the authors of this work [24, 25]. In the so-called "correlated speed-dependent asymmetric Rautian-Sobelman profile" (CSDARSP) the dispersion asymmetry (caused by the finite duration of collisions and/or the line mixing) and the correlation between velocity-changing and dephasing collisions were taken into account as well [25]. The dispersive asymmetry can be incorporated in the profile by replacing of $\rho(\omega,\vec{v})$ in Eq. (1) by $(1 - i\chi)\rho(\omega,\vec{v})$ where χ is the asymmetry parameter. The correlation between velocity-changing and dephasing collisions [1] can be considered in a phenomenological way by replacing of the original frequency of the velocity-changing collisions $v_{\text{diff}} = k_B T/(m_A D)$ (D is the diffusion coefficient) by the quantity $v_{diff} - \eta(\Gamma + i\Delta)$ in which η is the correlation parameter. Chaussard et al. [26, 27] have demonstrated that this new analytical profile can fit experimental data over a wide range of physical condition.

The analytical expressions describing the spectral line shape are important for analysis and interpretation of experimental data. However, when deriving such expressions we always pay a price: making simplifications and introducing phenomenological quantities which can affect our interpretation of the physical effects. To be free from some of these limitations we are forced to make numerical line shape calculations which allow us to describe collisional processes affecting the shape of spectral lines in a more realistic way. The numerical method described in the following section is in fact a simpler version of the approach presented by Blackmore [3].

MATRIX APPROACH TO LINE SHAPE CALCULATIONS

To solve Eq. (2) numerically we represent the function $\rho(\omega,\vec{v})$ using an infinite set of orthonormal functions $\varphi_s(\vec{v})$ where $s = 0, 1, 2, ...$, in which $\int d^3\vec{v} f_m(\vec{v}) \varphi_s^*(\vec{v}) \varphi_{s'}(\vec{v}) = \delta_{s,s'}$ and $\varphi_0(\vec{v}) = 1$ (c.f. Ref. [5]). Then we can write the function $\rho(\omega,\vec{v})$ as

$$\rho(\omega,\vec{v}) = f_m(\vec{v}) \sum_{s=0}^{\infty} c_s(\omega)\varphi_s(\vec{v}), \tag{7}$$

where the coefficients $c_s(\omega)$ depend only on the frequency ω. Inserting Eq. (7) into the transport/relaxation equation, Eq. (2), we obtain an infinite system of complex simultaneous linear equations which can be written in the matrix form as [5, 19]

$$\underline{b} = \mathbf{L}(\omega)\underline{c}(\omega), \tag{8}$$

where the column vector $\underline{b}$, contains elements $[\underline{b}]_s = \delta_{s,0}$. The column vector $\underline{c}(\omega)$ consists of the elements $[\underline{c}(\omega)]_s = c_s(\omega)$. The matrix $\mathbf{L}(\omega)$ can be given in the following form

$$\mathbf{L}(\omega) = -i(\omega - \omega_0)\mathbf{1} + i\mathbf{K} - \mathbf{S}^f, \tag{9}$$

where $\mathbf{1}$ is the unit matrix, i.e. $[\mathbf{1}]_{s,s'} = \delta_{s,s'}$. $\mathbf{K}$ is the matrix which represents the Doppler shift, i.e. $[\mathbf{K}]_{s,s'} = \int d^3\vec{v} f_m(\vec{v})(\vec{k} \cdot \vec{v}) \varphi_s^*(\vec{v}) \varphi_{s'}(\vec{v})$, and $\mathbf{S}^f$ is the matrix which represents the collision operator, i.e. $[\mathbf{S}^f]_{s,s'} = \int d^3\vec{v} \varphi_s^*(\vec{v}) \widehat{S} f_m(\vec{v}) \varphi_{s'}(\vec{v})$. In terms of coefficients $c_s(\omega)$ the line shape can be written in the following form [5, 19, 28]

$$I(\omega) = \frac{1}{\pi} \text{Re } c_0(\omega). \tag{10}$$

To calculate the line shape Eq. (8) should be solved at each frequency. In some cases it is more convenient to use a diagonalization technique (c.f. [19, 29, 30]). To do it in this way we need to find the full set of eigenvectors $\underline{e}_j$ and corresponding eigenvalues ε_j which fulfil the following equation

$$(i\mathbf{K} - \mathbf{S}^f)\underline{e}_j = \varepsilon_j \underline{e}_j. \tag{11}$$

Once the eigenvectors and eigenvalues are known the coefficient $c_0(\omega)$ can be computed from the following expression

$$c_0(\omega) = \sum_{j=0}^{\infty} \frac{\beta_j[\underline{e}_j]_0}{\varepsilon_j - i(\omega - \omega_0)}, \tag{12}$$

where the coefficients fulfil the relation $\underline{b} = \sum_{j=0}^{\infty} \beta_j \underline{e}_j$. In this case the time consuming diagonalization can be carried out once to calculate shape of a line for different frequencies.

In practice, the calculations are carried out using a finite set of basis functions what lead to approximate results. However, in cases where collisional processes have comparable or greater influence on the line shape than the Doppler broadening the computation can be done with sufficient accuracy using a small number of basis functions.

Using the approach presented above we are able to calculate the line shape in the case when velocity-changing collisions are described by the billiard-ball model (BB) [3, 28, 31]. In this model the velocity changing collisions are treated as collisions of rigid spheres. This model allows us to capture the dependence of velocity-changing collisions on the mass ratio α of the active atom or molecule and perturber. By changing the mass of the pertuber we can modify the nature of the collisions from mostly direction-changing collisions in the case of very heavy perturbers (corresponding in limit $\alpha = \infty$ to the Lorentz gas model) to collisions significantly changing speed in the case of very light perturbers (corresponding in limit $\alpha = 0$ to the soft collision model). Calculating the correlated speed-dependent asymmetric billiard-ball profile (CSDABBP) we also take into account the dispersion asymmetry and correlation between velocity-changing and dephasing collisions in the phenomenological way described above. In this case the matrix elements of the velocity-changing collision operator can be given by the analytical formula derived by Lindenfeld and Shizgal [28, 32].

In the following section both analytical and numerical approaches are applied to the analysis of experimental data.

APPLICATION

First, we present an example of the R(6) line from the fundamental band of HF perturbed by argon recorded with a linear-scan-controlled difference-frequency laser spectrometer [33]. The speed dependence of the broadening and shifting is computed from the realistic quantum close-coupling collision cross sections calculated by Green and Hutson [34] using an accurate HF-Ar van der Waals molecule potential. In Fig. 1, we compare several profiles to the 100 Torr trace fit simultaneously with the 200 and 500 Torr scans [35]. This multifitting procedure provides a test of the consistency of the lineshape model and parameters over the range of measured pressures [35, 36]. We obtain the best

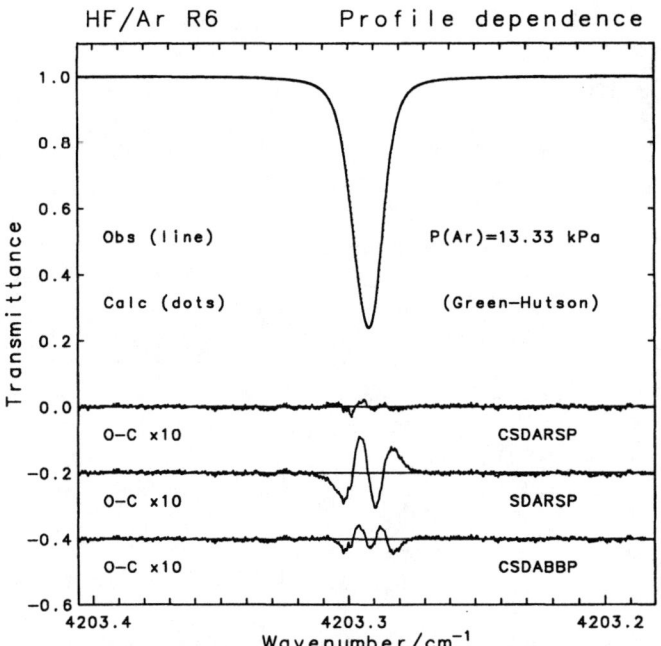

FIGURE 1. The measured shape of the R6 line of HF perturbed by argon compared to the correlated speed-dependent asymmetric Rautian-Sobelman profile (CSDARSP), the speed-dependent asymmetric Rautian-Sobelman profile (SDARSP) and the correlated speed-dependent asymmetric billiard-ball profile (SDABBP).

fit for the analytical CSDARSP, incorporating a linear combination of hard and soft collision models with a "hardness" fraction of 0.72. Correlation between the velocity-changing and dephasing collisions is crucial for explaining the observed asymmetry for these strongly shifted lines ($\Delta/\Gamma \simeq 2.6$) [33, 35]. If we neglect correlation, as seen for the SDARSP fit, we obtain large systematic residuals indicating that neither the speed dependence of the shift nor the collision duration effects have sufficient magnitude or the correct pressure dependence to account for this asymmetry. The residuals for CSDABBP are also worse than for the CSDARSP, though the latter does have one more parameter. This indicates that the billiard ball profile with a fixed mass ratio does not accurately interpolate between soft and hard collision models the way it does between the soft and the Lorentz gas models. In fact, the CSDABBP residuals are very similar to the pure soft collision profile for HF/Ar shown previously [35]. Generally though, billiard ball profiles are intermediate between hard and soft collision lineshapes and, when multifit over a range of pressures, exhibit discrepancies that could easily be distinguished experimentally. Because of convergence difficulties with the large number of basis functions needed for the CSDABBP in the Doppler regime, we have not included lower pressure traces [33, 35] in our fits here. Presented results demonstrate that we need to construct a realistic collision operator starting only from the intermolecular interaction

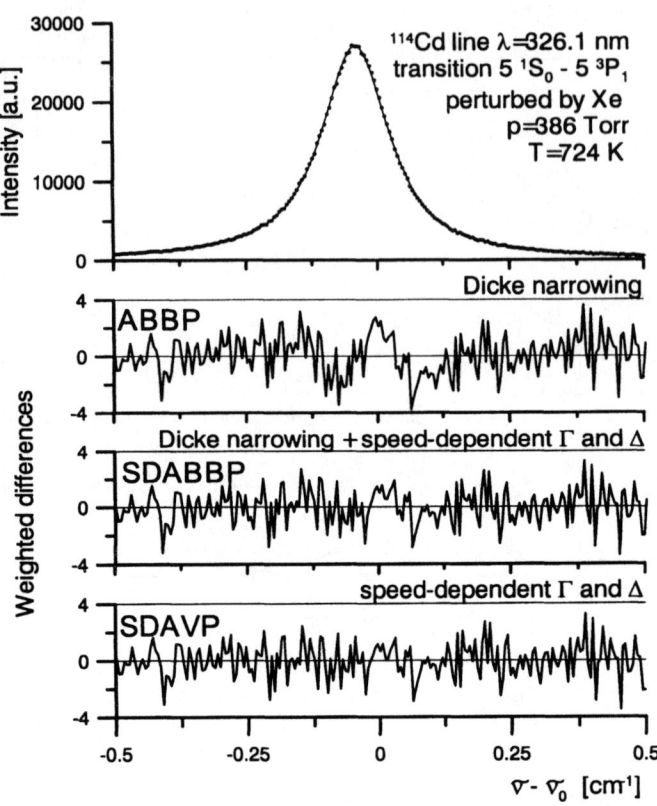

FIGURE 2. The measured shape of the Cd 326.1-nm line perturbed by xenon compared to the asymmetric billiard-ball profile (ABBP), the speed-dependent asymmetric billiard-ball profile (SDABBP) and the speed-dependent asymmetric Voigt profile (SDAVP). The weighted difference is defined as a difference of the measured and fitted intensity divided by the standard deviation of the measured intensity.

able to reach good description of measured profiles of HF perturbed by argon.

Now we turn to the intercombination line of ^{114}Cd perturbed by 386 Torr of xenon. Bielski *et al.* [37] observed that the Doppler width of this line obtained from the fit of the measured profile to the ordinary Voigt profile is significantly smaller than the value corresponding to the cell temperature. To verify the reason for this narrowing we have fitted the asymmetric billiard-ball profile (ABBP) in which the Dicke narrowing is taken into account using the billiard-ball model. As can be seen from Fig. 2 the Dicke narrowing itself cannot completely eliminate departures between the fitted and experimental profiles. Therefore, we have also fitted the data to the speed-dependent asymmetric Voigt profile (SDAVP) [38] which takes into account the speed dependence of collisional broadening and shifting but neglects the Dicke narrowing. We have used analytical speed dependence of collisional broadening and shifting given by the confluent hypergeometric function corresponding to the van der Waals potential [37, 6] which gives results close to those obtained using the potentials calculated by Czuchaj and Stoll [39]. As

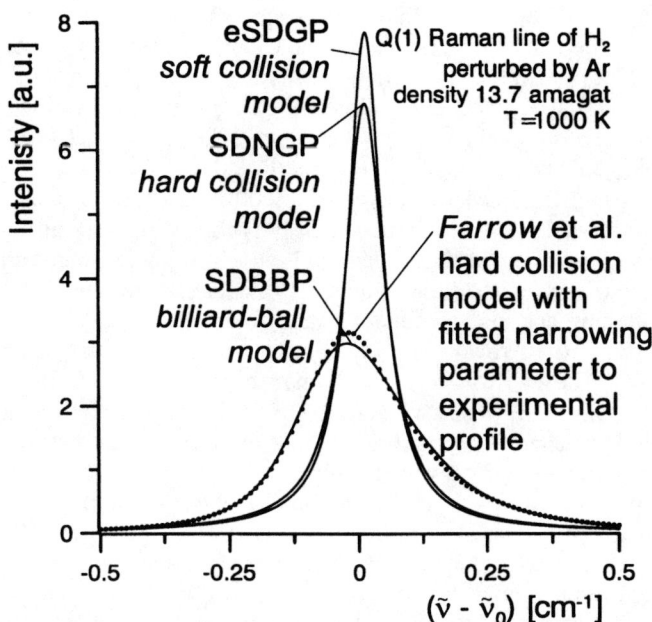

FIGURE 3. Comparison of the speed-dependent billiard ball profile (SDBBP), the speed-dependent Nelkin-Ghatak profile (SDNGP) and the "exact" speed-dependent Galatry profile (eSDGP) (solid lines) simulated for the Q(1) Raman line of H_2 perturbed by Ar and the profile fitted by Farrow *et al.* [48] (dotted) to their experimental data.

can be seen from Fig. 2 the differences are spread uniformly about zero which confirms the good quality of the SDAVP fit. We have also fitted the speed-dependent asymmetric billiard-ball profile (SDABBP) in which both the speed dependence of the collisional width and shift and the Dicke narrowing are taken into account. The quality of the fit is slightly worse than that for the SDAVP if the frequency of velocity changing collisions is fixed at the value expected for the Cd-Xe system based on the potentials calculated by Czuchaj and Stoll [39]. It would be necessary to improve the signal to noise ratio to conclude whether the Dicke narrowing is smaller than expected in the Cd-Xe system. However it also should be noted that Lewis and his coworkers [40] did not report any signs of the Dicke narrowing in their experiment with calcium perturbed by rare gases. Moreover, there is a increasing number of evidence for molecular systems for which narrowing of the line is caused by the speed dependence of collisional broadening rather than by Dicke narrowing [41]-[44]. Wehr *et al.* [45] discuss this problem for the P2 line of CO perturbed by argon [46].

Finally we show the results of simulations of the Q(1) Raman line of H_2 perturbed by argon at temperature 1000 K and the density 13.7 amagat which corresponds to the experiment of Farrow *et al.* [48]. We have calculated the speed-dependent billiard-ball profile (SDBBP) [3, 31, 47] taking into account the strong speed dependence of the shift of this line and the velocity-changing collisions described by the BB model. As can be seen from Fig. 3 the SDBBP appears much different than the speed-dependent Nelkin-Ghatak profile (SDNGP) [1, 17] and the exact speed-dependent Galatry profile (eSDGP) [16, 31] where velocity-changing collisions are described by the "hard" and "soft" collision models, respectively. Collisions of light hydrogen molecules with the much heavier argon atom change mostly the direction of the H_2 velocity. The speed is only slightly changed. On the other hand the "hard" and "soft" collision model describes collisions which significantly change the speed of colliding atoms or molecules. The difference between the SDBBP, SDNGP and eSDGP is caused only by different treatment of speed-changing collisions in these models. It should be noted that all three profiles were calculated with exactly the same value of parameters, assuming speed-independent collisional broadening and using a function describing speed dependence of the collisional shifting given by Farrow $et\ al.$ [48]. Similar variations of the shape of line were also modelled by Lance and Robert [22] who changed the phenomenological "hardness" parameter in the CSDARSP. When we fit the SDNGP to the simulated SDBBP the fitted frequency of velocity-changing collisions will be more than 10 times smaller than the value used in the simulation. This can explain the results reported by Farrow $et\ al.$ [48]. They fitted a profile based on the hard collision model to their experimental profiles of H_2 perturbed by argon and obtained a narrowing parameter more than 10 times smaller than expected. Recently, Hoang $et\ al.$ [49] modelled such behavior for a wider range of perturbers using the molecular dynamic simulation (MDS) technique. We believe that the billiard-ball model is a promising tool to analyze the experimental line shape of H_2 perturbed by foreign gases.

CONCLUSIONS

Concluding we would like to point out that the flexible analytical expression describing the line shape in the form of the correlated speed-dependent asymmetric Rautian-Sobelman profile was derived. This expression with parameters having reasonable physical interpretation alow us to analyse experimental data for different systems and in wide range of conditions. We have demonstrated that the fast numerical line shape calculation for any kind of the collision operator are possible. It opens the way to the line shape calculations which are $ab\ initio$ in spirit and are able to reach agreement with experimental results. Introducing the billiard-ball model to the treatment of the line shape allows us to describe the velocity-changing collisions in a way more realistic than other standard models used before. Especially, the variation between the direction- and speed-changing collisions caused by the change of the perturber-absorber mass ratio can be obtained as a natural consequence of application of the BB model.

The additional effort is necessary to reach a full agreement between theoretical and experimental results for the isolated lines. We need to have the realistic collision operators calculated starting from the intermolecular interaction and taking into account the speed dependence of collisional broadening and shifting, velocity-changing collisions and the correlations between them, in a unified way. In the case of the low density regime when the Doppler effect has dominant influence on the line shape the numerical method described above fails and another effective numerical approach should be developed. Further experimental investigations of the Dicke narrowing in atomic and molecular systems should lead to better understanding of the correlation between dephasing and velocity-changing collisions and their impact on the line shape.

ACKNOWLEDGMENTS

This work was supported by the Kosciuszko Foundation and grant No. 5 P03B 066 20 (354/P03/2001/20) from the Committee for Scientific Research.

REFERENCES

1. S. G. Rautian and I. I. Sobelman, Usp. Fiz. Nauk **90**, 209 (1966) [Sov. Phys. Usp. **9**, 701 (1967)].
2. G. Nienhuis, J. Quant. Spectrosc. Radiat. Transf. **20**, 275 (1978).
3. R. Blackmore, J. Chem. Phys. **87**, 791 (1987).
4. A. D. May, Phys. Rev. A **59**, 3495 (1999).
5. D. A. Shapiro, R. Ciuryło, J. R. Drummond, and A. D. May, Phys. Rev. A **65**, 012501 (2002).
6. J. Ward, J. Cooper, and E. W. Smith, J. Quant. Spectrosc. Radiat. Transf. **14**, 555 (1974).
7. R. M. Herman, *Spectral Line Shapes*, edited by J. Seidel (AIP Press, New York, 2001) Vol. 11, pp. 237.
8. P. R. Berman, J. Quant. Spectrosc. Radiat. Transf. **12**, 1331 (1972).
9. S. Chandrasekhar, Rev. Mod. Phys. **15**, 1 (1943).
10. L. Galatry, Phys. Rev. **122**, 1218 (1961).
11. M. Nelkin and A. Ghatak, Phys. Rev. **135**, A4 (1964).
12. R. H. Dicke, Phys. Rev. **89**, 472 (1953).
13. P. Duggan, P. M. Sinclair, A. D. May, and J. R. Drummond, Phys. Rev. A **51**, 218 (1995).
14. R. Ciuryło and J. Szudy, J. Quant. Spectrosc. Radiat. Transf. **57**, 411 (1997).
15. D. Priem, F. Rohart, J. M. Colmont, G. Wlodarczak, and J. P. Bouanich, J. Mol. Structure **517**, 435 (2000).
16. R. Ciuryło, R. Jaworski, J. Jurkowski, A. S. Pine, and J. Szudy, Phys. Rev. A **63**, 032507 (2001).
17. B. Lance, G. Blanquet, J. Walrand, and J. P. Bouanich, J. Mol. Spectrosc. **185**, 262 (1997).
18. D. A. Shapiro, R. Ciuryło, R. Jaworski, and A. D. May, Can. J. Phys. **79**, 1209 (2001).
19. D. Robert and L. Bonamy, Eur. Phys. J. D **2**, 245 (1998).
20. D. A. Shapiro, J. Phys. B **33**, L43 (2000).
21. J. Keilson and J. E. Storer, Quart. J. App. Math. **10**, 243 (1952).
22. B. Lance and D. Robert, J. Chem. Phys. **109**, 8283 (1998).
23. B. Lance and D. Robert, J. Chem. Phys. **111**, 789 (1999).
24. R. Ciuryło, Phys. Rev. A **58**, 1029 (1998).
25. R. Ciuryło, A. S. Pine, and J. Szudy, J. Quant. Spectrosc. Radiat. Transf. **68**, 257 (2001).
26. F. Chaussard, X. Michaut, R. Saint-Loup, H. Berger, P. Joubert, B. Lance, J. Bonamy, and D. Robert, J. Chem. Phys. **112**, 158 (2000).
27. F. Chaussard, R. Saint-Loup, H. Berger, P. Joubert, X. Bruet, J. Bonamy and D. Robert, J. Chem. Phys. **113**, 4951 (2000).
28. M. J. Lindenfeld, J. Chem. Phys. **73**, 5817 (1980).
29. S. Dolbeau, R. Berman, A. D. May, and J. R. Drummond, Phys. Rev. A **59**, 3506 (1999).
30. D. A. Shapiro and A. D. May, Phys. Rev. A **63**, 012701 (2001).
31. R. Ciuryło, D. A. Shapiro, J. R. Drummond, and A. D. May, Phys. Rev. A **65**, 012502 (2002).
32. M. J. Lindenfeld and B. Shizgal, Chem. Phys. **41**, 81 (1979).
33. A. S. Pine, J. Chem. Phys. **101**, 3444 (1994).
34. S. Green and J. Hutson, J. Chem. Phys. **100**, 891 (1994).
35. A. S. Pine and R. Ciuryło, J. Mol. Spectrosc. **208**, 180 (2001).

36. D. C. Benner, C. P. Rinsland, V. M. Devi, M. A. H. Smith, and D. Atkins, J. Quant. Spectrosc. Radiat. Transfer **53**, 705 (1995).
37. A. Bielski, R. Ciuryło, J. Domysławska, D. Lisak, R. S. Trawiński, and J. Szudy, Phys. Rev. A **62**, 032511 (2000).
38. M. Harris, E. L. Lewis, D. McHugh, and I. Shannon, J. Phys. B **17**, L661 (1984).
39. E. Czuchaj and H. Stoll, Chem. Phys. **248**, 1 (1999).
40. E. L. Lewis, *Spectral Line Shapes*, edited by J. Szudy (Ossolineum Publ. Wrocław, 1988) Vol. 5, pp. 485.
41. P. Duggan, P. M. Sinclair, R. Berman, A. D. May, and J. R. Drummond, J. Mol. Spectrosc. **186**, 90 (1997).
42. D. Priem, J. M. Colmont, F. Rohart, G. Wlodarczak, and R. R. Gamache, J. Mol. Spectrosc. **204**, 204 (2000).
43. J. M. Colmont, J. F. D'Eu, F. Rohart, G. Wlodarczak, and J. Buldyreva, J. Mol. Spectrosc. **208**, 197 (2001).
44. J. F. D'Eu, B. Lemoine, and F. Rohart, J. Mol. Spectrosc. **212**, 96 (2002).
45. R. Wehr, A. Vitcu, R. Ciuryło, F. Thibault, J. R. Drummond, A. D. May, (to be published).
46. R. Berman, P. M. Sinclair, A. D. May, and J. R. Drummond, J. Mol. Spectrosc. **198**, 283 (1999).
47. R. Ciuryło, D. Lisak, and J. Szudy, Phys. Rev. A (in press).
48. R. L. Farrow, L. A. Rahn, G. O. Sitz, and G. J. Rosasco, Phys. Rev. Lett. **63**, 746 (1989).
49. P. N. M. Hoang, P. Joubert, and D. Robert, Phys. Rev. A **65**, 012507 (2002).

Comparison of an *ab initio* calculation of the CO-Ar P(2) line shape with high-resolution measurements

R. Wehr*, A. Vitcu*, R. Ciurylo[†], F. Thibault[‡],
J.R. Drummond*, and A.D. May*

*Dept. of Physics, University of Toronto, Toronto, Canada, M5S 1A7
[†]Inst. Of Physics, Nicholas Copernicus University, Grudziadzka 5/7, 87-100 Torun, Poland
[‡]UMR 6627 du CNRS, Université de Rennes, Campus de Beaulieu,
34042 Rennes Cedex, France

Abstract. A practical matrix-based formalism for solving the master equation for a spectral line is applied to the P(2) transition in the fundamental band of carbon monoxide perturbed by argon. The method assumes that the effect of intermolecular collisions on the internal relaxation of the molecules is uncorrelated with the effect of those collisions on the translational motion of the molecules. Comparison with high-resolution line shape measurements reveals that at low pressures, the omission of statistical correlation leads to a miscalculation of the shape of the line.

INTRODUCTION

A master transport-relaxation equation for the shape of a spectral line was presented as equation (10) in [1], and a theoretical formalism for solving that equation to calculate real line profiles was presented in [2] and [3]. That formalism is applied here to the case of the P(2) line of carbon monoxide perturbed by argon in an attempt to make – for the first time to our knowledge – accurate calculations of an isolated spectral line at all densities. In order to make the calculation more tractable, one key simplifying assumption is made: it is assumed that the effect of intermolecular collisions on the internal relaxation of the molecules is statistically uncorrelated with the effect of those collisions on the translational motion of the molecules. The calculations are then compared to high-resolution experimental data obtained recently by [4].

CP645, *Spectral Line Shapes: Volume 12, 16th ICSLS*, edited by C. A. Back
© 2002 American Institute of Physics 0-7354-0100-4/02/$19.00

COLLISIONAL BROADENING

Because the internal and external degrees of freedom are assumed to be uncorrelated, it is possible to assume a different potential energy surface for each. The starting point for the calculation of the speed-dependent collisional broadening (i.e. the internal relaxation) is the potential energy surface of [5]. From there, fully quantal close coupled calculations of the broadening coefficient have been carried out using MOLSCAT and MOLCOL codes at the Université de Rennes. The dependence of the broadening coefficient on the relative molecular velocity can be fitted to an analytic function, which is then used to analytically determine the dependence of the broadening coefficient on the velocity of the active molecule. A dephasing collision matrix, $\mathbf{S}_D^f$, is then determined using equation (11) of [3].

DOPPLER BROADENING AND DICKE NARROWING

The potential energy surface used for determining the translational motion is that of a rigid sphere. The Doppler broadening is represented by the matrix $\mathbf{K}$, while the effect of Dicke narrowing is represented by the velocity changing collision matrix, $\mathbf{S}_{VC}^f$, as determined for the rigid sphere potential in [3].

MATRIX CALCULATIONS

As shown in [3], the line shape can now be determined from

$$I(\omega) = \frac{1}{\pi} \operatorname{Re} c_0(\omega),$$

where $c_0(\omega)$ is the first element of the matrix $\mathbf{c}(\omega)$, and

$$\mathbf{b} = \mathbf{L}(\omega)\mathbf{c}(\omega).$$

Here, $\mathbf{b}$ is a column matrix whose n^{th} element is $\delta_{n,0}$, and $\mathbf{L}(\omega)$ is given by

$$\mathbf{L}(\omega) = -i(\omega - \omega_0)\mathbf{1} + i\mathbf{K} - \mathbf{S}_D^f - \mathbf{S}_{VC}^f.$$

The size of these matrices is determined by the number of basis functions required to give the accuracy desired for the solution. In our case, it is always sufficient to use 121 basis functions. At higher pressures, only 36 basis functions are required.

COMPARISON WITH EXPERIMENTAL DATA

In the equation for the broadening coefficient as a function of the velocity of the active molecule, the half-width at half-maximum of the spectral line, Γ_0, appears as a multiplicative scaling factor. We are only able to calculate Γ_0 to an accuracy of 2%. We are able, however, to fit Γ_0 to an accuracy of 0.1% using the experimentally measured data. Since we are presently interested in testing our understanding of the physics and not in testing our ability to calculate Γ_0 precisely, we have chosen to fit Γ_0; that is, to determine it from a least squares minimization of the difference between our calculated and measured profiles.

At high densities, when the pressure is greater than about 0.2 atmospheres, the calculations agree with the measurements within the experimental noise. As the measurements were recorded with a signal-to-noise ratio of 3000:1 and a frequency resolution of 1.5 MHz, the calculations can be termed highly accurate.

As the pressure decreases below 0.2 atm, increasing disagreement is observed, as seen in the residuals plotted in Figure 1. Also shown is the result of removing Dicke narrowing from the calculations, which universally improves the agreement. The resulting profile is a speed-dependent Voigt profile (SDVP).

Figure 2 shows the behaviour of the fitted $\gamma_0 = \Gamma_0/P$ as a function of pressure. At high pressures, γ_0 is constant and within 1% of the calculated value, as should be expected. However, at lower pressures, γ_0 diverges, increasing drastically for our *ab initio* profile and decreasing slightly for the simplified SDVP. This picture is consistent with over-calculated Dicke narrowing in our *ab initio* profile and under-calculated Dicke narrowing (in fact, no Dicke narrowing at all) in the SDVP, for which the least squares minimization attempts to compensate by altering the overall half-width, Γ_0. The residuals in Figure 1 attest that altering the fitted Γ_0, even by these extreme amounts, is not capable of producing agreement between the calculated and measured profiles. This fact serves as an *a posteriori* justification for fitting Γ_0 in the first place.

There are two chief suspects in the failure of the *ab initio* calculation to reproduce real spectral profiles at low densities. The first is our choice of rigid spheres for the modelling of the translational motion. However, calculations have also been carried out using the speed-dependent Nelkin-Ghatak hard collision profile (SDNGP) [6] and the speed-dependent Galatry soft collision profile (SDGP) [7], and the same pattern of disagreement is observed, as shown in Figure 2. The interactions used in these two profiles are generally considered to be opposite, extreme cases, with most other models falling somewhere in between. Thus our results are not sensitive to our choice of interaction for the translational motion.

The second suspect is the assumption that the internal and external degrees of freedom are statistically uncorrelated. In fact, it seems likely that this assumption is the cause of the disagreement for two reasons. First, the CO-Ar system under study is known to be dominated by inelastic collisions, which by their definition involve a direct coupling between the internal and external degrees of freedom of the molecules. The second reason is that in a purely theoretical model proposed by [8] to account for

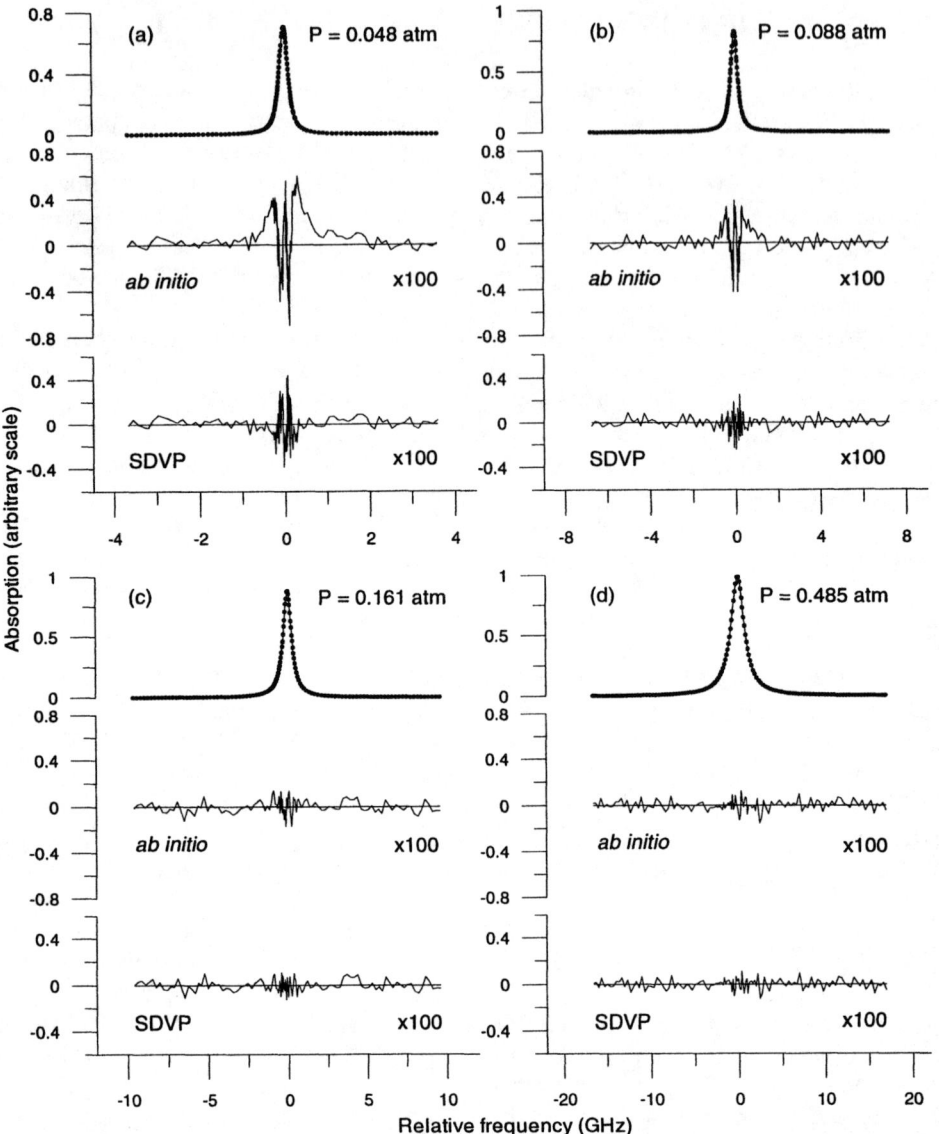

FIGURE 1. Line absorption profiles and residuals for the *ab initio* and speed-dependent Voigt (SDVP) profiles. Residuals are the calculated profile minus the measured profile, and are magnified 100 times. Four different pressures are shown in panels (a) – (d).

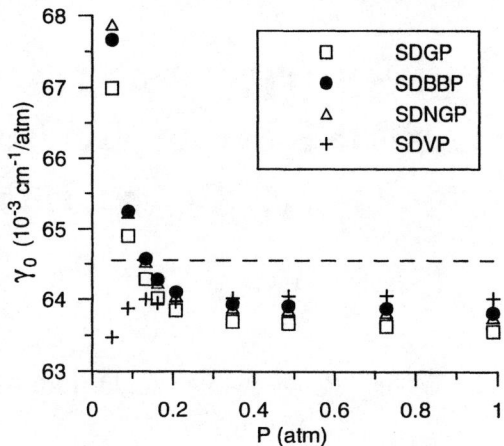

FIGURE 2. Fitted broadening coefficients versus pressure for the *ab initio* speed-dependent billiard ball profile (SDBBP), speed-dependent Galatry profile (SDGP), speed-dependent Nelkin-Ghatak profile (SDNGP), and speed-dependent Voigt profile (SDVP). The calculated value with only 2% accuracy is shown as a dashed line.

statistical correlation, the effect of including the correlation was to reduce the amount of Dicke narrowing, which is consistent with the results of these *ab initio* calculations.

CONCLUSION

It will be necessary to perform a fully correlated calculation in order to generate accurate line shapes *ab initio* at all densities for the CO-Ar P(2) line. Omission of statistical correlation between the effects of collisions on the internal relaxation and external motion of the molecules results in an apparent reduction in Dicke narrowing. This reduction is not evident at high densities, but is manifested at low densities, where Doppler broadening and Dicke narrowing dominate the spectral line shape.

REFERENCES

1. May, A. D. *Phys. Rev. A* **59**, 3495-3505 (1999).
2. Shapiro, D. A., Ciurylo, R., Drummond, J. R., and May, A. D. *Phys. Rev. A* **65**, 012501 (2001).
3. Ciurylo, R., Shapiro, D. A., Drummond, J. R., and May, A. D. *Phys. Rev. A* **65**, 012502 (2001).
4. Berman, R., Sinclair, P. M., May, A. D., and Drummond, J. R. *J. Mol. Spec.* **198**, 283-290 (1999).
5. Tockzylowski, R. R. and Cybulski, S. M. *J. Chem. Phys.* **112**, 4604-4612 (2000).
6. Nelkin, M. and Ghatak, A. *Phys. Rev.* **135**, A4-9 (1964).
7. Galatry, L. *Phys. Rev.* **122**, 1218-1223 (1961).
8. Rautian, S. G. and Sobelman, I. I. *Usp. Fiz. Naut.* **90**, 209-236 (1966), *Trans. In Sov. Phys. Usp.* **9**, 701-716 (1967).

HIGH RESOLUTION AND HIGH SIGNAL-TO-NOISE MEASUREMENTS IN THE $03^10 \leftarrow 01^10$ Q-BRANCH OF N_2O AT 1160 CM^{-1}

A. Vitcu*, R. Wehr*, R. Ciurylo[†], J.R. Drummond* and
A.D. May*

*Dept. of Physics, University of Toronto, Toronto, Canada, M5S 1A7
[†]Inst. Of Physics, Nicholas Copernicus University, Grudziadzka 5/7, 87-100 Torun, Poland

Abstract. High-resolution measurements of the $\Pi \leftarrow \Pi$ Q-branch of pure N_2O near 1160 cm^{-1} were made using a difference-frequency spectrometer with resolution of 5 x 10^{-5} cm^{-1} and a signal-to-noise ratio of 2000:1. Lines Q18F through Q12E have been recorded in a single scan, at room temperature and at pressures ranging from 1 to 130 torr. The spectra are analyzed up to 23 torr on a line-by-line basis using a hard collision profile including Dicke narrowing and line mixing. Since the separation of the central lines of this double-sided Q-branch is of the same order of magnitude with the collisional broadening, line mixing is considered in the analysis even at 1 torr and its dependence with pressure is studied.

INTRODUCTION

The process of a spectral line shape formation is far from being completely understood. Although Gaussian, Lorentzian and Voigt models have earned their places in the common knowledge about this field, the physics they carry is not enough to explain the details shown by high-resolution experiments. However more recent theories which include line mixing, speed dependent profiles and statistical correlation between the relaxation of the internal motion and the translational motion of the molecules require even better quality spectral information for their validation. Typical values of line mixing effects, for example, are no more than few tenths of a percent of the peak absorption value of a line, so signal-to-noise ratios in excess of 1000:1 are needed if one looks for experimental line-mixing signature in a spectrum with a minimum degree of accuracy.

One of the best ways to experimentally satisfy these demands is the Difference Frequency Spectroscopy technique. It consists of generating IR radiation by mixing

CP645, *Spectral Line Shapes: Volume 12, 16th ICSLS*, edited by C. A. Back
© 2002 American Institute of Physics 0-7354-0100-4/02/$19.00

two visible lasers (one of which is modulated) in a nonlinear crystal and extracting the difference frequency. The infrared signals are then detected using a Lock-in amplifier. Using single mode tunable dye lasers as visible radiation creates an important advantage to the technique, in that the single mode signature of the emerging IR beam improves the spectral resolution considerably.

EXPERIMENTAL

Figure 1 presents the current setup of the Difference-Frequency Spectrometer at the University of Toronto. It consists of two systems (**2.2-5.5 μm** and **6-11 μm**) , each requiring two visible lasers and a nonlinear crystal. The 2.2 to 5.5 μm spectrometer uses the (green) Ar laser and the (yellow) dye laser to generate infrared (IR) in a $LiIO_3$ crystal, while two (red and yellow) dye lasers combine into a $AgGaS_2$ nonlinear crystal to generate IR in the 6 to 11 μm region for the second spectrometer. The two spectrometers create one versatile unit, and switching from one IR region to the other proves to be rather easy, by removing a couple of kinematic-mounted mirrors.

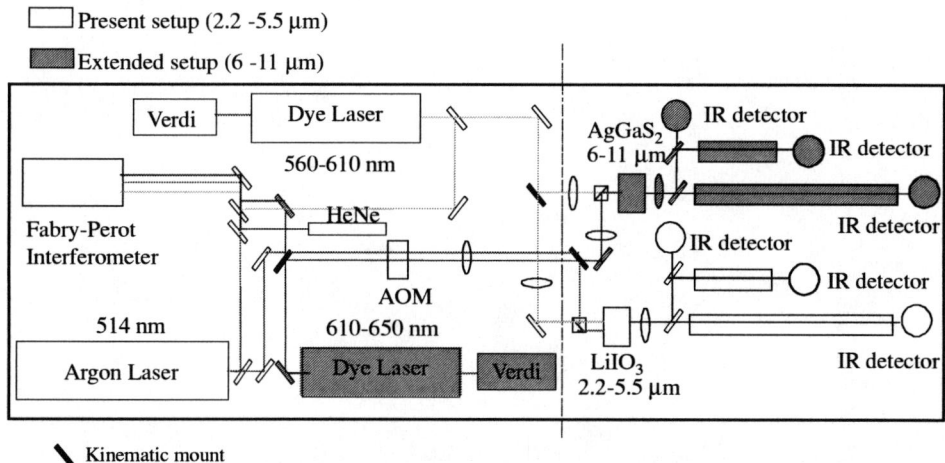

FIGURE 1. Experimental setup of the difference-frequency spectrometers

All the optics on the table to the right of each crystal are Germanium, which block the visible beams, and are AR coated for the required IR wavelengths. The emerging IR beam is split in two: 2/3 of it goes straight through the main cell to one of the detectors, while 1/3 is reflected towards a second (reference) detector for signal calibration. In the setup in figure 1, there is an extra arm in each spectrometer - created by a second beam splitter. A low pressure cell in this third arm acts as a frequency reference and allows us to include pressure shifts in our measurements.

The characteristics of the two spectrometers are similar:
- •S/N ratio in excess of 2000:1
- •Resolution: 1.5 MHz (5×10^{-5} cm^{-1})
- •Single mode operation
- •Contiunous tunability over more than 1 cm^{-1}

Data recording and processing is computer controlled. The analog infrared signals from each of the Lock-in amplifiers are sampled at 10 Ksamples/sec and fed into a 16 bit analog-to-digital converter. The signal is recorded only if the the frequency of the two visible lasers remains constant relative to a frequency stabilized HeNe laser. The transmission peaks of the three lasers going through a confocal Fabry-Perot interferometer are recorded and the relative position of the peaks determines if the stability condition is met at every IR frequency step measurement.

Figure 2 presents a typical recorded spectrum of the Q-branch of N_2O at 1160 cm^{-1} (8.6 µm).

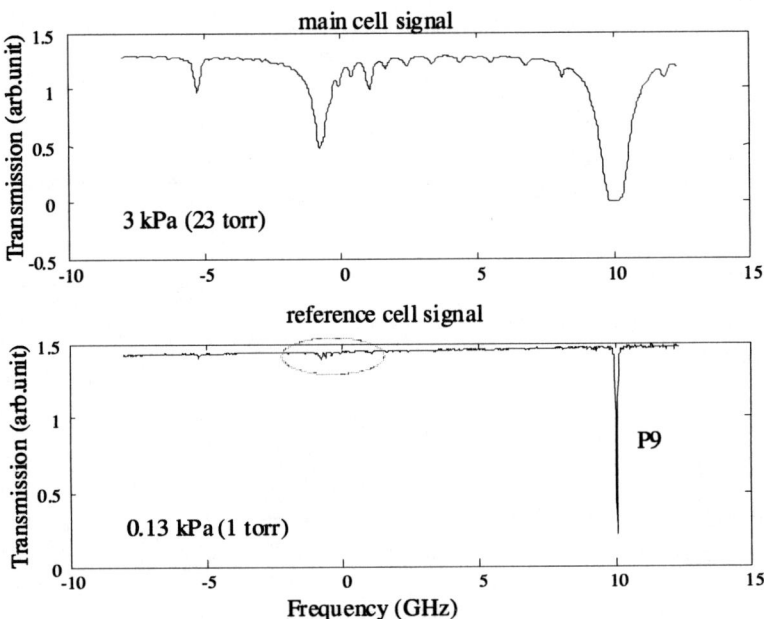

FIGURE 2. Typical experimental transmission spectrum for N_2O around 1160 cm^{-1}

The upper spectrum represents the main cell at 23 torr; the bottom one shows the reference cell N2O spectrum at 1.35 torr. The center of the strong P9 line is used as a reference for the pressure shift measurement.

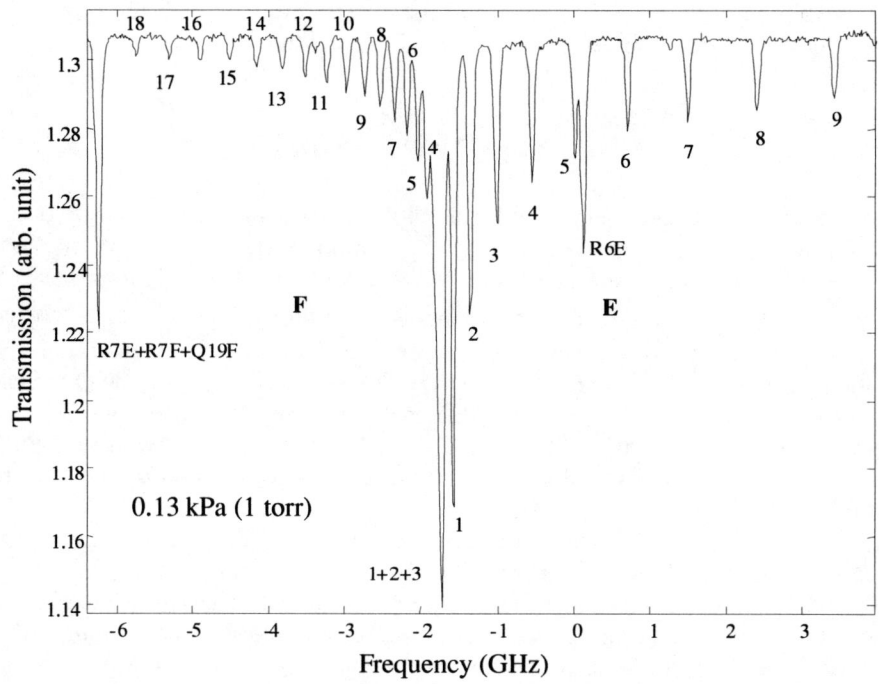

FIGURE 3. N_2O Q-branch at 1160 cm^{-1}: detail

Figure 3 provides a detailed look at the double-sided Q-branch at 1.35 torr (the circled area in the bottom spectrum of figure 2) and shows the signal-to-noise and spectral resolution capabilities of the spectrometer.

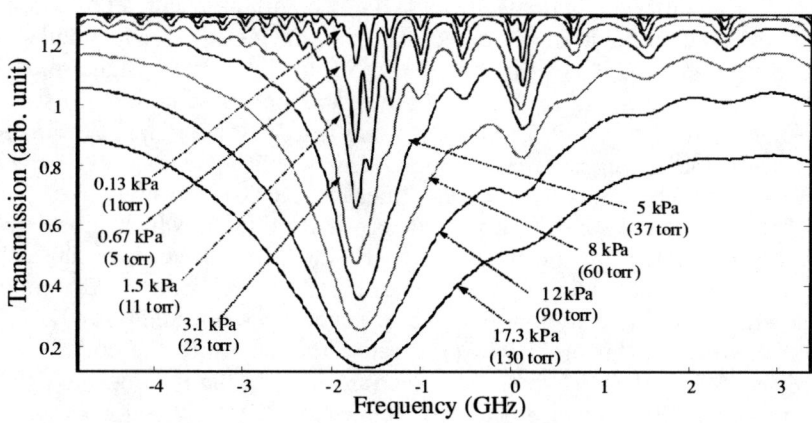

FIGURE 4. Q-branch collapsing as the pressure increases from 1 to 130 torr.

In figure 4, the experimental transmission spectrum for 8 different pressures are presented. The collapse of the Q-branch at high pressures is easily seen.

ANALYSIS AND RESULTS

In the absorption spectra of linear molecules, Q branches are obtained for $\Delta l=\pm 1$, where l is the vibrational angular momentum. Because the value of l is not zero, the upper or lower or both levels are split due to l-type doubling. In our case, the $03^{1}0 \leftarrow 01^{1}0$ band of N_2O is a $\Pi \leftarrow \Pi$ band ($l = 1$) and two close levels – labeled e and f, denoting the parity - are associated with each J value in both upper and lower vibrational states. Since the Q lines are of the $f \leftarrow e$ type or $e \leftarrow f$ type, the Q branch is composed of two subbands denoted as E and F according to the lower level of the transition. For the 1160 cm^{-1} Q branch of N_2O, the e levels are always below the f ones, so the E subband is positioned at the higher frequency end of the spectrum (figure 3).

The lines Q18F through Q10E have been recorded in a single scan, at room temperature and pressures ranging from 1 to 130 torr. The results presented in figures 5 to 7 are based on a preliminary analysis of the spectra recorded at the first four lowest pressures: 1, 5, 11 and 23 torr. Since the separation of the central lines in this Q-branch is of the same order of magnitude as the collisional width, line mixing is considered in the analysis even at the lowest pressures. In each of the figures, the negative values of the line numbers correspond to the F subband, while the positive ones belong to the E subband. The value zero has no line correspondent in the spectrum of the Q branch.

The spectra are analyzed on a line-by-line basis using a hard collision (HC) profile including Dicke narrowing and line mixing[1]. For each line, four parameters are fitted: the strength, the center, the collisional broadening and the profile asymmetry. The narrowing parameter is fitted as a common parameter to all lines.

Lines on either side of the branch (E or F) having the same J number should have equal strengths. This is verified at 1 and 5 torr. However at higher pressures, it is more difficult to fit lines near the band centre and more assumptions must be made. In each figure, the 1 torr and 5 torr spectra are fitted allowing the strengths to float independently; the 11 torr spectrum is fitted using two different assumptions: with and without the constraint that equal J number lines must have equal strength.

The results vary significantly at 11 torr, depending on which fitting routine is chosen. Most affected are the lines on the F side of the branch, because the line spacing is smaller and mixing is more important. The uncertainty in determining an accurate value of the line strength is a major concern for lines Q5F to Q1E. These 6 lines are confined within a 500 MHz spectral region, while the collisional width (FWHM) is ~100 MHz at 11 torr. As shown in figure 5, the mixing causes line Q2F to collapse and redistributes intensity from the central (Q1F, Q2F) lines towards Q5F line. A drop in line strength corresponds in figure 7 to a drop in half width for the central 1F, 2F lines.

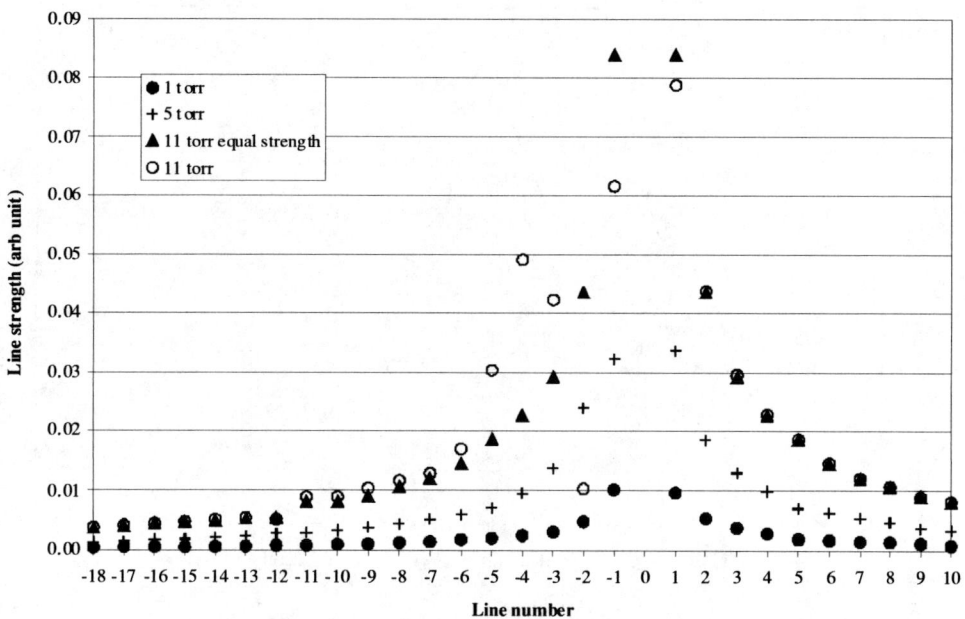

FIGURE 5. Line strength comparison for the three lowest pressures

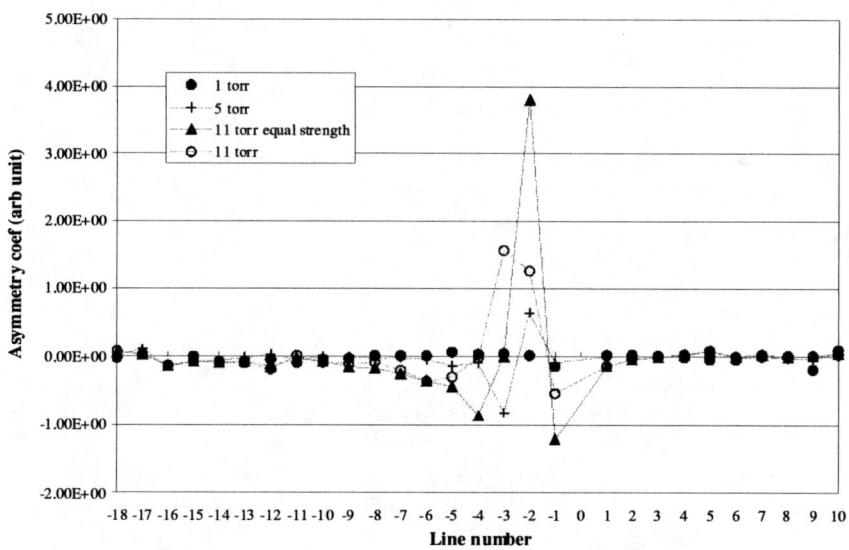

FIGURE 6. Fitted asymmetry coefficients as a function of J

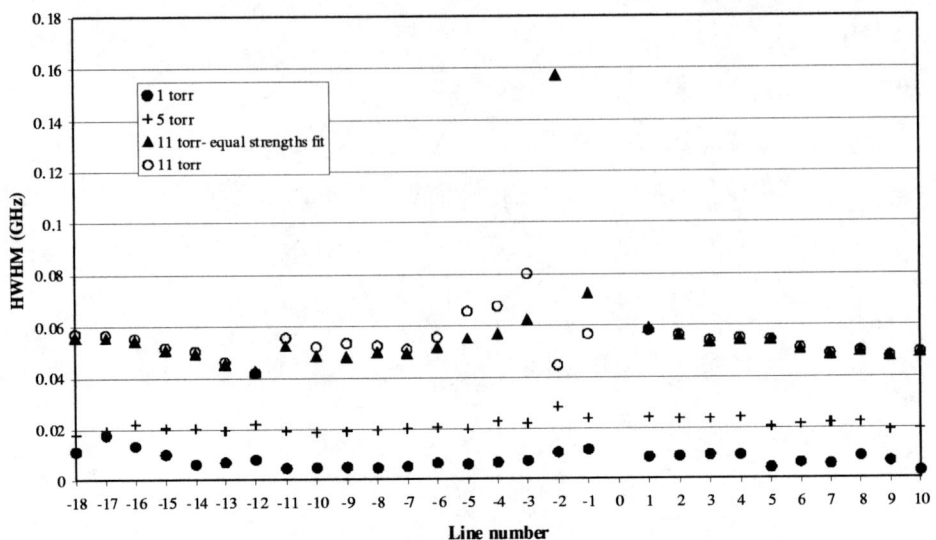

FIGURE 7. Collisional widths (HWHM) for 1, 5 and 11 torr

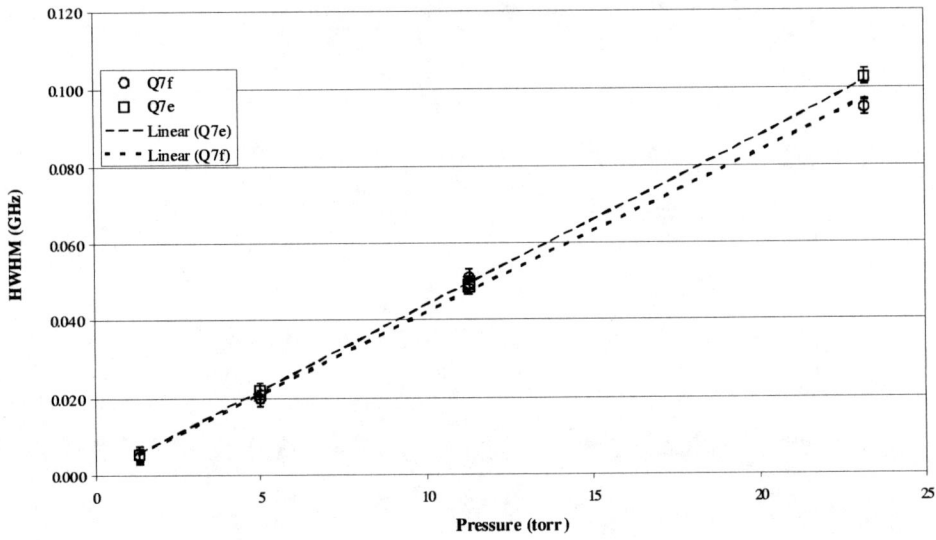

FIGURE 8. Linear pressure dependence of the collisional broadening for Q7F and Q7E lines

In figure 8, the linear pressure dependence of the broadening is presented for Q7E and Q7F lines. Using results from analysis of the spectrum up to 23 torr, the self-broadening coefficients for these lines were calculated and their values are 0.1115 cm^{-1}/atm for Q7E and 0.1065 cm^{-1}/atm for Q7F.

In terms of mixing asymmetry, the results in figure 6 are consistent for higher J values on either side of the branch, where the first order line mixing assumed in the fitting works much better. Although large discrepancies are obvious for the 4F to 1F lines in figure 6, there is a consistent trend that mixing coefficients are positive for low J and negative for higher J, a signature of the band collapse.

ACKNOWLEDGEMENTS

This work was supported by funding from the Natural Sciences and Engineering Research Council of Canada, COMDEV, ABB BOMEM Inc., Meteorological Service of Canada, University of Toronto Research fund, the Canadian Space Agency and the Industrial Research Chair in Atmospheric Remote Sounding from Space at the University of Toronto.

REFERENCES

1. Ciurylo, R. *Phys. Rev. A* **58**, p.1029 (2001).

Collision induced light scattering by gaseous neon

F. Rachet, M. Chrysos, and G. Lothon

Laboratoire des Propriétés Optiques des Matéraux et Applications, UMR CNRS 6136
Université d'Angers, 2 bd Lavoisier 49045 Angers, France

R. Moszynski[1] and A. Milet[2]

[1]Department of Chemistry, University of Warsaw, ul. Pasteura 1, 02-093 Warsaw, Poland
[2]Laboratoire d'Etudes Dynamiques et Structurales de la Sélectivité, UMR CNRS 5616
Université J. Fourier, 38041 Grenoble, France

Abstract. Binary collision induced light scattering by gaseous room temperature neon has been both experimentally and theoretically studied. Our experimental scattering cross sections have been calibrated on an absolute scale. The anisotropic spectrum and the depolarization ratio have been measured over the 5-300 cm^{-1} frequency domain, whereas the isotropic spectrum has been recorded over the domain 125-300 cm^{-1}. In comparison with any previous work in the topic, our measurements provide a substantial extension of the probed far wing. An exhaustive study of the light scattering process has been made from the theoretical/numerical viewpoint also. This has been made possible by quantum mechanically computing scattering intensities via a rigorous numerical treatment (previously developed by the authors) accounting both for free Ne pairs and for stable Ne_2 diatoms, which has been coupled to a novel *ab initio* incremental polarizability model based on Symmetry Adapted Perturbation Theory.

INTRODUCTION

The light which is scattered by an atomic gas yields — in addition to the well-known Rayleigh peak resulting from the atomic polarizabilities — a broad and weak Raman-like spectrum which is generated by incremental pair polarizabilities. While the Rayleigh peak is the spectral signature of atoms colliding elastically, the Raman-like spectrum clearly indicates the existence of inelastic collisions between atoms in the medium. The latter process is commonly referred to as the collision-induced light scattering (CILS). This kind of process involving optically inactive atomic systems is quite a simple one and relatively much studied since a long time. Discovered in the late 60s by Mc Tague and Birnbaum [1] and by Thibeau and Oksengorn [2], it has been thoroughly investigated by many research-workers until nowadays. As far as inert gases are concerned, the works by Frommhold's group in the early 80s remain pioneering works in the literature of this topic [3,4].

CP645, *Spectral Line Shapes: Volume 12, 16th ICSLS*, edited by C. A. Back
© 2002 American Institute of Physics 0-7354-0100-4/02/$19.00

Within this context, we recently accomplished an exhaustive study on low density gaseous helium both at room temperature [5-7] and at a low temperature [8]. From this study a complete picture of the dynamics of the helium pair was obtained and an exhaustive description of its CILS spectrum over an extremely large interval of detunings was made possible. A number of surprising, seemingly unique characteristic features of this system were reported, never revealed before by any other species studied by CILS spectroscopy.

In the present work, we turned our attention to the next atomic gas, neon. The interest in performing this study relies on that neon, due to its higher (compared to helium) degree of complexity, may serve as an effective prototype for checking the response of the *ab initio* approaches used to compute incremental polarizability invariants. Among such approaches, the symmetry adapted perturbation theory (SAPT) [9] has the reputation to be a powerful and promising tool as it is able to correctly account for tiny effects that contribute to the weak van der Waals mechanisms and which may drastically affect the spectral response of the system in the far wing. In order to provide evidence of these effects, most of the effort concerning our experiment was put on the substantial extension (compared to any previous work for this system) of the spectral domain probed.

EXPERIMENTAL SETUP

For this experiment a typical Raman scattering setup at a right angle scattering geometry was employed (Fig. 1). As a source, the green spectral line (at 514.5 nm) of an Ar^+ laser stabilized at a power of 2 W was used, whose beam was polarized in the direction perpendicular to the scattering plane, that is the plane defined by the incident and the scattered beams. Recorded in this configuration, the intensity of the scattered light, referred to as $I_\perp(v)$, constitutes the *polarized spectrum*. On the other hand, once the incident beam is polarized within the scattering plane, a complementary as well as independent spectral information can be obtained. This is often considered as one of the major advantages of Raman spectroscopy. In this case the intensity of the scattered signal is designated by $I_\parallel(v)$ and describes the so-called *depolarized spectrum*. In order us to acceede to this latter spectrum, the polarization axis of the incident beam was rotated by means of a half-wave plate followed by a glan polarizer.

The gaseous neon sample was

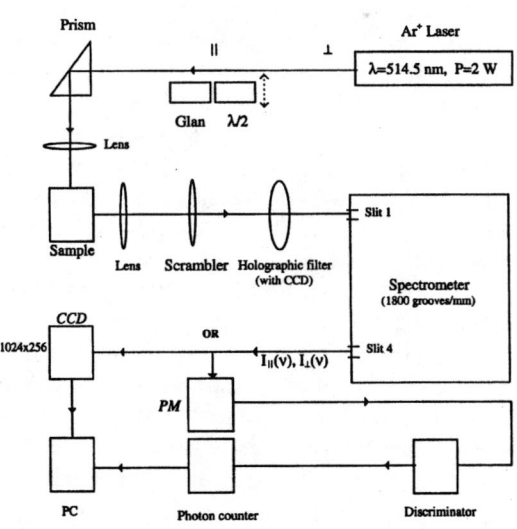

FIGURE 1. Experimental setup.

contained in a four-window high-pressure cell at room temperature. Given the dramatic decrease of the scattered signal with increasing wavenumber, densities up to 240 amagat were dealt with, that is a gas pressure going up to 300 bars. Depending on the gas pressure, fused silica or sapphire windows were used. It should be stressed that, in this kind of experiments, the use of a highly purified gas is crucial, as the contribution to the scattering intensities of parasitic signals due to impurities must remain as small as possible. In our experiment, highly purified neon gas with total residual impurities less than 10 ppm was used and a particular attention was put in the preparation of the cell. The scattered light was analyzed by a double monochromator supplied with two 1800 grooves/mm holographic gratings, after a scrambler crossed the scattered beam so that the different polarizations of the signal are mixed up. One has to keep in mind that for an aperture angle of the scattered beam close to zero, the anisotropic spectrum $I_{ani}(v)$ corresponds to the $I_{\parallel}(v)$ component, and that the isotropic spectrum $I_{iso}(v)$ can be deduced from the following linear combination of the two recorded components $I_{\perp}(v)$ and $I_{\parallel}(v)$:

$$I_{iso}(v) = I_{\perp}(v) - \frac{7}{6} I_{\parallel}(v)$$

Two different detection devices were employed. A low dark-current noise photomultiplier (PM) for low and intermediate frequencies, and a nitrogen-cooled CCD detector for the far spectral wing where the signal becomes extremely weak. Typically, with our CCD we were able to detect signals as low as one photon per week for every pixel of the CCD.

For the sake of completeness, we note that the extraction of the binary spectrum was possible by making a systematic study over density of the sample, for each of the frequencies probed. Then, having recorded the binary signal, the spectrum was calibrated on an absolute scale by taking as reference a rotational line of H_2.

THEORETICAL ASPECTS

In this study, we dealt with binary spectra, that is spectra probing pair atomic interactions. As the presence of any neon atom in the neighborhood of a single one breaks the spherical atomic symmetry, a Raman-like spectrum is induced, due to inelastic transitions between initial and final states of the *diatom*. This is schematically shown by Fig. 2. The incremental pair polarizability due to this interaction is represented by a second rank tensor whose invariants $\alpha(r)$ and $\beta(r)$ are functions of the interatomic separation r. They are called *incremental trace* and *anisotropy* respectively. Knowledge of the two invariants as a function of interatomic distance is crucial since it provides information on the collision atomic dynamics at close distances, as well as on macroscopic electro-optical properties of the probed medium, such as its refractivity index, the second dielectric virial coefficient and the constant of Kerr. It turns out that these invariants are straightforwardly related to the two spectral components $I_{iso}(v)$ and $I_{ani}(v)$ respectively, and they can therefore be checked via comparison of calculated spectra with experimental ones, as we show below.

The incremental polarizability models developed in this study were obtained by means of the *ab initio* SAPT, a powerful method able to separately account for effects

176

due to *polarization, exchange, dispersion* and *induction* interactions. For a review of this method, see Ref. [9]; for specific applications of the approach to the He_2 polarizability, see Refs [10,11]. In the present work we employed the same computational scheme as in Ref. [10]. A [6s5p4d3f1g] basis-set centered on the Ne atoms was used at first. No convergence of total incremental polarizabilities was obtained at short interatomic distances in spite of the fact that the 142 functions-set used seemed well spanning the phase space of the neon pair. Although absolute convergence is extremely difficult to ensure, a larger set, comprising 178 basis functions, was finally employed, obtained by supplementing the former one with a 3s3p3d2f set of bond functions. A more complete implementation and detailed description of the *ab initio* calculations, as well as a further extension of the spectral domain probed are in progress and will be reported in forthcoming publications.

Theoretical profiles were computed by using the Aziz and Slaman Ne_2 interatomic potential [12] and by employing a highly reliable quantum mechanical procedure developed by one of our groups [13,14]. The wave functions of the continuum were built step by step, according to the Fox-Goodwin propagative method, through outward propagation of the wave function ratio at every pair of adjacent points defined on a spatial grid ranging from 3 to 150 Bohr. A multitude of initial-state energies varied by steps of 15 cm^{-1} and attaining the value of 3000 cm^{-1} were considered. The maximum angular momentum quantum number was fixed at J_{max}=300. These numerical parameters provided convergency of total cross sections to within < 1% [13]. Contributions due to bound and predissociating states were computed by means of a DVR basis approach. The major advantage of this method relies on the fact that, within this DVR basis, the matrix representation of the effective potential interaction is diagonal while that of the kinetic operator is analytic [14].

$$\underline{\alpha} = \begin{pmatrix} \alpha_{\|} & 0 & 0 \\ 0 & \alpha_{\perp} & 0 \\ 0 & 0 & \alpha_{\perp} \end{pmatrix}$$

$$\underline{\alpha}_{ab}(\mathbf{r}) = \alpha(r)\delta_{ab} + \beta(r)\left\{\frac{ab}{r^2} - \frac{\delta_{ab}}{3}\right\}$$

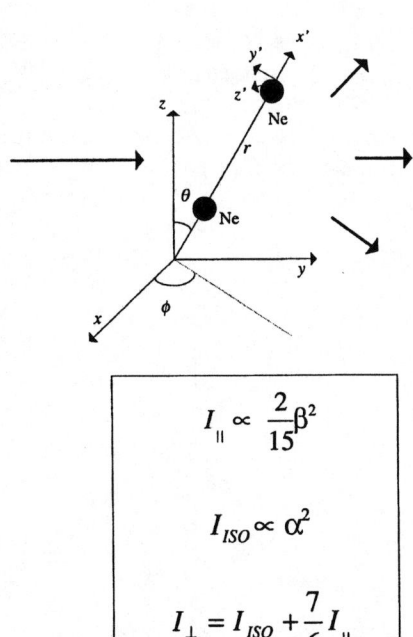

$$I_{\|} \propto \frac{2}{15}\beta^2$$

$$I_{ISO} \propto \alpha^2$$

$$I_{\perp} = I_{ISO} + \frac{7}{6}I_{\|}$$

FIGURE 2. Incremental polarizability tensor in both molecular-fixed and laboratory-fixed frames, schematic description of the neon diatom, and proportionality relationships between polarizability invariants and spectral components.

177

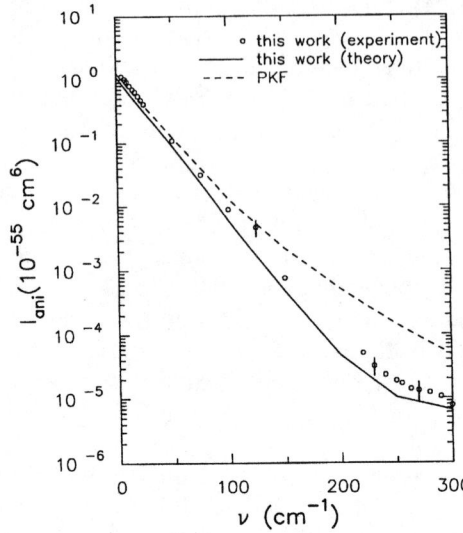

FIGURE 3. Anisotropic spectrum of (Ne)$_2$. Symbols: this work (experiment); solid-line curve: this work (theory); dashed-line curve: PKF

At the end of the 70s and at the beginning of the 80s two experimental groups reported on the neon binary anisotropic spectrum at room temperature. The measurements by Frommhold's group [4], performed over the interval 10-170 cm^{-1}, were found to agree well with those of Bérard and Lallemand [15] covering the domain 6-80 cm^{-1}. In the present work, we report on the neon anisotropic CILS spectrum in absolute units over an as yet unstudied spectral domain attaining 300 cm^{-1}. This extension was possible by employing the experimental setup described above, especially adapted to the detection of an extremely low photon flux. Figure 3 shows the binary anisotropic spectrum on an absolute logarithmic scale (cm^6) as a function of Raman shift ν. Our measurements, represented by circles, covered the spectral domain 5-300 cm^{-1}. The response of the semi-empirical PKF model, developed by Frommhold's group to fit their experimental data [4], is represented by a broken-line curve. Compared to our measurements, the latter model was found to be overall unable to account correctly for the lineshape, and only within the frequency domain common to the two experiments was it found to be adequate. The spectral signature of the SAPT anisotropy model proposed here, and obtained with the large basis-set, is illustrated by a solid-line curve. A very satisfactory agreement with our experiment is observed over the entire frequency domain probed, where the curvature of the lineshape is well reproduced as well.

For the sake of completeness, the

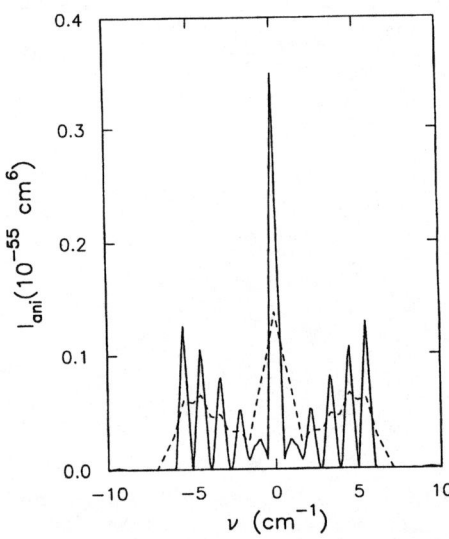

FIGURE 4. Anisotropic intensities by bound (Ne)$_2$ dimers convoluted with a triangular slit function. Solid-line curve: FWHM=0.5 cm^{-1}; dashed-line curve: FWHM=1.6 cm^{-1}.

spectrum due to stable Ne_2 dimers is shown in Fig. 4 for two triangular slit functions of full-width at half maximum (FWHM) fixed at 0.5 and 1.6 cm^{-1}, the latter value corresponding to the experimental conditions. For the neon gas, this spectrum was found to be weak and confined to a very short frequency domain. Finally, with this respect, figure 5 illustrates the non-convoluted stick-spectrum due to the bound dimers, arising from each of the possible transitions between initial and final states. The $\Delta J=0, \pm 2$ Raman branches are depicted on the figure.

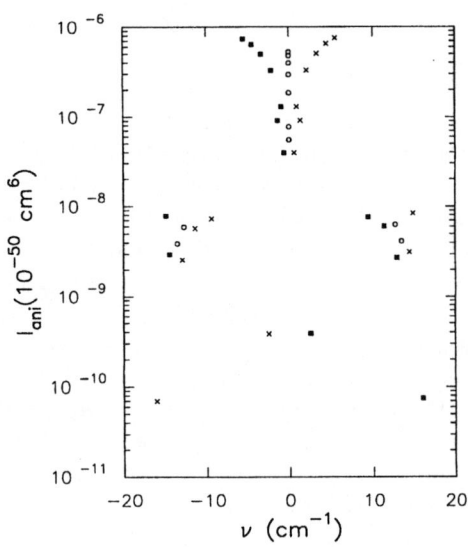

FIGURE 5. Anisotropic non-convoluted spectral lines by bound dimers. Circles: $\Delta J=0$; crosses: $\Delta J=2$; squares: $\Delta J=-2$.

ISOTROPIC SPECTRUM

The neon isotropic spectrum was first reported by Frommhold's group in the interval 10-170 cm^{-1} [3,4]. Their measurements, obtained for a single density of the sample (30.5 amagat), remain thus far the only experimental results available in the literature. Here we report new measurements over a large spectral range corresponding to an extension of the thus far explored domain by almost two times. Figure 6 illustrates our experimental isotropic spectrum on an absolute scale (cm^6), as well as the one obtained quantum mechanically by using our SAPT incremental trace model. A very satisfactory agreement is observed throughout the entire domain probed. This consistency, together with the one for the depolarized spectrum, strengthens our confidence to the quality of the *ab initio* invariants provided by SAPT, provided that the latter has been implemented within a sufficiently large basis set.

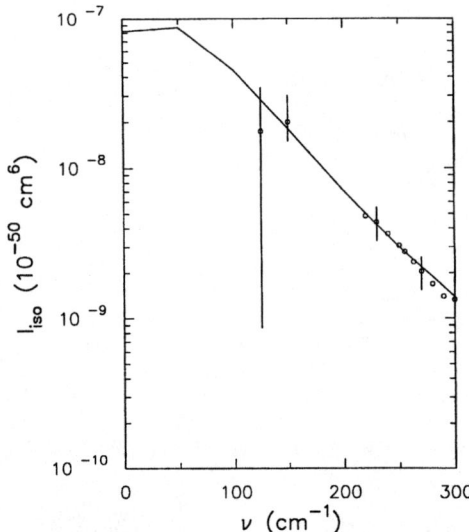

FIGURE 6. Isotropic spectrum of $(Ne)_2$. Symbols: experiment; solid-line curve: theory.

DEPOLARIZATION RATIO

The ratio between $I_\parallel(\nu)$ and $I_\perp(\nu)$ defines what is known as the *depolarization ratio* $\eta(\nu)$. Fig. 7 illustrates η up to $\nu = 300$ cm^{-1}. Experimental results are designated by circles; the solid line curve refers to quantum theory, where all computations have been carried out with the *ab initio* induced-polarizability incremental trace, α, and anisotropy, β, SAPT model, and with the Aziz and Slaman interatomic potential [12]. At low frequencies, $\eta(\nu)$ remains close to the value 6/7, a signature of depolarized scattering. As frequency increases, however, $\eta(\nu)$ acquires a rapidly decreasing behavior, a feature characterizing a quickly polarized system. The experimental η attains a minimum value of only ~0.06 in the probed far wing. Although the agreement between experimental and the theoretical η is satisfactory, improvement of the response of theory should be expected by further increasing the SAPT orbital basis. Work is in progress in our institutes.

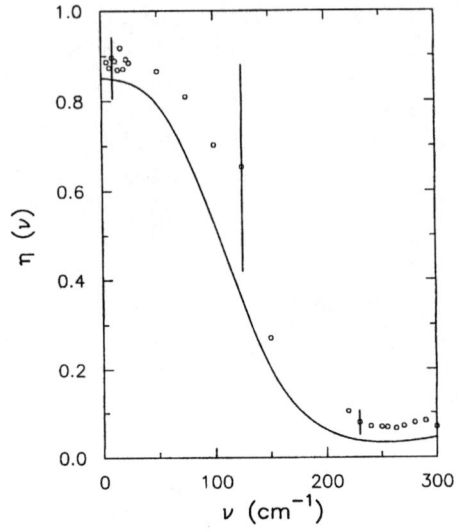

FIGURE 7. Depolarization ratio of (Ne)$_2$. Symbols: experiment; solid-line curve: theory.

REFERENCES

1. Mc Tague, J.P., and Birnbaum G., *Phys. Rev. Lett.* **21**, 661-664 (1968).
2. Thibeau M., and Oksengorn B., *Mol. Phys.* **15**, 579-586 (1968).
3. Frommhold L., and Proffitt M.H., *Phys. Rev. A* **21**, 1249-1255 (1980).
4. Proffitt M.H., Keto J.W., and Frommhold L., *Can. J. Phys.* **59**, 1459-1474 (1981).
5. Guillot-Noël C., Chrysos M., Le Duff Y., and Rachet F., *J. Phys. B: At. Mol. Opt.Phys.* **33**, 569-580 (2000).
6. Rachet F., Le Duff Y., Guillot-Noël C., and Chrysos M., *Phys. Rev. A.* **61**, 062501 (2000).
7. Rachet F., Chrysos M., Guillot-Noël C., and Le Duff Y., *Phys. Rev. Lett.* **84**, 2120-2123 (2000).
8. Guillot-Noël C., Le Duff Y., Rachet F., and Chrysos M., *Phys. Rev. A.* **66** (2002), in press
9. Jeziorski B., Moszynski R., and Szalewicz K., Chem. Rev. **94**, 1887 (1994).
10. Moszynski R., Heijmen T.G.A., Wormer P.E.S., and van der Avoird A., *J. Chem. Phys.* **104**, 6997-7007 (1996).
11. Heijmen T. G. A., Moszynski R., Wormer P. E. S., and van der Avoird A., Mol Phys **89**, 81 (1996).
12. Aziz A., and Slaman. M.J., *Chem. Phys.* **130**, 187-194 (1989)
13. Chrysos M., Gaye O., and Le Duff Y., *J. Phys. B: At. Mol. Opt. Phys.* **29**, 583-593 (1996).
14. Chrysos M., Gaye O., and Le Duff Y., *J. Chem. Phys.* **105**, 31-36 (1996).
15. Bérard M., and Lallemand P., *Mol. Phys.* **34**, 251-260 (1977).

Collision-Induced Absorption of H₂ Pairs in the High Frequency Wing of the Rototranslational Spectrum

D. Bailly[*], J.-P. Bouanich[*], C. Brodbeck[*], A. Borysow[#]

[*]Laboratoire de photophysique moléculaire, CNRS, bâtiment 350, campus d'Orsay,
91405 Orsay cedex, France
[#]Niels Bohr Institute for Astrophysics, Physics and Geophysics, Copenhagen University Observatory,
Juliane Maries vej 30 DK-2100 Copenhagen, Denmark

Abstract. The spectra of hydrogen in the 5 μm window of Jupiter have been recorded at temperatures of 297.5 and 77.5 K for gas densities ranging from 51 to 610 amagats. The binary absorption coefficient has been determined by extrapolation to zero density of the measured profiles. These extrapolated measurements are compared with calculations from the extended Birnbaum-Cohen model which represents the existing quantum profiles with full accuracy. Within the experimental errors, which are rather larger at low temperature, the agreement is satisfactory at room as well as low temperature.

INTRODUCTION

Molecular hydrogen and helium are the most abundant constituents in the upper atmospheres of the major planets and the pressure-induced spectra from H_2-H_2 and H_2-He collisions are an important source of thermal emission and absorption from these planets [1]. In particular, the 5 μm window of Jupiter (from 1900 to 2300 cm⁻¹) allows the determination of the abundance of several minor compounds [2], the continuum being partly due to the absorption of H_2-H_2 and H_2-He pairs. Therefore, the study of collision-induced spectra for these interacting systems is of considerable astrophysical interest. The experimental data can be useful as a benchmark for testing the theoretical models. The absorption arises from transient dipole moments induced by intermolecular interactions. The dominant induction mechanism is the polarization of the collisional partner in the electric field (mainly quadrupolar) of a hydrogen molecule.

In this work, we present the results of new measurements for collision-induced spectra of H_2 at room temperature and at 77.5 K in the frequency range 1900-2300 cm⁻¹. The values of the extrapolated binary absorption coefficient are compared with calculated results based on an extended six-parameter model that fits the results of the existing quantum calculations accurately.

CP645, *Spectral Line Shapes: Volume 12, 16ᵗʰ ICSLS*, edited by C. A. Back
© 2002 American Institute of Physics 0-7354-0100-4/02/$19.00

EXPERIMENTAL PROCEDURES

The set of infrared experiments has been carried out on a Bruker IFS 66V FT spectrometer with a resolution of 0.5 cm^{-1}. Each interferogram was averaged over 512 scans with a four-term Blackmann-Harris apodization. The high pressure absorption cell [3] is a 214.4 cm path length cell, cooled by liquid nitrogen for the low temperature recordings. The collision-induced absorption spectra of hydrogen – with high purity-grade (99.9999 %) - in the region between 1900 to 2300 cm^{-1} have been recorded at 297.5 and 77.5 K and for gas densities ranging from 167 to 610 amagats (Am) at low temperature and 51 to 204 amagats at room temperature.

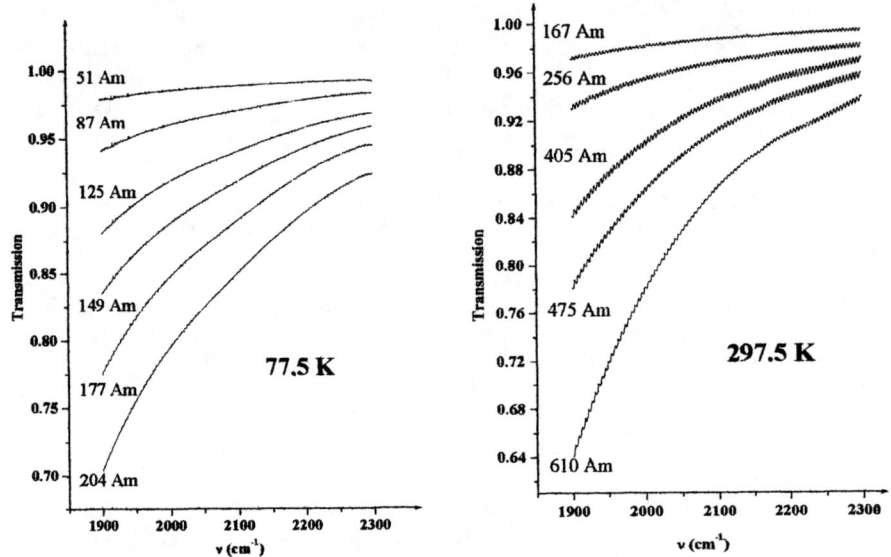

FIGURE 1. Measured absorption spectra of H$_2$ for several densities.

DATA REDUCTION

The absorption coefficient $\alpha(\nu)$ at the wavenumber ν is obtained from the usual Beer-Lambert law by:

$$\alpha(\nu) = - \ln \left[I(\nu)/I_0(\nu) \right] / L, \tag{1}$$

where I and I_0 are respectively the transmitted intensities through the cell filled with hydrogen and the cell under vacuum or filled by helium at the same pressure than H$_2$, respectively, L is the optical path length.

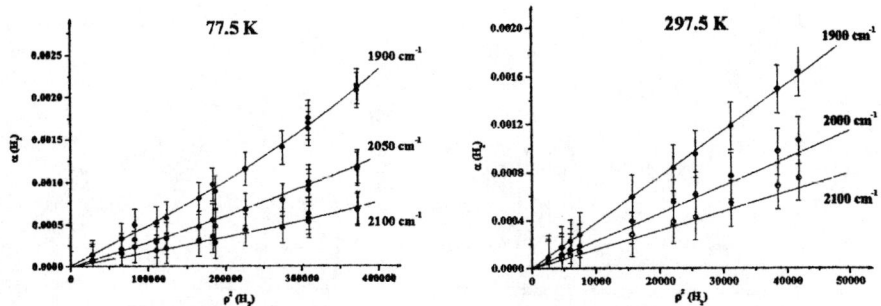

FIGURE 2. Quadratic density dependence of the measured absorption spectra.

At room temperature, the binary absorption coefficient of H_2 is defined as:

$$\alpha_0(\nu) = \alpha(\nu) / \rho^2, \tag{2}$$

where ρ is the gas density, and is obtained from a least-square procedure.

At low temperature, since higher densities of H_2 were used, the binary absorption coefficient of H_2 is derived through a third degree polynomial fit as:

$$\alpha(\nu) = \alpha_0(\nu)\,\rho^2 + \alpha_1(\nu)\,\rho^3. \tag{3}$$

In these two cases the absolute uncertainty of α_0 is estimated to be equal to three times the error of α_0 derived from least-squares fits.

THEORETICAL RESULTS AND COMPARISONS

The spectral range studied here corresponds to the high frequency wing of the rototranslational spectrum of H_2-H_2. The significant contributions in the 5 μm region are the anisotropic overlap components, the quadrupole-induced single and double transitions and the hexadecapole-induced transitions.

The calculations of line shapes for the collision-induced spectrum have been carried out on the basis of an extended six-parameter Birnbaum-Cohen model (EBC) [4] which represents the quantum profiles with full accuracy.

At low temperature, the measured binary absorption coefficient is larger than that derived from the EBC model, but in view of the large uncertainties in our measurements, the agreement is still reasonable.

At room temperature, the theoretical curve lies systematically under our measurements. The differences may be explained by experimental errors (uncertainties in the temperature − 0.5%, and in the position of the baseline − 5%). It is also possible that the theory based on the isotropic potential and neglecting the J-dependence starts being insufficient in the very far spectral wings studied here. By considering these uncertainties, an overall reasonable agreement is obtained between experimental data and theoretical results.

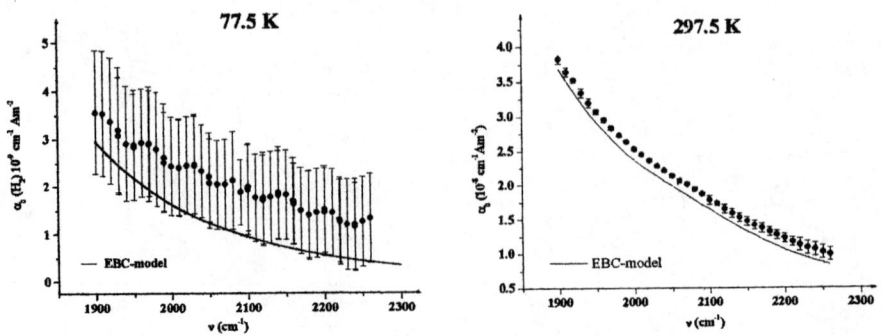

FIGURE 3. Comparison of our extrapolated measurements at zero density with results from the EBC model.

REFERENCES

1. Trafton, L. M., *Astrophys. J.* **146**, 558-571 (1966).
2. Lellouch, E., Drossart, P., Encrenaz, T., *Icarus* **77**, 457-465 (1989).
3. Brodbeck, C., Bouanich, J.-P., Nguyen-Van-Thanh, Fu, Y., Borysow, A., *J. Chem. Phys.* **110**, 4750-4756 (1999).
4. Borysow, J., Trafton, L; Frommhold, L., Birnbaum, G., *Astrophys. J.* **296**, 644-654 (1985).

Pressure effects on 326.1 nm cadmium line perturbed by neon

A. Bielski*, D. Lisak*, J. Szudy* and R. S. Trawiński*

*Institute of Physics, Nicholas Copernicus University, Grudziądzka 5, 87-100 Toruń, Poland

Abstract. Using a laser-induced fluorescence method a detailed analysis of profiles of the [114]Cd 326.1 nm line perturbed by neon was performed which revealed deviations from the ordinary Voigt profile. These deviations are shown to be consistent with fits of experimental profiles to an asymmetric Voigt profile. Coefficients of the pressure broadening, shift and collision-time asymmetry are determined and compared with those calculated in the adiabatic approximation for different interaction potentials.

INTRODUCTION

At low densities of perturbing atoms, the core of collisionally broadened and shifted line can be described by a Lorentzian profile [1, 2]. Such a line profile is justified only in the so–called impact approximation in which the duration of each collision is assumed negligibly short compared with the time between collisions. More detailed theoretical treatment showed, however, that the inclusion of the finite duration of collisions modifies the Lorentzian profile, the first order correction being the addition of a dispersion component to the Lorentzian distribution. Thus the resulting line shape becomes asymmetric and this type of asymmetry is usually referred to as collision-time asymmetry [3]-[11].

Spectral line shapes are also determined by thermal motion of the emitting (or absorbing) atom which alters the shape of the line through the Doppler effect, but it also determines the relative emiter–perturber velocity distribution, and through that the collisional broadening. In addition to the line shape asymmetries mentioned above there is also an asymmetry due to the correlation between Doppler and collisional broadening [12, 13] caused by the fact that in the Maxwellian average of the profile, the line shift is speed dependent. This asymmetry is small and depends on the value of the line shift. It must be emphasized that because of the similarity between contributions from collision duration time and speed-dependent effects an extreme care is needed in any quantitative interpretation of asymmetry of pressure broadened spectral line shapes.

In our systematic study [14]-[16] of collisional effects on the [114]Cd 326.1 nm ($5^1S_0 - 5^3P_1$) intercombination line perturbed by heavier noble gases (Xe, Kr, Ar) we used a laser-induced fluorescence (LIF) technique, and a digital laser spectrometer described elsewhere [18].

A good signal to noise ratio and negligible instrumental profile enabled us to fit the line shapes in considerable detail and in the case of the Cd-Xe [14] and Cd-Kr [15] systems we were able to detect the asymmetry of the 326.1 nm line shape caused by finite

CP645, *Spectral Line Shapes: Volume 12, 16th ICSLS*, edited by C. A. Back
© 2002 American Institute of Physics 0-7354-0100-4/02/$19.00

TABLE 1. Comparison of experimental values of the β, δ (in units 10^{-20} cm^{-1}/atom cm^{-3}) and κ (in units 10^{-21}/atom cm^{-3}) coefficients with those calculated for different interatomic potentials. For experimental data the values of standard uncertainty are given

Experimental values	β_{exp}	δ_{exp}	κ_{exp}
This work	0.715(4)	-0.090(5)	0.17(7)
Calculated values	β_{calc}	δ_{calc}	κ_{calc}
Czuchaj-Stoll [19]	0.750	-0.142	-0.025
Morse	0.734	-0.063	0.06
Lennard-Jones (SC)	0.783	0.048	0.07

duration of collision as well as that due to correlation between Doppler and collisional broadening.

As a sequel to our previous papers in this paper we report results of measurements of the pressure broadening, shift and asymmetry of the 326.1 nm ^{114}Cd line perturbed by neon. Profiles of this line were obtained by observing the total fluorescence resulting from excitation of Cd–Ne mixtures by a tunable Coherent CR 899-21 ring dye laser pumped by Innova Ar$^+$ laser. The Cd–Ne mixture was present in a quartz cell at temperature 724 K. The Ne-pressure varied between 5 and 442 Torr at room temperature.

RESULTS AND DISCUSSION

As was shown in preceding paper [17] in the case of light perturbers the measured line shapes can be fitted well to asymmetric Voigt profiles. Experimental values of pressure broadening β, shift δ and asymmetry κ coefficients determined from the best-fit procedure are listed in Table 1.

In order to interpret our experimental data we calculated the theoretical values of pressure broadening β and shift δ coefficients on the basis of the adiabatic phase-shift theory with straight-line trajectories using equation (2) from Ref. [17]. Our calculations were performed for the Czuchaj and Stoll [19] numerical potential as well as for two empirical potentials derived from experimental data in the form of the Morse and Lennard-Jones potentials with the spectroscopic constants determined recently by Koperski *et al.* [20].

The values of β and δ coefficients evaluated for these potentials are listed in Table 1 and marked as "Czuchaj-Stoll [19]", "Morse", "Lennard-Jones (SC)", respectively.

As can be seen from Table 1 there is generally poor agreement between the experimental and theoretical values both for β and δ coefficients.

In Table 1 we have also listed the value of asymmetry coefficient κ calculated for all of the above potentials using equation (3) from Ref. [17]. Let us note that our experimental asymmetry coefficient κ is positive in contradiction to the pressure shift coefficient which was found to be negative. This means that for the Cd 326.1 nm line perturbed by neon the red shift is associated with the higher intensity blue wing of this line.

ACKNOWLEDGMENTS

This work was supported by the grant No. 5 P03B 066 20 (354/P03/2001/20) from the Committee for Scientific Research.

REFERENCES

1. N. Allard and J. Kielkopf, Rev. Mod. Phys. 54 (1982) 1103.
2. J. Szudy and W. E. Baylis, Phys. Rep. 266 (1996) 127.
3. P. W. Anderson and J. D. Talman, Bell Teleph. Syst. Tech. Publ. No 3117 University of Pittsburg, USA, 1955.
4. G. Traving, Uber die Theorie der Druckverbreiterung von Spektrallinien, Verlag G. Braun, Kalsruhe, 1960.
5. J. Szudy and W. E. Baylis, J. Quant. Spectrosc. Radiat. Transf. 15 (1975) 641.
6. J. Szudy and W. E. Baylis, J. Quant. Spectrosc. Radiat. Transf. 17 (1977) 681.
7. G. Peach, J. Phys. B 17, (1984) 2599.
8. B. N. I. Al-Saqabi and G. Peach, J. Phys. B 20, (1987) 1175.
9. A. Royer, Acta Phys. Pol. A 54 (1978) 805.
10. P. S. Julienne and F. H. Mies, Phys. Rev. A 34 (1986) 3792.
11. R. Ciuryło, J. Szudy, and R. S. Trawiński, J. Quant. Spectrosc. Radiat. Transf. 57 (1997) 551.
12. P. R. Berman, J. Quant. Spectrosc. Radiat. Transf. 12 (1972) 1331.
13. J. Ward, J. Cooper, and E. W. Smith, J. Quant. Spectrosc. Radiat. Transf. 14 (1974) 555.
14. A. Bielski, R. Ciuryło, J. Domysławska, D. Lisak, R. S. Trawiński, and J. Szudy, Phys. Rev. A 62 (2000) 032511.
15. R. S. Trawiński, A. Bielski, D. Lisak, Acta Phys. Pol. A 99 (2001) 243.
16. A. Bielski, D. Lisak, R. S. Trawiński, Eur. Phys. J. D 14 (2001) 27.
17. A. Bielski, D. Lisak, J. Szudy and R. S. Trawiński, "Pressure broadening, shift and asymmetry of the 326.1 nm Cd line perturbed by N_2 and CH_4", in this book.
18. A. Bielski, R. Ciuryło, J. Domysławska, D. Lisak, R. S. Trawiński, and J. Wolnikowski, Acta Phys. Pol. A 97 (2000) 1003.
19. E. Czuchaj and H. Stoll, Chem. Phys. 248 (1999) 1.
20. J. Koperski and M. Czajkowski, Eur. Phys. J D 10 (2000) 363.

Pressure broadening, shift and asymmetry of the 326.1 nm Cd line perturbed by N_2 and CH_4

A. Bielski*, D. Lisak*, J. Szudy* and R. S. Trawiński*

*Institute of Physics, Nicholas Copernicus University, Grudziądzka 5, 87-100 Toruń, Poland

Abstract. The experimental values of pressure broadening, shift and collision-time asymmetry coefficients of the 326.1 nm ^{114}Cd line perturbed by N_2 and by CH_4 were determined at perturbing gas pressures between 1 and 350 Torr. All measurements were performed using a Laser Induced Fluorescence (LIF) technique. The experimental coefficients of pressure broadening, shift and collision-time asymmetry were compared with theoretical values calculated from the impact theory assuming van der Waals form of interaction potentials.

INTRODUCTION

In previous papers [1]-[4] precise studies of pressure broadening and shift of the ^{114}Cd 326.1 nm $(5^1S_0 - 5^3P_1)$ intercombination line perturbed by noble gases (Ar, Kr, Xe) and two non-polar molecular gases (H_2 and D_2) performed by means of a laser-induced fluorescence method were reported. In this paper we report results of measurements of the shapes of the 326.1 nm Cd line by use of N_2 and CH_4 as perturbing gas.

EXPERIMENTAL SETUP

Profiles of the 326.1 nm ^{114}Cd cadmium line were registered using a digital laser spectrometer described elsewhere [4]. A single-frequency Coherent CR 899-21 ring dye laser equipped with intracavity frequency doubler CR 8500, operating on DCM dye was pumped by INNOVA-400 argon-ion laser. The ring laser provided single mode UV output continuously tunable for up to 60 GHz. The line width of this laser line was about 1 MHz. The intensity of the fluorescence signal was measured by a photomultiplier working in the photon counting mode. The cell containing ^{114}Cd isotope and perturbing gas was situated in an oven at constant temperature 450 K.

LINE SHAPE

In the impact limit the combined influence of Doppler and pressure effects can be represented by the familiar Voigt profile $I_{VP}(\tilde{v})$ which is a convolution of the Lorentzian and Gaussian profiles. Beyond the impact limit, i. e. in the case when the finite duration of collisions has to be taken into account the resultant profile can be represented by

CP645, *Spectral Line Shapes: Volume 12, 16th ICSLS*, edited by C. A. Back
© 2002 American Institute of Physics 0-7354-0100-4/02/$19.00

the so-called asymmetric Voigt profile $I_{AVP}(\tilde{v})$ which is a convolution of the collisional and Doppler components. The collisional component of the line profile $I_C(\tilde{v})$ can be presented as a sum of the Lorentzian and dispersion profiles [5, 6]

$$I_C(\tilde{v}) = I_C^{(0)} \left(\frac{\gamma_L}{2} \right) \frac{(\gamma_L/2) + \chi(\tilde{v} - \tilde{v}_0 - \Delta)}{(\tilde{v} - \tilde{v}_0 - \Delta)^2 + (\gamma_L/2)^2} \ . \tag{1}$$

Here γ_L, Δ and χ denote the Lorentzian width (FWHM), pressure shift, and collision-time asymmetry parameter, respectively, $I_C^{(0)}$ is the intensity in the line peak and $\tilde{v}_0$ is the unperturbed wavenumber of emitted line. The Lorentzian width and shift can be evaluated from the following expression [7]-[9]:

$$\frac{\gamma_L}{2} + i\Delta = \frac{N}{c} \int d^3 \tilde{v} f(\tilde{v}) \, v \int_0^{+\infty} d\rho \, \rho \left\{ 1 - \left\langle S_{ii}(\rho, v) S_{ff}^{-1}(\rho, v) \right\rangle_{Ang.Av.} \right\}. \tag{2}$$

Here N is perturber density, ρ is the impact parameter, $f(\tilde{v})$ is the Maxwellian distribution of the relative velocities $\tilde{v}$ of the colliding atoms and $S_{ii}(\rho, v)$ and $S_{ff}(\rho, v)$ are the diagonal elements of the scattering matrix $\hat{S} = \hat{U}(-\infty, +\infty)$ for the initial and final states of the radiating atom. $\hat{U}(t_2, t_1)$ denotes the time evolution operator and the symbol $\langle ... \rangle_{Ang.Av.}$ means the average over angular coordinates.

The collision-time asymmetry parameter can be evaluated from the expression [9, 10]:

$$\chi = 2\pi N \int d^3 \tilde{v} f(\tilde{v}) \, v \int_0^{+\infty} d\rho \, \rho \int_{-\infty}^{+\infty} dt_0 \ \mathrm{Im} \Big\{ 1 + \left\langle U_{ii}(+\infty, -\infty) U_{ff}^{-1}(+\infty, -\infty) - \right.$$

$$\left. - U_{ii}(t_0, -\infty) U_{ff}^{-1}(t_0, -\infty) - U_{ii}(+\infty, t_0) U_{ff}^{-1}(+\infty, t_0) \right\rangle_{Ang.Av.} \Big\}. \tag{3}$$

In the classical limit time ordering in the time evolutions operator is ignored and these operators are replaced by exponential functions of the phase-shift functions

$$\eta(t_1, t_2) = \frac{1}{\hbar} \int_{t_1}^{t_2} \Delta V(r(t)) dt \ , \tag{4}$$

where $\Delta V(r) = V_u(r) - V_l(r)$ is the difference of adiabatic interaction potentials for the upper (u) and lower (l) levels of the emitting atom, respectively, expressed as a function of the distance $r(t)$ between emitter and the perturber at time t.

RESULTS AND DISCUSSION

The asymmetric Voigt profiles $I_{AVP}(\tilde{v})$ were fitted to our experimental data. The fitting parameters were: Gaussian γ_D and Lorentzian γ_L widths, line shift Δ, the asymmetry parameter χ, intensity in the line peak and constant background signal. The Doppler widths determined from the fits agree with a theoretical value 43.7×10^{-3} cm^{-1} corresponding to the cell temperature (450 K).

TABLE 1. Experimental (exp) and theoretical (vdW) values of pressure broadening β, shift δ and collision-time asymmetry κ coefficients of the ^{114}Cd 326.1 nm line perturbed by N_2 and CH_4

Perturber	β [10^{-20}cm^{-1}/molec. cm^{-3}]	δ [10^{-20}cm^{-1}/molec. cm^{-3}]	κ [10^{-21}/molec. cm^{-3}]
N_2 (exp)	1.054(9)	-0.281(7)	-0.36(8)
N_2 (vdW)	0.933	-0.338	-0.60
CH_4 (exp)	1.548(5)	-0.446(4)	-0.81(7)
CH_4 (vdW)	1.255	-0.455	-0.66

For both Cd – N_2 and Cd – CH_4 systems the experimental values for the pressure broadening $\beta = \gamma_L/N$, shift Δ/N and the collision-time asymmetry coefficients $\kappa = \chi/N$ coefficients were determined and listed in Table 1. Note that both for N_2 and CH_4 the parameter κ is negative which corresponds to red asymmetry of the profiles.

The same sign of pressure shift δ and asymmetry κ coefficients agree with theoretical predictions for the van der Waals potential that a red shift should be associated with a higher intensity red wing [11, 12].

Pressure broadening and shift coefficients have been calculated from Eq. (2) and asymmetry coefficient from Eq. (3) assuming the van der Waals form of the interaction potential and were listed in Table 1. We used the approximate formula given by Unsöld [13] to calculate C_6 force-constants: $C_6 = \alpha e^2 \langle r^2 \rangle$, where α is polarizability of the perturber, e is the elementary charge and $\langle r^2 \rangle$ is the expectation value of r^2 in a given state of radiating atom. $\langle r^2 \rangle$ were calculated using the Coulomb approximation:

$$\langle r^2 \rangle = \frac{a_0^2 \, n_{eff}^2}{2} \left[5n_{eff}^2 + 1 - 3l(l+1) \right] \tag{5}$$

where n_{eff} denotes the effective quantum number for a given (upper or lower) state.

As can be seen the theoretical values β, δ and κ are in reasonable agreement with experiment except the asymmetry coefficient for Cd – N_2 system which is about two times greater than the experimental value. The cause of this disagreement may come from the model assumed for the interaction potential. Unfortunately quantum mechanical interaction potentials for Cd – N_2 and Cd – CH_4 have not been calculated as yet.

ACKNOWLEDGMENTS

This work was supported by the grant No. 5 P03B 066 20 (354/P03/2001/20) from the Committee for Scientific Research.

REFERENCES

1. A. Bielski, R. Ciuryło, J. Domysławska, D. Lisak, R. S. Trawiński, and J. Szudy, *Phys. Rev. A* **62**, 032511 (2000).
2. R. S. Trawiński, A. Bielski, D. Lisak, *Acta Phys. Pol. A* **99**, 243 (2001).
3. A. Bielski, D. Lisak, R. S. Trawiński, *Eur. Phys. J. D* **14**, 27-31 (2001).

4. A. Bielski, R. Ciuryło, J. Domysławska, D. Lisak, R. S. Trawiński and J. Wolnikowski, *Acta Phys. Pol. A* **97**, 1003 (2000).
5. J. Szudy and W. E. Baylis, Phys. Rep. **266**, 127 (1996).
6. J. Szudy and W. E. Baylis, J. Quant. Spectrosc. Radiat. Transf. **15**, 641 (1975); **17**, 681 (1977).
7. M. Baranger, Phys. Rev. **111**, 494 (1958)
8. N. Allard and J. Kielkopf, Rev. Mod. Phys. **54**, 1103 (1982).
9. R. Ciuryło, Phys. Rev. A **58**, 1029 (1998).
10. R. Ciuryło, J. Szudy, and R. S. Trawiński, J. Quant. Spectrosc. Radiat. Transf. **57**, 551, (1997).
11. J. Szudy and W. E. Baylis, *Phys. Rep.* **266**, 127 (1996).
12. N. Allard and J. Kielkopf, *Rev. Mod. Phys.* **54**, 1103 (1982).
13. A. Unsöld, *Physik der Sternatmospharen*, Springer, Berlin 1955.

On the Impact Parameters of the Broadening and Shift of Spectral Lines

M. S. Helmi and G. D. Roston[*]

Department of physics - Faculty of science - Alexandria University - Alexandria , Egypt
* gamal_daniel@yahoo.com

Abstract. In a previous work , the authors obtained simple analytical approximate formulas for the broadening β_c and shift δ_c coefficients in case of Lennard-Jones potential. The obtained formulas are based on the broadening and phase shifts $\pm\,\eta_{ob}$ and $\pm\,\eta_{o\delta}$ respectively. The correct signs of these phases are obtained and defined. When these phases are applied with   their correct signs in the approximate formulas, the broadening β_c and shift δ_c coefficients for some interactions of (Th, Hg, Cd, Zn, Ar and Ne) with inert gases and self interactions are in agreement with the corresponding values obtained   numerically by other authors. The limit at which the shift changes its sign is also obtained . It depends on the ratio ($\Delta C_{12} / \Delta C_6$). New impact parameters which are not known up to now have been discussed and obtained.

1. INTRODUCTION

It has long been appreciated that studying of the collision broadening and shift of spectral lines contain information concerning the interatomic potentials between the radiated and perturbed atoms. The theoretical treatment   of this process is greatly interested for the region of low densities at which the two particles interactions are predominant, where the impact approximation take place . In this case the broadening and shift coefficients β and δ respectively are specified for such interactions. For that, in the previous paper [1], the authors obtained simple analytical formulas for calculating the coefficients β and δ in case of Lennard – Jones potential . These formulas are based on the assumption that the ranges of the impact parameters ρ_{ob} and $\rho_{o\delta}$ responsible for the broadening and shift of the spectral line are different, and then the phase shift for the broadening η_{ob} and shift $\eta_{o\delta}$ are also different. In that paper the authors obtained that $\eta_{ob} = \pm\,\pi / 5 = \pm\,0.63$ and $\eta_{o\delta} = \pm\,\pi / 2 = \pm\,1.57$.

The aim of this work is to obtain the correct sign of the broadening and shift phases η_{ob} and $\eta_{o\delta}$ and how to apply them to calculate the correct values of β and δ.

The authors also obtained the critical value of the impact parameter ρ_δ which separates between the red and blue shifts of the spectral line.

CP645, *Spectral Line Shapes: Volume 12, 16th ICSLS*, edited by C. A. Back
© 2002 American Institute of Physics 0-7354-0100-4/02/$19.00

2. THEORETICAL BACKGROUND

The approximated formulas for the broadening β_c and shift δ_c obtained by the authors [1] in case of Lennard – Jones potential are given by:

$$\beta_c = \overline{V} \ \rho_{ob}^2$$

and

(1)

$$\delta_c = \frac{\alpha_{12}\Delta C_{12}}{9} \rho_{o\delta}^{-9} - \frac{\alpha_6 \Delta C_6}{3} \rho_{o\delta}^{-3}$$

Here ρ_{ob} and $\rho_{o\delta}$ are respectively the broadening and shift impact parameters, and ΔC_6 , ΔC_{12} are the potential parameters for the Lennerd – Jones potential. $\overline{V}$ is the mean relative velocity which is related to the temperature T and the reduced mass μ of the radiator and perturber by the formula:

$$\overline{V} = \sqrt{\frac{8KT}{\pi \mu}}$$

To test the validity of the obtained formulas (1), the values of β_c and δ_c obtained with the broadening and shift phase values $\eta_{ob} = \pm 0.63$ and $\eta_{o\delta} = \pm 1.57$ are compared with the corresponding Hindmarsh et al [2] values β_H and δ_H which are given by:

$$\beta_H = 4 \ (3\pi/8)^{2/5} \overline{V}^{3/5} \ (\Delta C_6)^{2/5} \ B(\alpha)$$

(2)

$$\delta_H = (3\pi/8)^{2/5} \overline{V}^{3/5} \ (\Delta C_6)^{2/5} \ S(\alpha)$$

where the broadening and shift functions $B(\alpha)$ and $S(\alpha)$ are defined by the following integrals:

$$B(\alpha) = \int_0^\infty x \ \sin^2 0.5(\alpha \ x^{-11} - x^{-5}) \ dx$$

(3)

$$S(\alpha) = \int_0^\infty x \ \sin(\alpha \ x^{-11} - x^{-5}) \ dx$$

where

$$\alpha = 0.536 \ \overline{V}^{6/5} \ (\Delta C_6)^{-11/5} \ \Delta C_{12}$$

It was shown [1] that there are a good agreement between the two values for some interactions when η_{ob} and $\eta_{o\delta}$ are positive and other interactions when they are negative, but we did not clarify the reason of the different signs, and when these signs can be applied.

The approximated formulas (1) are based on the phase shift $\eta(\overline{V}, \rho)$, which is defined for any impact parameter ρ and velocity $\overline{V}$ as:

$$\eta(\overline{V},\rho) = (\frac{\alpha_{12}}{\overline{V}} \frac{\Delta C_{12}}{\overline{V}}) \, \rho^{-11} - (\frac{\alpha_6}{\overline{V}} \frac{\Delta C_6}{\overline{V}}) \, \rho^{-5} \qquad (4)$$

Let ρ_o, ρ_E, ρ_{ob} and $\rho_{o\delta}$ are the impact parameters corresponding respectively to the phase shifts $\eta = 0$, $\eta = \eta_E$ (the phase shift well depth), $\eta_{ob} = \pm 0.63$ and $\eta_{o\delta} = \pm 1.57$. When $\eta(\overline{V}, \rho)$ is plotted against ρ for constant values of $\overline{V}$, the plotted curves has the form shown in Figs.(1-3). The parameters ρ_o, ρ_E and η_E are given from (4) by:

$$\rho_o = \left[\frac{21 \Delta C_{12}}{32 \Delta C_6} \right]^{1/6}, \qquad \rho_E = 1.14 \, \rho_o$$

$$\eta_E = \left(\frac{\alpha_{12} \Delta C_{12}}{\overline{V}} \right) \rho_E^{-11} - \left(\frac{\alpha_6 \Delta C_6}{\overline{V}} \right) \rho_E^{-5} \qquad (5)$$

It is also seen from (1) that the impact parameter ρ_δ which separates between the negative and positive signs of the shift coefficients δ is given by:

$$\rho_\delta = \left[\frac{7 \Delta C_{12}}{32 \Delta C_6} \right]^{1/6} = 1.2 \, \rho_o \qquad (6)$$

So that if $\rho_{o\delta} < \rho_\delta$, then δ has a positive sign, but if $\rho_{o\delta} > \rho_\delta$, then it has a negative sign.

RESULTS AND DISCUSSION

To obtain the appropriate sign of $\eta_{ob} = \pm 0.63$ and $\eta_{o\delta} = \pm 1.57$, which should be applied in (4), to give the correct values for the parameters ρ_{ob} and $\rho_{o\delta}$, we proceed as follows:

Knowing ΔC_6 and ΔC_{12}, η_E can be obtained using (5).

1- If $\eta_E > -0.63$ then, ρ_{ob} and $\rho_{o\delta}$ are obtained with the positive sign of η_{ob} and $\eta_{o\delta}$ [see Fig. (1)]. In this case $\rho_{o\delta} < \rho_{ob} < \rho_E$.

2- When $\eta_E \leq -1.57$, then ρ_{ob} and $\rho_{o\delta}$ will be taken with the negative sign of η_{ob} and $\eta_{o\delta}$ [see Fig. (2)]. In this case $\rho_{ob} > \rho_{o\delta} > \rho_E$.

3- If $\eta_E < -0.63$, but > -1.57, then ρ_{ob} will be taken with the negative sign of η_{ob} so that, $\rho_{ob} > \rho_E$, but $\rho_{o\delta}$ will be taken with the positive sign of $\eta_{o\delta}$, so that $\rho_{o\delta} < \rho_E$. [see Fig. (3)].

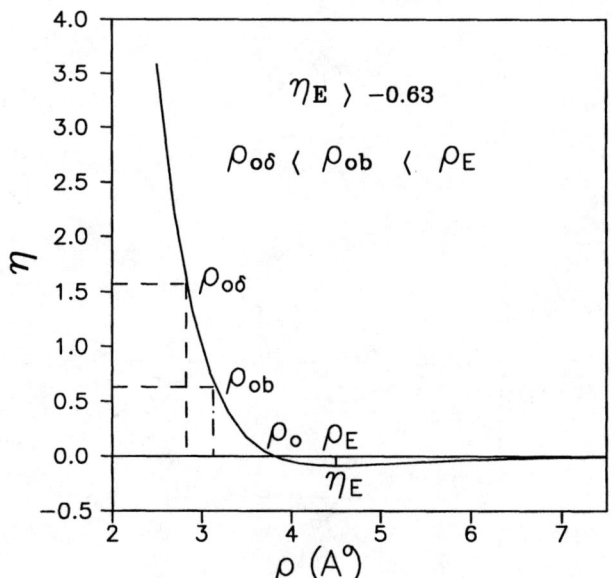

FIGURE 1. The phase shift η as a function of the impact parameter ρ. when $\eta_E > -0.63$.

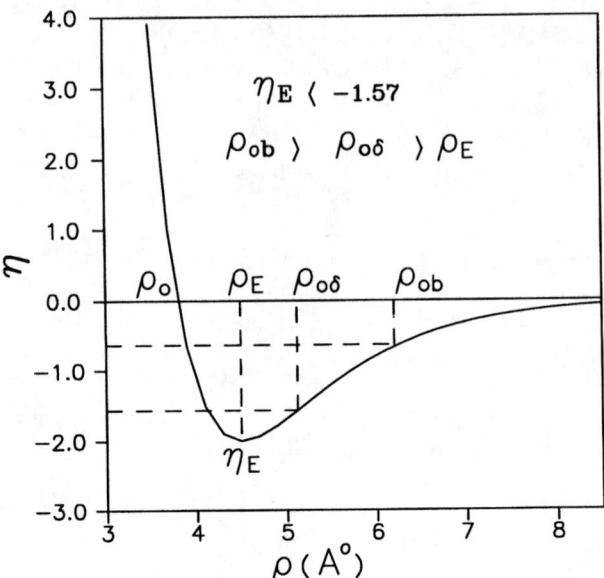

FIGURE 2. The phase shift η as a function of the impact parameter ρ, when $\eta_E \leq -1.57$.

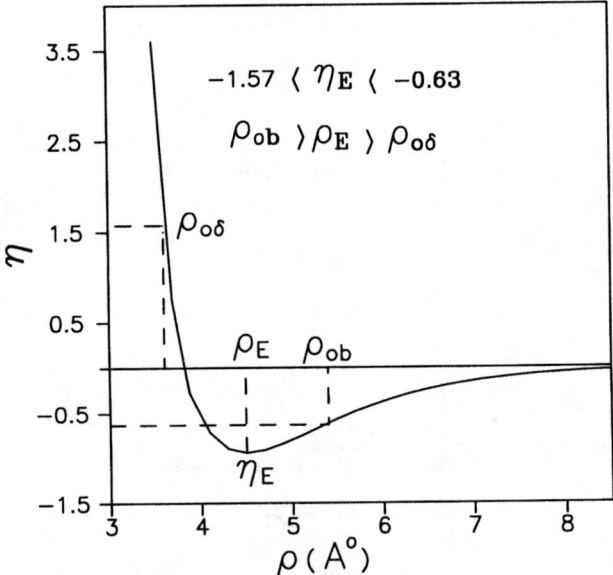

FIGURE 3. The phase shift η as a function of the impact parameter ρ, when $-1.57 < \eta_E < -0.63$.

The calculated coefficients β_c and δ_c for different interactions are illustrated in the table with the corresponding Hindmarsh values β_H and δ_H.

TABLE I. The calculated values of pressure broadening β_c and shift δ_c coefficients with the corresponding Hindmarsh values β_H and δ_H in units 10^{-20} cm^{-1}/atom cm^{-3} for Ar, Ne, Th, Hg, Cd and Zn perturbed by inert gases. The values of ρ are in (Å), ΔC_6 in units 10^{-32}cm^6 rad s^{-1} and ΔC_{12} in units 10^{-74} cm^{12} rad s^{-1} are taken from the labeled references.

Perturber	Potential Parameters		Phase	Impact Parameters				Broadening		Shift	
	$\Delta C6$	ΔC_{12}	η_E	ρ_δ	ρ_E	ρ_{ob}	$\rho_{o\delta}$	β_H	β_C	δ_H	δ_C
Ar line 703 nm , T = 330 K											
Ar[3]	531.8	83750	-0.060	18.0	24.7	18.8	17.6	7.040	6.930	0.200	0.200
Ne[3]	130.0	55870	-0.005	21.3	29.2	18.4	17.0	8.250	8.170	0.990	1.000
He[3]	70.50	48190	-0.001	23.0	31.7	17.3	16.0	13.89	13.75	1.890	1.780
Ne line 540 nm , T = 330 K											
Ne[4]	5.400	1.2400	-0.100	6.09	8.34	6.52	6.15	1.153	1.170	-0.028	-0.028
He[4]	2.900	0.8800	-0.025	6.37	8.73	6.22	5.79	1.870	1.870	0.150	0.150
Th line 377.68 nm , T = 860 K											
Xe[5,6]	145.2	30.910	-5.290	5.99	8.21	14.1	11.7	3.270	3.160	-1.100	-1.170
Kr[5,6]	49.40	20.610	-1.900	6.08	9.49	12.4	10.0	2.840	2.840	-1.400	-1.100
Ar[5,6]	58.70	15.250	-1.170	6.20	8.49	7.72	7.12	1.900	1.949	-1.180	-1.200
Ne[5,6]	14.30	6.5800	-0.130	6.82	9.34	7.42	7.00	1.640	1.800	-0.080	-0.090
Hg line 253.6 nm , T = 860 K											
Hg[7]	11.700	0.0572	-10.98	3.19	4.38	8.76	7.29	1.079	1.085	-0.396	-0.394
Xe[8]	0.2970	0.0010	-0.270	3.01	4.28	3.52	3.36	0.207	0.202	-0.037	-0.036
Kr[9]	0.7920	0.0080	-0.300	3.61	4.94	4.08	3.93	0.304	0.314	-0.068	-0.067
Ar[10]	0.0630	0.4970	-0.003	5.09	6.97	4.20	3.87	0.438	0.443	0.055	0.059
Ne[10]	0.7450	0.0640	-0.027	5.16	7.05	5.07	4.72	0.850	0.847	0.064	0.066
Cd line 326.1 nm , T = 860 K											
Cd[11]	7.2200	0.0451	-4.080	3.33	4.56	7.45	6.15	1.140	1.070	-0.365	-0.396
Xe[12]	1.3900	0.0110	-0.670	3.46	4.75	4.50	3.90	0.359	0.370	-0.154	-0.156
Kr[13]	0.2190	0.0127	-0.018	4.85	6.61	4.58	4.26	0.428	0.437	0.039	0.041
Ar[12]	0.2197	0.0306	-0.458	3.82	5.21	4.37	4.22	0.549	0.510	-0.180	-0.175
Ne[12]	0.3740	0.0388	-0.760	3.63	4.96	4.63	4.10	0.775	0.753	-0.355	-0.370
He[12]	0.3170	0.0106	-0.010	4.43	6.03	4.04	3.75	1.190	1.200	0.123	0.127
Zn line 307.6 nm , T = 860 K											
Zn[14]	1.7875	0.0199	-0.475	3.67	5.02	4.22	4.07	0.491	0.441	-0.163	-0.161
Xe[15]	10.532	0.1926	-2.558	3.98	5.46	9.36	7.71	1.220	1.188	-0.665	-0.705
Kr[16]	2.6630	0.0028	-5.369	2.47	3.39	5.91	4.95	0.829	0.817	-0.268	-0.291
Ar[16]	0.1650	0.0072	-0.012	4.61	6.31	4.25	3.95	0.514	0.515	0.051	0.053
Ne[17]	0.1337	0.0030	-0.014	4.12	5.63	3.83	3.56	0.529	0.531	0.052	0.054

4. CONCLUSIONS

The following conclusion are based on the fact that the calculated coefficients β_C and δ_C by simple analytical formulas obtained by the authors before [1], when the Lennard – Jones potential is applied are in good agreement with the corresponding coefficients obtained numerically by the other authors. This has led to the following consideration:

1- New impact parameters ρ_o, ρ_E, ρ_{ob}, $\rho_{o\delta}$ and ρ_δ which are firstly defined and obtained.

2- The impact parameters ρ_{ob}, and $\rho_{o\delta}$ responsible receptively for the broadening and shift of spectral lines are different, and depend strongly on the values of ΔC_6, ΔC_{12} and the values and signs of the phase shifts η_{ob} and $\eta_{o\delta}$ due to the broadening and shift respectively.

3- The phases (η_{ob} and $\eta_{o\delta}$) at which the broadening and shift respectively start to occur are also different in values. They may have positive or negative signs depending on the colliding particles.

4- The sign of phases η_{ob} or $\eta_{o\delta}$ depends on the value of the phase η_E at the equilibrium position of phases as follows:

(a) If $\eta_E > -0.63$ then , the broadening and shift impact parameters ρ_{ob} and $\rho_{o\delta}$ are obtained with the positive sign of η_{ob} and $\eta_{o\delta}$. In this case $\rho_{o\delta} < \rho_{ob} < \rho_E$.

(b) If $\eta_E \leq - 1.57$, then ρ_{ob} and $\rho_{o\delta}$ will be taken with the negative sign of η_{ob} and $\eta_{o\delta}$. In this case $\rho_{ob} > \rho_{o\delta} > \rho_E$.

(c) If $\eta_E < - 0.63$, but $> - 1.57$, then ρ_{ob} will be taken with the negative sign of η_{ob}, so that $\rho_{ob} > \rho_E$, but $\rho_{o\delta}$ will be taken with the positive sign of $\eta_{o\delta}$, so that $\rho_{o\delta} < \rho_E$.

5- The impact parameter ρ_δ which separates between the positive and nagative signs of the shift of spectral line is obtained and given by a simple formula (6), so that if $\rho_{o\delta} < \rho_\delta$, then the line is shifted to the blue wing (δ = positive value) , but if $\rho_{o\delta} > \rho_\delta$, then it is shifted to the red wing (δ = nagative value).

6- From the foregoing conclusions, the simple analytical formulas obtained in [1] with the mentioned values of η_E can be used for calculating the coefficients β and δ instead of the numerical calculations of Hindmarsh et al. [2].

REFERENCES

1- Helmi M. S. and Roston G. D., Physica Scripta, **62**, 36 , (2000).
2- Hindmarsh W.R., Petford A. D., Smith G., Proc. Roy. Soc. **297A**, 296, (1967).
3- Bielski A., Wawrzynski J., and Wolnikowski J., J. Acta Phys. Polonica, **67A**, 621, (1985).
4- Bielski A., Dokurno W., Szudy J., and Wolnikowski J., J. Physica **101C**, 113, (1980).
5- Dygdała R. S., J. phys. B: At. Mol. Opt. Phys. **21** , 2039, (1988).
6- Dygdała R. S., Bobkowski R. and Lisicki E., J. phys. B: At. Mol. Opt. Phys. **22** , 1563, (1989).
7- Czuchaj E., Rebentrost F., Stoll. H, Preuss H., Chem.Phys., **214**, 277, (1997).
8- Okunishi M., H., Nakazawa, Yamanouchi K. and Tsuchiya S., J.Chem.Phys., **93**, 7526, (1990).
9- Kurosawa T., Ohmori K., Chiba H., Okunishi M., Ueda K. and Sato Y., J. Chem. Phys. , **108** , 8101, (1998).
10- Petzold H. C. and Behmenburg W., Z. Naturf., **33a**, 1461, (1978).
11- Helmi M. S. , Grycuk T. and Roston G. D., Spectrochimica Acta , **B 51**, 633, (1996).
12- Czuchaj E. and Stoll H., Chem.Phys., **248**, 1, 1999.
13- Czajkowski M., Bobkowski R. and Krause L.. Phys. Rev. **A44**, 9, 5730, (1991).
14- Czajkowski M. and KoperskJ J. Spectrochimica Acta, A **55**, 2221, (1999).
15- Inguar Wallace, John G. Kaup and Breckenridge W. H., **95**, 8060, (1991).
16- Inguar Wallace, Jarral Ryter and Breckenridge W. H., **96**, 1, (1992).
17- Koperski J. and Czajkowski M., Phys. Rev. A , **62**, 012505, (2000).

Broadening And Shift of The Intercombination Spectral Line of The Second Group Elements (Hg, Cd and Zn) Perturbed by Inert Gases

G. D. Roston , Z. F. Ghatass* and M. S. Helmi

Department of Physics - Faculty of Science - Alexandria University - Alexandria , Egypt
gamal_daniel@yahoo.com

*Institute of Graduate Studies and Research - Alexandria University - Alexandria , Egypt

Abstract. The Van der Waals and Lennard - Jones potential parameters for the ground and exited states of the second group elements (Hg, Cd and Zn) perturbed by inert gases (Xe, Kr, Ar, Ne and He) have been calculated using coulomb approximation. The basis of the Lindholm –Foley impact theory and the calculated potential parameters are used for calculating the collision broadening and shift coefficients β and δ for the intercombination lines 253.7, 326.1 and 307.6 nm of Hg, Cd and Zn atoms respectively when they are colliding with inert gas atoms . The calculating coefficients of β and δ are compared with their corresponding experimental values, taken from literatures.

1. INTRODUCTION

The study of pressure broadening and shift of spectral lines is an important fundamental process strongly related to the interaction potentials between the radiating and perturbing atoms. It is well known that at low densities the impact approximation is valid and the line shape can be described by the Lorentzian profile with the half width and shift proportional to the density of the perturbing gas [1].

The aim of this work is to calculate the interatomic potential parameters C_6 and C_{12} between the radiating atoms (Hg, Cd and Zn) and the perturbing inert gas atoms (Xe, Kr, Ar, Ne and He) for the ground and excited states of the radiating atoms, and apply them to obtain the broadening and shift coefficients β and δ for the intercombination lines 253.7, 326.1 and 307.6 nm of Hg, Cd and Zn atoms respectively using Lindholm – Foley impact theory. The calculated theoretical data for β and δ in case of Van der Waals and Lennerd – Jones potentials are compared with their corresponding experimental results.

CP645, *Spectral Line Shapes: Volume 12, 16th ICSLS*, edited by C. A. Back

2. THEORETICAL BACKGROUND

2. 1. Van der Waals Potential

The Van der Waals potential is given by:

$$V(R) = -C_6 / R^6 \qquad (1)$$

Where C_6 is the Van der Waals force constant in the given state and R is the interatomic separation between the radiating and perturbing atoms. To calculate the Van der Waals force constant C_6, the approximate formula given by Unsöld [2] has been used:

$$C_6 = \alpha e^2 \langle r^2 \rangle \qquad (2)$$

Where e is the electron charge, α is the dipole polarizability of the perturbing atom (The values of α for the inert gases are taken from [3 - 4]) and $\langle r^2 \rangle$ is the quantum mechanical mean value of r^2 in a given state of the radiating atom. The values of $\langle r^2 \rangle$ have been calculated for both the ground and exited states using the Coulomb approximation [5]:

$$\langle r^2 \rangle = (1/2) a_o^2 n_{eff}^2 [5 n_{eff}^2 + 1 - 3\lambda(\lambda + 1)] \qquad (3)$$

where a_o is the Bohr radius , n_{eff} and λ are the effective and azimuthal quantum numbers. The calculated values of n_{eff} and $\langle r^2 \rangle$ for the ground and excited states of Hg, Cd and Zn respectively are presented in table I.

TABLE I. The effective quantum numbers n_{eff} and the mean values of $r^2 (10^{-20} m^2$) For the ground and excited states of Hg. Cd and Zn, leading to the intercombination lines of these atoms. The Hindmarsh radii R_A for these states.

Radiating atom	Hg			Cd			Zn		
	n_{eff}	$\langle r^2 \rangle$	R_A	n_{eff}	$\langle r^2 \rangle$	R_A	n_{eff}	$\langle r^2 \rangle$	R_A
Ground state	1.141	1.371	5.474	1.23	1.815	4.617	1.204	1.672	3.946
Excited state	1.566	2.496	6.177	1.62	2.983	5.210	1.594	2.741	4.454

2. 2. The Lennerd - Jones Potential

The Lennard - Jones potential is given by:

$$V(R) = \frac{C_{12}}{R^{12}} - \frac{C_6}{R^6} \qquad (4)$$

Where C_6 and C_{12} are the attractive and repulsive potential parameters and R is the interatomic separation between the radiating and perturbing atoms. For calculating the c_{12} constant, Hindmarsh [6] formula has been used as:

$$C_{12} = q(R_A + R_B)^2 \qquad (5)$$

Where $q = (0.9 \pm 0.3)10^{-16}$ erg and R_A or R_B denotes the so –called Hindmarsh radius which is defined as the distance from the nucleus of atom A or B at which the unperturbed radial charge density has the value 0.012 a.u. Those are calculated using the Froese-Fischer program [7] for the Hartree-Fock wave functions (see [5, 8]). The calculated values of R_A for the states of Hg, Cd and Zn are presented in table I. The calculated values of C_6 and C_{12} for the ground (g) and excited states (e) with the potential difference ΔC_6 and ΔC_{12} are presented in table II.

TABLE II. The Van der Waals (C_6 x 10^{-32} cm^{-1} cm^6) and Lennard-Jones (C_{12} x 10^{-74} cm^{-1} cm^{12}) potential parameters for the ground and exited states of Hg, Cd and Zn perturbed by inert gases (Xe, Kr, Ar, Ne and He), with the potential parameters differences ΔC_6 and ΔC_{12}.

Radiated Atoms	Perturbed Atoms	Ground State Parameters		Excite State Parameters		ΔC_6	ΔC_{12}
		C_6^g	C_{12}^g	C_6^e	C_{12}^e		
Hg	He	0.097804	0.011719	0.178041	0.038631	0.080241	0.026912
	Ne	0.188183	0.021318	0.342575	0.066484	0.154392	0.045166
	Ar	0.782976	0.060699	1.425370	0.172834	0.642379	0.112145
	Kr	1.182252	0.091494	2.152128	0.251887	0.969952	0.160398
	Xe	1.928464	0.158104	3.510715	0.416813	1.582101	0.258708
Cd	He	0.129433	0.002286	0.212753	0.007255	0.083325	0.004968
	Ne	0.249034	0.004528	0.409372	0.013512	0.160323	0.008986
	Ar	1.036191	0.014789	1.703139	0.039995	0.667069	0.025207
	Kr	1.564594	0.023441	2.571687	0.061153	1.007229	0.037712
	Xe	2.552068	0.043189	4.19514	0.107625	1.642922	0.064436
Zn	He	0.119256	0.000533	0.195457	0.001630	0.076212	0.001096
	Ne	0.229460	0.001150	0.376094	0.003289	0.146642	0.002137
	Ar	0.954724	0.004306	1.564745	0.011075	0.610127	0.006769
	Kr	1.441579	0.007167	2.362813	0.017748	0.921250	0.010581
	Xe	2.351494	0.014510	3.854211	0.033162	1.502747	0.018652

2. 3. Calculation of the Broadening and Shift Coefficients β and δ

The collision broadening and shift coefficients β and δ (in angular frequency units) for the intercombination lines 253.7, 326.1 and 307.6 nm of Hg, Cd and Zn respectively using both Van der Waals and Lennerd – Jouns potentials paramters have been calculated as follows:

1- In case of the Van der Waal's potential the pressure broadening β and shift δ coefficients are calculated using Behmenburg and Dygdała formula [1, 9] as:

$$\beta = 8.16 \ \overline{V}^{-3/5} (\Delta C_6)^{2/5}$$

and

$$\delta = -0.357 \ \beta$$

(6)

Where $\overline{V}$ is the mean relative velocity which is related to the temperature T, Boltezmann constant K and the reduced mass μ of the radiator and perturber by the formula:

$$\overline{V} = \sqrt{\frac{8KT}{\pi\mu}}$$

(7)

2- In case of Lennard - Jones potential, if ΔC_6 and ΔC_{12} are the potential parameters difference between the ground and excite states then, β and δ are calculated using Hindmarsh et al. [6]:

$$\beta = 8\pi \ (3\pi/8)^{2/5} \overline{V}^{3/5} \ (\Delta C_6)^{2/5} \ B(\alpha)$$

(8)

$$\delta = 2\pi \ (3\pi/8)^{2/5} \overline{V}^{3/5} \ (\Delta C_6)^{2/5} \ S(\alpha)$$

(9)

where the broadening and shift functions B(α) and S(α) are given by :

$$B(\alpha) = \int_0^\infty x \ \sin^2 0.5(\alpha \ x^{-11} - x^{-5}) \ dx$$

(10)

$$S(\alpha) = \int_0^\infty x \ \sin(\alpha \ x^{-11} - x^{-5}) \ dx$$

(11)

and

$$\alpha = 0.536 \ \overline{V}^{6/5} (\Delta C_6)^{-11/5} \ \Delta C_{12}$$

(12)

The two functions B(α) and S(α) are tabulated for different values of α from 1×10^{-4} to 1×10^4 by Hindmarsh et al. [6].

3. RESULTS AND DISCUSSION

The calculated values of the collision broadening and shift coefficients β and δ for the intercombination lines 253.7, 326.1 and 307.6 nm of Hg , Cd and Zn respectively based on Van der Waals and Lennerd - Jones potentials with the corresponding experimental values for Hg [3, 5] and Cd [10, 11] are presented Figs.[1-3].

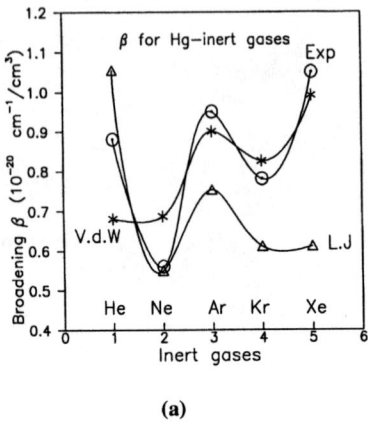
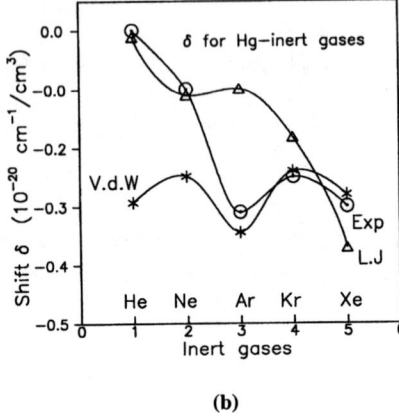

(a)   (b)

Figure. 1. The calculated values of the broadening β and shift δ coefficients (a) and (b) respectively (in units of 10^{-20} cm^{-1}/ atom cm^{-3}) for the intercombination line of Hg with the corresponding experimental values.

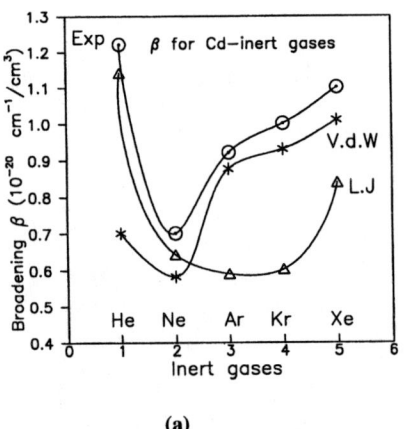
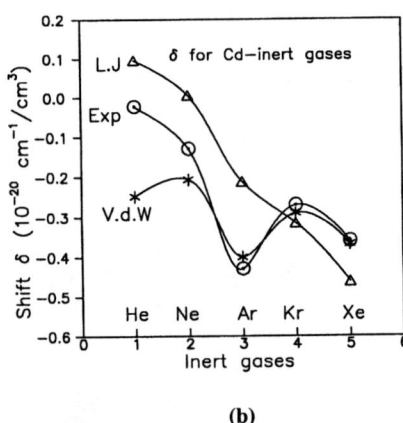

(a)   (b)

Figure. 2. The calculated values of the broadening β and shift δ coefficients (a) and (b) respectively (in units of 10^{-20} cm^{-1}/ atom cm^{-3}) for the intercombination line of Cd with the corresponding experimental values.

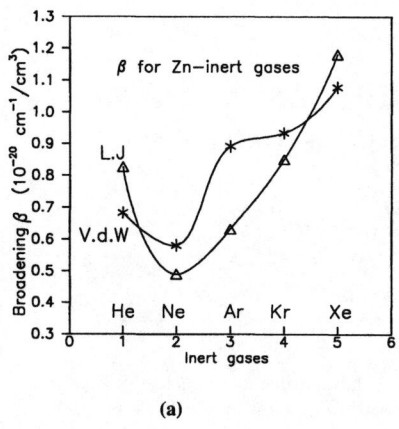 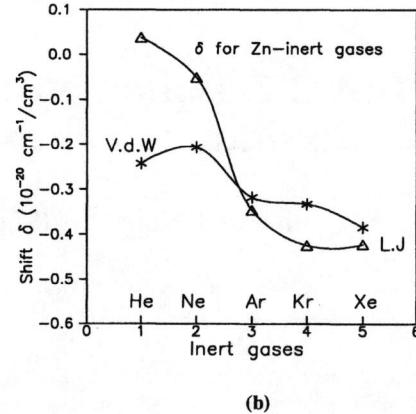

(a)   (b)

Figure. 3. The calculated values of the broadening β and shift δ coefficients (a) and (b) respectively (in units of 10^{-20} cm^{-1}/atom cm^{-3}) for the intercombination line of Zn.

It is seen from the figures that the calculated values of the collision broadening and shift coefficients β and δ due to Van der Waals potential are in agreement with the experimental results for the heavy perturbing atoms but the results due to Lennard – Jones potential are in agreement for light perturbing atoms. This means that Van der Waals potential can be used to describe the interaction of heavy perturbing atoms while Lennard – Jones potential is used for light perturbing atoms.

REFERENCES

1- Behmenburg W. and Schuller F. Phys. Rev. **12C**, 274, 1974.
2- Unsöld A., Physik der sternatmospharen, Springer, Berline, 1955
3- Bielski A. , Brym S. , Ciuyto R. , Domyslawaka J. and Trawiniski R., J.Phys. B:At.Mol.Opt.Phys. **27** , 5863, 1994.
4- Dietz K.J., Dabkiewic P., Kluge H.J., Kuhl T. and Schuessler H.A., J. Phys. B: A. Mol. Phys. **13**, 2749, 1980.
5- Brym S. ,and Domyslawaka J., Physica Scripta. Vol. **52** , 511, 1995.
6- Hindmarsh W. R., Petford A. D. and Smith G., Proc. Roy. Soc.A **297** , 296, 1967.
7- Froese - Fischer C., Comput. Phys. Commun., **4**, 107, 1972.
8- Bielski A. and Wasilewski J., Z. Naturforsch. **35a**, 1249, 1980
9- Dygdała R. S., J. phys.B:At.Mol.Opt.Phys. **21** , 2039, 1988
10- Bautax J. and Lennuier R. , C. D. Acad. Sci., Ser. **B261** , 671, 1965.
11- Bautax J., Schuller F. and Lennuier R.J., J.Phys. **33**, 635, 1972 and **35** , 361, 1974.

Direct Observation of Collisions by Laser Excitation of the Collision Complex

Frank Rebentrost* and Joachim Grosser†

*Max-Planck-Institut für Quantenoptik, 85748 Garching, Germany
†Institut für Atom- und Molekülphysik, Universität Hannover, 30167 Hannover, Germany

Abstract. Laser excitation of atomic collision pairs has developped into a powerful tool for the study and manipulation of atomic and molecular collision pairs. The interference structure in the differential cross sections has been measured for different alkali-rare gas systems, allowing tests of attractive as well as repulsive potential curves with an accuracy of $10 \ \mathrm{cm}^{-1}$ or better. Polarization experiments give access to geometric properties of collision pairs ("collision photography"). For Na + Ne, the variation of the differential cross section with the laser polarization has been used to determine the pair of Condon vectors (the direction of the internuclear axis at the moment of the optical transition) with high accuracy. For atom-molecule collisions, polarization data reveal details of the electronic structure of the short-lived collision complex, for instance the presence of a conical intersection. Polarization can also be used to switch trajectories on or off.

INTRODUCTION

The study of optical collisions has originated from the investigation of the collisional broadening of spectral lineshapes in the quasistatic wing of an atomic transition. A typical case is the optical collision of an alkali atom with a rare gas atom perturber

$$A(n^2 S_{1/2}) + X + h\nu \rightarrow A^*(n^2 P_{1/2,3/2}) + X$$

The motivation followed here is however not so much the spectral lineshape itself which apart from nonadiabatic details is well understood in the single collision limit. Rather we use the fact that the study of optical collisions allows to extract information on the collisional event itself. This information pertains to the dynamics evolving on a subcollisional time scale and provides an unique access to features occurring during the collision.

The basic principle of an optical collision is simple and can be understood by the Condon points associated with a given laser detuning from the atomic resonance. Thus via the detuning one has control over the initial internuclear separation and the electronic potential term, Fig. 1a. This fact has been the basis for the investigation of nonadiabatic interactions occurring after the optical excitation. In earlier experiments using gas cell conditions the spin-orbit coupling determines the final $^2P_{1/2}/^2P_{3/2}$ ratio measured by the emitted fluorescence. Rotational couplings related to the orbital decoupling/locking mechanism of the electronic angular momentum to the molecular axis are seen in the corresponding multipoles (orientation and alignment) of the fluorescence. In these experiments the optical collisions occur isotropically in space and averages over the impact conditions and thermal collision energies are observed.

CP645, *Spectral Line Shapes: Volume 12, 16th ICSLS*, edited by C. A. Back
© 2002 American Institute of Physics 0-7354-0100-4/02/$19.00

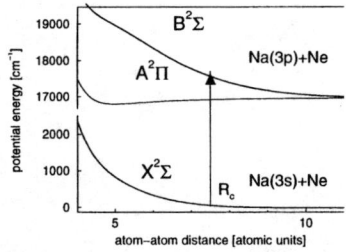 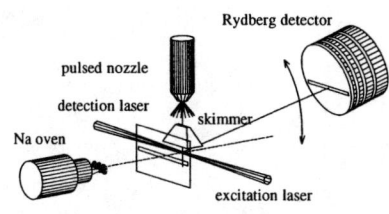

FIGURE 1. a) (left) Potential scheme for an optical collision in NaNe. b) (right) Experimental setup for differential detection of optical collisions.

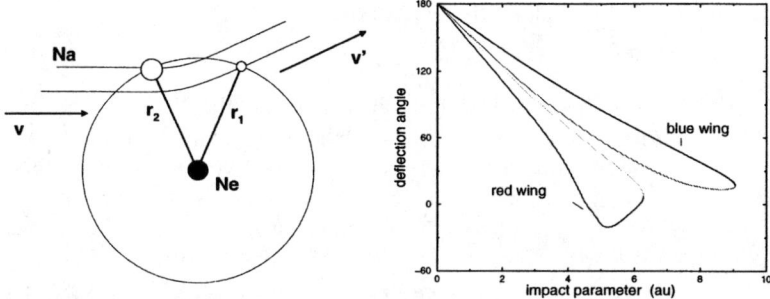

FIGURE 2. a) (left) Condon vector geometry for an optical excitation of the collision pair. The deflection is in the direction of v'. b) (right) Classical deflection function for excitation in the blue and red wing. In the latter case up to four trajectories may occur with the same scattering angle.

In 1994 it became possible to investigate optical collisions under the conditions of a molecular beam experiment. The experimental setup which since then has been considerably refined is shown in Fig. 1b. Essentially it consists of two atom or molecule beams and the two lasers used for pumping and detection. Of course such an experiment must be designed to overcome the low signal intensity related to the short collision duration (1 ps) as well as the low absorption probability in the far wing of the atomic resonance. Due to its short lifetime (Na*: 16 ns) a direct differential detection of A* is not possible. The differential detection scheme is therefore based on a conversion to a high Rydberg level which lives long enough to travel from the collisional region to the detector.

The advantages of the molecular beam method (well-defined initial and final velocities or energies, control of impact parameter by the scattering angle) have been widely used in many fields [1]. In combination with the optical excitation scheme used here one can even go beyond this and gets access to geometric properties of a collisional system. This new aspect will be discussed in Sec. 3 for atomic and molecular colliders.

Using the situation of a single Condon radius that is characteristic for a repulsive upper state as in Fig. 1a the following simple geometric picture arises for an optical collision with differential detection, Fig. 2a. The optical collision is described by classical trajectories which switch at the Condon points from the lower to the upper electronic term.

207

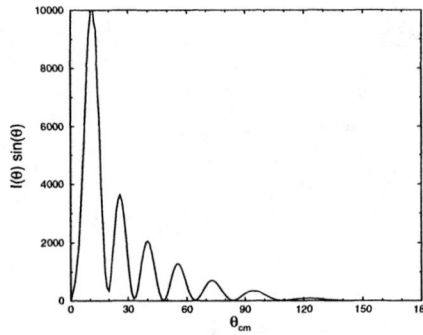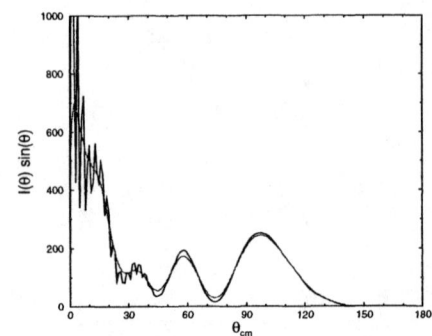

FIGURE 3. Differential optical collision cross section for NaNe. a) (left) Excitation of the $B^2\Sigma$ state (blue wing). b) (right) Excitation of the $A^2\Pi$ state (red wing).

These have the typical deflection functions shown in Fig. 2b. With differential detection and energy resolution of the collision this may lead for a given detuning to just a single pair of trajectories scattered into a given direction. These two paths have different impact parameters, phases and excitation weights corresponding to optical excitation occurring either in ingoing or outgoing direction. This picture forms the basis of an almost quantitative description of the differential optical collision cross sections [2] and the geometric analysis discussed in Secs. 2 and 3. A complete quantum description of optical collisions is also possible for atomic collisions and is capable to treat in addition the nonadiabatic interactions occurring after excitation [2].

At present the experimental research using the molecular beam approach to optical collisions is along the following lines

Differential optical collisions.
The oscillatory structure in the differential cross sections may be used to manipulate or even control the collision [3]. Probing of potentials with high accuracy of $\sim 10\,\mathrm{cm}^{-1}$ or better has been demonstrated for repulsive [4] and attractive potentials.

Collision geometry using the polarization of the excitation laser.

Final state probing using the detection laser.
This involves the study of spin-orbit or rotational induced nonadiabatic couplings occurring after optical excitation [5, 6, 7]. Examples for the investigation of molecular fine-structure changing collisions involve optical collisions with N_2, N_2O and C_2H_2 [8].

DIFFERENTIAL OPTICAL COLLISIONS

The typical feature of the differential collision cross sections is their oscillatory structure, Fig. 3. These Stueckelberg oscillations are due to the interference from the two paths shown in Fig. 2a which have different phases. Note that the paths also differ in their im-

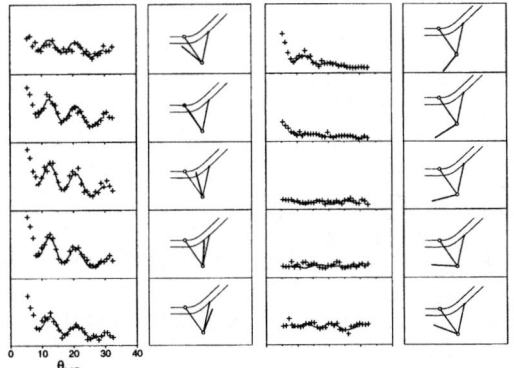 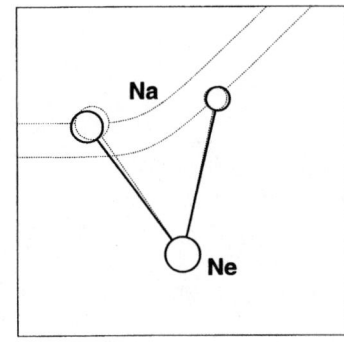

FIGURE 4. a) (left) Variation of the interference structure with the laser polarization. b) (right) Condon vector geometry obtained from experiment, comparison with theory.

pact parameters and sample different regions of the potentials. The simple oscillatory structure seen in Fig. 3a is however related to the interference of just two trajectories with a common Condon point under the conditions chosen in the experiment. The situation gets immediately more complex if more than two trajectories are possible, as *e.g.* for excitation in the red wing where the transition involves the attractive $A\Pi$ state. As an example Fig. 3b shows a calculated differential cross section in this case. The rapid oscillations seen in the range of smaller angles are due to interference from trajectories with different sign of the deflection function and will be washed out under the resolution of the actual experiment. The resulting smoothened structures are then comparable with the experiment.

COLLISION GEOMETRY

The molecular beam approach to optical collisions with differential detection and energy resolution of the colliders implies a drastic reduction of the actually observed collision events. This selection is the key feature for the possibility to observe the oscillatory structure and from this obtain access to the collision geometry at the instant of the optical excitation. For the situation of Fig. 2a the collision plane is in the laboratory plane and also the Condon vectors have well-defined orientation to this plane. In particular for a $\Sigma - \Sigma$ transition they are in the collision plane with given angle to the direction of the colliders.

In Fig. 4a the variation of an experimental differential cross is shown when one rotates the laser polarization in the collision plane. Clearly it is seen that the oscillatory structure disappears when the polarization is perpendicular to one of the Condon vectors. From the interference pattern the actual geometric arrangement of the Condon vectors can be reconstructed with high accuracy, Fig. 4b [5]. This also includes the relative transition probabilities associated with a given Condon vector.

A very promising aspect of this geometric investigation of collision complexes is the

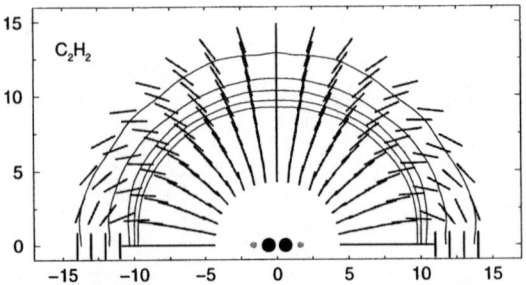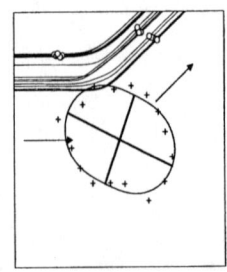

FIGURE 5. a) (left) Na+C_2H_2: Transition dipoles for X$\rightarrow$ 2A'. These show strong variation with the orientation of the molecule due to the long-range quadrupole-quadrupole interaction and the presence of a conical intersection for collinear geometry. b) (right) Geometry of molecular collision pairs: Experimental excitation tensor for Na+C_2H_2 optical collisions (obtained by rotating the laser polarization in the scattering plane). Also shown are calculated trajectory bundles with the Condon points o related to an optical transition on either the ingoing or outgoing direction.

fact that a similar polarization experiment is also useful for molecules. In this case the observed signal will also reflect the average orientation of the molecule at the instant of the optical excitation. This is because the transition dipole depends on this orientation as *e.g.* shown for C_2H_2 in Fig. 5a. In the same way the simple trajectory pairs existing for atomic collision pairs have to be replaced by trajectory bundles. This is indicated in Fig. 5b, where also an example for an experimental excitation tensor is shown. These additional complexities characteristic for molecular colliders lead to the absence of any oscillatory structure in the differential cross sections. Despite of this the measured excitation tensor can be used to derive specific information on the Condon vectors. By varying the detuning the excitation occurs in different regions of the intermolecular potentials. This is reflected in the excitation tensor through the very different behaviour of the transition dipoles for the long-range quadrupole-quadrupole and the short-range exchange interaction. The interplay of these interactions leads to a conical intersection that can be visualized by the present experiment.

REFERENCES

1. "Atomic and Molecular Beams: The State of the Art 2000", ed. R. Campargue, Springer, 2001.
2. F. Rebentrost, S. Klose and J. Grosser J, Eur. Phys. D **1**, 277 (1998).
3. J. Grosser, O. Hoffmann and F. Rebentrost, see ref. [1], p. 485.
4. J. Grosser, O. Hoffmann and F. Rebentrost, J. Phys. B **33**, L577 (2000).
5. J. Grosser, O. Hoffmann and F. Rebentrost, Europhys. Lett. **58**, 209 (2002).
6. R. Goldstein, J. Grosser, O. Hoffmann, V. Schumann, D. Wößner, M. Jungen and M. Lehner, J. Chem. Phys. **114**, 2144 (2001).
7. J. Grosser, O. Hoffmann, F. Schulze Wischeler and F. Rebentrost, J. Chem. Phys. **111**, 2853 (1999).
8. C. Figl, A. Grimpe, R. Goldstein, J. Grosser, O. Hoffmann and F. Rebentrost, to be published.

Absorption spectra of broadened sodium resonance lines in presence of rare gases

H.-K. Chung*, M. Shurgalin† and J. F. Babb**

*Lawrence Livermore National Laboratory, Livermore CA 94550
†OmniGuide Communications Inc., Cambridge, MA 02139
**ITAMP, Harvard-Smithsonian Center for Astrophysics, Cambridge, MA 02138

Abstract. The pressure broadening of alkali-metal lines is a fundamental problem with numerous applications. For example, the sodium resonance lines broadened by xenon are important in the production of broad spectra emitted in the HPS (High-Pressure Sodium) lamp and they potentially can be used for gas condition diagnostics. Broadened absorption lines of alkali-metal atoms are prominent in the optical spectra of brown dwarfs and understanding the broadening mechanism will help elucidate the chemical composition and atmospheric properties of those stars.

The far-line wing spectra of sodium resonance lines broadened by rare gases are found to exhibit molecular characteristics such as satellites and hence the total absorption coefficients for vapors of Na atoms and perturbing rare gas atoms can be modeled as Na-RG (rare gas) molecular absorption spectra. In this work, using carefully chosen interatomic potentials for Na-RG molecules we carry out quantum-mechanical calculations for reduced absorption coefficients for vapors composed of Na-He, Na-Ar, and Na-Xe. Calculated spectra are compared to available experimental results and the agreement is good in the measured satellite positions and shapes.

INTRODUCTION

Line broadening of sodium resonance lines in presence of perturbing atoms have been extensively studied over several decades. Due to interatomic interaction between Na atom and perturbers, far line wings of the broadened lines exhibit characteristics of molecular spectra. Therefore, experimental methods such as laser-induced fluorescence, emission and absorption spectroscopy have been used to study properties of interatomic interactions such as interaction potentials. In addition to the fundamental studies of molecular physics, thermal emission and absorption spectra of sodium resonance lines are studied in the context of various applications such as the modeling of high-pressure sodium lamps and optical diagnostics of combustion and chemical reactions.

Theoretically, the far line wing spectra has been studied with classical and semi-classical methods. These methods give a good description of gross features of absorption and emission spectra. However, recent high-precision measurements of absorption coefficients [1] reveal more detailed features such as ro-vibration lines and satellite structures in the far line wings of sodium resonance lines when broadened by the same or different perturbing atoms requiring more precise theoretical descriptions. Accurate quantum-mechanical methods in molecular spectroscopy are not usually applied to the spectral calculations for high temperature gas partly because the uncertainty of available molecular potentials is comparable to that of the classical and semi-classical methods and partly

CP645, *Spectral Line Shapes: Volume 12, 16th ICSLS*, edited by C. A. Back
© 2002 American Institute of Physics 0-7354-0100-4/02/$19.00

TABLE 1. Molecular potentials used in the calculations

Rare gas	$X^2\Sigma^+$, $A^2\Pi$ and $B^2\Sigma^+$ potentials
He	Theodorakopoulosis and Petsalakis [8]
Ar	Kerner *et al* [9]
Xe	Baumann *et al.* ($X^2\Sigma^+$ and $A^2\Pi$) [10] Düren-type potential function ($B^2\Sigma^+$) [11, 7]

because of the computational limitations resulting from the difficulty of accounting for all the partial waves contributing to the total spectra. Recently, however, accuracies for atomic and molecular data have been greatly improved by utilizing various methods such as laser spectroscopy, photoassociation spectroscopy, Feshbach resonance, and ultracold collisions, and also recent advances in computer speed and techniques of parallel and distributed processing made it possible to consider the convergent number of partial waves involved in calculations. In this work, using the PVM (Parallel Virtual Machine) technique, we carry out quantum-mechanical calculations of absorption coefficients for far line wing spectra of sodium resonance lines broadened by rare gases.

QUANTUM-MECHANICAL METHOD

When broadened by rare gases, absorption coefficients of sodium resonance lines arising from 3s to 3p transitions can be modeled as molecular spectra of transitions within dimolecules composed of sodium and rare gas atoms. Absorption transitions occur from the lower molecular states which approach the absorbing sodium (3s) and the perturbing rare gas atomic states at a long-range limit to the upper molecular states which approach the excited sodium (3p) and the spectator rare gas atomic states. Depending on whether the molecular potentials have bound states formed by sodium and rare gas atoms, the total spectra consist of four possible transitions; bound-bound, bound-free, free-bound, free-free transitions. Details of quantum-mechanical formulation of each transition are described in our earlier papers [2, 3].

Most important are the molecular potentials involved in the radiative transitions and a lot of care was taken into using the most accurate molecular potentials of sodium-rare gas di-molecules. Molecular potentials are roughly divided into short-range potentials which can be obtained by *ab initio* calculations or experimental methods such as laser spectroscopy or line broadening measurements and long-range potentials normally described as $V(R) = -C_6/R^6$ where van der Waals forces are dominant. We investigated the potentials available in the literature by calculating theoretical values of thermal gas parameters such as diffusion coefficients and index of refraction of sodium matter waves using those potentials and comparing with the measured values [4, 5]. Measured satellite positions [1, 6, 7] were applied to select and modify molecular potentials as well. Molecular potentials of our choice are listed in Table. 1–2 for Na-He and Na-Ar and Na-Xe molecules. Details of our studies on molecular potentials will be published elsewhere shortly [7].

212

TABLE 2. C_6 values in atomic units for long-range molecular potentials $V(R) = -C_6/R^6$ used in calculations

Rare gas	$X^2\Sigma^+$ potential	$A^2\Pi$ potential	$B^2\Sigma^+$ potential
He	26.2 [12]	49.2 [13]	71.4 [13]
Ar	190.0 [12]	534.7 [9]	717.8 [9]
Xe	455.9 [12]	897.4 [14]	1095.4 [14]

COMPARISONS WITH MEASUREMENTS

Experimentally, the absorption coefficients of sodium vapor at $900(^+10^-50)$K in presence of various rare gases were measured in the range from 400 nm to 850 nm. With the sodium atomic densities obtained accurately using the anomalous dispersion method [1], density-independent reduced absorption coefficients of Na-Na and Na-Rare gases were obtained simultaneously. The measured reduced absorption coefficients (upper curves) for Na-He, Na-Ar and Na-Xe are compared with the calculated values (lower curves) in Fig. 1. The blue wing absorption arises from transitions between the $X^2\Sigma^+$ molecular state to the $B^2\Sigma^+$ molecular state and the red wing from transitions between the $X^2\Sigma^+$ molecular state to the $A^2\Pi$ molecular state. Since the ground electronic states of Na and rare gas di-molecules are weakly bound, most absorption occurs from free-bound and free-free transitions and the absorption spectra should be smooth without ro-vibrational structures[1].

In the blue wings notable features are the primary peaks occurring at 530, 555 and 560 nm for Na-He, Na-Ar and Na-Xe respectively. Quantum-mechanical calculations reproduce the primary peaks at the measured positions. In addition to the primary peaks, there are weaker secondary peaks shown for Na-Ar and Na-Xe and they are also reproduced in calculations. The semi-classical method suggested by Bieniek and Streeter [15, 16] also yields results similar to the quantum-mechanical method, however, other classical and semi-classical methods [17] fail to describe the primary and the secondary peaks altogether. While our results give a overall good agreement, we found that the absolute values of calculated intensities are slightly lower than the measured values and the sources of discrepancies are not known yet. We also found that the secondary peaks require more stringent tests of molecular potentials than primary peaks and the shapes and the positions of the secondary peaks can be applied to refine molecular potentials.

---

[1] The structures in the experimental red wing spectra are believed to be merely the remnants of sodium dimer spectra appearing because reduced absorption coefficients are extracted from spectra of total absorption coefficients for the gas mixtures.

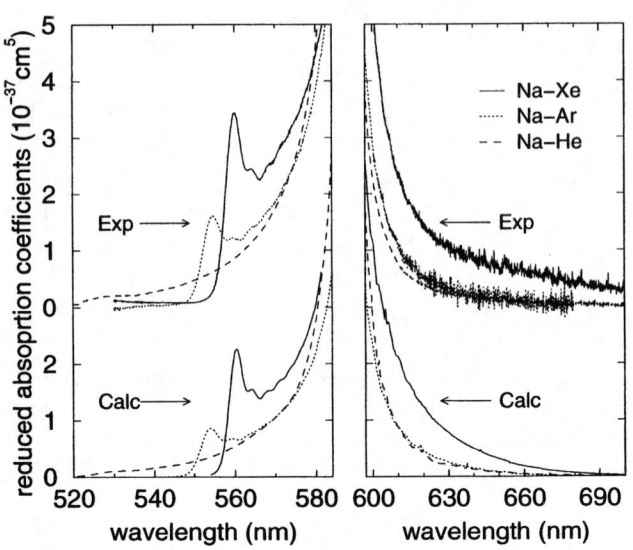

FIGURE 1. Comparisons for Na-rare gas molecules of reduced absorption coefficients between experiment at temperature of 900+10 − 50K (upper curves) and theory at temperature of 900K (lower curves).

APPLICATIONS

Pressure broadening of atomic resonance lines is a key ingredient of various lighting sources. In particular, our research is germane to the spectra of high-pressure sodium lamps [18]. Our measured and calculated absorption coefficients might be utilized for improved emission models.

Another intriguing application of the present research is found in astrophysics. The spectra of brown dwarfs contains alkali-metal atom resonance lines pressure-broadened predominantly by the perturbing gases He and molecular hydrogen [19]. The resonance lines of K and Na are particularly prominent and improved modeling of their broadening may lead to better knowledge of atmospheric composition, effective temperature and gravity [20, 21].

SUMMARY

High precision measurements were used to obtain the absorption spectra of sodium resonance lines in presence of rare gases. The absorption coefficients show interesting peak structures in the blue wings. We demonstrated that quantum-mechanical methods can be applied to describe the peak and the overall features of far wing spectra of sodium resonance lines when broadened by rare gases. Our results show that the structures provide a good test of molecular potentials of sodium-rare gas di-molecules.

ACKNOWLEDGMENTS

This work is supported in part by the NSF, grant PHY-9724713 and a partial support for H.K.Chung is provided under the auspices of the U.S. Department of Energy by University of California Lawrence Livermore National Laboratory under contract No. W-7405-Eng-48.

REFERENCES

1. Shurgalin, M., Parkinson, W. H., Yoshino, K., Schoene, C., and Lapatovich, W. P., *Meas. Sci. Technol.*, **11**, 730–737 (2000).
2. Chung, H.-K., Kirby, K., and Babb, J. F., *Phys. Rev. A*, **60**, 2002 (1999).
3. Chung, H.-K., Kirby, K., and Babb, J. F., *Phys. Rev. A*, **63**, 032516 (2001).
4. Ager III, J. W., and Howard, C. J., *J. Chem. Phys.*, **85**, 3469 (1986).
5. Schmiedmayer, J., Chapman, M. S., Ekstron, C. R., Hammond, T. D., Whinger, S., and Pritchard, D. E., *Phys. Rev. Lett.*, **74**, 1043 (1995).
6. Ch'en, S. Y., and Jr., R. A. W., *Physica*, **27**, 497 (1961).
7. Chung, H.-K., Shurgalin, M., and Babb, J. F., Experimental and theoretical studies of sodium resonance line broadening in presence of rare gases (2002), in progress.
8. Theodorakopoulos, G., and Petsalakis, I. D., *J. Phys. B: At. Mol. Opt. Phys.*, **26**, 4367 (1993).
9. Kerner, C., and Meyer, W., *Ph. D. thesis*, Universität Kaiserslautern, Kaiserslautern, Germany, 1995.
10. Baumann, P., Zimmermann, D., and Brühl, R., *J. Mol. Spectros.*, **155**, 277 (1992).
11. Düren, R., Hasselbrink, E., and Hillrichs, G., *J. Chem. Phys.*, **89**, 2822 (1988).
12. Derevianko, A. (2000), private communication.
13. Leo, P. J., Peach, G., and Whittingham, I. B., *J. Phys. B: At. Mol. Opt. Phys.*, **33**, 4779 (2000).
14. Mahan, G. D., *J. Chem. Phys.*, **50**, 2755 (1969).
15. Bieniek, R. J., and Streeter, T. J., *Phys. Rev. A*, **28**, 3328 (1983).
16. Sato, Y., Nakamura, T., Okunishi, M., Ohmori, K., Chiba, H., and Ueda, K., *Phys. Rev. A*, **53**, 867 (1996).
17. Szudy, J., and Baylis, W. E., *J. Quant. Spectrosc. Radiat. Transfer*, **15**, 641 (1975).
18. de Groot, J. J., and van Vliet, J. A. J. M., *The high pressure sodium lamp*, Philips Technical Library, Kluwer Techn. Boeken, Deventer, 1986.
19. Burrows, A., Hubbard, W. B., Lunine, J. I., and Liebert, J., *Rev. Mod. Phys.*, **73**, 719 (2001).
20. Schweitzer, A., Gizis, J., Hauschildt, P., Allard, F., and Reid, I., *Ap.J.*, **555**, 368 (2001).
21. Burrows, A., Burgasser, A. J., Kirkpatrick, J. D., Liebert, J., Milsom, J. A., Sudarsky, D., and Hubeny, I. (2002), astro-ph/0109227 Ap.J. in press 2002.

Effects of the Anisotropy of the Intermolecular Potential on the Collision-induced Spectra of H_2-H, H_2-He, H_2-H_2, and HD-He

Magnus Gustafsson and Lothar Frommhold

Physics Department, University of Texas at Austin, Texas 78712-1081

Abstract. We calculate collision-induced absorption spectra by including a (weak) electromagnetic radiation field in the molecular scattering Hamiltonian and solving the close-coupled Schrödinger equation numerically. Advanced anisotropic intermolecular potential surfaces and *ab initio* dipole surfaces are used to obtain rototranslational and rotovibrational spectra at various temperatures of H_2 colliding with He, H, and H_2. We also consider the case of HD colliding with He to study interesting interference effects of the permanent dipole of HD with the induced dipole of the supramolecular system HD–He near the $R(j)$ and $P_1(1)$ lines of HD.

INTRODUCTION

Good laboratory measurements of collision-induced absorption (CIA) spectra are known only for a limited number of gases and gas mixtures; at few temperatures; and usually over a limited range of frequencies. Studies of stellar and planetary atmospheres require, however, dependable information of the kind at many temperatures; for rather complex gas mixtures; and over wide frequency bands. Reliable quantum calculations of binary CIA spectra can significantly support applied spectroscopists in their efforts. In the past such calculations were often done in the isotropic potential approximation, which neglects the anisotropy of the intermolecular energy surfaces. This approximation leads to a substantial simplification of the computational task [1], but raises questions concerning the reliability of such calculations, especially when applications call for substantial temperature or frequency extrapolations relative to the laboratory data. We attempt in this work is to account for the anisotropy of the intermolecular potential. This has been done previously in low temperature calculations for H_2–H_2 by Schäfer and McKellar [2]. These focus on sharp spectral features due to the bound $(H_2)_2$ van der Waals molecule rather than the effects of the anisotropy on the CIA continuum. Estimates of the zeroth and first spectral moments using semi-classical expressions that account for the anisotropy of the potential have been reported for H_2–He CIA for a range of temperatures [3, 4]. These showed that the anisotropy indeed affects the CIA spectrum and the corrections due to the anisotropy are on the order of 10%.

In general, for accurate computations of spectral line shapes, broadened by collisions, or collision-induced, a theoretical formulation involving couplings of the colliding particles' internal degrees of freedom is necessary. Specifically, in a calculation of diatomic scattering off any target, without radiation being involved, rotational excitation can only

CP645, *Spectral Line Shapes: Volume 12, 16th ICSLS*, edited by C. A. Back

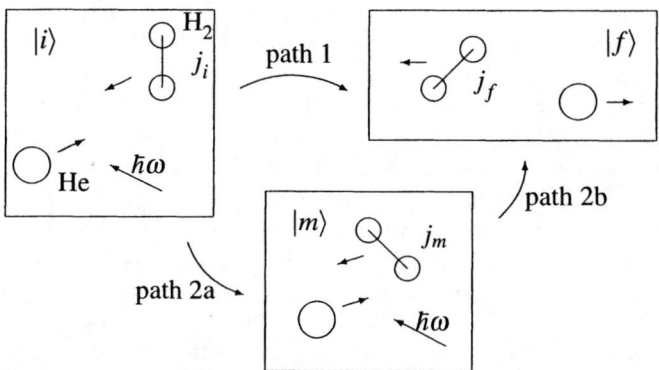

FIGURE 1. An example of an absorption process made possible through the inclusion of the anisotropy of the intermolecular potential for the H_2–He pair. The letter j indicates the rotational quantum number of the hydrogen molecule. The transition can go either direct from the initial channel $|i\rangle$ to the final channel $|f\rangle$ (path 1), or via an intermediate channel $|m\rangle$ (path 2). In this specific example the transition via the intermediate channel implies an inelastic or elastic collision, $j_m \neq j_i$ or $j_m = j_i$, respectively, (path 2a) followed by absorption of the photon (path 2b). The incident photon, which has energy $\hbar\omega$, is unaffected by the transition $|i\rangle \rightarrow |m\rangle$.

occur if the anisotropy of the intermolecular potential is included. Figure 1 illustrates how the anisotropy can affect the absorption process. An isotropic potential does not couple rotations and it only supports radiative transitions via path 1 in figure 1.

Arthurs and Dalgarno [5] developed the formulation of rigid rotor scattering by a structure less particle, e.g., H_2–He collisions, by means of the close-coupling scheme. That theory is essentially what is needed for CIA line shape computations, if one somehow includes the radiation field. The general method for inelastic scattering in radiation fields by means of close-coupling theory was formulated by DeVries and George [6] and later reviewed by Mies [7]. A formulation for atomic line shape calculations was made by Julienne [8], which we have extended for the case of collision-induced absorption line profiles [9] and allowed rotational line profiles [10]. The method, which is commonly called the *radiative close-coupling scheme*, is an approach that includes the quantized radiation field in the Hamiltonian.

CLOSE-COUPLING LINE SHAPE CALCULATIONS

Consider the system of two colliding particles and one incident photon. Limiting oneself to *one* incident photon is the same as saying that only processes with probability linear in the incident light intensity are considered. The colliding particles have a set of quantum numbers corresponding to their internal degrees of freedom. Considering an atom-diatom collision these quantum numbers are the diatomic vibrational and rotational quantum numbers v and j with corresponding energy levels, E_{vj}, which are considered unperturbed during the collision. In other words, the colliding partners are assumed to have strong internal interaction while they interact weakly with each other.

For computation of the collision-induced spectra the Schrödinger equation is integrated using the close-coupling scheme that combines the radiative coupling with a non-perturbative scattering formalism [7]. In this way multi-photon processes may be treated, which however is here of no interest since we focus on single-photon absorption processes. It is indeed a perturbative treatment of the radiation that will be presented, a Fermi's Golden Rule equivalent if you will. Admittedly one can instead adopt the more traditional approach and treat the atom-diatom scattering separate, store the wave functions and then calculate radiative transition probabilities from the dipole matrix elements [2]. In this work the concern is mainly the broad, free-free features of collision-induced spectra which allow for a rather sparse frequency grid, so that the computational overhead resulting from the higher number of coupled channels is acceptable.

The Hamiltonian describing interaction of He with a rotating and vibrating H_2-molecule in the presence of a radiation field is

$$H(\mathbf{r}, \mathbf{R}) = H^{H-H}(\mathbf{r}) - \frac{\hbar^2}{2m}\nabla_{\mathbf{R}}^2 + V(\mathbf{r}, \mathbf{R}) + V^{\text{rad}}(\mathbf{r}, \mathbf{R}) + H^{\text{rad}}, \tag{1}$$

where V is the intermolecular potential of the H_2-He complex which has the reduced mass m. The Hamiltonian of the isolated H_2-molecule is H^{H-H}, the radiative coupling is V^{rad}, and the Hamiltonian for the isolated photon field is H^{rad}. Now, take the Schrödinger equation and integrate out five of the six spatial variables. The evaluation yields the radial equation,

$$\left(E_{v''j''} - \frac{\hbar^2}{2m}\frac{d^2}{dR^2} + \frac{\hbar^2 l''(l''+1)}{2mR^2} + n''\hbar\omega - E \right) F_{\alpha''\alpha}^{n''n}(R;E)$$
$$+ \sum_{\alpha'n'} \left[V_{\alpha''\alpha'}(R)\delta_{n''n'} + W_{\alpha''\alpha'}(R)\delta_{n''\pm 1,n'} \right] F_{\alpha'\alpha}^{n'n}(R;E) = 0, \tag{2}$$

which forms a coupled system of differential equations in the intermolecular spacing R. The set of quantum numbers is $\alpha = (v, j, l, J, M)$ where $\mathbf{J} = \mathbf{j} + \mathbf{l}$ is the total angular momentum, and n is the number of photons of frequency ω. The coupling matrix elements $V_{\alpha''\alpha'}$ and $W_{\alpha''\alpha'}$ will be defined below. Equation (2) can be solved for $F_{\alpha''\alpha}^{n''n}$. The asymptotic boundary condition for $R \to \infty$ yields the scattering matrix element, $S_{\alpha'\alpha}^{n'n}$, from which the desired quantities related to the transition probability between two channels (α, n) and (α', n') are calculated. In principle the summations over α' and n' in equation (2) are infinite, but the close-coupling scheme prescribes a truncation of the basis set. We have limited ourselves to one-photon absorption so that the n'-summation runs only from 0 to 1. We have also truncated the sum over α' by excluding those (closed) channels which hardly affect the radiative transition probabilities at the kinetic energies under consideration.

For an atom-diatom system it is appropriate to expand the intermolecular potential in Legendre polynomials, according to

$$V(\mathbf{r}, \mathbf{R}) = \sum_{\gamma} V_{\gamma}(R, r) P_{\gamma}(\hat{\mathbf{r}} \cdot \hat{\mathbf{R}}) \tag{3}$$

where the subscripts γ are even integers for a homonuclear diatom due to its inversion symmetry. The angular and radial integrals resulting in equation (2) yield the intermolecular potential matrix element

$$V_{\alpha\alpha'}(R) = \sum_{\gamma} V_{\gamma}^{vjj'}(R)\, f_{\gamma}(j,l,j',l';J)\, \delta_{JJ'}\, \delta_{MM'}\, \delta_{vv'}\,. \tag{4}$$

where the kinetic energy of the collisions of interest is assumed to be small enough to prevent non-radiative transitions between different vibrational states. The coefficients f_{γ} originate in the angular integrals and are given in [9, 10].

For the H_2–He complex the radiation-dipole interaction operator is given by

$$W(\mathbf{r},\mathbf{R}) = \sqrt{\frac{2\pi\hbar\omega\phi}{c}}\, \mu_z(\mathbf{r},\mathbf{R}) \tag{5}$$

where $\bar{\mu}$ is the collision-induced dipole moment and ϕ is the radiation flux in photon number per unit area and time. The standard series expansion of $\bar{\mu}$ [1] produces angular and radial integrals similar to those for the potential energy above. We assume linearly polarized light along the z-axis so that only the z-component of the dipole is at work. The angular integral can be evaluated and the desired matrix elements are

$$W_{\alpha\alpha'}(R) = \sqrt{\frac{2\pi\hbar\omega\phi}{c}} \sum_{\lambda L} B_{\lambda L}^{vjv'j'}(R)\, d_{\lambda L}(j,l,J,j',l',J')\, \delta_{MM'}\,, \tag{6}$$

with $d_{\lambda L}$ defined in reference [9]. For HD–He $B_{\lambda L}^{vjv'j'}(R)$ is replaced by $B_{I;\lambda L}^{vjv'j'}(R)\,+$ $B_{A}^{vjv'j'}\delta_{\lambda 1}\delta_{L0}$ where the I indicates the induced dipole and A indicates the permanent (allowed) dipole.

The binary collision-induced absorption coefficient of a gas mixture of the species with densities ρ_1 and ρ_2, respectively, at the temperature T, is calculated from the S-matrix as shown in Ref. [8]. Letting the subscripts i and f designate the initial and final channels, respectively, i.e., channels with one or no photon, yields

$$\alpha(\omega,T) = \rho_1\rho_2\lambda_0^3\, \frac{1-e^{-\beta\hbar\omega}}{\phi h} \sum_i P_{0j_i}(T) \int_0^{\infty} dE_i e^{-\beta E_i} \sum_f \left| S_{fi}(E_i) \right|^2 \tag{7}$$

where $f = (j_f, l_f, J_f)$, $i = (j_i, l_i, J_i)$. λ_0 is the thermal de Broglie wavelength, P_{vj} is the rotovibrational occupancy function for the diatom and E_i is the initial kinetic energy.

Note that, in order to make the calculations of the absorption coefficient (7) less time consuming, three decouplings of the equations (2) can be applied as described in reference [9]. Various trial calculations have been performed for each case to confirm a numerical accuracy at the 1% level or better. For the integration of the coupled Schrödinger equation (2) the COUPLE computer program developed by Mies, Julienne and Sando [13] has been used.

RESULTS

H_2–He: In figures 2 and 3 the heavy, smooth solid lines represent the spectra that we calculate with the *ab initio* dipole surface [9] and the anisotropic intermolecular potential

219

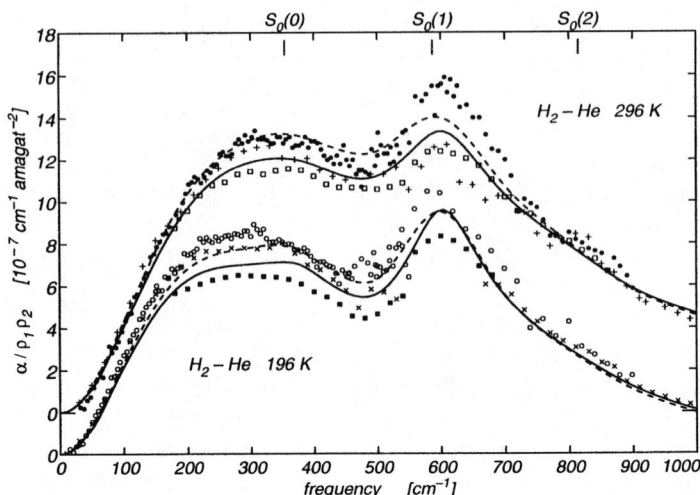

FIGURE 2. The absorption coefficient α, normalized by the helium and hydrogen gas densities, ρ_1 and ρ_2, respectively, as function of frequency in the rototranslational band, at the temperatures of 296 K (upper trace) and 196 K (lower trace, shifted downward one step for better readability of the figure). Solid and dashed curves represent calculations [9] with and without the anisotropy of the intermolecular potential accounted for, respectively. Also shown are the measurements by Birnbaum [11] at 293 K (•) and 195 K (○); Birnbaum et al. [12] at 296 K, with 35% He (+) and 10% He (open boxes) and at 196 K, with 35% He (×) and 10% He (■).

[14, 15]. In order to study the effects of the anisotropy, we also show calculations obtained with the same dipole surface, but with the anisotropic potential components ($\gamma = 2$ and 4) suppressed (dashed curves). The comparison shows that the spectroscopic effects of the anisotropy are generally small, corroborating the assumption we made when using the isotropic approximation in previous work. Still, especially in the region of the lower S lines but also in the region of the Q lines, the anisotropy generally reduces the intensities by up to 10% as may be seen in the figures below. In the far wings of both the rototranslational and the fundamental bands of H_2, on the other hand, intensities are enhanced by the anisotropy by up to 15% for the data shown [9] and presumably more at even higher frequencies. To a large extent, these effects are due to the mixing and the resulting interference of the various dipole components λL. We note that in the isotropic interaction approximation these dipole components do not mix [1].

The spectra of dissimilar pairs such as H_2–He are often referred to as enhancement spectra. These must be obtained as the difference between the total absorption spectra of the gas mixture (helium and hydrogen) and those of the pure gas (hydrogen). The measurements of the collision-induced absorption spectra of dissimilar pairs (e.g., H_2–He) are therefore typically more uncertain than those of like pairs (H_2–H_2) — a fact that cannot be ignored when detailed comparisons of theory and measurements are attempted.

Figure 2 shows the computed rototranslational spectra of H_2–He pairs in the far infrared, at room temperature and 196 K. The available laboratory measurements are

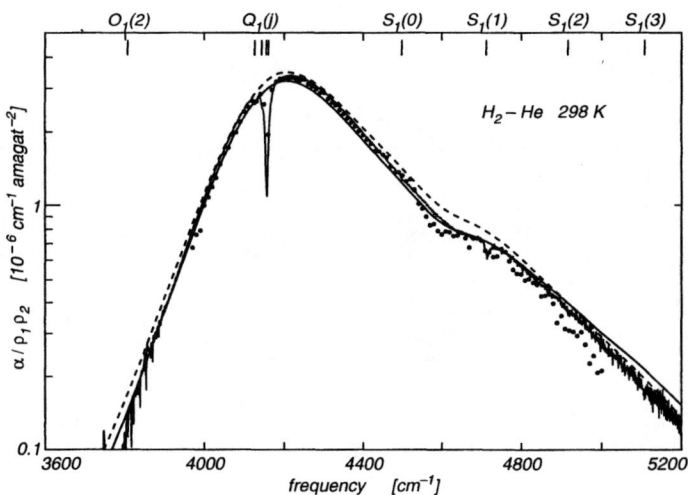

FIGURE 3. Rotovibrational collision-induced absorption spectrum of H_2–He pairs at 298 K in the fundamental band of H_2; Solid and dashed curves [9]: as in figure 2; measurement: thin line (with discernible noise) [16]; measurement (●) [17].

also shown for comparison. These are rather similar with respect to their shapes but they vary by up to ±10% in terms of their absolute intensity scale. The calculated spectra (solid lines) are well bracketed by the experimental data and may be considered a reasonable "average" of the existing experiments. Mean deviations of the experiments from theory are smaller than the variations of the measurements among themselves. We conclude that nearly complete agreement of measurements and fundamental theory can be obtained at 195 K and room temperature by rescaling the measured intensities by factors 0.94 and 1.04, respectively. The same conclusion may be drawn for the far wing of the spectra as well as spectra for lower temperatures [9].

Figure 3 shows the rotovibrational spectrum in the fundamental band of H_2. We note that the striking dips of absorption seen in the measurements near $4155 \, \text{cm}^{-1}$ and a much lesser one near $4700 \, \text{cm}^{-1}$ are due to intercollisional interference [19, 20, 21] which is a many-body feature which the binary line shape theory cannot reproduce. We compare theory with two of the most recent measurements of Brodbeck et al. [16] and Birnbaum et al. [17]. Since the thin (noisy) trace of the measurement [16] is often covered by the thick (smooth) line representing our calculation, we state here that from about 4100 to $4400 \, \text{cm}^{-1}$ the measured trace and the calculated curve practically coincide. Up to about $4600 \, \text{cm}^{-1}$ the experiment is just three percent or so above theory, but the traces coincide again at higher frequency, up to the point where the noise of the experiment becomes excessive. The agreement of the two measurements [16, 17] with theory is excellent.

H_2–H: In figure 4 is presented the rototranslational spectrum at 200 K that we calculate with the *ab initio* dipole surface [18] and the intermolecular potential [22]. We notice that the spectroscopic effects of the anisotropy are at the most 1-2 % which is much smaller than in the previous case of H_2–He. This can be understood from the

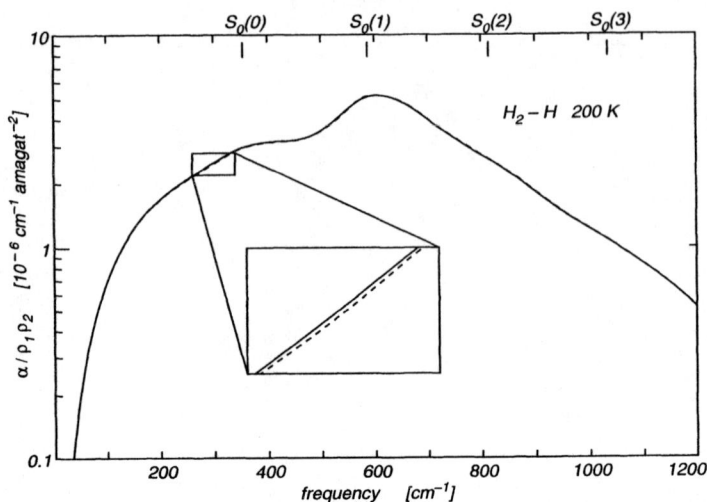

FIGURE 4. Collision-induced absorption spectrum of H_2–H pairs at 200 K in the rototranslational band; Solid and dashed curves [18]: as in figure 2

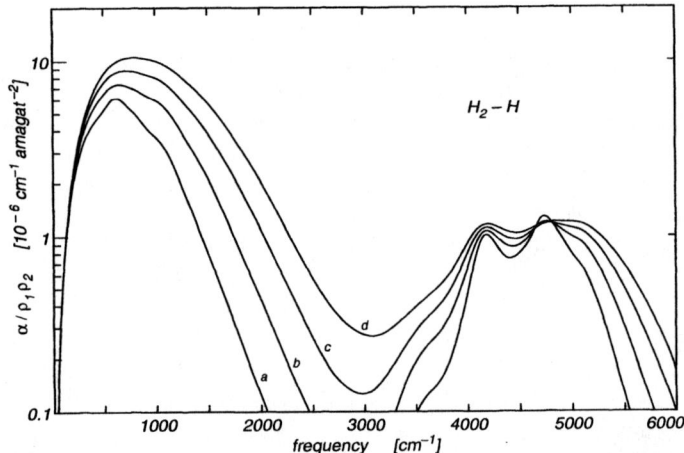

FIGURE 5. Collision-induced absorption spectra in the rototranslational and the H_2 fundamental band [18]. The line shapes are computed without the anisotropy of the interaction accounted for. The temperatures are: (a) 400 K, (b) 600 K, (c) 800 K, and (d) 1000 K.

characteristics of the intermolecular potential. In the region outside the classical turning point the anisotropic components, V_2 and V_4, are much smaller for the H_2–H potential than for H_2–He. This results in less rotational level mixing for the case at hand.

Similarly to reference [4] we have computed the correction to the zeroth moment as a function of temperature from 200 K to 1000 K [18]. Even at a temperature of 1000 K the

222

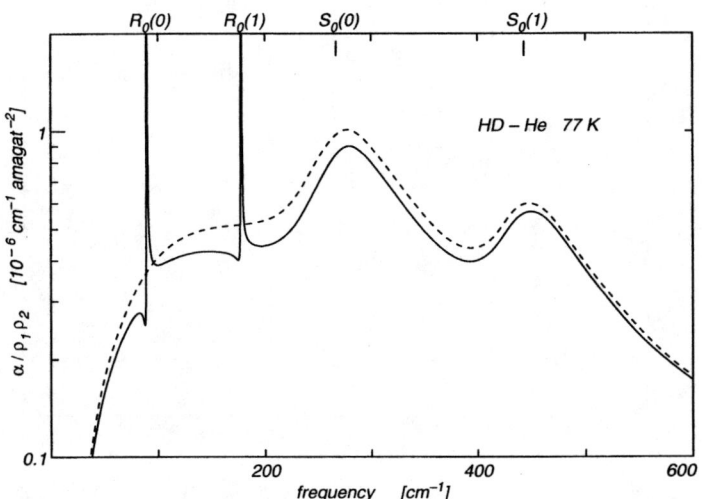

FIGURE 6. Collision-induced absorption spectrum of HD–He pairs at 77 K in the rototranslational band [23]; Solid and dashed curves: as in figure 2.

correction is no more than 2% in both the translational and the H_2 fundamental band. This is an indicator for that the isotropic potential approximation is good for that range of temperatures. Figure 5 shows spectra at four temperatures, from 400 K to 1000 K, covering the translational and the H_2 fundamental band. These are all computed using the isotropic interaction approximation.

HD–He: The computed absorption profile for the rototranslational band is shown in figure 6. The permanent dipole, interaction-induced dipole, and intermolecular potential come from references [24, 23, 15], respectively. We note that the allowed lines do not appear in the isotropic potential model because the allowed lines are delta functions (not shown); the anisotropy of the potential causes the broadening of the rotational lines. The collision-induced backgrounds calculated with and without the anisotropic parts of the intermolecular potential have similar shapes, but intensities differ more or less. While in the rotational band these differences are generally less than 10%, with a few exceptions near the $R(j)$ lines, we notice differences of up to 25% or so in the fundamental band (see Ref. [23]).

The region close to the $R(0)$ line in the HD fundamental band is shown in figure 7. We notice that the computed points agree well with the measured profile for the lowest He density. For an analysis of the allowed lines we use the low density formula for the Fano profile reported in reference [10],

$$\frac{\alpha(v)}{\rho_1\rho_2} = \frac{c_0 v}{2\pi}\left\{\frac{B_0}{(v-v_0(0))^2} + \frac{4\Delta''I}{v-v_0(0)}\right\} + a + bv + dv^2 \qquad (8)$$

to fit the calculated points in the vicinity of the P and R lines. The six parameters B_0, $\Delta''I$, a, b, d, $v_0(0)$, in equation (8) are obtained by a nonlinear fit [23] and reported in table 1 (Radiative CC). For c_0 the theoretical values are used [23].

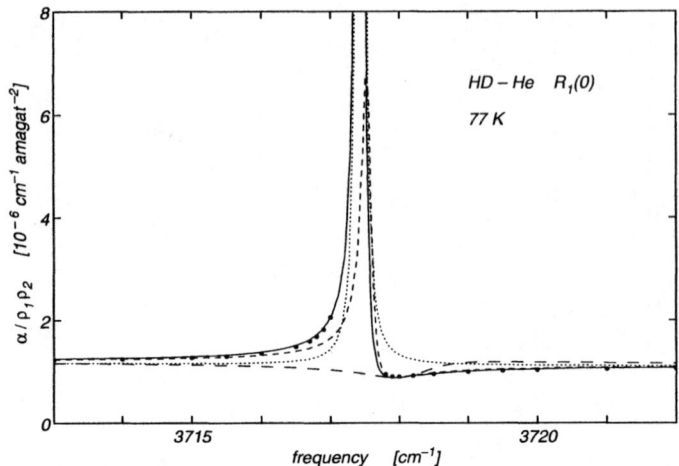

FIGURE 7. The absorption coefficient α, normalized by the He and HD gas densities, ρ_1 and ρ_2, respectively, as function of frequency at the $R_1(0)$ line, and at the temperature of 77 K. Dots represent radiative close-coupling calculations [10] using the interaction-induced dipole of Ref. [25]. Also shown is the Fano profile, determined experimentally in Ref. [26], for three different helium densities; solid line: 1 amagat; short dashes: 29.4 amagats; long dashes: 205 amagats. The dotted curve is the line shape obtained with the impact theory.

The broadening parameter obtained with the impact approximation is also shown in table 1. Agreement is observed between these two theoretical results for the broadening parameter, especially for the $R(0)$ lines. The deviation from experimental values is slightly larger: especially in the rototranslational band. The standard impact theory, which predicts a Lorentzian line shape, is based on accounting for many subsequent elastic binary collisions. This makes it different from the radiative close coupling theory which takes into account isolated inelastic binary collisions. Typically, the impact approximation is better suited for small detuning relative to the transition frequency, while the radiative close-coupling scheme is good for computation of the far wings of allowed lines.

H_2–H_2: The collision-induced absorption profile for equilibrium hydrogen at a temperature of 77 K is shown in figure 8. The solid line and the crosses represent values calculated with the radiative close-coupling theory. The dipole and potential are taken from references [33, 34], respectively. It is clear that including the anisotropy of the intermolecular potential does not have a very significant effect on the CIA profile in figure 8. However, in the region around 200 cm^{-1} there is an enhancement of the absorption by about 5%. This seems to correct the theoretical prediction to, almost perfectly, fit the measurement by Birnbaum [11]. The effect on the absorption due to dimers has been investigated previously [35, 2] and here only the free–free contribution is included.

In the rototranslational band the isotropic dipole component, i.e., the one with $\lambda_1 \lambda_2 \lambda L = 0001$, is almost zero due to the approximate inversion symmetry in intermolecular coordinates [33]. However, in the H_2 fundamental band this is not the case and there the isotropic dipole component is significant for the CIA spectrum. This may

TABLE 1. Comparison of calculated and measured line shape parameters of HD-He at 77K. The semi-classical calculations are made using the theory of Ref. [31]. Square brackets indicate powers of ten.

		B_0 [cm^{-1} am^{-1}]	Δ'_l [am^{-1}]	a [cm^{-1} am^{-2}]	b [am^{-2}]	d [cm am^{-2}]	$v_0(0)$ [cm^{-1}]
$R_0(0)$	Radiative CC	0.00356	0.00104	-0.208977[-6]	0.876095[-8]	-0.306704[-10]	89.23
	Impact theory	0.00357					
	Measurement of Ref. [27]	0.00293(12)	0.0015(3)				
	Semi-classical of Ref. [28]	0.0062	0.00060				
$R_0(1)$	Radiative CC	0.00464	0.00081	0.628021[-6]	-0.220323[-8]	0.611017[-11]	177.84
	Impact theory	0.00479					
	Measurement of Ref. [27]	0.00391(11)	0.0013(3)				
	Semi-classical of Ref. [28]	0.0039	-0.000027				
$P_1(1)$	Radiative CC	0.00412	0.00419	0.419582[-3]	-0.241736[-6]	0.348191[-10]	3542.82
	Impact theory	0.00407					
	Semi-classical of Ref. [29]	0.0064	0.0100				
$R_1(0)$	Radiative CC	0.00477	-0.00476	-0.331911[-3]	0.185684[-6]	-0.258495[-10]	3717.44
	Impact theory	0.00478					
	Measurement of Ref. [26]	0.0056(2)	-0.0059(1)				
	Semi-classical of Ref. [29]	0.0064	-0.00266				
	Quantum [30]		-.0011				
$R_1(1)$	Radiative CC	0.00593	-0.00377	0.275607[-3]	-0.140382[-6]	0.179002[-10]	3798.39
	Impact theory	0.00617					
	Measurement of Ref. [26]	0.0062(3)	-0.0055(1)				
	Semi-classical of Ref. [29]	0.0040	0.00079				
	Quantum [30]		-0.0003				

lead to a bigger influence of the anisotropy of the potential due to the mixing between that dipole component and the others.

CONCLUSION

A formalism for calculation of collision-induced absorption profiles including molecular degrees of freedom has been developed. The wings of allowed line shapes can be

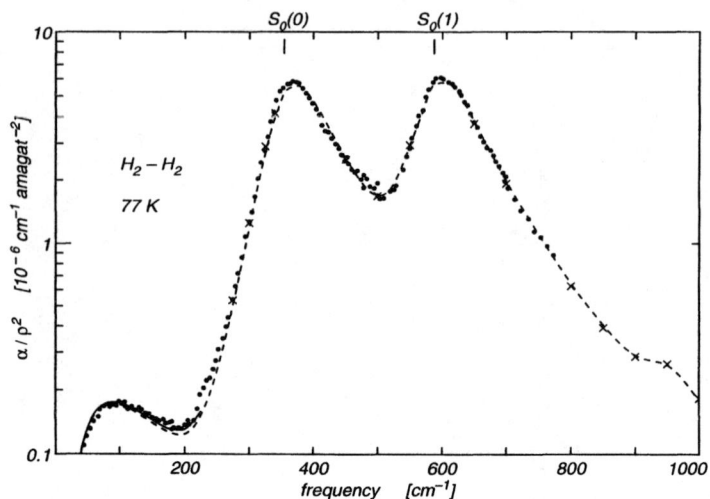

FIGURE 8. Collision-induced absorption spectrum of hydrogen at 77 K in the rototranslational band. The solid curve and the crosses represent the calculation with the anisotropy of the intermolecular potential accounted for [32]. The dashed curve represent the calculation in the isotropic potential approximation. A measurement by Birnbaum [11] is presented as dots.

computed with the same method. Two things turns out to be essential for the anisotropy of the intermolecular potential to significantly affect the collision-induced absorption spectrum. First, the anisotropic components of the intermolecular potential must be relatively strong. Above was reported how this requirement is not fulfilled for the H_2–H complex and the effect of the anisotropy on the CIA spectra of that system was indeed almost insignificant. Second, the effects of the anisotropy originate mainly in interferences between the various dipole components. Thus several dipole components must contribute to the absorption with comparable strength for the anisotropy have an effect. If one dipole component dominates the spectrum the interference with other components become weak.

The collision-induced absorption of HD–He is computed, including the wings of the allowed R and P lines, taking into account single binary collisions between HD and He at the temperature of 77 K. The inclusion of the anisotropy leads to a mixing of the rotational levels, i.e., pressure broadening of the allowed rotational lines. It also couples the various induced dipole components and the permanent dipole giving rise to interferences between those dipole moments, and an asymmetry of the allowed spectral lines. Agreement between theory and measurements is observed in the low-helium-density limit of the measured absorption line shapes.

Other things which could be investigated with the radiative close-coupling scheme are, for example, rotational mixing in H_2–H_2 collision-induced light scattering and vibrational mixing in high temperature H_2–He CIA [32].

226

ACKNOWLEDGMENTS

We want to express our sincere thanks to Dr. Paul Julienne for kindly providing us with a copy of his COUPLE computer code which we have used in this work. The R. A. Welch Foundation supported this work through grant F-1346.

REFERENCES

1. Frommhold, L., *Collision-induced Absorption in Gases*, Cambridge University Press, Cambridge, New York, 1993.
2. Schäfer, J., and McKellar, A. R. W., *Z. Physik D*, **15**, 51 – 65 (1990), erratum: Z. Physik D 17 (1990) 231.
3. Moraldi, M., Borysow, A., and Frommhold, L., *Phys. Review*, **A 35**, 3679 – 3687 (1987).
4. Moraldi, M., and Frommhold, L., *Phys. Review A*, **52**, 274 – 281 (1995).
5. Arthurs, A. M., and Dalgarno, A., *Proc. Roy. Soc. (London)*, **Ser. A, 256**, 540 – 551 (1960).
6. DeVries, P. L., and George, T. F., *Molec. Phys.*, **36**, 151 (1978).
7. Mies, F. H., "Quantum theory of atomic collisions in intense laser fields," in *Theoretical Chemistry: Advances and Perspectives*, edited by D. Henderson, Academic Press, New York, 1981, vol. 6B, pp. 127 – 198.
8. Julienne, P. S., *Phys. Review A*, **26**, 3299 (1982).
9. Gustafsson, M., Frommhold, L., and Meyer, W., *J. Chem. Phys.*, **113**, 3641 – 3650 (2000).
10. Gustafsson, M., and Frommhold, L., *Phys. Review A*, **63**, 052514 (2001).
11. Birnbaum, G., *J. Quant. Spectroscopy and Rad. Transfer*, **19**, 51 – 62 (1978).
12. Birnbaum, G., Bachet, G., and Frommhold, L., *Phys. Review*, **A 36**, 3729 – 3735 (1987).
13. Mies, F. H., Julienne, P. S., and Sando, K. M., A close coupling code (1993).
14. Meyer, W., Hariharan, P. C., and Kutzelnigg, W., *J. Chem. Phys.*, **73**, 1880 – 1897 (1980).
15. Schaefer, J., and Koehler, W. E., *Physica A*, **129**, 469 – 502 (1985).
16. Brodbeck, C., Nguyen-van-Thanh, Bouanich, J. P., and Frommhold, L., *Phys. Review A*, **51**, 1209 – 1213 (1995).
17. Birnbaum, G., Borysow, A., and Orton, G. S., *Icarus*, **123**, 4 – 22 (1996).
18. Gustafsson, M., Frommhold, L., and Meyer, W., To be published (2002).
19. van Kranendonk, J., *Can. J. Phys.*, **46**, 1173 (1968).
20. Lewis, J. C., "Intercollisional interference—Theory and experiment," in *Phenomena Induced by Intermolec. Interactions*, edited by G. Birnbaum, Plenum Press, New York, 1985, pp. 215 – 258.
21. Poll, J. D., Hunt, J. L., and Mactaggart, J. W., *Can. J. Phys.*, **53**, 954 (1975).
22. Partridge, H., C. W. Bauschlicher, j., Stallco, J. R., and Levin, E., *J. Chem. Phys.*, **99**, 5951 – 5960 (1993).
23. Gustafsson, M., and Frommhold, L., *J. Chem. Phys.*, **115**, 5427 – 5432 (2001).
24. Thorson, W. R., Choi, J. H., and Knudsen, S. K., *Phys. Review A*, **31**, 34 – 42 (1985).
25. Borysow, A., Frommhold, L., and Meyer, W., *J. Chem. Phys.*, **88**, 4855 – 4860 (1988).
26. McKellar, A. R. W., and Rich, N. H., *Can. J. Phys.*, **62**, 1665 – 1672 (1984).
27. Ulivi, L., Lu, Z., and Tabisz, G. C., *Phys. Review A*, **40**, 642 – 651 (1989).
28. Gao, B., Cooper, J., and Tabisz, G. C., *Phys. Review A*, **46**, 5781 – 5788 (1992).
29. McQuarrie, B., and Tabisz, G. C., *J. Molec. Liquids*, **70**, 159 – 168 (1996).
30. Herman, R. M., and Lewis, J. C., "Scalar collisional interference parameters for the HD $R_1(0)$ and $R_1(1)$ lines in mixtures with He," in *Spectral Line Shapes* 11, edited by J. Seidel, AIP Conference Proceedings, vol.559, Am. Inst. Physics, New York, 2000, pp. 397 – 399.
31. Gao, B., Tabisz, G. C., Trippenbach, M., and Cooper, J., *Phys. Review A*, **44**, 7379 – 7391 (1991).
32. Gustafsson, M., Dissertation (2002).
33. Meyer, W., Borysow, A., and Frommhold, L., *Phys. Review A*, **40**, 6931 – 6949 (1989).
34. Schäfer, J., and Köhler, W. E., *Z. Physik D*, **13**, 217 – 229 (1989).
35. Meyer, W., Frommhold, L., and Birnbaum, G., *Phys. Review A*, **39**, 2434 – 2448 (1989).

FAR-WING LINE SHAPES: APPLICATION TO THE WATER CONTINUUM

R. H. Tipping[*] and Q. Ma[†]

[*]Department of Physics and Astronomy, University of Alabama, Tuscaloosa, AL 35487
[†]Department of Applied Physics, Columbia University, and Institute for Space Studies, Goddard Space Flight Center, 2880 Broadway, New York, NY 10025

Abstract. A far-wing line shape theory that satisfies the detailed balance principle has been developed invoking only the binary-collision and quasistatic approximations. This first-principles theory has been applied to calculate the far-wing line shapes and the corresponding absorption for H_2O-H_2O and H_2O-N_2 which, for historical reasons, are called the self- and foreign-broadened water continua. Using sophisticated interaction potentials that give good agreement with other transport data and the coordinate representation, in which the required traces become multidimensional integrals over the angular coordinates necessary to specify the positions before and after the collision (11- dimensional for the self and 9-dimensional for the foreign continuum, respectively), we can obtain converged results using moderate computational resources. Results obtained are in very good agreement with existing laboratory data, and comparisons with an empirical continuum will be presented.

INTRODUCTION

In the Earth's atmosphere both in the far infrared and infrared regions there are two main sources for the continuous absorption of radiation over a range of frequencies: the self and foreign water continua [1,2] arising from the far wings of the allowed pure rotational and vibration-rotational dipole transitions, and collision induced absorption (CIA) arising from transient dipole moments induced during binary collisions. Both these absorptions scale quadratically with the number density, ρ^2 for a pure gas, or as the product of the ρ's for a binary mixture [3]. In the case of zenith measurements, the former continua dominate and are included in most atmospheric radiative transfer programs [4]. However, for low humidity conditions such as long-path limb measurements through the stratosphere, the collision-induced fundamental absorptions of O_2 and N_2 can be important and are also included.

During the past few years, we have developed a first principles far-wing line shape theory making only the binary collision and quasistatic approximations, both of which are valid for the low density gases of interest. The theory explicitly satisfies the detailed balance principle [5]. In earlier work, we utilized for the basis set the product of the two unperturbed molecular wave functions to describe the internal degrees of

CP645, *Spectral Line Shapes: Volume 12, 16th ICSLS*, edited by C. A. Back
© 2002 American Institute of Physics 0-7354-0100-4/02/$19.00

freedom; the translational motion was treated classically and the average over the separation r between the centers of mass of the two molecules was done analytically using conservation of energy during the collisions. In this representation, the anisotropic part of the interaction potential is off-diagonal and had to be diagonalized in order to obtain the eigenvalues and eigenvectors needed to carry out the trace operation for calculating the line shape. For simple systems such as CO_2-Ar, this presented no difficulties, but for more complicated systems, such as H_2O-H_2O, the size to the matrix increases very quickly as more rotational states are included. As a result, the matrix had to be truncated, which resulted in convergence problems and a loss of accuracy. In recent work we have avoided this problem by adopting the coordinate representation where the internal state $| \zeta >$ is expressed as

$$| \zeta > = |\delta(\Omega_a - \Omega_{a\zeta}) \, \delta(\Omega_b - \Omega_{b\zeta})> , \qquad (1)$$

where the subscript a or b denote the absorber and bath molecules, respectively. For linear molecules, Ω_ζ represents the two angles θ_ζ and φ_ζ needed to specify the orientation of the axis with respect to a space-fixed coordinate system, while for an asymmetric top molecule it represents the three Euler angles α_ζ, β_ζ and γ_ζ. The advantages of this representation are that the anisotropic interaction potential, $V(r, \Omega_a, \Omega_b)$, is diagonal and one can include as many state as desired in the calculations. The disadvantages are that the density operator is off-diagonal and that the sums over internal quantum numbers become multi-dimensional integrals. The diagonalization of the density operator is straightforward, and although it is time consuming, it only has to be done once for a fixed temperature T. The dimensionality of the integrals depends on the types of molecules: for two linear molecules, one linear and one asymmetric top, or for two asymmetric tops, it is 7, 9, and 11, respectively. For the latter two cases, one has to use a Monte Carlo method, and even then, in order to obtain converged results, a large amount of CPU time is required. This, of course, depends on the complexity of the interaction potential model.

For the H_2O-H_2O case, we tested several of the interaction potentials available in the literature, but none gave good results for the far-wing absorption. In particular, the far-wing line shapes depend very sensitively on the short-range anisotropic gradients. Consequently, we adopted the following model potential:

$$V(r, \Omega_a, \Omega_b) = \sum_{i \in a} \sum_{j \in b} \frac{q_i q_j}{r_{ij}} + \sum_{k \in a} \sum_{l \in b} V_{kl} \, e^{-\frac{r_{kl}}{\rho_{kl}}} - \frac{B}{r^6} \qquad (2)$$

where q_i, V_{kl}, and B are adjustable parameters. We assume that for each H_2O, there are two point charges +q located on the H atoms and one −2q at a position along the symmetry axis a distance δ from the O atom. We require that these charges give the correct permanent dipole moment and good values for the quadrupole moment components. We assume that there are three repulsive force centers for each molecule: two located on the H atoms and one on the O atom. In addition to yielding good agreement with laboratory data for the self continuum at T = 296 K, we also require this potential to reproduce the measured differential scattering cross-sections and T-dependent second virial coefficients. Once we obtain the parameters, we use the same potential to calculate the line shapes and absorption at other temperatures [1,2].

RESULTS

Once the line shapes are determined, we calculate the corresponding absorption coefficient using the known data for the line centers and intensities for comparisons with experiment. In Fig.1 we present results for H_2O-H_2O at T = 296 K for the high-frequency region of the pure rotational band.

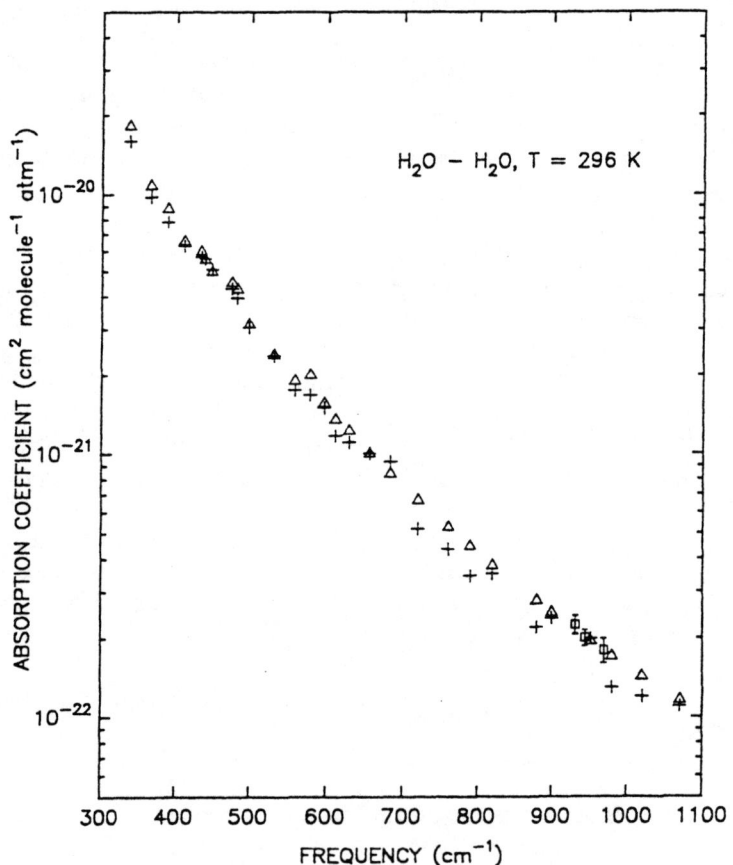

FIGURE 1. The calculated self-broadened absorption coefficient (in units of cm^2 molecule^{-1} atm^{-1}) at T = 296 K in the 300-1100 cm^{-1} spectral region is represented by Δ. For comparison, the experimental data of Burch et al. [6] are denoted by + and those of Cormier et al. [7] along with their error bars by $\square$.

In Fig.2 we present a similar comparison for the foreign continuum at T = 296 K. The results of the empirical CKD 2.4 continuum [8] are also shown. As can be seen from these figures, the present formalism gives very good agreement with the experimental results over several orders of magnitude for both the self- and foreign continuum absorptions, whereas the empirical results for the foreign continuum are significantly below the measurements in the spectral region between 700 – 1,000 cm^{-1}.

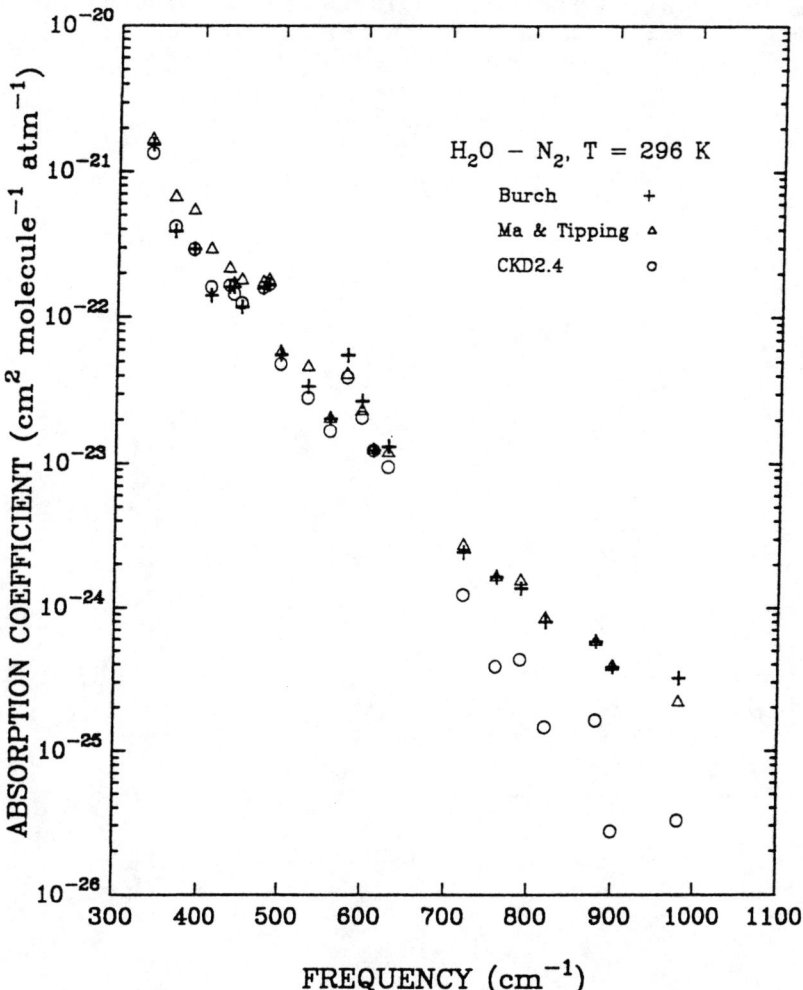

FIGURE 2. The calculated N_2-broadened absorption coefficient (in units of cm^2 molecule^{-1} atm^{-1}) at T = 296 K in the 300-1100 cm^{-1} spectral region is represented by Δ. For comparison, the experimental data of Burch et al. [6] are denoted by + and the empirical CKD 2.4 values are represented by O.

REFERENCES

1. Ma, Q, and Tipping, R. H., J. Chem. Phys. **112**, 574-584 (2000).
2. Ma, Q, and Tipping, R. H., J. Chem. Phys. **116**, 4102-4115 (2002).
3. Moreau, G., Boissoles, J., Le Doucen, R., Boulet, C., Tipping, R. H., and Ma, Q., JQSRT **69**, 245-256 (2001).
4. Edwards, D. P., GENLN2, NCAR Technical Note, January 1992.
5. Ma, Q., and Tipping, R. H., JQSRT **51**, 751-757 (1994).
6. Burch, D. E., and Gryvnak, Report Number AFGL-TR-79-0054, 1979; Burch, D. E., and Alt, R. L., Report Number AFGL-TR-84-0128, 1984.
7. Cormier, J. G., Ciurylo, R., and Drummond, J. R., J. Chem. Phys. **116**, 1030-1034 (2002).
8. Clough, S. A., Kneizys, F. X., and Davies, R. W., Atmospheric Research **23**, 229-241 (1989).

Dipole Spectra of H₂ and HD in Interstitial Channels in Carbon Nanotube Bundles

R. M. Herman[a] and J. Courtenay Lewis[b]

[a]Department of Physics, The Pennsylvania State University, University Park, PA 16802
[b]Department of Physics and Physical Oceanography, Memorial University of Newfoundland,
St. Johns, NF A1B 3X7, Canada

Abstract. Quadrupole induced dipole fundamental band transition lines are predicted and characterized for H_2 and HD trapped within threefold symmetry interstitial channels bounded by sets of three carbon nanotubes.

We have calculated transition dipoles and positions of H_2 quadrupole-induced dipole spectra, and have estimated absorption strengths for the fundamental band of H_2 adsorbed into interstitial channels (IC's) between sets of three carbon nanotubes within nanotube bundles. Such sites may be important for hydrogen storage applications, and the spectra that we are reporting could be used to shed important information on adsorption dynamics and adsorbate locations within such systems. While the estimated optical absorption cross sections are quite large, the relatively low density of sites and short transmission path lengths will require some effort to observe the spectra.

We are assuming the nanotubes to be nonconducting, as opposed to a recent study by Kostov, *et al* [1] in which the surrounding triplet of nanotubes were envisioned as being electrically conducting. As a result of this and other refinements, the orientation energies calculated here are substantially lower than those reported in ref. [1].

To find the dipole moments induced by the H_2 quadrupole moments in the surrounding C atoms of the nanotube nets, we first consider the result that would be obtained if all C atoms were replaced by He atoms. Then the x-component of the transition dipole would be given by

$$p_x = -\sqrt{2}\ \mathrm{Re}\{p_1\} \tag{1}$$

where the ν-th component of the transition dipole for H_2-He is [2]

$$p\nu = \frac{4\pi}{\sqrt{3}} \sum_{\Lambda L} B^{01}_{\Lambda L}(R) \sum_M C(\Lambda L1; M(\nu - M)) Y_{\Lambda M}(\vartheta\varphi) Y_{L(\nu-M)}(\Theta\Phi) \tag{2}$$

We envision the z-axis as being parallel to the tube axes, with x being chosen as a convenient perpendicular direction. (In practice, the x-and y- directions are entirely equivalent.) The notation is standard, ($\vartheta\varphi$) representing the orientation angles of the H_2 figure axis and ($\Theta\Phi$) specifying the angular position of each C atom, with H_2 lying at the origin and the tube axes specifying the $\vartheta = 0$ and $\Theta = 0$ directions. In the above threefold

CP645, *Spectral Line Shapes: Volume 12, 16th ICSLS*, edited by C. A. Back
© 2002 American Institute of Physics 0-7354-0100-4/02/$19.00

symmetry situation, only the $\Lambda=2$, L=3 component of the above expression survives. Because at the distances of interest (equal to or larger than about 5.5 atomic units, a.u.) the 2,3 component represents the quadrupole-induced dipole, it is proportional to the polarizability of He. $B_{23}^{01}(R)$ is known very precisely, from *ab initio* calculations [3]. To relate this result to that for C nanotubes, one might replace the He polarizability by the C atom polarizability. The latter, for field components perpendicular to the carbon nets, is 3.72 a.u. (as opposed to 1.39 a.u. for He). A complication arises in that the C atom polarizability, for field components parallel to the net surfaces, is about 3.5 times the value for perpendicular field components. A somewhat lengthy calculation reveals that ultimately one achieves the correct result by replacing α_{He}, as it would appear in an expression for the long-range behaviour of B_{23}^{01}, simply by the average of the two carbon polarizabilities. This results from the threefold symmetry nature of the present situation, and would not be true for any particular C atom by itself. The threefold symmetry situation is such that only $Y_{30}(\Theta\Phi)$ and $Y_{33}(\Theta\Phi)$ operators can possibly contribute to a transition. However, the odd symmetry behaviour of Y_{30} in $\cos(\Theta)$ leads to the failure of this component to contribute. Therefore, for present purposes,

$$ p_1 = \frac{4\pi}{\sqrt{3}}\left(\frac{\alpha_{c\perp}+\alpha_{c\parallel}}{2\alpha_{He}}\right)B_{23}^{01}(R)C(231;-23)Y_{2-2}(\vartheta\varphi)Y_{33}(\Theta\Phi) \tag{3} $$

For a $\nu=1$ transition, while one unit of angular momentum is removed from the radiation field, three units are delivered to the nanotube system, while -2 units are given to the H_2 molecule itself! We now sum this (trivially) over equivalent points in the three-tube system, then integrate over surface elements in a single tube, assuming that the sum on C atoms can be replaced by a surface integral of a smoothed distribution of C atoms. The results are presented in Table 1 below.

We also have calculated the energy perturbations for the various relevant states, through assuming that

$$ V(R,\gamma) = \sum_L V_L(R)P_L(\cos\gamma) \tag{4} $$

with γ being the angle between the H_2 figure axis and the intermolecular displacement. The Lennard-Jones parameters for H_2-C are well known, and we assume that V_2 and V_4 bear the same relationship to V_0 as they do in H_2-He interactions. (This would seem to be justified in view of the fact that the Lennard-Jones radii and the short range forces are quite comparable for the two situations.) The functions V_L for L=0, 2, 4 are well known for H_2-He [4]. This idea seems to work reasonably well, when tested on the experimentally known V_2 and V_0 functions for H_2-Ar interactions, even though there, the dynamical similarity is less evident. It is of note that in the H_2-He and H_2-Ar cases, the induction energy has been included implicitly, so that there is no need to separately consider it as being distinct from overlap forces. This leads to a potential function which, including vibrational dependencies, has the form

$$V(R, \gamma) \cong B_e[-(0.554 + 0.167v) \qquad (5)$$
$$+ (0.180 - 0.857\cos^2 \gamma + 0.527\cos^4 \gamma) + (0.192\cos^2 \gamma - 0.064)v]$$

Adopting the energy shifts as a measure of perturbability, we estimate the linewidths at 77K as being due to inhomogeneous broadening (arising from slight differences in trapping sites) at a level of, say, 10% of the energy shifts themselves. Through standard calculations, the peak absorption for the transitions can be estimated for passage through 50 nanotube layers, with a loading of one H_2 molecule per 10 a.u. within the IC's. The results are shown below.

Table 1. Fundamental Band Transitions of H_2 and HD in Carbon Nanotube Interstitial Channels at 77K

Transition	Quantum Number	Transition Dipole (10^{-3} a.u.)	Shift from Free Molecule Line (cm^{-1})	Estimated Width (cm^{-1})	Estimated Peak Absorption (%)
H_2: $Q_1(1)$	$(01\pm1 \rightarrow 11\mp1)$	6.65	12.3	1.2	0.14
$S_1(0)$	$(000 \rightarrow 12\pm2)$	4.30	18.8	1.9	0.13
$S_1(1)$	$(010 \rightarrow 13\pm2)$	2.30			Weaker
	$(01\pm1 \rightarrow 13\mp1)$	1.02			Weaker
	$(01\pm1 \rightarrow 13\pm3)$	3.98			Weaker
HD: $R_1(0)$	$(000 \rightarrow 11\pm1)$	2.58	17.5	1.8	0.07
$R_1(1)$	$(010 \rightarrow 12\pm1)$				Weaker
	$(01\pm1 \rightarrow 120)$				Weaker
	$(01\pm1 \rightarrow 12\pm2)$				Weaker
$P_1(1)$	$(01\pm1 \rightarrow 100)$				Weaker

Also shown in Table 1 is a set of transitions which would be absent in H_2 but which exist by virtue of the displacement between mass and electrical centers in HD. It corresponds to the Λ, L=1,0 dipole component in an expression corresponding to eq. 2 for HD. This is an overlap component, simply proportional to $\cos(\gamma)$. Its absorption strength has been estimated through comparison with known H_2-He and H_2-Ar results.

There will also be a $Q_1(J)$-vibron band associated with the Λ,L = 0,1 overlap dipole term in eq. 2. However this would seem to be too diffuse to observe. While the above absorption strengths appear to be almost impossibly small to be of practical importance, it should be noted that recently an observation of dipole spectra of H_2 trapped in C_{60} cubic lattices has been made [5]. Noting the odd-L, even-Λ form of eq. 2, one would reason that the phonon bands may arise from the largest Λ,L = 0,1 dipole term, though even at the higher temperature of those observations, it is somewhat surprising that they show up with the strength that was observed. But of more interest in the present context is the fact that sharp features at expected line positions also were recorded. These could come only from H_2 trapped in sites lacking inversion symmetry (in order for the odd-L terms to survive). The potential of H_2 trapped in cubic C_{60} lattices appears to support H_2

molecules in vibrational states symmetrically oriented about the centers of the cubic lattice faces. So the above analysis, applied to that situation would depend upon the H_2 molecules lying in sites lacking such an inversion symmetry, which normally would be quite a rare occurrence. A rough calculation leads to the prediction of transition dipoles comparable in magnitude to those calculated for the IC's for each asymmetric site. However, the bandwidth shown in ref. 5 appears to be on the order of 25 cm^{-1}. Even at the higher temperature than those considered in this paper, the intensities of the sharp features reported in ref. 5 would seem to greatly exceed our expectations. In view of this situation, it becomes doubly interesting to see if one can observe similarly anomalously high intensity sharp lines associated with H_2 trapped in nanotube IC sites.

REFERENCES

1. Kostov, M. K., Cheng, M., Herman, R. M., Cole, M. W. and Lewis, J. C., *J. Chem. Phys.* **116**, 1720 (2002).
2. Poll, J. D. and Hunt, J. L., *Can. J. Phys.* **54**, 461 (1976).
3. Borysow, A. and Frommhold, L., *J. Chem. Phys.* **88**, 4855 (1988).
4. Schaffer, J. and Kohler, W. E., *Physica* **129A**, 460 (1985).
5. Fitzgerald, S. A., Forth, S. and Rinkowski, M., *Phys. Rev. B* **65**, 140302(R) (2002).

Light emitting processes of sonoluminescence

Dominik Hammer and Lothar Frommhold

Department of Physics, The University of Texas at Austin, Austin, TX 78712, USA

Abstract. We consider elementary radiative processes that play a role under conditions such as encountered in sonoluminescence studies. These are electron-atom and electron-ion bremsstrahlung; electron-atom polarization bremsstrahlung; electron-ion recombination radiation; radiative attachment of electrons to atoms; ion-atom bremsstrahlung and charge exchange collisions; and ion-atom association radiation. We include these processes in a hydrodynamic-chemical model of the expanding and collapsing bubble and compute spectral profiles, spectral intensities, and duration of the light pulses for the rare gas bubbles. We generally obtain good agreement with observations for photon numbers, pulse durations, and in most cases for the spectral shapes. Some calculated spectral profiles agree, however, less well with the experiment, especially if water temperatures near freezing are involved, and in the case of helium bubbles. We show that by accounting for a somewhat non-uniform heating across the bubble diameter, agreement with the observed spectral profiles may be obtained even in these latter cases.

INTRODUCTION

Sonoluminescence (SL) is the conversion of sound into light in tiny bubbles levitated in a liquid [1, 2]. The gas bubbles expand during the tensile phase of sound and implode violently during the compression phase. We are concerned here with "single bubble sonoluminescence," where the experimental conditions are such that a single bubble is maintained over hours, expanding and collapsing with the frequency of sound ($\approx$ 20 kHz). The present paper is an extension of the work presented at the previous ICSLS [3], where calculations of electron-rare gas atom bremsstrahlung spectra were combined with a simple hydrodynamical model of the bubble dynamics [4]. Here this model is significantly enhanced. In addition to the rare gas, water vapor is considered a part of the bubble contents, with an amount that varies during the acoustic cycle. The accounting for water vapor modifies bubble dynamics; it also causes chemical reactions (dissociation) to occur in the bubble and significantly affects the light emission. We again use our *ab initio* calculations of electron-atom bremsstrahlung spectra. In addition, we consider electron-ion bremsstrahlung; electron-atom polarization bremsstrahlung; radiative attachment of electrons to hydrogen and oxygen atoms; and charge exchange, ion-atom bremsstrahlung and ion-atom association radiation in helium.

The Rayleigh-Plesset equation [4, 5, 6, 7] gives the radius of the bubble as function of time, internal pressure and external pressures. The external pressure is given by the sum of the ambient pressure and the sound pressure. The internal pressure is given by the equation of state, which requires knowledge of the number and types of particles, the volume of the bubble, and the temperature. These are all functions of time and, in general, functions of the radial coordinate r.

CP645, *Spectral Line Shapes: Volume 12, 16th ICSLS*, edited by C. A. Back

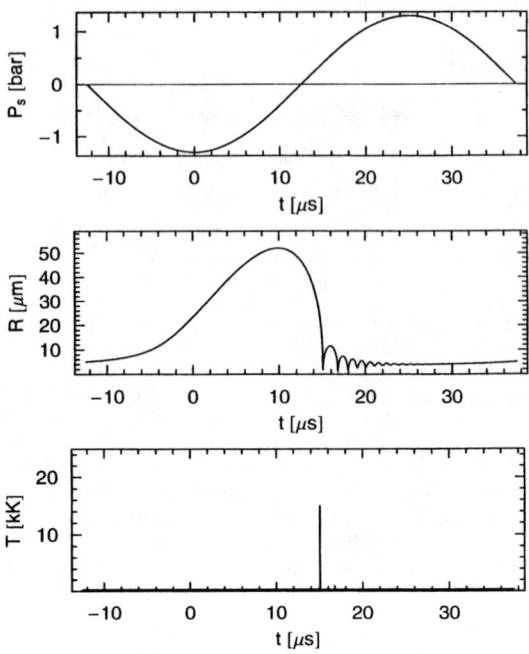

FIGURE 1. Driving pressure P_s, bubble radius R, and temperature T, as functions of time t, for one complete cycle of the driving pressure.

To compute such a non-uniform model of sonoluminescent bubbles a set of partial differential equations must be solved. Initially it was assumed that strong shocks exist in the bubble which cause a very inhomogeneous heating of the gas and thus require a non-isotropic treatment of the bubble content [8, 9, 10]. However, it has been shown later [11] that in these early computations several effects were neglected and that, at best, mild compression waves may heat the bubble slightly non-uniformly.

Our main interest here is the light emission. We may therefore assume a uniform bubble interior, with temperatures, densities, etc. that do not much depend on the radial position r. This approximation results in an enormous simplification of the computational task; it is generally satisfactory and should suffice for much of our work. But we will see that in certain cases that approximation leaves something to be desired.

The energy content and heat capacity of the bubble determine the temperature. Energy gains and losses of the environment are taken into account. During the rapid collapse the gas in the bubble may be thought of being heated adiabatically. However, since the bubble loses energy due to the endothermic reactions, such as ionization and the dissociation of water molecules, the compression is actually "quasi-adiabatic" [7, 12], with temperatures high enough so that a small fraction of the gas is ionized. The degree of ionization is calculated i)from rate equations and ii)from the Saha equation, with consistent results. Once the temperature, composition and degree of ionization in the bubble has been calculated for a complete cycle of the driving sound wave we use

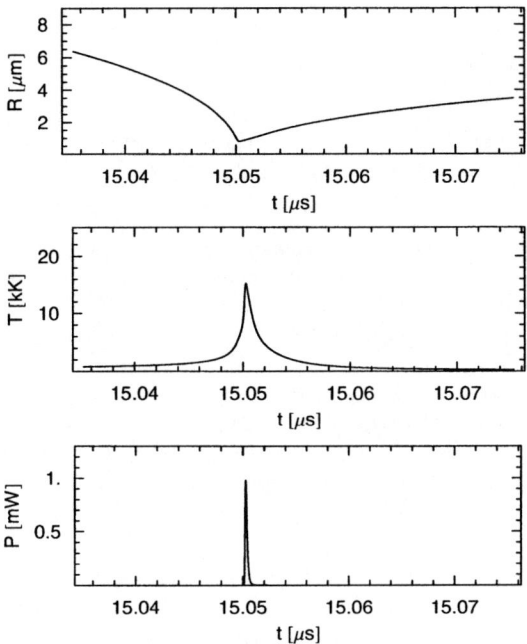

FIGURE 2. Bubble radius and temperature as in Fig. 1, but with a one-thousand-fold expanded time scale; bottom: the light pulse.

this information to compute the light emission due to the various radiative processes mentioned.

Figure 1 shows the driving pressure P_s, the bubble radius R and the temperature in the bubble T as functions of time t for one complete acoustic cycle. Note the bubble expansion at negative acoustic pressures, and the sudden collapse as the driving pressure changes sign, in agreement with experiment. At the end of the collapse, for a very brief moment, the temperature T rises to about 15,000 K. In single bubble sonoluminescence studies the bubble dynamics repeats itself through many million cycles of the sound wave with virtually no variations.

In Fig. 2 the time scale is expanded by roughly three orders of magnitude near the point of maximum compression. Now the radius as function of time barely changes, while the temperature *versus* time curve has a discernible width. Yet, even on this expanded time scale, the light emission *versus* time curve is still very sharp. This is a result of the roughly exponential dependence of the degree of ionization, and thus of the emitted intensity, on temperature.

Figure 3 shows the light pulse on a further expanded time scale. Along with the emitted intensity, integrated over the entire optical window of water from 200 nm to 800 nm, we also show the intensity integrated over the ultraviolet (200 nm to 300 nm) and the red (700 nm to 800 nm) parts of the spectrum. As observed in the experiment [13], the pulse lengths of the red and the UV pulses practically coincide.

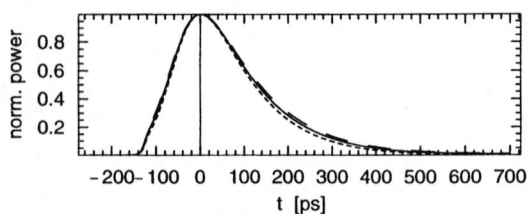

FIGURE 3. The emission as function of time, *i*.integrated over the entire optical window of water (solid line); *ii*.the ultraviolet part of the spectrum (200 nm to 300 nm; dotted line); and *iii*.the red part of the spectrum (700 nm to 800 nm; dashed line).

The results presented in these figures are for one particular set of experimental parameters. Similar results may be expected in general for single bubble sonoluminescence. Our model predicts peak temperatures exceeding 10,000 K at densities of a few hundred amagat ($\approx 10^{28}$ particles per m^3). The gas mixture in the bubble is weakly ionized.

RADIATIVE PROCESSES

The spectral intensities of most of the radiative processes included in our model are computed from a quantum mechanical line shape code [14], assuming binary collisions. We solve the Schrödinger equation for the two-body scattering processes in a partial wave expansion, for a given interaction potential $V(r)$. We determine wave functions for a given initial kinetic energy ε of the scattering pair and a final kinetic energy $\varepsilon \pm hc/\lambda$. The h and c have their usual meanings and λ is the wavelength of the light, the upper sign referring to absorption and the lower to emission. Intensities are determined from Fermi's Golden Rule using a dipole surface $\mu(r)$ that again characterizes the specific process under investigation. Finally for each process initial kinetic energies are thermally averaged.

Electron-neutral bremsstrahlung. For this process we use the Hartree-Fock-Slater potentials [15] for electron-ion interaction with semi-empirical corrections for the electron-neutral case [16, 17]. We use the acceleration form of the dipole operator for fast convergence.

Figure 4 shows the results for the case of electron-argon bremsstrahlung: The emission spectra are "white" at low frequencies, i.e., they are of nearly constant intensity, up to the point where the duration of the interaction is comparable to the reciprocal frequency of the emitted light. At that point a sharp fall-off is seen which at the temperatures considered occurs somewhere in the optical window of water.

Polarization bremsstrahlung. Polarization bremsstrahlung is also emitted during the collision of electrons with neutral atoms: the time-varying dipole due to polarization of the atom in the electric field of the electron generates these spectra [18]. The Hartree-Fock-Slater interaction potential is again used [15], but now with the dipole operator

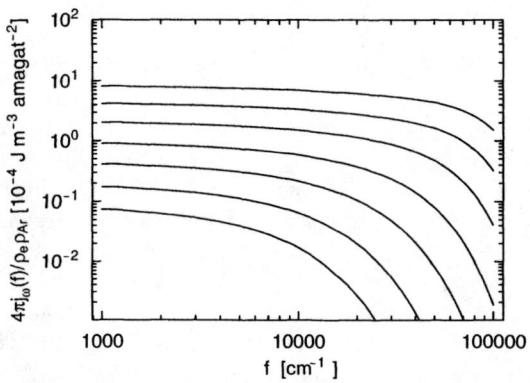

FIGURE 4. Electron-argon bremsstrahlung emission coefficient; quantum calculations for 5, 7, 10, 14, 20, 28 and 40 kK (bottom to top).

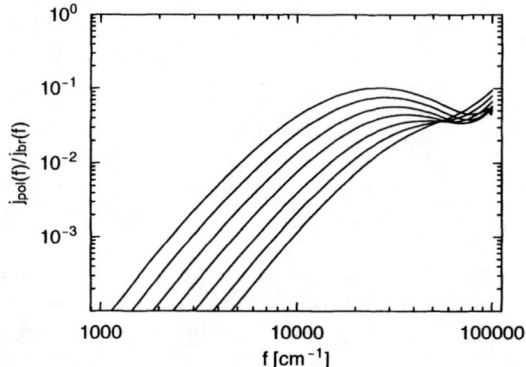

FIGURE 5. Ratio of the emission coefficient of polarization bremsstrahlung j_{pol} to electron-neutral bremsstrahlung j_{br} of electron-argon collisions at 5, 7, 10, 14, 20, 28 and 40 kK, from top to bottom.

$\mu(r) = -\alpha/r^2$ where α is the polarizability of the atom.

Figure 5 compares the polarization bremsstrahlung intensities to those of ordinary electron-atom bremsstrahlung for the case of electron-argon collisions. We see that at low frequencies ordinary bremsstrahlung is many orders of magnitude more efficient, but at the higher frequencies polarization bremsstrahlung intensities amount to roughly 10% of the electron-atom bremsstrahlung. Similar results are obtained for the other rare gases. At temperatures of up to 20 kK and frequencies smaller than 50,000 cm^{-1}, polarization bremsstrahlung amounts to at most a few per cent [19, 20].

Electron-ion bremsstrahlung. Electron-ion bremsstrahlung spectra can be approximated by semiclassical formulas [21], but quantum calculations may differ substantially from such estimates. The ratio of the quantal and semiclassical results are commonly expressed by (temperature-averaged) Gaunt factors $\xi(\lambda, T)$,

$$j_{\text{quant}}^{\text{e}-\text{i}}(\lambda, T) = \xi(\lambda, T) j_{\text{class}}^{\text{e}-\text{i}}(\lambda, T). \tag{1}$$

Our quantum computations are based on the Hartree-Fock-Slater ionic potentials; for the dipole operator we use again the acceleration form. The Gaunt factors thus computed for hydrogen and the light rare gas atoms are of order unity, but for argon, krypton and xenon Gaunt factors amount to 5 or 8 at the shortest wavelengths and highest temperatures [22].

Radiative attachment. Furthermore we consider radiative attachment of electrons to hydrogen and oxygen atoms that originate from the dissociation of H_2O. The intensities of these processes are computed for hydrogen from theoretical radiative attachment cross sections [23]), and for oxygen from measured cross sections of the inverse processes, absorption due to radiative detachment [24] and Kirchhoff's law [21].

Charge exchange radiation and related processes in helium. In He–He$^+$ collisions emission of radiation may be quite efficient, e.g., by radiative ion-atom association

$$\text{He}^+ + \text{He} \longleftrightarrow h\nu + \text{He}_2^+, \tag{2}$$

ion-atom bremsstrahlung, and charge exchange,

$$\text{He}^+ + \text{He} \longleftrightarrow h\nu + \begin{cases} \text{He}^+ + \text{He} \\ \text{He} + \text{He}^+ \end{cases}. \tag{3}$$

Absorption and emission coefficients for these processes are known for helium and other gases [25]. Charge exchange and ion-atom association radiation show strong emission in the blue and ultraviolet parts of the spectrum. For our purposes we use the results as presented in Ref. [25].

RESULTS

With our hydrodynamic-chemical model and the radiative processes discussed above we compute sonoluminescence spectra, pulse lengths and emitted photon numbers for a variety of rare gases and experimental conditions. We generally observe good agreement of emitted photon numbers (on the order of 10^5 for helium bubbles in water at room temperature and up to 10^7 for xenon bubbles in water near freezing temperature) as well as for the lengths of the emitted light pulses, about 100 to 350 ps, depending on the circumstances; see Fig. 3.

Figure 6 shows a sonoluminescence spectrum for an argon bubble in water at room temperature. The experimentally controlled parameters were approximated as closely as possible for our computation. Spectral shape and integrated absolute intensity agree well with experiment. Along with the total emitted intensities, in Fig. 6 we also show the contributions to the total spectrum due to different radiative mechanisms. One notices that electron-neutral argon bremsstrahlung gives the most significant contribution, with

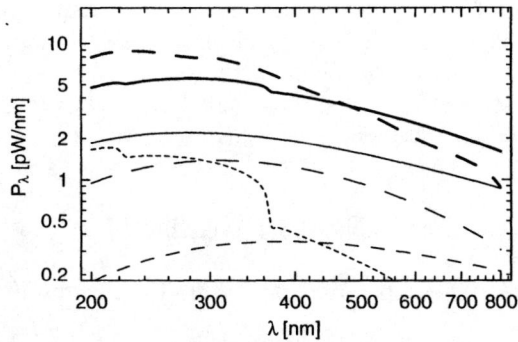

FIGURE 6. The spectrum and its composition for argon bubbles in water at room temperature. Shown are the total spectrum (thick solid line) and contributions due to electron-neutral argon bremsstrahlung (thin solid line), O⁻ radiation (dotted), H⁻ radiation (long dashed) and e-H neutral bremsstrahlung (short dashed). For comparison, the measured spectrum is indicated by the thick dashed line [26, 27].

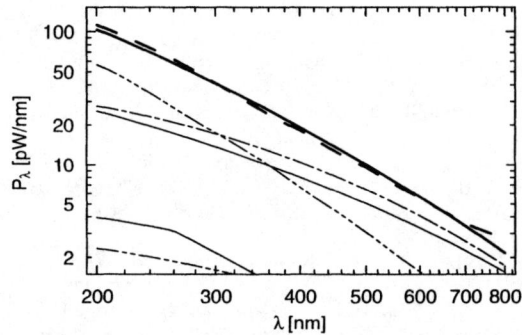

FIGURE 7. Spectrum of a helium bubble in water near 0°. For the computation it was assumed that the interior of the bubble with radius $R'_{min} = 0.160 \ \mu$m is heated to a maximum temperature of about 59,000 K. All radiative processes are indicated as in Fig. 6 above; electron-helium ion bremsstrahlung is indicated by the dashed-dotted line and the emission due to charge exchange and related processes, Eqs. (2) and (3), is indicated by the dash-triple dotted line.

the radiative attachment processes also being important. Electron-ion bremsstrahlung is negligible, owing to the small degree of ionization (less than 1%).

At higher bubble temperatures the degree of ionization increases. Higher temperatures in the bubble may be obtained in water at freezing temperatures: In that case the amplitude of the acoustic pressure field is typically larger than at room temperature which leads to a more dramatic collapse and thus more efficient quasi-adiabatic heating. However, even in that case our model predicts that electron-ion bremsstrahlung contributes little to the total intensity. In other words, even the largest Gaunt factors seen above do hardly affect the total spectra. But we note that at freezing water temperatures computed and measured spectral shapes show the greatest deviations. We attribute this to our assumption of a homogeneous bubble interior [20].

Simple extensions of our model, which model temperature gradients by a hotter core and a cooler area surrounding the core, give very good agreement of computed and measured spectral shapes [20]. Figure 7 shows the spectrum obtained from such a computation. In that case electron-ion bremsstrahlung is important, along with charge exchange radiation and related processes discussed above.

ACKNOWLEDGMENTS

The support of the Welch Foundation, Grant No. 1346, is gratefully acknowledged.

REFERENCES

1. Hammer, D., and Frommhold, L., *J. Mod. Opt.*, **48**, 239 – 277 (2001).
2. Brenner, M. P., Hilgenfeldt, S., and Lohse, D., *Rev. Mod. Phys.*, **74**, 425 – 484 (2002).
3. Hammer, D., and Frommhold, L., "The spectra of sonoluminescent rare gas bubbles," in *Spectral Line Shapes, Vol. 11*, edited by J. Seidel, AIP Press, N.Y., 2001, pp. 202 – 204.
4. Hilgenfeldt, S., Grossmann, S., and Lohse, D., *Phys. Fluids*, **11**, 1318 – 1330 (1999).
5. Brennen, C. E., *Cavitation and Bubble Dynamics*, Oxford University Press, New York, 1995.
6. Prosperetti, A., and Lezzi, A., *J. Fluid Mech.*, **168**, 457 – 478 (1986).
7. Toegel, R., Gompf, B., Pecha, R., and Lohse, D., *Phys. Rev. Lett.*, **85**, 3165 – 3168 (2000).
8. Wu, C. C., and Roberts, P. H., *Phys. Rev. Lett.*, **70**, 3424 – 3427 (1993).
9. Wu, C. C., and Roberts, P. H., *Proc. Roy. Soc. (London)*, **A 445**, 323 – 349 (1994).
10. Moss, W. C., Clarke, D. B., White, J. W., and Young, D. A., *Phys. Fluids*, **6**, 2979 – 2985 (1994).
11. Vuong, V. Q., Szeri, A. J., and Young, D. A., *Phys. Fluids*, **11**, 10 – 17 (1999).
12. Yasui, K., *Phys. Rev. E*, **56**, 6750 – 6760 (1997).
13. Gompf, B., Günther, R., Nick, G., Pecha, R., and Eisenmenger, W., *Phys. Rev. Lett.*, **79**, 1405 – 1408 (1997).
14. Frommhold, L., *Collision-Induced Absorption in Gases*, Cambridge Univ. Press (1993).
15. Herman, F., and Skillman, S., *Atomic Structure Calculations*, Prentice-Hall, Englewood Cliffs, N.J., 1963.
16. Frommhold, L., *Phys. Rev. E*, **58**, 1899 – 1905 (1998).
17. Geltman, S., *J. Quant. Spectros. Rad. Transfer*, **13**, 601 – 613 (1973).
18. Astapenko, V. A., and Bureyeva, L.A., and Lisitsa, V. S., *JETP* **90**, 434 (2000).
19. Hammer, D., and Frommhold, L., *Phys. Rev. A*, **64**, 024705 (2001), erratum: Ref. [28].
20. Hammer, D., and Frommhold, L., *Phys. Rev. E*, **65**, 046309 (2002).
21. Zel'dovich, Y. B., and Raizer, Y. P., *Physics of Shock Waves and High Temperature Hydrodynamic Phenomena*, vol. 1, Academic Press, New York, 1966.
22. Hammer, D., and Frommhold, L., *Phys. Rev. E* (2002), to appear in the July issue.
23. Geltman, S., *Astrophys. J.*, **136**, 935 – 945 (1962).
24. Bates, D. R., editor, *Atomic and Molecular Processes*, Acad. Press, New York, 1962.
25. Ermolaev, A. M., Mihajlov, A. A., Ignjatović, L. M., and Dimitrijević, M. S., *J. Phys. D*, **28**, 1047 – 1057 (1995).
26. Barber, B. P., Hiller, R. A., Löfstedt, R., Putterman, S. J., and Weninger, K. R., *Phys. Rep.*, **281**, 65 – 143 (1997).
27. We divided the intensity of the experimental spectra by 4π since this factor was reported to be missing in plots of the measured spectral intensity [29, 30].
28. Hammer, D., and Frommhold, L., *Phys. Rev. A*, **64**, 059901 (2001).
29. Weninger, K. R., Camara, C. G., and Putterman, S. J., *Phys. Rev. E*, **63**, 016310 (2001).
30. Vazquez, G., Camara, C., Putterman, S., and Weninger, K., *Opt. Lett.*, **26**, 575 – 577 (2001).

HIGH-DENSITY SOURCES

Frequency fluctuation model survey

A. Calisti, S. Ferri, C. Mossé, B. Talin

PIIM, Université de Provence, 13397 Marseille cedex 20, France.

Abstract. Developed since the late eighties the frequency fluctuation model line shape code (today called PPP, formerly PimPamPoum) is recognized as a flexible fast and accurate state of the art plasma spectroscopy tool altogether with a powerful link between charge dynamics theory in hot and dense matter and spectroscopy.

INTRODUCTION

The lineshape code [1] has been developed at the origin with the objective of synthesizing spectral lines emitted by plasmas for any emitter and plasma conditions. This work has been motivated both for plasma diagnostic purposes and non linear charge dynamic investigations. The numerical code [1] based on original models involves a few steps that realize the best compromise between accuracy and speed. Particularly, the perturber charge motion is accounted for within the so-called frequency fluctuation model (FFM) that post-processes the spectra from static to dynamic. The code has played an active role convincing people to use shapes instead of simple line widths. Over a decade the lineshape code has been improved mostly under the control of lineshape molecular dynamics simulation which is an indissoluble part of it. The whole model has been the source of several other specific developments. Today, a new portable version more accurate and flexible is on the way. The purpose of this paper is to recall the main features and capabilities of the model and outline its short term expected evolution.

STARK BROADENING SUMMARY

Static ions

The preliminary formalism rigorous but somewhat useless in this case ends where start most of the Stark broadening models i.e. the following formula, already involving the main necessary approximations

$$I(\omega) = \frac{1}{\pi} Re \int_0^\infty e^{i\omega t} C(t)dt \qquad (1)$$

---

[1] A well interfaced version of the PPP code named TOTAL is used at the LLNL. Some calculations cited in reference below mention this name.

CP645, *Spectral Line Shapes: Volume 12, 16th ICSLS*, edited by C. A. Back
© 2002 American Institute of Physics 0-7354-0100-4/02/$19.00

Most of the work about modelling Stark broadened lineshapes has to be carried out to calculate the dipole auto-correlation function $C(t)$. Generally the framework of line-shape models is a semi-classical one i.e. limited to a description of the behavior of a quantum emitter submitted to the perturbation of a time dependent classical electric field. The basic ingredients needed for synthesizing plasma lineshapes are of three types i) relevant quantum atomic data for the problem under interest, ii) static and dynamic characteristics of the external classical perturbation depending on the plasma constituents, the electron temperature Te and the electron density Ne iii) ionization rates and atomic level populations whose knowledge relies on both the previous quantum and classical data. A fourth ingredient, namely computing, should be also mentioned. Owing to their complexity, such problems are solved numerically implying a discretization of the whole data involved. In the so-called quasi-static approximation two key hypothesis are used: static ions and impact electrons. This results in solving the lineshape problem for quantum systems with imaginary energy levels. Numerically, lineshapes are expressed with weighted sums running over a discrete microfield probability function. This stage is used to collect the Stark components by diagonalizing the Liouville operator for each discrete microfield. Ultimately, the lineshape can be written as a scalar product in Liouville space

$$I(\omega) = Re << P|(i\omega \cdot 1 - i\Omega)^{-1}|Q >>$$
$$= \frac{1}{\pi} Re \sum_{k=1,N} \frac{\xi_k}{\omega - \zeta_k} \qquad (2)$$

where Ω is a complex diagonal matrix formed with the distinct generalized complex frequencies of a set of observables called the radiative channels (or components). The process allowing to characterize these objects from the Stark resonances is equivalent to a numerical renormalisation of a very large data set in order to extract from it reduced relevant information which, contrary to the Stark resonances, involve a clear physical meaning. The detailed calculation of the two complex vectors P and Q partially depending on dipole operators and level populations will not be given here. This formalism limited to static ion perturbers makes use of homogeneous and inhomogeneous broadening notions respectively related to electron and ion perturbers. In the following section an intermediate profile status will be encountered due to ion motion.

Ion dynamics

To account for ionic field fluctuations with an average rate v [2] it is postulated that the components are mixed according to a simple stationary Markov process whose transition rate matrix vW obeys the relations:

$$\sum_j W_{i,j} p_j = 0$$

---

[2] In general v is taken to be a plasma characteristic rate v_t/r_0 where r_0 is the mean particle distance and v_t the thermal velocity

248

$$\sum_i W_{i,j} = 0 \tag{3}$$

where p is the component weight vector $\sum_i p_i = 1$. The lineshape with ion dynamics takes the simple form

$$I(\omega) = Re << P|(i\omega \cdot 1 - i\Omega - vW)^{-1}|Q >> \tag{4}$$

which in turn can be transformed into a diagonal form similar to 2.

It is straightforward to note that the fluctuation process satisfies two limit conditions, first, the slow fluctuation limit $(v \to 0)$ useful for nearly static conditions thus, mainly relevant for ion broadening and second, the fast fluctuation limit $(v \to \infty)$ that matches nearly impact conditions, mainly relevant for electron broadening. This will be commented again below.

Molecular dynamics simulation

As mentioned, lineshape simulations based on molecular dynamics (MDLS) providing benchmark spectra is an indispensable part of the model development. MDLS allows to carry out classical lineshape calculations accounting for all the correlations between charges. First a standard simulation process is used whose details are given elsewhere. The sample microfields are calculated with molecular dynamics that also provides all relevant quantities about microfield statistical properties and about charge density and charge motion around the emitter. Second, for each sample, the dipole autocorrelation function is obtained by integrating an evolution equation involving the quantum system whose dipole is coupled to the external perturbation. This problem is equivalent to solve numerically a stochastic equation that can be written:

$$\frac{dU_f(t)}{dt} = -iL_f(t)U_f(t)$$
$$\mathbf{U}(t) = \{U_f\}_{f \in F} \tag{5}$$

where the mean evolution operator $\mathbf{U}$ results from an average over a well chosen sampling set F of time dependent microfield sequences generated by molecular dynamics. MDLS has been used mainly to benchmark ion broadening i.e. to study moderate fluctuation effects on line. This corresponds to conditions such as the fluctuation rate v is small compared with some characteristic ionic broadening of the line. An overall good agreement has been found comparing MDLS with PPP results that validates the fast code. More recently, MDLS has been extended to study line broadening for neutral emitters in electrons. For this case v is no longer small compared with the average static broadening.

APPLICATIONS AND PROSPECT DEVELOPMENTS

Because it is based on the definition of physical observables the lineshape code results very flexible. Recently it has proven to be very useful for investigating radiative properties beyond the standard Stark broadening for moving ions and impact electrons. Many aspects have to be considered as the radiative properties are the tip of plasma physics.

Applications to plasma spectroscopy

A few necessary interpretations of measured spectra have been a part of the work associated to the code development[2]. Several applications have been performed for plasma conditions ranging from $Ne = 10^{13}cm^{-3} \rightarrow 10^{26}cm^{-3}$ and $Te = 10^4 K \rightarrow 10^7 K$ for moderate plasma coupling parameters. A feedback is obtained with a few experiments providing indications for tentative model improvements. This has been particularly the case with neutral hydrogen lines in tokamaks and helium like hydrogen lines in various laboratory devices.

Resonant diffusion

Owing to the physical observable concept linked to the line homogeneities defined for the static limit, it has been possible to extend the FFM model to second order optical mechanisms. This extension is carried out accounting for the whole aspects of line broadening including ion dynamics and fine structure[3]. In addition this quite general approach can be extended also to Doppler broadening as a typical inhomogeneous broadening mechanism. Two main objectives are considered for modelling resonant diffusion. The first one is related to radiation transport in plasmas, the second one to active spectroscopy i.e. to model the response of a plasma perturbed by a monochromatic radiation. For both cases the spectral emissivity of a uniform plasma sample perturbed by an harmonic external field has to be modelled.

Electron broadening

The capability of the model for describing emitter dipole relaxation in plasmas accounting for stochastic ionic field fluctuations has been extended naturally to model electron broadening. On a formal point of view the only difference between ion and electron field fluctuations is the fluctuation rate magnitude. Thus, as for moving ions, dipole relaxation by electrons is modelled in terms of a markovian mixing of the components. As noted before, for the ion case the fluctuation rate is generally small compared to the average static ion broadening, however, for electrons the fluctuation rate can be large and can force the lineshape to tend to a Lorentz shape. Because the need to work with classical electrons the study relies on a regularized classical Coulomb ion-electron potential that prevents divergences at short distances[4]. This potential is designed to account approximatively for quantum features related to the Coulomb potential. Here, one of objectives is to make possible the discussion of some unfounded hypothesis used in classical line broadening models designed for ion emitters in plasmas.

ACKNOWLEDGMENTS

The authors would like to thank several people for their contribution to the code development and for helpfull discussions and collaborations: S.Alexiou, J.Dufty, L.Godbert-Mouret, M.Gigosos, C.Iglesias, C.J.Keane, L.Klein, M.Koubiti, H.-J.Kunze, R.W.Lee, V.Lisitsa, J.K.Nash, F.Rosmej, R.Stamm.

REFERENCES

1. A.Calisti, F.Khelfaoui, R.Stamm, B.Talin and R.W.Lee, *Phys Rev*A**42** 5433 (1990).
   B.Talin, A.Calisti, L.Godbert, R.Stamm, R.W.Lee and L.Klein,*Phys. Rev.*A**51** 1918 (1995).
   A.Calisti, S.Ferri, M.Koubiti, T.Meftah, C.Mossé, L.Mouret, R.Stamm, S.Alexiou, R.W.Lee and L.Klein,*JQSRT* **58** 953 (1997).
2. L.Godbert, A.Calisti, R.Stamm, B.Talin, S.Glenzer, H.J.Kunze, J.Nash, R.W.Lee and L.Klein *Phys. Rev.* E**49**, 5889 (1994).
   N.C.Woolsey, A.Asfaw, B.Hammel, C.Keane, C.A.Back, A.Calisti, C.Mossé, R.Stamm, B.Talin, J.S.Wark, R.W.Lee and L.Klein *Phys Rev* E**53** 6396 (1996).
   N.C.Woolsey, B.Hammel, C.Keane, A.Asfaw, C.A.Back, J.C.Moreno, J.K.Nash, A.Calisti, C.Mossé, R.Stamm, B.Talin, L.Klein and R.W.Lee *JQSRT* **58**, 975 (1997).
   S.Ferri, A.Calisti, R.Stamm, B.Talin, R.W.Lee and L.Klein Phys Rev E**58**, R6943 (1998).
   L.Godbert-Mouret, T.Meftah, A.Calisti, R.Stamm, B.Talin, M.Gigosos, V.Cardenoso, S.Alexiou, R.W.Lee and L.Klein. *Phys Rev Lett* , 5568 (1998).
   F.B.Rosmej, D.H.H.Hoffmann, W.Süss, M. Geibel, A.Ya.Faenov, T.A.Pikuz, A.Calisti, R.Stamm and B.Talin *JQSRT* **71**, 631 (2001).
3. C.Mossé, A.Calisti, R.Stamm, B.Talin, R.W.Lee, J.Koch, A.Asfaw, J.Seely, J.Wark et L.Klein, *JQSRT* **58**, 803 (1997).
   C.Mossé, A.Calisti, R.Stamm, B.Talin, R.W.Lee and L.Klein *Phys Rev* A **60**, 1005 (1999).
4. B.Talin, A.Calisti and J.W.Duffy, *Phys Rev* E **65**, 056406(2002).
   A.Calisti, E.Dufour B.Talin and J.W.Dufty, this proceedings, page 252.
   E.Dufour, A.Calisti, B.Talin, M.A.Gigosos, M.González and J.W.Dufty, this proceedings page 268.

Models For Electronic Electric Field Distribution Function At a Positive Ion

A. Calisti[*], E. Dufour[*], B. Talin[*] and J.W. Dufty[+].

[*]Laboratoire de Physique des Interactions Ioniques et Moléculaires, Université de Provence, 13397 Marseille cedex 20, France.
[+]Department of Physics, University of Florida, Gainesville, USA.

Abstract. A single positive ion is imbedded in an electron gas with overall charge neutrality. A classical statistical mechanics is considered using an electron-ion Coulomb potential regularized at distances within the de Broglie length. The electric field distribution at the ion is studied as a function of ion-electron coupling using molecular dynamics simulation and theoretical model (APEX, Adjustable-Parameter Exponential Approximation). Agreement between theory and simulation is quite good in general, although differences are observed for very strong ion-electron coupling due to enhanced importance of close electron-ion configurations.

INTRODUCTION

The description of a system of same sign charges (with a uniform neutralizing background) is a well-studied problem. Because of the repulsive Coulomb interactions close encounters are rare and for a wide range of state conditions a description via classical mechanics can be justified. In this way classical molecular dynamics (MD) and many-body methods have been applied to describe both structure and dynamics of such system with considerable success. In contrast, systems with charges of opposite sign are inherently quantum mechanical, due to unbounded Coulomb attraction at short distances, and have no simple classical counter part. Nevertheless, there may be state conditions (Landau length and average inter-particle distance large compared to the de Broglie length) such that most properties of interest are not sensitive to the quantum domain. In this case, a representative classical mechanics can be postulated using a Coulomb potential that is regularized to be finite for distances smaller than the de Broglie length [1]. Classical molecular dynamics and many-body methods then can be applied for such systems as well [2]. Here, a single positive ion embedded in an electron gas with overall charge neutrality is considered. For this simple impurity system there are only three dimensionless parameters: the charge number of the ion Z, the electron-electron coupling constant Γ, and the de Broglie wavelength relative to the inter-electron distance δ. The electron-ion coupling is measured by the value of the regularized ion-electron potential at the origin, $\sigma = Z\Gamma/\delta$. The application of semi-classical methods requires $\sigma > 1$ and $\delta < 1$. The electric field distribution at the ion is studied as a function of ion-electron coupling using molecular dynamics simulations and theoretical models. The theoretical models considered here are based on the hypernetted chain approximation (HNC) [3] integral equation for the electron-electron

CP645, *Spectral Line Shapes: Volume 12, 16th ICSLS*, edited by C. A. Back
© 2002 American Institute of Physics 0-7354-0100-4/02/$19.00

and electron-ion charge densities. Properties of the electron electric field at the ion (covariance and microfield distribution) are obtained from approximations that are determined from these charge densities. The microfield distribution and its nearest neighbour approximation are calculated from MD, the Baranger-Mozer (BM) approximation [4] and its extension to strong coupling APEX [5] adapted to the case of opposite sign charges.

THEORY AND SIMULATION

The system considered here consists of N_e electrons with charge $-e$, an infinitely massive positive ion with charge Ze placed at the origin, and a rigid uniform positive background for overall charge neutrality. The electron-electron interactions are taken to be Coulomb while the electron-ion interactions occur via a regularized Coulomb potential $V(r)$.

$$V(r) = Z\Gamma e^{-r/\lambda}(1 - e^{-r/\delta})/r \qquad (1)$$

with

$$\delta = h/\sqrt{2\pi m_e k T_e} \qquad (2)$$

and

$$\Gamma = e^2/r_0 k T_e. \qquad (3)$$

The regularized potential prevents for divergences at short and long distances. The long range box-size screening $\lambda \gg r_0$ must be compatible with MD simulation and large enough in order to not affect the short range structure of the electron gas and not mask the electron screening mechanisms. The structural properties of interest here are the electronic electric field distribution functions at the ion.

The electron electric field at the ion due to all electrons is obtained by:

$$\mathbf{E} = -\nabla_0 V(\{\mathbf{r}_{io}\}) = \sum_{i=1}^{N_e} \mathbf{e}(\mathbf{r}_{io}) \qquad (4)$$

where $\mathbf{r}_{io} = \mathbf{r}_i - \mathbf{r}_0$ is the position of the ith electron relative to the ion and,

$$V(\{\mathbf{r}_{io}\}) = \sum_{i=1}^{N_e} V(\mathbf{r}_{io}), \qquad (5)$$

$$\mathbf{e}(\mathbf{r}_{i0}) = \frac{\hat{\mathbf{r}}_{i0}}{r_{i0}^2}(1 - (1 + \frac{r_{i0}}{\delta})e^{-r_{i0}/\delta}). \qquad (6)$$

The covariance of the electric field is related to the electron charge density through:

$$C = \frac{r_0^4}{e^2}\langle \mathbf{E}.\mathbf{E} \rangle = -\frac{1}{Z\Gamma}\langle \nabla_0.\mathbf{E} \rangle = \frac{3}{4\pi Z\Gamma}\int d\mathbf{r}\, g_{ie}(r)\nabla.\mathbf{e}(r) \qquad (7)$$

The probability density for electric field magnitudes is defined by:

$$P(\varepsilon) = 4\pi\varepsilon^2\langle \delta(\varepsilon - \mathbf{E}) \rangle = \frac{\varepsilon^2}{2\pi^2}\int d\lambda e^{-i\lambda.\varepsilon}\langle e^{-i\lambda.\mathbf{E}} \rangle \equiv \frac{\varepsilon^2}{2\pi^2}\int d\lambda e^{-i\lambda.\varepsilon}e^{G(\lambda)} \qquad (8)$$

The second equality defines the generating function $G(\lambda)$.

$$G(\lambda) = \sum \int d\mathbf{r}_1...d\mathbf{r}_m g^m(\mathbf{r}_1...\mathbf{r}_m)\phi(\lambda,\mathbf{r}_1)...\phi(\lambda,\mathbf{r}_m) \qquad (9)$$

$$\phi(\lambda, \mathbf{r}) = e^{i\lambda.e(\mathbf{r})} - 1 \tag{10}$$

Here, $g^m(\mathbf{r}_1, \ldots, \mathbf{r}_m)$ is a cluster function representing the correlations among m electrons and the ion. At weak coupling it is expected that the series can be truncated at m=1, leading to the Baranger-Mozer (BM) approximation.

$$P(\varepsilon) = \frac{\varepsilon^2}{2\pi^2} \int d\lambda e^{-i\lambda.\varepsilon} \exp \frac{3}{4\pi} \int d\mathbf{r} g_{ie}(\mathbf{r}) \phi(\lambda, \mathbf{r}) \tag{11}$$

In this approximation, the microfield distribution is determined from the electron charge density alone. But it is easy to verify that the covariance of the electric field differs from the exact result (7).

$$C^{BM} = \frac{3}{4\pi} \int d\mathbf{r} g_{ie}(\mathbf{r}) e(\mathbf{r}).e(\mathbf{r}) \tag{12}$$

To account for multielectron correlations it is necessary to include higher order terms in the series. An improvement on the BM microfield distribution is given by the APEX approximation.

This approximation can be obtained by defining a function $\phi^*(\lambda, \mathbf{r})$ in terms of a 'renormalized field' $e^*(\mathbf{r})$ [6].

$$\phi^*(\lambda, \mathbf{r}) = e^{i\lambda.e^*(\mathbf{r})} - 1 \tag{13}$$

with

$$e^*(\mathbf{r}) = \ln g_{ie}(\mathbf{r}) \approx \nabla \frac{\overline{Z}}{Z} \frac{1}{(1 - (\frac{\delta}{\overline{\lambda}})^2)} \frac{1}{r} (e^{-r/\overline{\lambda}} - e^{-r/\delta}) \tag{14}$$

This effective field has two effects due to correlations, a long range screening through $\overline{\lambda}$ and an overall intensity change through $\overline{Z}$. For same charge perurbers the correlations are long range and the standard method is to vary $\overline{\lambda}$ with $\overline{Z} = Z$ introduced and adjusted to get the correct covariance. Here, the effective charge number $\overline{Z}$ is used to require the correct covariance with $\overline{\lambda}$ constant and of the order of the system size. This appears to represent better the short range correlations, particularly at stronger electron-ion coupling.

The generating function is, then, given by:

$$G^{APEX} = \frac{3}{4\pi} \int d\mathbf{r} g_{ie}(\mathbf{r}) R(\mathbf{r}) \phi^*(\lambda, \mathbf{r}) \tag{15}$$

with R(r)=e(r)/e*(r); the corresponding covariance being:

$$C^{APEX} = \frac{3}{4\pi} \int d\mathbf{r} g_{ie}(\mathbf{r}) e(\mathbf{r}).e^*(\mathbf{r}) \tag{16}$$

The field $e^*(\mathbf{r})$ is chosen such that the result (16) is exact, i.e. $C^{APEX} = C$. As for BM, the electric field distribution is then determined in term of the electron charge density for which a standard theoretical description is given by the resolution of the hyper-netted-chain set of equations:

$$\begin{cases} \widetilde{h}_{ie}(k) = \widetilde{c}_{ie}(k)/(1 - \rho \widetilde{c}_{ee}(k)) \\ \widetilde{h}_{ie}(k) = \widetilde{c}_{ie}(k)/(1 - \rho \widetilde{c}_{ee}(k)) \\ h_{ie}(r) = g_{ie}(r) - 1 \end{cases} \tag{17}$$

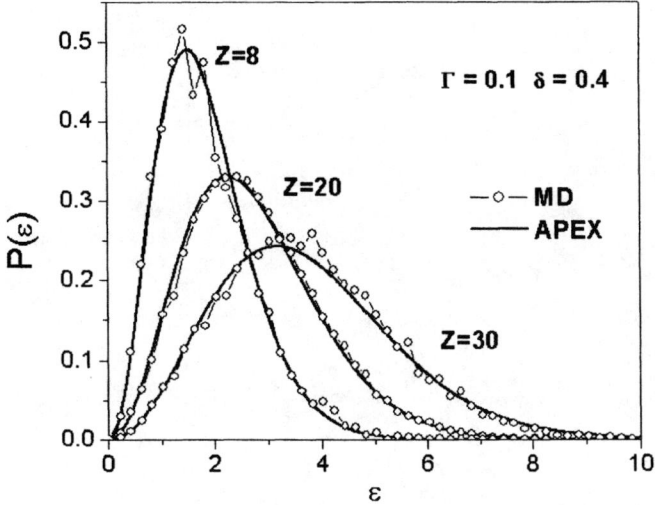

FIGURE 1. Comparison of microfield distributions from MD and APEX. A dimensionless microfield in units of e/r_0^2 is used. $n_e = 2.5 \ 10^{22} \text{cm}^{-3}$, $T_e = 7.9 \ 10^5$ K

where $c_{ee}(r)$ is the direct correlation function for the electron OCP.

A primary motivation for associating the classical Hamiltonian with the underlying quantum system is to allow application of molecular dynamics simulation. MD accounts for all correlations and many-body effects for both static and dynamic properties of classical systems. To allow the most efficient MD simulations, periodic boundary conditions are used and the potential is screened at distances of the order of the system size. This screening is always much larger than the correlation length, so the results are essentially the same as would be obtained with an Ewald sum to correct for small finite system size.

DISCUSSION

The microfield distributions from MD and APEX are compared in figure 1 for $n_e = 2.5 \ 10^{22} \ \text{cm}^{-3}$ and $T_e = 7.9 \ 10^5$ K, Z=8, 20 and 30 corresponding respectively to coupling values of $\sigma = 2$, 5 and 7.5. It is seen that APEX gives quite good approximation even at the strongest coupling values (the Z=40 case, not shown here, shows a similar good agreement). The same results are shown in figure 2, including the BM approximation and the corresponding nearest neighbor distributions, in the case of Z=20. Similar results are obtained in all cases. The microfield distribution differs significantly from the nearest neighbor distribution, showing that collective electron effects dominate. In all cases APEX is a significant improvement upon BM although the latter is qualitatively correct.

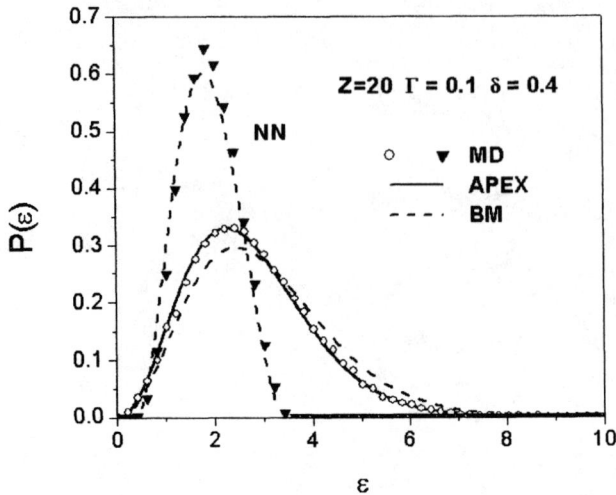

FIGURE 2. Comparison of microfield distributions from MD, BM and APEX for Z=20. Also shown is the nearest neighbor microfield distribution from MD. n_e=2.5 10^{22}cm^{-3}, T_e=7.9 10^5 K.

Now consider lower temperature cases. Figure 3 shows a comparison of MD, APEX and BM for Z=3 corresponding to a coupling value of σ=7.5. In contrast to the higher temperature cases, the nearest neighbor distribution becomes more similar to the full distribution as the coupling strength is increased.

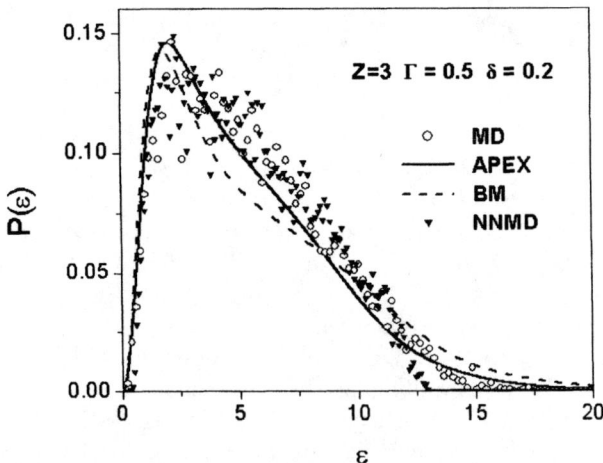

FIGURE 3. Comparison of microfield distributions from MD, BM and APEX for Z=3. Also shown is the nearest neighbor microfield distribution from MD. n_e=3.2 10^{18}cm^{-3}, T_e=7.9 10^3 K

This shows the increasing dominance of the nearest neighbor. A shoulder and a broadening around the most probable value appear. The BM approximation shows these general new features and APEX provides corrections for a good quantitative representation except at the strongest coupling case.

Figure 4 shows comparisons between MD and APEX for more realistic plasma conditions. The results correspond to $Z=1$ for $n_e=1.\ 10^{18}$ cm^{-3} and $T_e=60000$K which lead to $\Gamma=4.49\ 10^{-2}$. Here again APEX gives a quite good agreement compared to MD.

For the results presented in Figure 5, the APEX calculations have been performed in two ways, corresponding to two different choices for the renormalized field e*. In the case labeled APEX Z, $\overline{Z}$ varies and $\overline{\lambda}$ is constant (like in all the other cases, here) and in the case labeled APEX λ, the inverse is done. The plasma conditions ($Z=1$, $n_e=1.\ 10^{18}$ cm^{-3} and $T_e=20000$K) are such that the coupling parameter is $\Gamma=0.134$. It can be noticed that both the two APEX calculations give qualitatively correct results but MD lies in between. It seems that for APEX the good answer would be to vary the two parameters together.

CONCLUSION

Both MD and current theoretical methods developed for repulsive potentials have been shown to apply as well for the attractive classical electron-ion system. The accuracy of these methods extends even to strong attraction where electron densities increase by

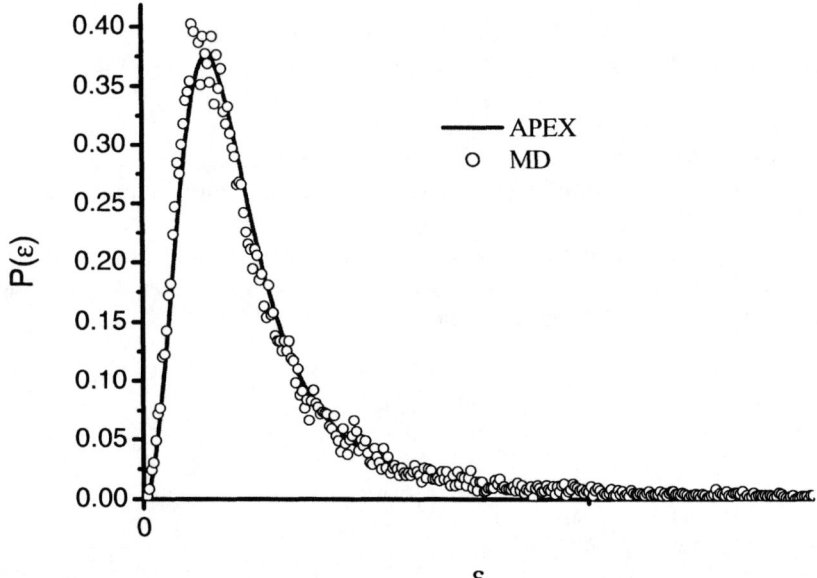

FIGURE 4. Comparison between APEX and MD for $Z=1$ for $n_e=1.\ 10^{18}$ cm^{-3} and $T_e=60000$K.

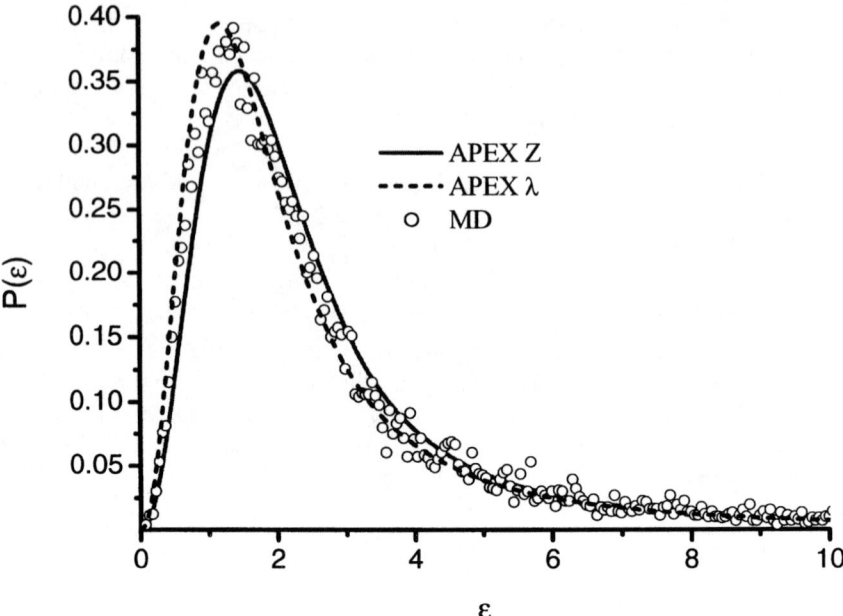

FIGURE 5. Comparison between the two different APEX calculations and MD for Z=1 for $n_e=1.10^{18}cm^{-3}$ and $T_e=20000K$.

several orders of magnitude. These results justify the application of these methods to more practical conditions of experimental interest. These include the low Z ions produced at low temperatures by short pulse lasers and the high Z ions for diagnostics in hot, dense laser fusion plasmas.

The main results of this study can be summarized by: -the electron electric fields at the ion can be studied by MD simulation, even at strong electron-ion coupling, except for conditions where close electron-ion configurations become important. In this domain, changes in the implementation of APEX are required for the microfield distribution, - both BM and APEX show qualitative as well as quantitative differences from the MD results whenever the nearest neighbor field becomes important.

REFERENCES

1. Minoo H., Gombert M.-M. and Deutsch C., Phys. Rev. A**23**, 924 (1981).
2. Talin B., Calisti A. and Dufty J., Phys. Rev. E **65**, 056406 (2002).
3. Rogers F.J., Phys. Rev. A**29**, 868 (1984).
4. Baranger M. and Mozer B., Phys. Rev. **115**, 521 (1959); Mozer B. and Baranger M., Phys. Rev. **118**, 626 (1960).
5. Iglesias C.A. and Lebowitz J.L., Phys. Rev. A**30**, 2001 (1984).
6. Dufty J.W., Boercker D.B. and Iglesias C.A., Phys. Rev. A **31**, 1681 (1985).

Stark Broadening Model for Ionic Emitter Spectroscopy and Application to Li-like Rydberg Satellite Series in Laser Produced Plasmas.

A.Calisti[a], F.B. Rosmej[a,b], B. Talin[a], C. Mossé[a], S. Ferri[a],
M. Geissel[c], D.H.H. Hoffmann[b,c],
A.Ya. Faenov[d], T.A. Pikuz[d].

[a]*Laboratoire de Physique des Interactions Ioniques et Moléculaires, Université de Provence, 13397 Marseille cedex 20, France.*
[b]*GSI-Darmstadt, Plasmaphysik, Germany.*
[c]*Technische Universität Darmstadt, Institut für Kernphysik, Germany.*
[d]*Multicharged Ions Spectra Data Center of VNIIFTRI, Mendeleevo, Russia.*

Abstract. Dense plasmas have been created at the GSI with the one hundred joules **nhelix**-laser under well-defined hydrodynamics conditions. The application of two-dimensional x-ray optics enabled the registration of the Li-like satellite series $1s2lnl'-1s^22l''$ with simultaneous high spectral and spatial resolution with excellent signal to noise ratio. The data show their emission is confined to the area of the laser spot thereby providing a localized emission source of the highest density region. The satellite transitions are also shown to be optically thin. Therefore, outstanding properties for Stark broadening analysis are met. The fast computer code PPP is capable to synthesise line shapes for experiment interpretation in a wide range of emitters and plasma conditions. Full Stark broadening calculations carried out with the PPP code involving the configurations $1s2lnl'$ with n=2-7 demonstrate their advantageous properties.

INTRODUCTION.

In hot, dense plasmas created in high power laser experiments, spectroscopy has been a very powerful tool for the determination of important plasma parameters. In particular bound-bound spectral lines originating from autoionizing and singly excited states are known to be very efficient to infer temperature, density and charge states through the simulation (kinetics and line broadening theories) of the observed spectral distribution. Obviously, the capability to treat complex experimental data strongly depends on the theoretical models and experimental boundary conditions (e.g. opacity, transient evolution). The rapid development of high intensity laser and ions beams allows to create high energy density plasmas in parameter regimes which stimulate new efforts in theory and simulation.

Numerous experiments request to diagnose large-scale extremely dense plasmas (e.g., future GSI-experiments employing the kilo-joule PHELIX-laser and the intense heavy ion beams). Estimations show that not only resonance lines but dielectronic

CP645, *Spectral Line Shapes: Volume 12, 16th ICSLS*, edited by C. A. Back
© 2002 American Institute of Physics 0-7354-0100-4/02/$19.00

satellites will show opacity effects [1] preventing the application of well-known temperature diagnostics. Further serious obstacles will be related to the plasma inhomogeneity. In this issue, test bed plasmas have been created with the **nhelix**-laser under well-defined hydrodynamics conditions. The application of two-dimensional x-ray optics enabled the registration of the Li-like satellite series $1s2lnl'-1s^221''$ with simultaneous high spectral and spatial resolution with excellent signal to noise ratio [2]. The data show their emission is confined to the area of the laser spot thereby providing a localized emission source of the highest density region. The highest satellite series transitions are shown to be optically thin. These very favourable properties suggest comparisons with theory in the way to perform diagnostics. In this context, the fast computer code PPP, developed a few years ago [3,4,5], is capable to provide Stark broadening spectral profiles for the radiation emission of multiply charged ions of even complex autoionizing states.

After a brief description of the laser-target interaction experiment, the outstanding properties of the observed line spectra are demonstrated. Then, the specific characteristics of the PPP code allowing to deal with very complex atomic structure are developed. Finally, full Stark broadening calculations carried out with the PPP code involving the configurations $1s2lnl'$ with $n=2-7$ demonstrate their advantageous diagnostic properties.

EXPERIMENTAL SETUP.

Dense plasmas have been created at GSI in Darmstadt, Germany, with the one hundred joules **nhelix**-laser facility irradiating massive silicon targets. The laser is a Nd-glass/Nd-Yag laser operating at 1.064 μm with a pulse duration of 15 ns. A focusing lens with 100 mm diameter and 150 mm focal length has been used. In the present experiments, targets have been irradiated with 30-60 J and spot sizes of about 100 until 500 μm. Space resolved X-ray spectra have been recorded by means of spherically bent mica crystals [6] with curvature radius of 150 mm and Kodak DEF-5 X-ray film (grain size of 1.6 μm). Two 1 μm polypropylene foils coated with 0.2 μm Al were used to protect the film from visible light. Typical distances target-crystal and crystal-film were 30 cm and 10 cm, respectively. The central Bragg angle was 35 degrees. Plasma images were digitised with 5000 dpi (dots per inch) employing a EUROCORE 10.000 dpi drum scanner.

This arrangement enabled a spatial resolution in the z direction of 12.8 μm per pixel and a spectral resolution of about 3500. (z is the direction of the expanding plasma perpendicular to the target surface). Spectra have been corrected for filter transmissions, crystal reflectivity, film response and non linear dispersion scale.

Figure 1(a) shows the plasma X-ray image and figures 1(b) and 1(c) depict the corresponding K-shell emission spectra at different target distances. The Rydberg resonance line series $1snp$ $^1P_1-1s^2$ up to about $n=10$ and the high energy Rydberg satellite transitions $1s2lnl'-1s^221''$ until the series limit are observed.

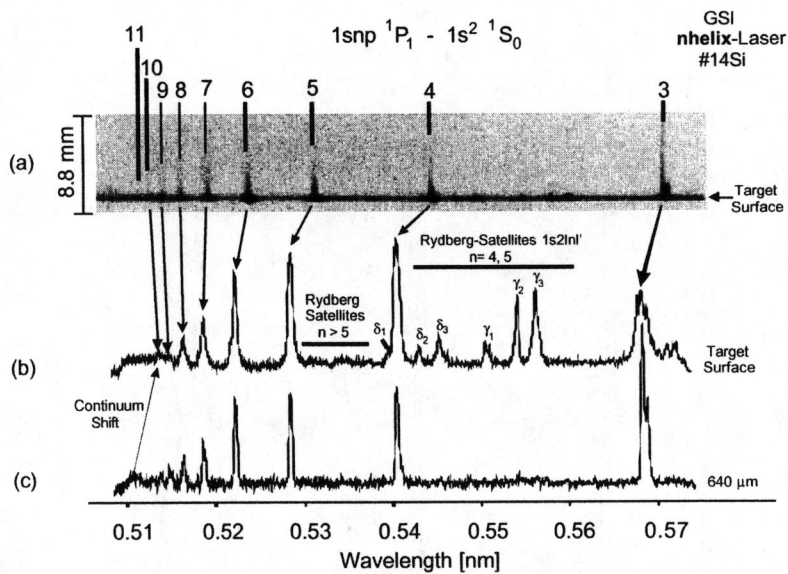

FIGURE 1. (a) is the space resolved plasma image of silicon K-shell X-ray line emission. The He-like resonance lines until the series limit are indicated. Two spectra are presented: (b) from the target surface showing the Rydberg-series of the He-like resonance lines and the associated dielectronic satellites and (c) in the expanding plasma taken 640 μm above the surface.

OUTSTANDING PROPERTIES FOR DIAGNOSTICS.

Figures 1(b) and 1(c) show spectra taken at the target surface and 640 μm above the target, respectively. The comparisons of these spectra clearly demonstrate that the satellite transitions are almost entirely confined to the spatial area of the laser spot size (which is about equal to the extension of the base line of Fig.1(a) created by bremsstrahlung). Prominent satellite groups γ_1, γ_2, γ_3 and also δ_1 and δ_2 are well resolved. Moreover, these transitions have favourable properties against opacity. Table 2 depicts the line-center opacities τ_0 of various resonance lines and satellites for $n_e = 10^{21}$ cm^{-3}, $kT_e = 300$ eV and $L_{eff} = 500$ μm (L_{eff} is the effective photon path length). Helium alpha and helium beta have a line center optical thickness of several hundred and even the helium alpha satellites are optically thick. In contrast, helium gamma and helium delta lines show moderate opacities and entirely optically thin satellite structure. This, first, enables stable temperature diagnostics based essentially on the mechanism of dielectronic recombination. The MARIA suite of codes [7] has been employed to perform multilevel collisional radiative opacity calculations and subsequent spectra simulations.

TABLE 1. Line-center opacities of various transitions.

	He$_\alpha$	He$_\beta$	He$_\gamma$	He$_\delta$	Sat$_\alpha$	Sat$_\beta$	Sat$_\gamma$	Sat$_\delta$
τ_0	980	170	50	16	O(1)	O(-1)	O(-2)	O(-2)

O(-1) means that τ_0 is of the order of 10^{-1}.

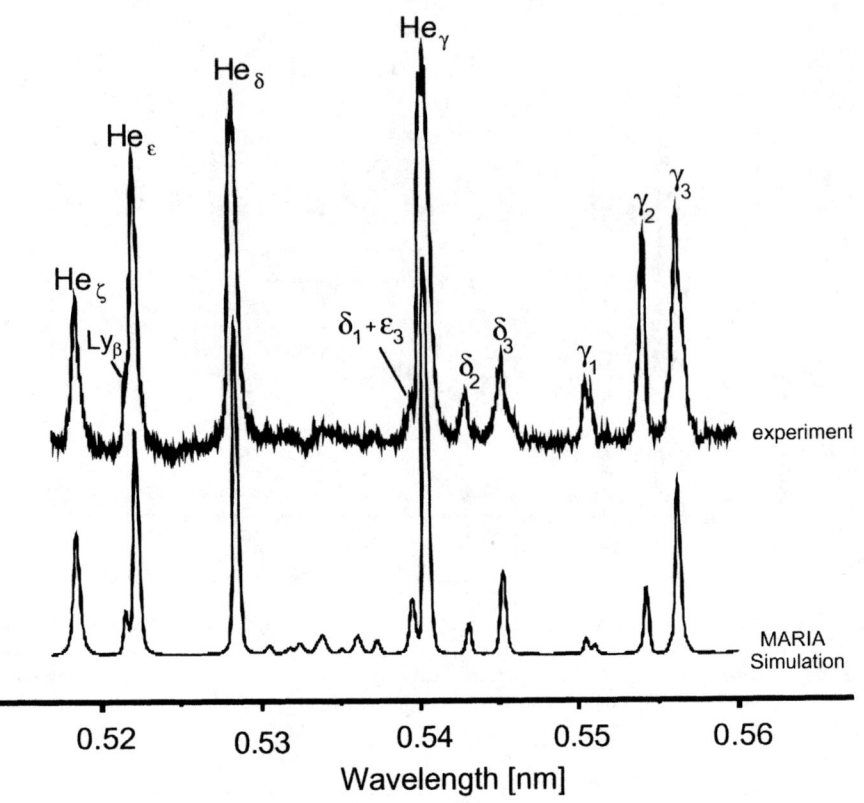

FIGURE 2. Comparison of the experimental spectrum and the MARIA simulations in the spectral interval from the He$_\zeta$ until the γ_3 satellites.

Simulations of spectra for different electron temperatures show a large sensitivity to the electron temperature over one order of magnitude for the He$_\delta$ and He$_\gamma$ satellites. Figure 2 shows the application of this Rydberg-satellite temperature diagnostic to the present data. Simulation parameters are kT$_e$=350 eV, n$_e$=10^{21} cm^{-3}, L$_{eff}$=100 μm and $\lambda/\delta\lambda_{instrument}$=3500. Comparisons of the upper (experiment) and lower (MARIA simulation) spectrum show excellent agreement with the line positions and quite good overall agreement. Due to the fact that spatially restricted emission data are analysed and entirely low or opacity free transitions are taken, it should be emphasized that this procedure is highly advanced. Moreover, to go further, it is possible to envisage density diagnostic using the Rydberg satellite series for Stark broadening analysis.

STARK BROADENING MODEL.

A fast computer code capable to provide spectral profiles for the radiation emitted by multiply charged ions imbedded in hot and dense non-equilibrium plasmas, with particular attention paid to Stark broadening due to ionic perturbers, has been developed a few years ago. The capability for the code to model radiation emitted by complex quantum structures and its capability to be joined with other plasma codes, require a systematic use of the methods and algorithms giving the best compromise between accuracy and fast calculations.

The code is mostly concerned with situations where line shapes are influenced by Stark splitting due to random microfield related to ionic perturbers. In plasma spectroscopy, Stark splitting is frequently concealed by Doppler effects or other broadening mechanisms. However, when Stark broadened lines are involved they are of fundamental interest for diagnostics. A line shape calculation is related to the study of the linear response of a quantum system immersed in a bath of moving charged particles. As usual we start with the relation giving the radiated intensity as the Fourier transform of the dipole correlation function C(t).

$$I(?) = \frac{1}{p} R \int_0^\infty e^{i\omega t} C(t) dt .$$ (1)

With the help of the Liouville formalism the following scalar product [8], equivalent to the usual trace over quantum states, can be written:

$$I(\omega) = \frac{1}{\pi} \left\langle\!\!\left\langle \vec{d}^t \left| \sum_{q=1}^{N} W_q \int_0^\infty e^{i\omega t} e^{-i(L_q - i\Phi)t} dt \right| \rho.\vec{d} \right\rangle\!\!\right\rangle$$ (2)

where, the discrete sum stands for the weighted mean value on the microfield distribution, L_q is the Liouville operator for the field q with the probability W_q, Φ is an approximate constant diagonal relaxation operator accounting for electron impact and spontaneous emission and ρ is the density matrix. Interference terms are taken into account in the electronic collision operator representing the effect of the electronic microfield component on the radiator. The APEX method for the calculation of a microfield distribution valid for all conditions from weakly to strongly coupled plasmas [9] has been chosen. The data set W_q is chosen in order to optimise computer time and accuracy for the final profile. Working in well suited basis for which the field dependent evolution operators are diagonal, and after a little algebra, one obtains:

$$I(\omega) = \sum_{k=1}^{M} \frac{c_k(\omega - f_k) + a_k g_k}{(\omega - f_k)^2 + g_k^2} .$$ (3)

The pure static profile appears as a sum of static Stark components each one being characterized by two complex numbers. These coefficients appearing in the sum represents the whole information available to further post processing. The number of Stark components is about the number of eigenstates that can be found diagonalizing the evolution operator with a constant external field times the sampling number of microfields used for the microfield average. This number can be huge and a process that corresponds to a statistical coarse-graining analysis of the components is needed to lower it to the order of a few hundreds. The radiative channels are then built appearing to be the true inhomogeneity that could be observed if the perturbers were static. The frequency fluctuation model (FFM) is based on the hypothesis that the microfield fluctuations affect these observable objects and that the dynamic spectrum results from a stationary Markovian stochastic process which mixes together the radiative channels leading to an additional broadening on each components and a shifting of the components towards the position of the unperturbed line.

RESULTS AND DISCUSSION.

In the case of the present experiment, Stark profiles of the He-like Si resonance lines until n=7 and the associated satellites until n=7 have been worked out with the PPP code. The calculation of the line profile requires knowledge of the ionic energy levels and their associated quantum numbers. This calculation also supposes that quantities like the Liouville, dipole and homogeneous operators are connected to the data available in a standard atomic structure code. All the spectra have been obtained by using the atomic data calculated by a multi-configuration Dirac-Fock code developed by Grant et al. [10]. A typical calculation with the model requires a large data set to be processed at a time. This data set is extracted from the atomic database, it involves: i)all the states belonging to upper and lower levels of an allowed radiative transition whose energy fall into the spectral interval of interest, ii)all the upper and lower atomic states linked to the previous ones through at least one dipole transition and whose role in Stark effect is non negligible. iii)all the data necessary for implementing the dipole operator in the calculation. iiii)all the data entering in the electron collision operator including the interference terms. The larger upper and lower subset size, the larger the matrices to be processed for each electric field are.

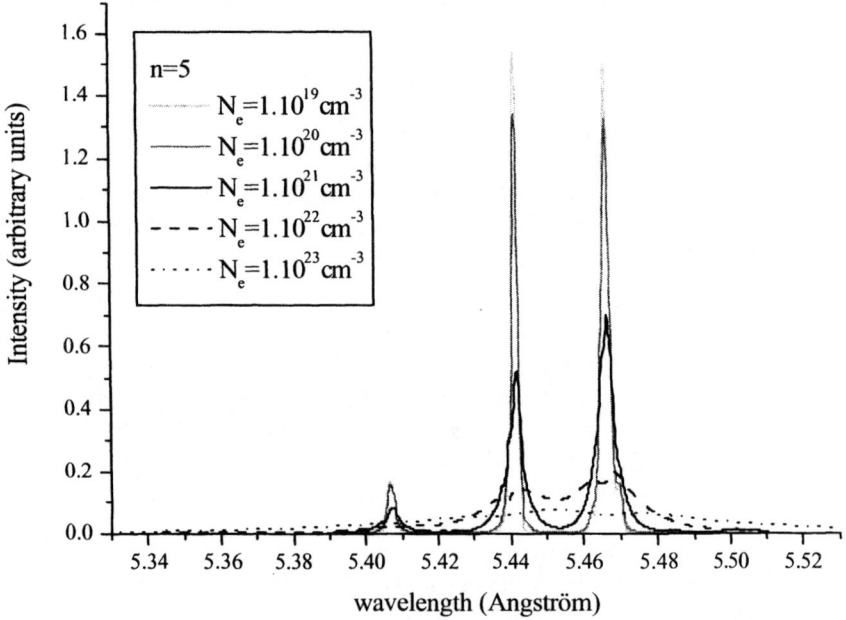

FIGURE 3. PPP Stark broadening calculations for the 1s2l5l'-complex of lithiumlike silicon for various densities and $kT_e = kT_i = 350$ eV.

For example, in the 1s2l5l'-complex case presented on Figure 3, the selected system involves 90 radiative transitions, 3 lower levels and 132 upper levels corresponding respectively to 8 and nearly 800 states leading to matrices of rank 6400. Typically 20 or 30 field values are used to perform the microfield average, so the Stark component number is of the order of 200000.

Detailed Stark broadening calculations of the entire 1s2l5l'-complex for various electronic densities are shown on Figure 3. All the calculations have been performed with a statistical population of the 1s2l5l' levels in order not to mask broadening effects with collisional redistribution of populations. Electronic and ionic temperatures were $kT_e=kT_i=350$ eV. Ion dynamics effects being negligible, the spectra are calculated in the quasi-static approximation. Doppler effect is included. For the lowest densities $n_e=10^{19}$ cm^{-3} and $n_e=10^{20}$ cm^{-3}, the profiles are almost unaffected by Stark effect whereas for the highest density $n_e=10^{23}$ cm^{-3}, Stark effect is the dominant broadening. The dependence of the width of the total shape with the density is clearly seen.

Figure 4 shows the Stark broadening calculations for the silicon Rydberg satellite series using the PPP code. Indicated are emissions arising from different configurations. Increasing width of single transitions and/or groups with increasing n (principle quantum number) is clearly seen.

The diagnostic developed above has been applied to the present experimental data. The direct comparison between experimental data (dots) and the PPP results (full line) including all resonance lines until n=8 and associated satellites until n=7 is shown on Figure 5. This illustrates the capability of the PPP code to accurately synthesize very complex configurations and spectra. Discrepancies appearing for the δ_3 satellites are

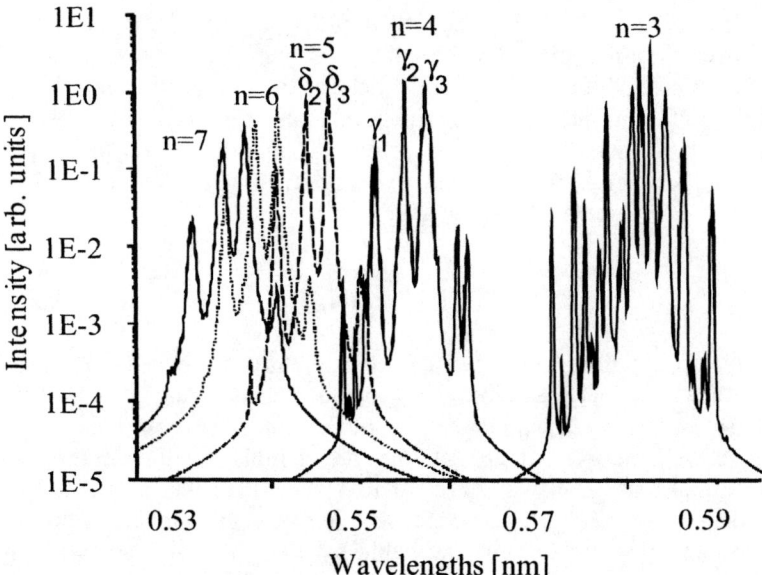

FIGURE 4. PPP Stark broadening calculations for the high energy Rydberg-satellite series 1s2lnl'-1s^{2}2l'', $Z_n=14$ $n_e=4.10^{20}$ cm^{-3} and $kT_e=kT_i=350$ eV.

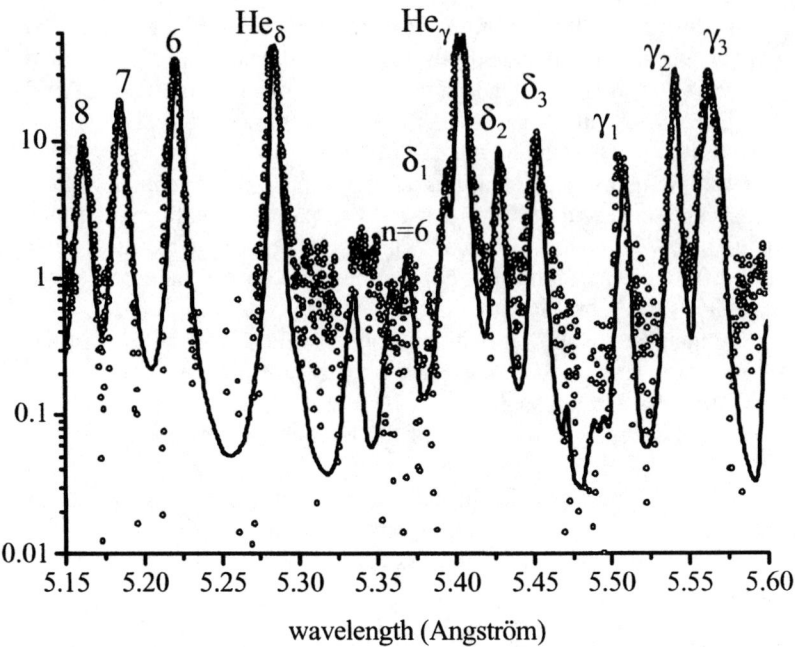

FIGURE 5. Comparison between experimental data (dots) and PPP calculations (full line) for $n_e=4.10^{20}$ cm-3. An apparatus function and Doppler effect are taken into account.

due to a corruption of 1s3l4l' satellites not taken into account in the calculations. For the series an overlap takes place by the 2l3l'-satellites. Apparatus function and Doppler effect are taken into account with the help of the narrow line emission from Figure 1(c). On Figure 6(a) and 6(b), each line group has been taken by itself. Results are presented for the γ and δ groups respectively. The calculations are made for $n_e=4.10^{20}$ cm^{-3} and $kT_e=kT_i=350$ eV. The apparatus function and Doppler effect are taken into account. The agreement is quite good.

CONCLUSION.

An essentially opacity free, space resolved temperature and density probe by the introduction of the Rydberg satellite concept to plasma diagnostics has been developed. This represents a significant improvement over common diagnostic methods because high-energy Rydberg satellites are essentially optically thin even in large-scale dense plasmas. The two codes MARIA and PPP have proven to be very powerful for diagnostics. The Stark broadening analysis of Rydberg autoionizing levels are rather complex and became available by using the PPP code which has proven its capability to accurately synthesize very complex spectra for data analysis.

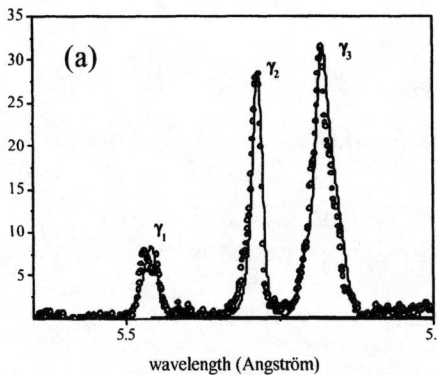
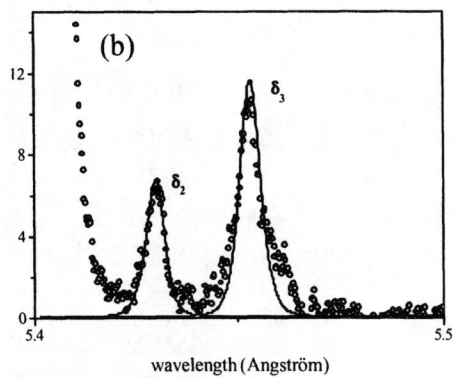

wavelength (Angström)   wavelength (Angström)

FIGURE 6. Comparisons between experimental data (dots) and PPP calculations for the gamma satellites (a) and the delta satellites (b) each group being taken by itself.

REFERENCES

1. Rosmej F.B., Faenov A.Ya., Pikuz T.A., Flora F., DiLazzaro P., Bollanti S., Lizi N., Letardi T., Reale A., Palladino L., Batani O., Bossi S., Bornardinello A., Scafati A., Reale L., Zigler A., Fraenkel M. and Cowan R.D., J. Quant. Spectrosc. Radiat. Transf. **58**, 859 (1997).
2. Rosmej F.B., Hoffmann D.H.H., Geissel M., Roth M., Pirzadeh P., Faenov A.Ya., Pikuz T.A., Skobelev I.Yu. and Magunov A.I., Phys. Rev. A **63**, 063409 (2001).
3. Calisti A., Khelfaoui F., Stamm R., Talin B. and Lee R.W., Phys. Rev. A **42**, 5433 (1990).
4. Talin B., Calisti A., Godbert L., Stamm R., Klein L. and Lee R.W., Phys. Rev. A **51**, 1918 (1995).
5. Calisti A., Godbert L., Stamm R. and Talin B., J. Quant. Spectrosc. Radiat. Transf. **51**, 59 (1994).
6. Skobelev I.Yu., Faenov A.Ya., Bryunetkin B.A. and Dyakin V.M., Zh. Eksp. Teor Fiz. **108**,1263 (1995) [JETP **81**,692 (1995)].
7. Rosmej F.B., J. Phys. B. **30**, L819 (1997).
8. Fano U., Phys. Rev. **131**, 259 (1963).
9. Iglesias C.A., Lebowitz J.L. and MacGowen D., Phys. Rev. A **28**, 1667 (1983).
10. Grant I.P., McKenzie B.J., Norrington P.H., Mayers D.F. and Pyper N.C., Comput. Phys. Commun. **21**, 207 (1980).

Spectral Line Shape Simulation for Electron Stark-Broadening of Ion Emitters in Plasmas

Emmanuelle Dufour[*][†], Annette Calisti[*], Bernard Talin[*],
Marco A. Gigosos[**], Manuel A. González[**] and Jim W. Dufty[‡]

[*]*Laboratoire de Physique des Interactions Ioniques et Moléculaires, UMR 6633,*
Université de Provence, 13397 Marseille cedex 20, FRANCE
[†]*Laboratoire de Physique Atomique dans les Plasmas Denses, LULI, UMR 7605,*
CNRS-Ecole Polytechnique-CEA-Université Paris VI, 75252 Paris cedex 05, FRANCE
[**]*Departamento de Óptica y Física Aplicada, Facultad de Ciencias,*
Universidad de Valladolid, 47071 Valladolid, ESPAÑA
[‡]*Department of Physics, University of Florida, Gainesville, Florida 32611, USA*

Abstract. Electron broadening for ions in plasmas is investigated in the framework of a simplified semi-classical model involving an ionic emitter imbedded in an electron gas. A regularized Coulomb potential that removes the divergence at short distances is postulated for the ion-electron interaction. Line shape simulations based on Molecular Dynamics for the ion impurity and the electrons, accounting for all the correlations, are reported. Comparisons with line shapes obtained with a quasi-particle model show expected correlation effects. Through an analysis of the results with the line shape code PPP, it is inferred that the correlation effect results mainly from the microfield dynamic properties.

INTRODUCTION

Most of the models designed to use plasma spectroscopy for diagnostic purposes are semi-classical ones, i.e. based on the study of quantum emitters perturbed by classical microfields. Both electrons and ions contribute to line broadening by Stark effect and most often they have been studied separately. In this work, attention is focussed on electron broadening for ion emitters. Electron broadening is usually modeled by means of an impact approximation relying on binary collisions based on quasi-particle kinetics. Recently, discrepancies noted between observed spectra and usual theory have suggested that improved electron broadening models would be necessary to obtain a better agreement between theories and experiments [1][2]. Such a work has been carried out successfully for neutral hydrogen. Molecular Dynamics line shape simulation (MDLS) has proven to be very useful in this case [1]. For ion emitters, beyond the quasi-particle approach, a few semi-classical electron broadening models - based on questionable or insufficiently validated hypothesis - have been proposed. This motivates the study of opposite charge systems reported below. The ion impurity in electrons considered in this work requires quantum mechanics because of unbounded Coulomb attraction at short distances. Nevertheless, plasma conditions exist for which classical mechanics can be postulated using an ion-electron regularized Coulomb potential allowing to carry out MDLS calculations. This approach is chosen here to investigate electron broadening ac-

CP645, *Spectral Line Shapes: Volume 12, 16th ICSLS*, edited by C. A. Back
© 2002 American Institute of Physics 0-7354-0100-4/02/$19.00

counting for all the correlations between charged particles. It is expected that the local changes in the electron gas properties around the impurity, due to the attractive potential, will induce a non negligible effect on electron broadening. After an outline of the used techniques, we discuss MDLS calculations of hydrogen-like helium Balmer-alpha lines in contrast with similar quasi-particle results.

FORMALISMS

Modeling the radiative response of an atomic system embedded into a gas of moving charges is a complex problem that cannot be solved analytically. The main difficulty is encountered when one attempts to solve the stochastic equation (SE) [3] satisfied by the emitter evolution operator $U(t)$. This equation can be written as follows:

$$\frac{dU_f(t)}{dt} = -i\left(L_0 - \vec{D}.\vec{E}_f(t)\right) \ , \ f \in F \tag{1}$$

Here, L_0 is the Liouville operator representation of the atomic Hamiltonian and $\vec{D}$ is the atomic dipole operator. The SE solution is defined as the average

$$U(t) = \left\langle U_f(t)\right\rangle_{f \in F} \tag{2}$$

over functions of time f, that characterize the time dependence of the microfield perturbation $\vec{E}_f(t)$. The simulation process gives a straightforward example of such a solution.

First, a sample set of N independent microfield configurations, $\vec{E}_k(t)$, is generated by using either a Molecular Dynamics technique (MD) [4] or independent quasi-particle simulation [5][6][7]. Second, the evolution operators, $U_k(t)$, are calculated by a stepwise integration of the Schrödinger equation for each configuration. Third, the mean evolution operator $U(t)$ is calculated by averaging over the configuration set. The SE solution becomes:

$$U(t) = \frac{1}{N}\sum_{k=1}^{N} U_k(t) \tag{3}$$

Although the simulation process is the method to treat the dynamic many-body problem with the best theoretical basis, the long calculation times necessary to obtain relevant results and the necessary restriction to not too complicated atomic systems make simulations definitely inappropriate for general purpose. Nevertheless, several models have been developed which provide an approximate solution to the problem with success [8][9][10] [11][12][13].

The starting point of the impact theory generally used to describe the electron-emitter interaction is to consider that the emitter is individually coupled with every plasma particle. The whole effect of the electrons may then be represented by an electronic collision operator, usually calculated in the framework of a binary collision relaxation theory. An element of this collision operator may be written as the sum of three terms [14]:

$$\Phi_{\alpha\alpha'\beta\beta'} = \sum_{\alpha''} \delta_{\beta\beta'} \vec{d}_{\alpha\alpha''}.\vec{d}_{\alpha''\alpha'} G(\Delta\omega_{\alpha''\beta})$$

$$+ \sum_{\beta''} \delta_{\alpha\alpha'} \vec{d}_{\beta'\beta''}.\vec{d}_{\beta''\beta} G(-\Delta\omega_{\alpha\beta''})$$

$$- \vec{d}_{\alpha\alpha'}.\vec{d}_{\beta'\beta} [G(\Delta\omega_{\alpha\beta'}) + G(-\Delta\omega_{\alpha'\beta})] \qquad (4)$$

with $\Delta\omega_{\alpha\beta} = \omega - \omega_{\alpha\beta}$. The two first terms are sums over perturbing states belonging respectively to the upper - a - and the lower - b - subset of states, and the last term represents the interference effect between the subsets a and b. The function $G(\omega)$ depends on the density and temperature of the plasma and is calculated to second order in the radiator-electron interaction. The line shapes, shown here, have been calculated with an expression for $G(\omega)$ [15] based on a modified semiclassical model, in which we have omitted the strong-collision term C usually added to the semi-classical term [16].

The capability of the FFM model [13][17] for describing emitter dipole relaxation in plasmas accounting for stochastic ionic field fluctuations has been extended to model electron broadening. Briefly, the FFM relies on the idea that an atomic system in an electric field behaves like a set of dressed two-level systems involved in a collision type mixing process. The electron dynamics which induces fluctuations of the microfield is taken into account through a mixing process modeled by a Markovian exchange between the dressed two-level systems. A unique rate is used : $v = v_{th}/d$ where v_{th} is the thermal velocity and d is the average distance between electrons. Note that this fluctuation rate can be rather large for electrons and can force the line shape to tend to a Lorentz shape.

SPECTRAL LINE SHAPE SIMULATION

Molecular Dynamics technique

MD technique [4] is the best tool for studying static and dynamic statistical properties of classical plasmas accounting for all charge correlations. One considers an isolated system that contains a finite number (a few hundred) of particles interacting through a given potential in a finite volume (a cubic box), and whose total energy is constant. Periodic boundary conditions are used, maintaining the system in a stationary state of temperature and density. A natural consequence of the use of periodic boundary conditions is the introduction of the "minimum image convention" which insures that the interaction volume is centered on the particle on which the force is calculated. The particles in the box obey classical mechanics laws. At each time step, the position and the velocity of the particles are obtained by integrating the equations of motion using the Verlet algorithm.

In the case of opposite charge systems, two different inter-particle potentials are used. The electron-electron one is the Coulomb potential

$$V_{\alpha\beta}(r) = \frac{z_\alpha z_\beta e^2}{r} \qquad (5)$$

and, in order to avoid a divergence at the origin, the ion-electron potential is a regularized Coulomb potential that is finite for distances smaller than the de Broglie length δ : the Deutsch potential [18]

$$V_{\alpha\beta}(r) = \frac{z_\alpha z_\beta e^2}{r} \left(1 - e^{-r/\delta}\right) \tag{6}$$

$$\delta = \frac{h}{\sqrt{2\pi m_e k T_e}} \tag{7}$$

Moreover, the two potentials are screened at distances of the order of the system size, chosen large enough in order to not affect the short range structure of the electron gas and not mask the electron screening mechanisms.

MD simulation allows to obtain static and dynamic statistical properties as density distribution functions around ion or electron, electron electric field distribution function at the ion, field autocorrelation function... Comparisons with theoretical models presented elsewhere show excellent agreement [19][20].

In the context of this paper, MD simulation is used to build the time dependent electron field sequences, $\vec{E}_k(t)$, necessary for the Schrödinger equation integration.

Spectral line shape calculation

This equation is solved by a technique [5] based on the SO(4) symmetry group of hydrogen atom (that allows to reduce the matrix size and then the calculation duration) and using the Euler-Rodrigues parameters (that guarantees the evolution operator unitarity and the solution accuracy). The field sequence is considered as a stepped function in which the electric field is constant in the interval between two temporary steps. The step size used allows to guarantee the quality of the numerical integration.

The mean evolution operator $U(t)$ is calculated by averaging over the whole set of electron field configurations. The spectral line shape is then obtained by :

$$I(\omega) = \frac{1}{\pi} Re \int_0^{+\infty} e^{i\omega t} C(t) dt \tag{8}$$

where $C(t)$ is the dipole autocorrelation function in Liouville space :

$$C(t) = \left\langle\left\langle \vec{d}^* | U(t) | \vec{d}.\rho_E \right\rangle\right\rangle \tag{9}$$

with ρ_E the emitter density matrix.

RESULTS AND DISCUSSION

The MDLS model outlined in the previous part has been applied to the case of the He$^+$ Balmer-α line ($\lambda = 1640,474$ Å) for impact conditions ($T_e = 60000\ K$,

$n_e = 10^{18} \ cm^{-3}$ and $\Gamma = 4,5.10^{-2}$).

Electron broadening is investigated through comparisons of three types : i) check the capability of the simulation procedure to reproduce impact broadening; ii) look for evidence of correlation effects; iii) interpret the origin of these effects.

In order to carry out comparisons for checking the first point, a model of independent electrons (IE) moving on straight trajectories and coupled with the ion impurity via a Debye screened Coulomb potential is used to generate microfield at the ion, accounting for all the electrons simultaneously. [1] A line shape is obtained using the integration procedure implemented for the MDLS calculations. On Figure 1, this line shape corresponding to a total field approach is compared to a standard impact limit based on complete binary collisions. From the good agreement observed, it can be concluded that first, both the plasma conditions and the studied Balmer-α line agree with impact theory hypothesis. Second, the numerical simulation is proved to be compatible with the impact limit.

In Figure 2, a comparison is performed to show a coupling effect when all the correlation between charged particles are accounted for. The line shape obtained from the MDLS procedure is plotted with the corresponding quasi-particle line shape resulting from a similar IE calculation than for Figure 1 but using the regularized ion-electron potential instead of the Coulomb potential. Full Width at Half Maximum (FWHM) is calculated for these lines and the IE line is about 20% broader than the MD line. A

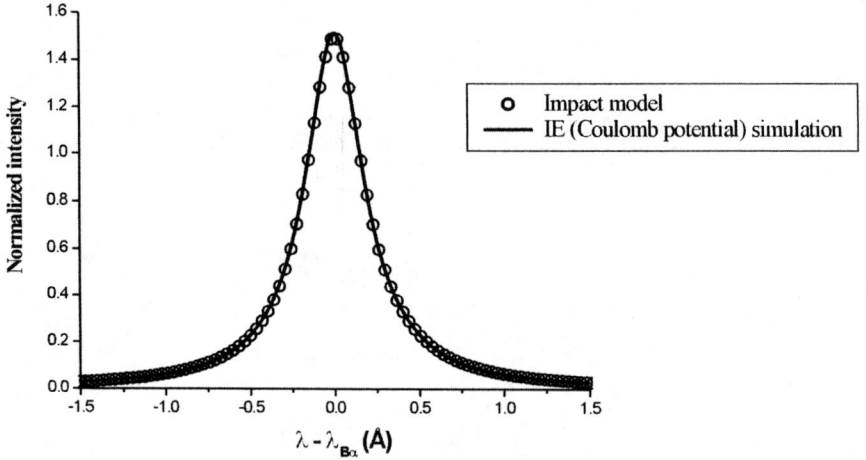

FIGURE 1. Comparison of Bα lines from impact model and IE (Coulomb potential) simulation

---

[1] Only a tiny radius of minimum approach is used in the simulation in order to avoid integration overflows.

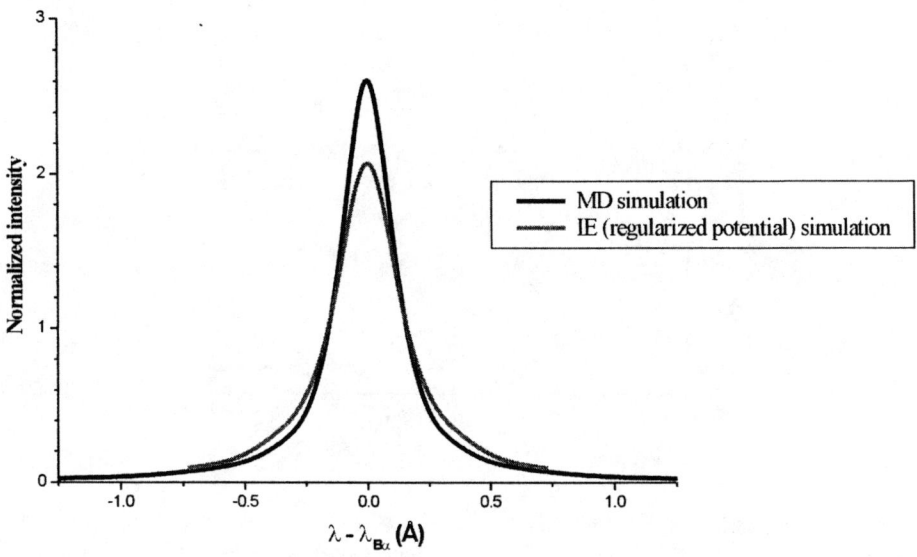

FIGURE 2. Bα lines from MD simulation and IE (regularized potential) simulation

discrepancy can be noted between the IE results in Figure 1 and Figure 2 due to the differences between the Coulomb and the regularized potentials at short distances.

On Figure 3, the field autocorrelation functions $C_{EE}(t)$ calculated from the two microfield sampling sets used in Figure 2 have been plotted. Using exponential fits, the net difference between the two curves is measured in terms of correlation times. In order to interpret the correlation effect shown on Figure 2, line shape calculations are carried out with the frequency fluctuation model code with two fluctuation rates in the same ratio than the correlation rates deduced from Figure 3. The two line shapes obtained in this way, plotted on Figure 4, show the same behavior than the simulated ones.

From this preliminary attempt to investigate correlation effect due to electron-ion and electron-electron couplings in electron broadening, it can be concluded that this effect can be large enough to affect line shape plasma diagnostics. The next step in this study will be to consider neutral systems involving several ions and electrons, in order to obtain total simulated profiles accounting for both ion and electron broadening.

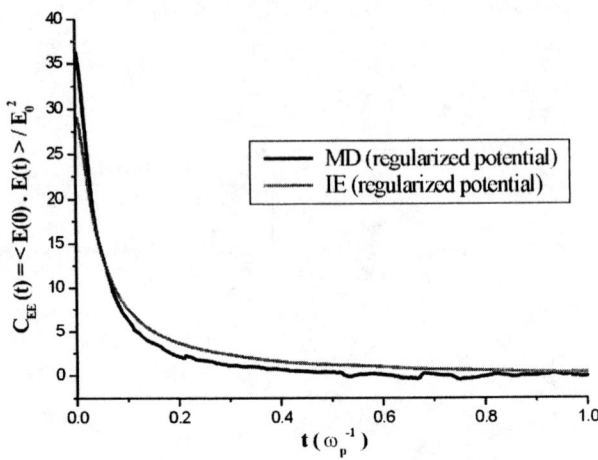

FIGURE 3. Comparison of field autocorrelation functions from MD and IE (regularized potential) simulations. $E_0 = e/r_0^2$ is the average field, r_0 the average distance between electrons and ω_p the electronic plasma frequency

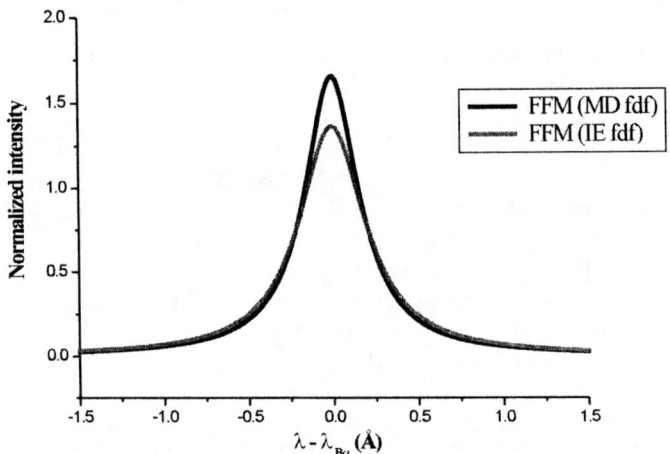

FIGURE 4. Bα lines from FFM : MD fdf and IE fdf

REFERENCES

1. S.Ferri, A.Calisti, R.Stamm, B.Talin, S.Büscher, Th.Wrubel, and H.J.Kunze, "Proceedings of the 15th ICSLS - Berlin (Germany) - 2000," in *Spectral Line Shapes XI*, edited by J.Seidel, AIP Conference Proceedings 559, American Institute of Physics, New York, 2001.
2. E.Oks, *JQSRT*, **65**, 405 (2000).
3. N.G.VanKampen, *Fundamental Problems in Statistical Mechanics III*, 1974 Wageningen Summer School North Holland, American Elsevier, E.G.D.Cohen, 1975.
4. M.A.Berkovsky, J.W.Dufty, A.Calisti, R.Stamm, and B.Talin, *Phys. Rev. E*, **54**, 4087 (1996).
5. M.A.Gigosos, J.Fraile, and F.Torres, *Phys. Rev. A*, **31**, 3509 (1985).
6. M.A.Gigosos, and V.Cardeñoso, *Phys. Rev. B*, **20**, 6005 (1987).
7. M.A.Gigosos, and V.Cardeñoso, *Phys. Rev. B*, **29**, 4795 (1996).
8. M.Baranger, *Phys. Rev.*, **112**, 855 (1958).
9. A.C.Kolb, and H.R.Griem, *Phys. Rev.*, **111**, 514 (1958).
10. A.Brissaud, and U.Frisch, *JQSRT*, **11**, 1767 (1971).
11. J.Seidel, *Z. Naturforsch.*, **32a**, 1207 (1977).
12. C.Stehlé, *Astron. Astrophys.*, **292**, 699 (1994).
13. B.Talin, A.Calisti, L.Godbert, R.Stamm, L.Klein, and R.W.Lee, *Phys. Rev. A*, **51**, 1918 (1995).
14. E.W.Smith, and C.F.Hooper, *Phys. Rev.*, **157**, 126 (1967).
15. A.Calisti, F.Khelfaoui, R.Stamm, B.Talin, and R.W.Lee, *Phys. Rev. A*, **42**, 5433 (1990).
16. H.R.Griem, M.Blaha, and P.C.Kepple, *Phys. Rev. A*, **19**, 2421 (1979).
17. B.Talin, A.Calisti, S.Ferri, and C.Mossé (this issue).
18. H.Minoo, M.Gombert, and C.Deutsch, *Phys. Rev. A*, **23**, 924 (1981).
19. B.Talin, A.Calisti, and J.W.Dufty, *Phys. Rev. E*, **65**, 056406 (2002).
20. A.Calisti, E.Dufour, B.Talin, and J.W.Dufty (this issue).

Shifting and Merging of Ar^{+17} Spectral Lines Due to Electron Collision Broadening

Mark A. Gunderson[1] and Charles F. Hooper, Jr[2]

[1]*Los Alamos National Lab, Los Alamos, NM 87545*
[2]*Department of Physics, University of Florida, Gainesville, FL 32611*

Abstract. As the electron density in hot, dense plasmas increases above 1×10^{24}/cm^3, the spectral lines of argon from argon-doped deuterium ICF capsules undergo a significant amount of shifting and broadening due to the effects of electron collisions. The shifts of these spectral lines are large enough so that adjacent spectral lines in a spectral line series begin to significantly merge together before the lines have been broadened beyond recognition or have been ionized. With the significant amount of merging of the spectral lines as the electron density increases, the width of these merged features requires that the validity of electron broadening models to second-order in the perturbing electron-radiator interaction be examined because these models are valid out to about twice the plasma frequency from line center. Thus, we present the results from an electron-broadening model to all-order in the perturbing electron-radiator interaction and compare them to results from a second-order model. Comparisons of line spectra from the two models show that for isolated lines the only significant differences occur in the line wings beyond twice the plasma frequency from line center. For spectral lines which are in the process of merging, significant and possibly observable differences between the two models emerge as the electron density approaches 5×10^{24} /cm^3.

INTRODUCTION

By placing a small amount of a dopant gas, eg. Argon or Krypton, in the deuterium fuel of an ICF microballoon target, we can determine the electron density and temperature of the core of the implosion through a Stark broadening analysis and a line intensity ratio analysis of the spectral lines emitted by the dopant gas. In recent years, this ICF electron density and temperature diagnostic has proven to be quite useful [1-5], and with some of our more recent experiments, we have achieved electron densities up to approximately 3×10^{24} /cm^3 as determined by analyzing Ar^{+17} spectral lines. With electron densities of this magnitude, we have noticed some interesting effects, the most notable being the shifting of spectral lines to lower energy due to collisions in which the plasma electrons penetrate the corresponding orbitals of the radiating ion. With the availability of more powerful multi-beam laser facilities in the future, eg. the National Ignition Facility or NIF, we expect to see higher electron densities during ICF implosion experiments, and thus we must have theoretical models that can accurately portray the high density effects, eg. line-merging effects, that we expect to see.

Our current electron broadening models, contained within or used in conjunction with our line shape code MERL [6,7], predict that as the electron density increases

CP645, *Spectral Line Shapes: Volume 12, 16th ICSLS*, edited by C. A. Back
© 2002 American Institute of Physics 0-7354-0100-4/02/$19.00

above 3×10^{24} /cm^3, the shifts of the Ar^{+17} spectral lines result in the merging of higher members of the spectral line series before their associated levels become ionized. However, there are some differences in the predictions of the electron broadening models as to how rapidly the spectral lines merge together as the electron density increases. To investigate further the differences between the electron broadening models in question, specifically the full Coulomb, quantum mechanical model to second-order in the plasma electron-radiator interaction and the full-Coulomb, semiclassical model to all-order in the plasma electron-radiator interaction, we also look at comparisons of individually calculated line spectra. These comparisons demonstrate that the two models differ the most in frequency regions of the line spectra beyond twice the electron plasma frequency from line center. Therefore, when considering calculations of merging line spectra and line wings at higher electron densities, it is important to perform calculations using both models in order to check the validity of cutting off the perturbation expansion at second-order.

THEORETICAL MODELS

Our motivation in using Stark-broadened spectra as an electron density diagnostic arises from the strong density dependence of the width, shift, and shape of these spectral features. Although weakly dependent on the temperature, we can also use these spectra as a temperature diagnostic when used in conjunction with a line intensity ratio analysis. We begin with the evaluation of the line shape equation [8,9]

$$I(\omega) = \frac{4\omega^4}{3c^3} \int_0^\infty d\vec{E}\, P(\vec{E}) J(\omega, \vec{E}) \tag{1}$$

where we integrate the microfield dependent line shape function

$$J(\omega, \vec{E}) = \frac{1}{\pi} \Re \mathrm{Tr}_r \left\{ \vec{d} \cdot R(\omega, \vec{E}) \rho_r \vec{d} \right\} \tag{2}$$

over the possible range of ion microfield values weighted by the ion microfield probability distribution function, calculated in this research work using the adjustable parameter exponential approximation (APEX). The trace is calculated over the relevant radiator states, and $\vec{d}$ is the radiator dipole operator. Line broadening due to radiator motion is included by convolving the line shape expression with a Doppler profile based on a Maxwellian velocity distribution [10].

The resolvent $R(\omega, \vec{E})$ in the absence of any lower state broadening has the form [2,11]

$$R(\omega, \vec{E}) = \frac{G(\omega, \vec{E})}{1 + iv \int d\vec{E}'\, P(\vec{E}') G(\omega, \vec{E})} \tag{3}$$

$$G(\omega, \vec{E}) = \frac{1}{\hbar\Delta\omega - H_{ir}(\vec{E}) - iv - M'(\omega, \vec{E})} \tag{4}$$

where $H_{ir}(\vec{E})$ is made up of the ion Stark splitting term (dipole term) and the ion quadrupole term, the terms with v comprise the effects of ion dynamical broadening using the BID method, and $M'(\omega, \vec{E})$ represents the shifting and broadening effects of the perturbing electrons on the line shape.

The electron broadening or width and shift operator as contained in the expression above has the form [12]

$$M'(\omega, \vec{E}) = n_e \text{Tr}_{1e} \left\{ V_{1e,r} f_{1e,r} + V_{1e,r} \frac{1}{\Delta\omega - H_{1e} - V_{1e,r}} f_{1e,r} V_{1e,r} \right\} f_r^{-1} \qquad (5)$$

where H_{1e} is the perturbing electron Hamiltonian, $f_{1e,r}$ is the reduced distribution operator for the perturbing electron-radiator subsystem, and the interaction between the perturbing electron and the radiator electronic structure is given by

$$V_{1e,r} = e^2 \sum_{i=1}^{N_r} \sum_{t=0}^{\infty} \sum_{q=-t}^{t} A^t(r_i, r_{1e}) C_q^{(t)}(\theta_i, \phi_i) C_q^{(t)*}(\theta_{1e}, \phi_{1e}) \qquad (6)$$

$$A^t(r_i, r_{1e}) = \left(\frac{r_<^t}{r_>^{t+1}} - \frac{\delta_{t,0}}{r_{1e}} \right) \qquad (7)$$

$$C_q^{(t)}(\theta, \phi) = \left(\frac{4\pi}{2t+1} \right)^{\frac{1}{2}} Y_{tq}(\theta, \phi) \qquad (8)$$

Note that the monopole interaction between the perturbing electron and the radiator has been removed from $V_{1e,r}$ and ha been included into the perturbing electron Hamiltonian. The first term in $M'(\omega, \vec{E})$ is a mean field term and gives rise to a large red shift in the calculated line shapes. This shift is occasionally referred to as the plasma polarization shift and is a result of perturbing electrons penetrating the orbitals of the radiator. The second term contains all of the dynamical effects of the collisions between the perturbing electrons and the radiator.

As the main focus of this paper deals with the shifts of the spectral lines and the effect of these shifts on line merging, we concentrate specifically on the calculation of the electron width and shift operator utilizing two different electron broadening and shifting models.

The Quantum Mechanical Second-order Model

When considering a model to second-order in the interaction, there are two conditions that must hold for the corresponding line shape. The first condition requires that most of the line shape lie within the frequency range $\Delta\omega \leq 2\omega_{pe}$ about line center [13]. Therefore, when using this model, the intensity of the wings and the relative level of symmetry between the red and blue wings of the line shape need to be examined carefully. The second condition is given by [14]

$$kT > \frac{Z(Z-1)}{n^2} \text{Ryd} \qquad (9)$$

where Z is the radiator charge and n is the principle quantum number of the upper radiator state associated with the line transition. This simply states that most of the collisions between the perturbing electrons and the radiator are weak in the sense that the perturbing electrons receive a small impulse relative to their momentum.

278

The width and shift operator in the second-order model has the form

$$M'(\omega,\vec{E})=n_e\text{Tr}_{1e}\left\{V_{1e,r}f_{1e}f_r\left[1+O(V_{1e,r})\right]+V_{1e,r}\frac{1}{\Delta\omega-H_{1e}}f_{1e}f_rV_{1e,r}\right\}f_r^{-1} \quad (10)$$

where we have expanded the distribution function in the first term to first-order in the interaction and removed the dependence of the distribution function on the interaction in the second term. For a quantum mechanical model, the one electron trace is written in terms of a trace over momentum scattering states of the perturber

$$\text{Tr}_{1e}[\ ]=\int d\vec{k}\langle\vec{k}\,|[\ ]|\,\vec{k}\rangle$$

In the case of the charged radiator, these states are expressed in terms of Coulomb wavefunctions. Also, note that electron-electron correlations are included through the use of a Debye static screening model. For a more detailed description of this model, please look at reference [14].

The Semiclassical All-order Model

In the calculation of the semiclassical all-order model [12], we utilize the width and shift operator as it is expressed in Equation 5. The one electron trace in terms of the classical path formalism is given by

$$\text{Tr}_{1e}[\ ]=\int_0^\infty dv_\infty\int_{b_{min}}^{b_{deb}}b\,db\int_0^{t_{col}}dt_0\int d\Omega[\ ] \quad (11)$$

where v_∞ is the velocity of the electrons before they enter the interaction region about the radiator defined by the Debye radius, b is the impact parameter, t_0 is a time parameter which tracks the location of the perturbing electron in its hyperbolic collision trajectory as it passes by the charged radiator, and the last integration is over the solid angle Ω. The minimum impact parameter is determined by including quantum effects at small radiator-electron perturber separations through the use of a Deutsch potential. The maximum impact parameter b_{deb} is an angular momentum cutoff corresponding to the Debye radius and approximates the screening in the plasma due to electron-electron correlations. t_{col} is the amount of time required for the perturbing electron to traverse the interaction region.

SHIFT EFFECTS ON LINE SHAPES

Almost all of the contribution to the red shift seen in line shape calculations results from the calculation of the $t=0$ term in the interaction $V_{1e,r}$ as is shown in Equation 6. Because the $t=2$ and higher terms give only minor contributions to both the width and the shift, we can demonstrate the effects of the shift on the line shape by comparing line shapes with only the $t=1$ term (dipole calculation) in the interaction with line shapes in which all relevant terms in the interaction are included (full-Coulomb calculation). Figures 1 and 2 show the Lyman-β line shape of Ar^{+17} evaluated at three different electron densities with the dipole and full-Coulomb treatment of the interaction, respectively. Note that while the dipole calculations show little or no shift and very little asymmetry with increasing density, the full-Coulomb calculations show a significant red shift and asymmetry as the density increases. This asymmetry occurs because upper level states with differing orbital angular momentum shift by different amounts, as can be seen in Figure 3.

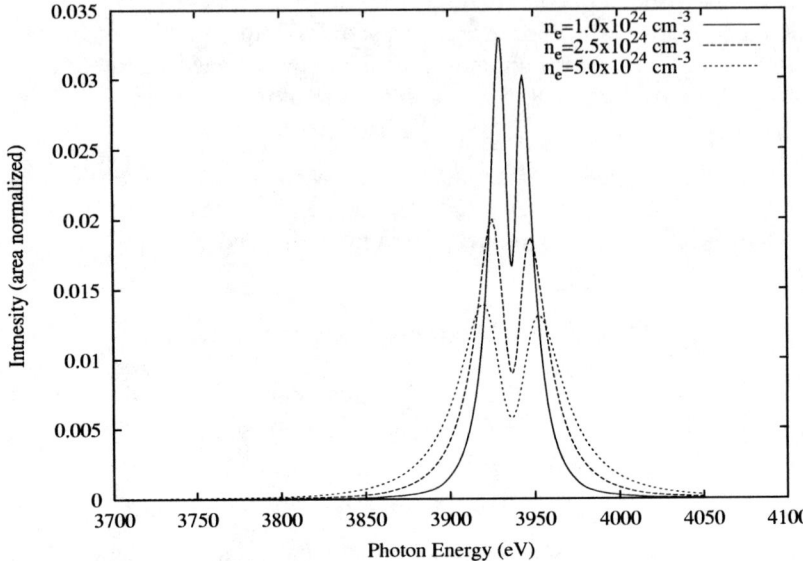

FIGURE 1. This plot shows the calculation of the argon Lyman-β line for three different electron densities and a temperature of 1keV utilizing only the t=1 term in the interaction given in Equation 6.

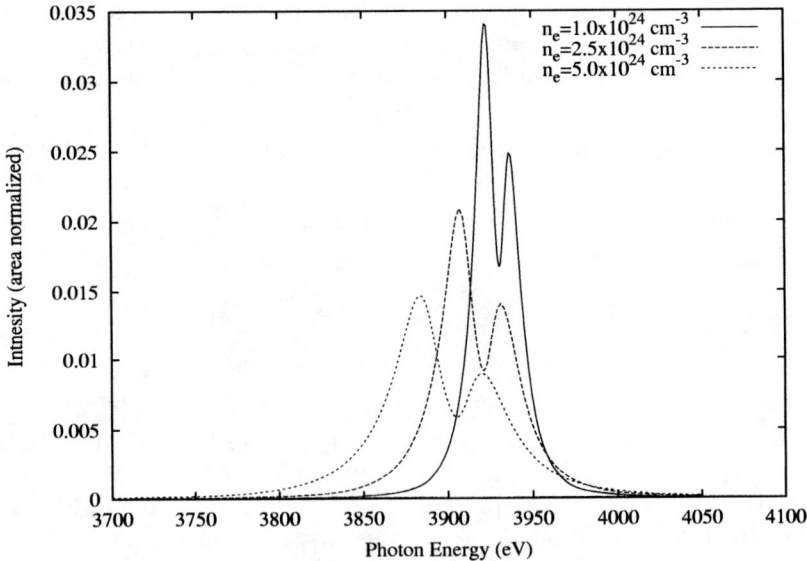

FIGURE 2. This plot shows the calculation of the argon Lyman-β line for three different electron densities and a temperature of 1keV utilizing all of the nonzero terms in the interaction given in Equation 6.

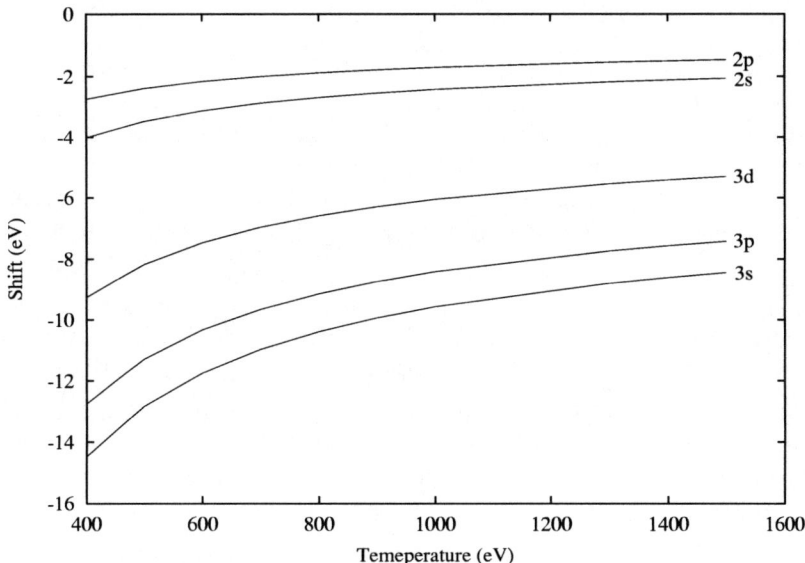

FIGURE 3. This plot shows the shifts of the upper state levels of both the argon Lyman-α and -β lines as a function of electron temperature at an electron density of $1 \times 10^{24}/\text{cm}^3$.

Let us examine Figure 3 in more detail. First, we see that as the electron temperature decreases, the magnitude of the shift of each level increases. The physical reasoning for this is relatively simple. As the electron temperature decreases, the electrons move more slowly in their hyperbolic trajectories as they undergo collisions with the charged radiator, thus giving these electrons more time to interact with and perturb the orbital levels of the radiator. We also note that as the orbital angular momentum quantum number of the orbital level increases, the shift associated with the orbital level decreases. One way to reason this out is to consider the spatial extent and volume of the orbital levels. As the orbital angular momentum of the levels increase, the spatial volume of the orbital decreases and thus results in a smaller amount of interaction with the perturbing electrons. Finally, note that the shift increases with increasing principle quantum number. This behavior is reasonable because the levels become less tightly bound with increasing principle quantum number and can therefore be more easily perturbed.

There are other ways in which the shift can affect line shapes used in the fitting of experimental data. The shifting of the energy levels associated with the line shapes could provide a possible modification to the Inglis-Teller limit [15]. Also, even though the shift is much smaller than the electron temperature, level shifts on the order of the observable line shifts can affect temperature inferences from line ratios through the use of a degeneracy-lowering model [15].

ELECTRON BROADENING MODEL COMPARISONS

Before we discuss the process of line merging, we should first examine and compare line shapes from the two electron broadening models presented, specifically the full-Coulomb second-order quantum mechanical model and the full-Coulomb all-order semiclassical model. From previously performed research and calculations, we expect most of the difference between the two models to occur in the far wings of the line shape beyond $2\omega_{pe}$ from line center. In Figure 4, we demonstrate that near the line center of the Ar^{+17} Lyman-β line at an electron density of and temperature of $5\times10^{24}/cm^3$ and 1keV, respectively, the two models agree quite well as was to be expected. In the semilog version of this plot shown in Figure 5, we see that the two models begin to diverge once you go beyond $2\omega_{pe}$ from line center. The difference in the far wing intensity of the line shape results mainly from a difference in a blue wing/red wing asymmetry between the two models. We plan to investigate further to gain a better understanding of this difference in the asymmetry.

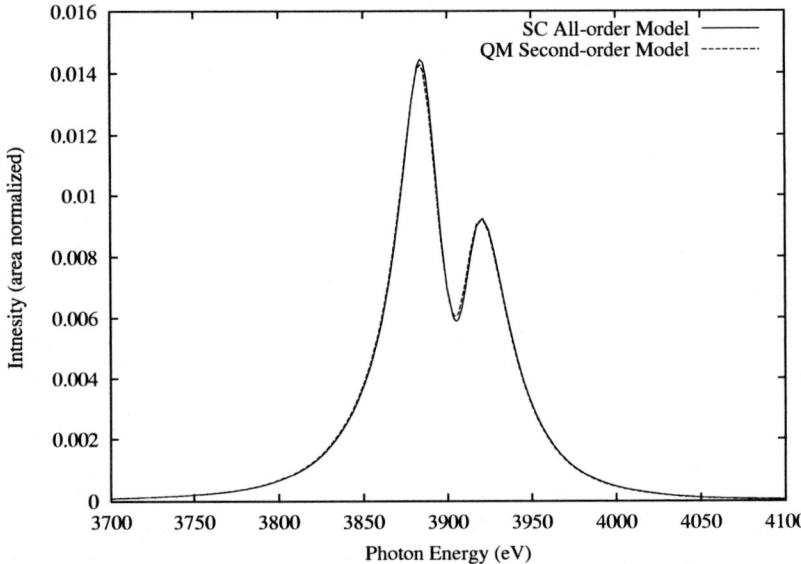

FIGURE 4. This plot shows the comparison of the argon Lyman-β line calculation from both the full Coulomb all-order semicalssical model and the full Coulomb second-order quantum mechanical model at an electron density and temperature of $5\times10^{24}/cm^3$ and 1keV, respectively.

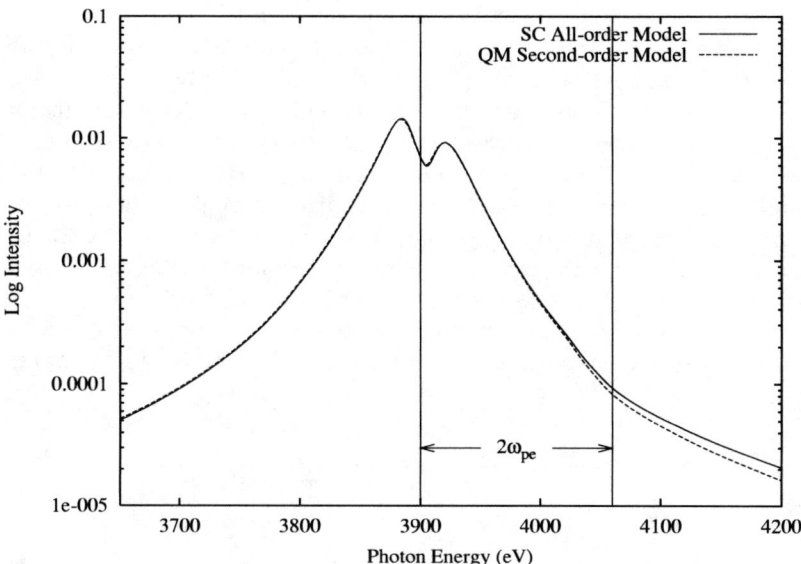

FIGURE 5. This semilog plot shows the comparison of the argon Lyman-β line calculation from both the full Coulomb all-order semicalssical model and the full Coulomb second-order quantum mechanical model at an electron density and temperature of $5\times10^{24}/cm^3$ and 1keV, respectively. This plot shows that most of the difference between the two models occurs in the wings of the line shape.

LINE MERGING

As was stated above, the shift of a level increases with increasing principle quantum number. For example, for the Lyman series of Ar^{+17}, a γ-line will shift more than a β-line, a δ-line will shift more than a γ-line, etc. This behavior, in fact, leads to a merging of these spectral lines as the density increases, and this merging occurs before the levels associated with these lines become ionized.

We consider two methods in calculating merging line spectra. In one method, the spectra are calculated independently and then are combined within the appropriate frequency range using appropriate line intensity ratios. Since the shift of the independently calculated line spectra is nearly linearly dependent on the electron density, we refer to this method as a linear line-merging model. However, there is a serious drawback to this method. If the electron density increases to a large enough value, line spectra with larger upper state principle quantum numbers, eg. a δ-line, may actually shift past line spectra with smaller upper state principle quantum numbers, eg. a γ-line. To avoid this problem, calculations have been done where the upper states of several adjacent spectral series members have been included in essentially the same "upper state manifold." In doing so, we have included mixing between the upper states of all of the spectra in the merged spectral feature. This mixing prevents spectral lines from shifting past one another, and thus we refer to this method as a nonlinear line-merging model.

In Figure 6, we show a comparison of the merging of Ar^{+17} Lyman-β, -γ, and -δ lines at an electron density and temperature of $5 \times 10^{24}/cm^3$ and 1keV from a linear line merging model, a second-order quantum mechanical nonlinear line merging model, and an all-order semiclassical nonlinear line merging model. We note that there are some differences in intensity over various portions of the merged spectral feature. However, the largest notable difference between the various models occurs in the all-order semiclassical nonlinear model where the gamma-delta portion of the spectral feature does not merge into the beta line as much as is predicted by the other two models. The difference in the energy location of the gamma-delta feature seems to be about 75 eV. Most of this difference is believed to come from mixing in the higher order terms of the all-order model that gives rise to a larger repulsive effect. At lower densities, this effect is much less noticeable and would be hard to pick out in any experimental data.

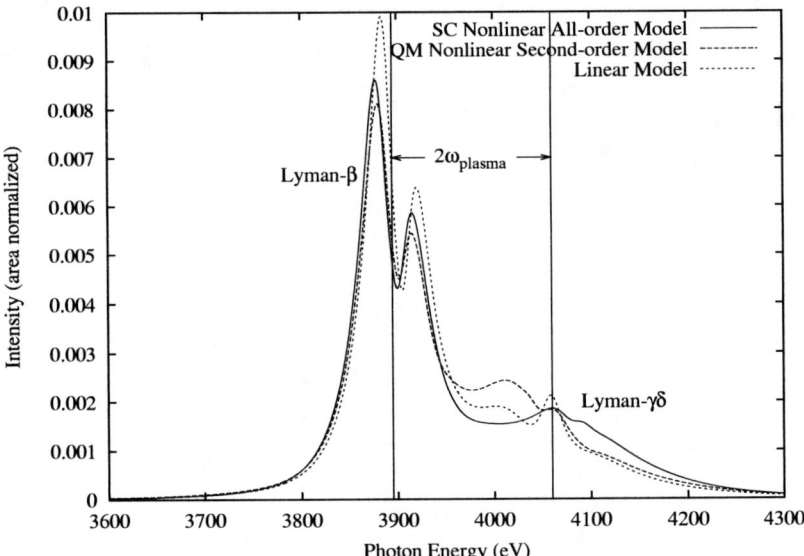

FIGURE 6. This plot shows the comparison of the merging argon Lyman-β, -γ, and -δ lines from an all-order semi-classical nonlinear line-merging model, a second-order quantum mechanical nonlinear line merging model, and a linear line-merging model at an electron density and temperature of $5 \times 10^{24}/cm^3$ and 1keV, respectively.

EXPERIMENTAL CAMPAIGN

Currently, we have an ICF experimental campaign underway through the National Laser User Facility (NLUF) program and Stockpile Stewardship program using the laser system at the Laboratory of Laser Energetics (LLE) to obtain higher core electron densities. To perform pre-experiment simulations of these argon spectroscopy implosions, we are utilizing the HYADES 1-D code [16], a three-temperature, three-geometry, Lagrangian hydrodynamics and energy transport code.

The electron and ion components are treated separately in a fluid approximation and are loosely coupled to each other, each in thermodynamic equilibrium and described by Mawell-Boltzmann statistics. The simulations use multi-group radiation transport in the diffusion approximation and LTE average atom ionization. The thermodynamic and equation of state quantities are derived from realistic (SESAME [17] or other theoretical models) tables. The HYADES code also includes thermonuclear burn of hydrogen isotopes and transports the fusion products using a particle tracking prescription. Since this is a 1D code, the simulations have a tendency to over-predict the achievable electron density. However, these simulations have still proven useful in the design of the ICF targets.

A comparison fit to data from NLUF shot 22534 using theoretical spectra from both the linear and non-linear second-order quantum mechanical line merging models is shown in Figure 7. It is clear visually and from Q^2 minimization techniques used in the fitting program that the nonlinear model tends to give a better fit to the data. With the achievement of higher electron densities in our upcoming NLUF shots, we plan to use theoretical spectra from the all-order semiclassical version of the nonlinear line-merging model in fitting the data.

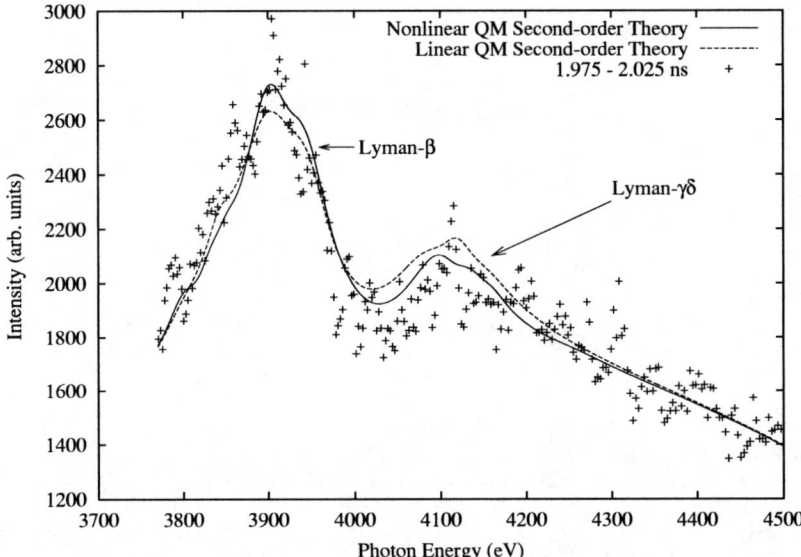

FIGURE 7. A comparison of theoretical fits of data from NLUF shot 22534 utilizing both the quantum mechanical linear and nonlinear line-merging models demonstrates a slightly better fit by the nonlinear line merging. The fitted electron density and temperature was $1.9 \times 10^{24}/cm^3$ and 1.15keV, respectively.

CONCLUSIONS

As higher electron densities are achieved through our ongoing NLUF campaign, it is quite apparent that the effects of line shifts on line spectra and on line merging is becoming an important tool in the analysis of experimental data. We have shown that line spectra utilizing the all-order semiclassical and second-order quantum mechanical electron broadening models differ the most beyond twice the plasma frequency from line center. Also, we have shown that as electron densities approach $5 \times 10^{24}/cm^3$, we begin to see differences between the two models in the calculation of merging spectral lines that may be observable in experimental data from ICF experiments that achieve this density. Thus, it is important that we have available to us an accurate and well-tested spectral model for the plasma conditions we expect to see in future experiments.

ACKNOWLEDGMENTS

We would like to thank Sean Regan, Paul Jaanimagi, and Jacquez Delletrez for their invaluable collaboration in fielding our NLUF experimental campaign at the Laboratory for Laser Energetics at the University of Rochester. We would also like to thank Don Haynes from the Fusion Technology Institute at the University of Wisconsin-Madison for his invaluable input and help on this research work. Finally, we would like to acknowledge Jeff Wrighton of the Physics Department at the University of Florida for his efforts in developing a user-friendly data analysis and fitting program. This research is being funded through NLUF grant DE-FG03-01SF22224.

REFERENCES

1. Hooper Jr., C. F., Kilcrease, D. P., Mancini, R. C., Woltz, L. A., Bradley, D. K., Jaanimagi, P. A., and Richardson, M. C., *Phys.Rev.Letters* **63**, 267-270 (1989).
2. Haynes Jr., D. A., Garber, D. T., Hooper Jr., C. F., Mancini, R. C., Lee, Y. T., Bradley, D. K., Delettrez, J., Epstein, R., Jaanimagi, P. A., *Phys. Rev. E* **53**, 1042-1050 (1996).
3. Bradley, D. K., et al., *Phys. Plasmas* **5**, 1870 (1998).
4. Woolsey, N. C., et al., *Phys. Rev. E* **57**, 4650-4652 (1998).
5. Regan, S. P., et al., *Laboratory for Laser Energetics (LLE) Review* **86**, 47 (2001).
6. Woltz, L. A., and Hooper Jr., C. F., *Phys. Rev A* **38**, 4766-4771 (1988).
7. Mancini, R. C., Kilcrease, D. P., Woltz, L. A., and Hooper Jr., C. F., *Comput. Phy. Commun.* **63**, 314-322 (1991).
8. Hooper Jr., C. F., *Phys. Rev.* **149**, 77-91 (1966).
9. Smith, E., and Hooper Jr., C. F., *Phys. Rev.* **157**, 126-137 (1967).
10. Woltz, L. A., and Hooper Jr., C. F., *Phys. Rev A* **30**, 468-473 (1984).
11. Boercker, D. B., Iglesias, C. A., and Dufty, J. W., *Phys. Rev. A* **36**, 2254-2264 (1987).
12. Gunderson, M. A., Junkel-Vives, G. C.,and Hooper Jr., C. F., *JQSRT* **71**, 373-382 (2001).
13. Tighe, R. J., and Hooper Jr., C. F., *Phys. Rev. A* **17**, 410-413 (1994).
14. Junkel, G. C., Gunderson, M. A., Hooper Jr., C. F., and Haynes Jr., D. A., *Phys. Rev E* **62**, 5584-5593 (2000).
15. Haynes Jr., D. A., Junkel, G. C., Gunderson, M. A., and Hooper Jr., C. F., "Plasma Induced Line Shifts and Their Effects on Line Merging and Population Kinetics" in *12th Atomic Processes in Plasmas Conference-2000*, edited by R. C. Mancini et al., AIP Conference Proceedings 547, New York: American Institute of Physics, 2000, pp. 227-237.
16. Larsen, J. T., *HYADES Manual and Users Guide*, Boulder, CO: Cascade Applied Sciences, 1997.
17. Lyon, S. P., and Johnson, J. D., *SESAME: The Los Alamos National Laboratory Equation of State Database*, LA-UR-92-3407 (1992).

Stark Broadening of Titanium K-shell Line Absorption Transitions

D.L. McCrorey and R.C. Mancini

Department of Physics, University of Nevada, Reno, NV 89557, USA

Abstract. We have computed detailed absorption line shapes for Ti K-shell transitions that include the effects of natural, thermal Doppler and Stark broadening, and we determined the density range where Stark broadening dominates. The combination of the density sensitivity of Stark broadened absorption line shapes and the temperature and density sensitivity of the level population kinetics results in an optical depth that is also sensitive to plasma temperature and density. In particular, the Stark broadening of the absorption line shapes removes the ambiguity with respect to density and temperature that the ionization balance brings into the optical depth calculation. Optical depth and transmissivity results are shown for several temperature and density conditions, and we discuss the potential for diagnostic applications.

INTRODUCTION

The determination of plasma conditions in the compressed shell of Inertial Confinement Fusion (ICF) implosions is important for diagnosing core-shell integrity and fuel-pusher mix [1]. To this end, Ti-doped tracer layers are embedded in the shell at various distances from the core. Radiation from the hot core backlights the tracer layer and characteristic Ti K-shell line absorption features can be observed in Ti ions from F- to He-like Ti [2]. Detailed modeling and analysis of these absorption features can yield Ti ionization state, areal-density, and electron temperature and density conditions in the absorbing layer. In particular, the density sensitivity of Stark-broadened absorption line shapes removes the ambiguity with respect to density and temperature that the ionization balance brings into the optical depth calculation [3].

ABSORPTION LINE SHAPE CALCULATIONS

Absorption line shapes for n=1 to n=2 line transitions in F-, O-, N-, C-, B-, Be-, Li- and He-like Ti ions were calculated using a multi-electron radiator line profile model and code that takes into account the effects of natural, thermal Doppler and Stark broadening [4,5]. The Stark broadening is due to the microfields of the plasma ions and electrons. The effect of the ions was calculated using a static ion approximation, while that for the electrons was calculated using a quantum-mechanical second-order relaxation theory approximation. The ion microfield distribution function was computed using the APEX model [6]. The atomic physics data needed for the line

CP645, *Spectral Line Shapes: Volume 12, 16th ICSLS*, edited by C. A. Back
© 2002 American Institute of Physics 0-7354-0100-4/02/$19.00

shape calculations (i.e. energy level structure and electric dipole matrix elements) were calculated with Cowan's atomic structure code [7]. Energy level widths due to radiative and autoionization decay rates, for the natural broadening effect, were calculated with the HULLAC code [8]. For each ionization stage, all fine structure n=1 to n=2 transitions considering all configurations with electrons in n=1 and n=2 were included in the calculation. For example, for the case of Be-like Ti the configurations for the initial levels were $1s^2 2s^2$, $1s^2 2s 2p$ and $1s^2 2p^2$, and the configurations for the final levels were $1s 2s^2 2p$, $1s 2s 2p^2$ and $1s 2p^3$. This group of initial and final configurations gives rise to a problem with 10 initial and 30 final J-energy levels, and 84 fine structure line transitions.

As an illustration of the absorption line profile calculations, Fig. 1 shows area-normalized (i.e. area=1) absorption line shapes for the case of Be-like Ti. Fig.1a displays line shapes for several values of the electron density N_e. For values of N_e below 1×10^{23} cm^{-3}, the line shapes show little change and become independent of N_e. This is an indication that the line broadening is dominated by natural and thermal Doppler broadening effects. A clear transition in the line shape features takes place between 1×10^{23} cm^{-3} and 1×10^{24} cm^{-3}, suggesting that the Stark broadening effect is gradually becoming more important. For electron densities above 1×10^{24} cm^{-3} the Stark effect clearly dominates the shape and broadening of the line profile, making them dependent on N_e. Fig. 1b shows the N_e dependence of the line profiles for several densities between 1×10^{24} cm^{-3} and 1×10^{25} cm^{-3}. This property is important for spectroscopic diagnostic applications. We note that mass densities on the order of 10 gcm^{-3} (equivalent to N_e of 3×10^{24} cm^{-3}) are expected in the compressed portion of the plastic shell of an ICF implosion. Hence, the N_e range of Fig. 1b is relevant for diagnostic applications in these experiments. On the other hand, the temperature sensitivity of the line shapes at these high electron densities is negligible. This fact is illustrated in Fig. 1c for several relevant electron temperatures T_e and $N_e=2\times10^{24}$ cm^{-3}. Similar results are obtained for other electron densities larger than 1×10^{24} cm^{-3}.

OPTICAL DEPTH AND TRANSMISSIVITY

The photon-energy-dependent optical depth due to n=1 to n=2 transitions in Ti ions, in a slab of plasma with physical thickness ΔR, is given by

$$\tau_v = \frac{\pi e^2}{mc} (\sum_{i,j} f_{ij}\phi_v^{ij} F_i)N\Delta R,$$ (1)

where e, m and c stand for the electron charge and mass, and the speed of light, respectively. The sum goes over all line transitions $i{\rightarrow}j$ in all ions included in the model, with i and j labeling the initial and final states of the transitions. f_{ij} is the absorption oscillator strength, ϕ_v^{ij} is the area-normalized line shape, F_i is the fractional population of the initial state, and N is the total Ti atom number density.

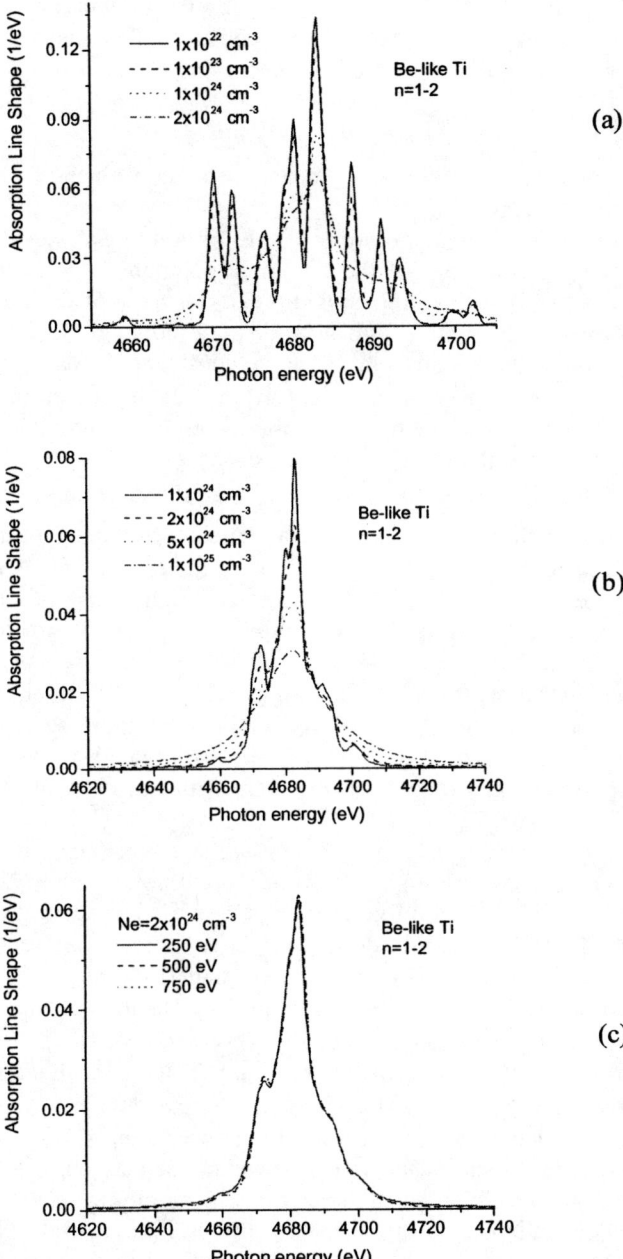

(a)

(b)

(c)

FIGURE 1. Area-normalized (area=1) absorption line shapes for Be-like Ti n=1 to n=2 transitions, for several temperature and density conditions.

If we neglect the effect of self-emission of the line, and denote incident and transmitted intensities by I_o and I_v, respectively, then the transmission through a uniform absorbing layer can be modeled according to,

$$I_v = I_o e^{-\tau_v}, \tag{2}$$

which is a solution of the radiation transport equation for the case of uniform opacity and negligible emissivity.

In addition to detailed area-normalized absorption line shapes, the evaluation of Eq. (1) also requires knowledge of the fractional populations F_i of the initial states (i.e. lower energy states) of the transitions. In this first stage of development of the Ti line absorption model for dense plasmas, these state populations are computed using an equilibrium model, which takes into account continuum lowering effects.

The photon-energy dependent optical depth τ_v is dependent on the electron density N_e and temperature T_e of the plasma through the temperature and density dependence of the fractional populations F_i and the density dependence of the Stark-broadened, area-normalized absorption line profiles ϕ_v^{ij}. To illustrate this functional dependence, Fig. 2 shows calculations of τ_v for several densities and temperatures. Fig. 2a displays τ_v for an electron density of 3×10^{24} cm^{-3} and three electron temperatures: 270 eV, 340 eV and 420 eV. The change in τ_v as the temperature rises reflects the shift in the ionization balance. For $T_e = 270$ eV the optical depth is dominated F-, O- and N-like Ti ions, while for $T_e = 420$ eV C-, B- and Be-like Ti ions are more important. We note that, for the latter case, the maximum values of τ_v for absorption by the dominant ions are larger than in the former case. This is due to the increase in absorption oscillator strength with ionization stage. Indeed, the average absorption oscillator strength increases monotonically by almost an order of magnitude from 0.085 for F-like Ti to 0.82 for He-like Ti.

More interesting from the point of view of Stark broadening is the dependence of τ_v on electron density N_e. Changing N_e affects both the ionization balance and the Stark broadening of the absorption line profiles. On one hand, for a given T_e, increasing N_e will shift the ionization balance downward, as an increase in three-body recombination will preferentially populate lower ionization stages. On the other hand, an increase in N_e will also increase the Stark broadening of the absorption line shapes. To focus on the broadening effect, Fig. 2b shows τ_v for several combinations of T_e and N_e that keep the ionization balance approximately constant. Thus, in each case τ_v is dominated by the same ions. As the density rises from 1×10^{24} cm^{-3} to 1×10^{25} cm^{-3}, the optical depth becomes less structured and the width of the absorption peaks become broader due to the density dependence of the line shapes. In fact, as density increases there is a significant decrease of the optical depth of the absorption peaks.

The Stark broadening effect on the absorption line profiles removes the ambiguity of the ionization balance with respect to electron temperature and density. As seen in Fig. 2b, several combinations of temperature and density can produce approximately the same ionization distribution, but the corresponding optical depths are different due to the density dependence of the Stark-broadened absorption line shapes.

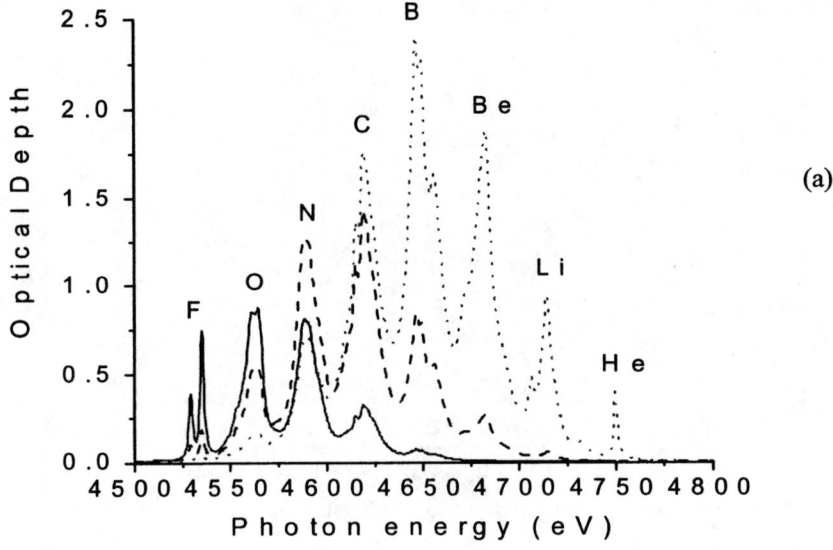

(a)

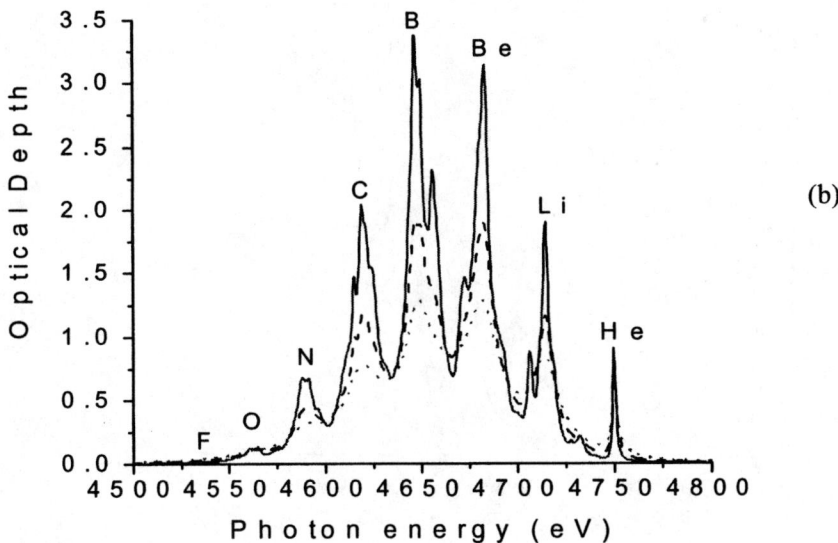

(b)

Figure 2. Optical depth of 1-2 K-shell transitions in a uniform absorbing slab of Ti with areal density $N\Delta R=1\times10^{19}$ cm^{-2}, for several electron density and temperature conditions. (a): Ne= 3×10^{24} cm^{-3} and Te=270 eV (—), 340 eV (– –) and 420 eV (…). (b): Ne/Te=2×10^{24} cm^{-3}/400 eV (—), 5×10^{24} cm^{-3}/505 eV (– –), 1×10^{25} cm^{-3}/620 eV (…). Each peak is labeled according to the contribution of the dominant absorbing ion.

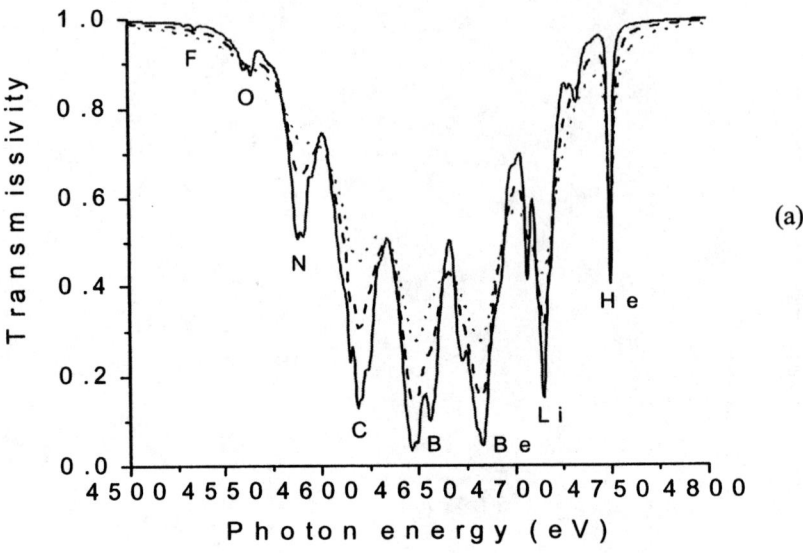

(a)

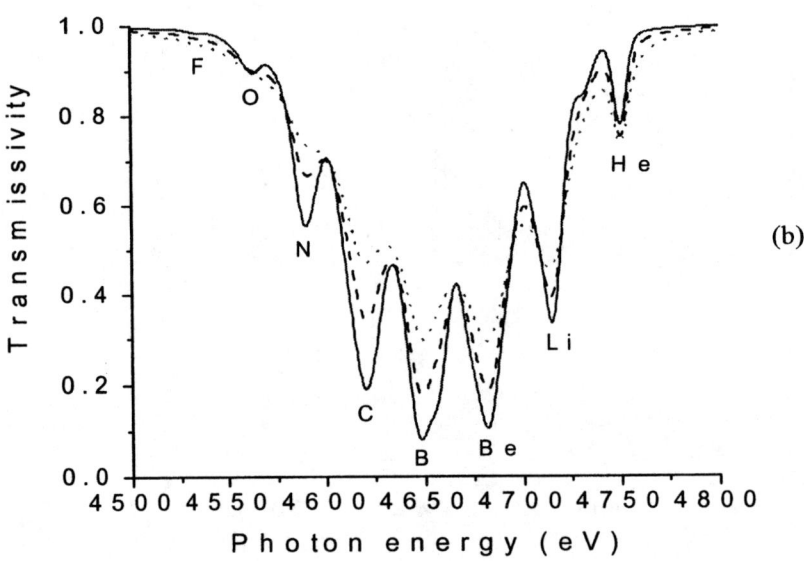

(b)

Figure 3. Trasmissivity of a uniform absorbing slab of Ti with areal density $N\Delta R=1\times10^{19}$ cm^{-2}, for electron density and temperature conditions Ne/Te=2×10^{24} cm^{-3}/400 eV (—), 5×10^{24} cm^{-3}/505 eV (– –), and 1×10^{25} cm^{-3}/620 eV (...). (a): no instrumental broadening effect, (b): including instrumental broadening effect by convolving the transmissivity with a Gaussian of FWHM=7 eV (i.e. $\lambda/\Delta\lambda=660$).

For the same combinations of temperature and density displayed in Fig. 2b, Fig. 3 shows the transmissivity (I_ν/I_o in Eq. 2). Fig. 3a shows that the Stark broadening effect of the absorption line shapes is also observed in the transmissivity. The question arises as to how much of this intrinsic broadening effect "will survive" instrumental broadening. In this connection, Fig. 3b displays the same results as Fig. 3a, but the transmissivity is now convolved with an area-normalized Gaussian of FWHM=7eV. This is done in an effort to approximate instrumental broadening effects on the transmissivity spectrum. This value of FWHM is equivalent to a resolution power of 660 in this photon energy range. This resolution power is suggested by current direct-drive implosion experiments with shell-embedded Ti-doped tracer layers at the OMEGA laser facility of the Laboratory for Laser Energetics, University of Rochester, where Ti K-shell line absorption spectra is being recorded [9]. As illustrated by Fig. 3b, the density sensitivity of the transmissivity persists when instrumental broadening is included in the model calculation.

CONCLUSIONS

We have developed a model of Ti K-shell line absorption spectra, which includes the effects of Stark broadening as well as natural and thermal Doppler broadening in determining the absorption line shapes. A database of these line shapes has been developed for all Ti ions, from F- to He-like Ti, and the density range where Stark broadening dominates has been determined. The results for the transmissivity spectrum are encouraging, and indicate that detailed modeling of Ti K-shell line absorption spectra in dense plasmas has significant potential as a diagnostic tool for compressed-shell conditions in ICF implosions. Work is in progress in order to address and include in the model the effects of self-emission of the lines, non-equilibrium population kinetics, n=1 to n=2 line transitions with spectator electron in n=3, and line shifts.

REFERENCES

1. C.J. Keane, R.C. Cook, T.R. Dittrich, B.A. Hammel, W.K. Levedahl, O.L. Landen, S. Langer, D.H. Munro and H.A. Scott, Rev. Sci. Instrum. **66**, 689 (1995).
2. B. Yaakobi, F.J. Marshall, D.K. Bradley, J.A. Delettrez, R.S. Craxton and R. Epstein, Phys. Plasmas **4**, 3021 (1997).
3. R.C. Mancini, C.F. Hooper, Jr., and R.L. Coldwell, J. Quant. Spectrosc. Radiative Transfer **51**, 201 (1994).
4. L.A. Woltz and C.F. Hooper, Jr., Phys. Rev. A **38**, 4766 (1988).
5. R.C. Mancini, D.P. Kilcrease, L.A. Woltz and C.F. Hooper, Jr., Comp. Phys. Commun. **63**, 314 (1991).
6. C.A. Iglesias, H.E. DeWitt, J.L. Lebowitz, D. Mac Gowan and W.B. Hubbard, Phys. Rev. A **31**, 1698 (1985).
7. R.D. Cowan, *The Theory of Atomic Structure and Spectra* (University of California Press, Berkeley, 1981).
8. A. Bar-Shalom, M. Klapisch and J.Oreg, J. Quant. Spectrosc. Radiative Transfer **71**, 169 (2001).
9. V. Smalyuk, private communication.

Spectroscopic Determination of Gradients in Indirect-Drive OMEGA Implosion Cores

L.A. Welser[a], R.C. Mancini[a], I.E. Golovkin[b], J.A. Koch[c], H.E. Dalhed[c], R.W. Lee[c], F.J. Marshall[d], J.A. Delettrez[d] and L. Klein[e]

[a]Department of Physics, University of Nevada, Reno, NV 89557, USA
[b]Prism Computational Sciences, Madison, WI 53703, USA
[c]Lawrence Livermore National Laboratory, Livermore, CA 94550, USA
[d]Laboratory for Laser Energetics, University of Rochester, Rochester, NY 14623, USA
[e]Department of Physics and Astronomy, Howard University, Washington, D.C. 20059, USA

Abstract. We report here on a spectroscopic method for determining the gradient structure in implosion cores based on the self-consistent analysis of simultaneous X-ray monochromatic images and X-ray line spectra. This technique is applied to a series of stable and low convergence indirect drive experiments where Ar-doped D_2-filled plastic shells were imploded with the OMEGA laser system. Argon K-shell X-ray line spectra were measured with streak crystal spectrometers, and X-ray monochromatic imagers recorded Ar Heβ based images of the core. The analysis self-consistently determines the temperature and density gradients that yield the best fits to the monochromatic spatial emissivity profiles and spectral line shapes. This measurement is critical for understanding the atomic kinetics, radiation transfer and plasma dynamics associated with the implosion process. In addition, since the results are independent of hydrodynamic simulations they are important for the verification and benchmarking of detailed fluid dynamic models of hot dense plasmas.

INTRODUCTION

Analysis of time-resolved, space-integrated X-ray line spectra is frequently used to determine the time history of spatially-averaged temperature and density in Inertial Confinement Fusion (ICF) implosion cores [1-5]. The next step in the X-ray spectroscopy of implosion cores is the bracketing of the spatial gradient structure as a function of time [6]. Knowledge of core gradients will result in improved diagnostics of core plasma conditions achieved at the collapse of the implosion, and will supply new data for detailed benchmarks of hydrodynamic simulation codes.

Our previous work in core gradient determination considered the analysis of data from Ar Heβ line ($1s^2$ ^1S – 1s 3p ^1P) spectra and monochromatic images recorded in direct drive ICF implosion experiments performed at the GEKKO XII laser system of the Institute of Laser Engineering at Osaka University [6]. Here, we focus on indirect drive experiments performed at the OMEGA laser facility of the University of Rochester's Laboratory for Laser Energetics. In these implosions, deuterium-filled argon-doped plastic microballoon targets were imploded inside Au hohlraums. Argon

CP645, *Spectral Line Shapes: Volume 12, 16th ICSLS*, edited by C. A. Back
© 2002 American Institute of Physics 0-7354-0100-4/02/$19.00

Heβ and Lyβ (1s ^{2}S – 3p ^{2}P) based X-ray line spectra and Heβ monochromatic images of the implosion core were simultaneously recorded.

The X-ray spectral and image data were processed using IDL [7] (Interactive Data Language), a powerful tool for image data visualization and graphical user interface. Space-integrated, time-resolved X-ray line spectra extracted from the data provided information on the time history of the Ar line emission. The standard method for the spectroscopic analysis of Ar K-shell X-ray line spectra from doped implosion cores is based on searching for the density and temperature values that yield the best fit to the X-ray line spectra [1-5]. This method relies on a uniform model approximation, and on the fact that the line spectrum (i.e. resonance and associated satellite lines) is temperature and density dependent through the temperature and density sensitivity of the level population kinetics, and the density sensitivity of the Stark-broadened line shapes. If the line spectrum is optically thin, the temperature and density obtained from the analysis can be interpreted as emissivity-weighted averages of the spatial distribution of temperature and density in the core [8]. In the experiments discussed here, the optical depth of the Heβ and Lyβ lines is small enough so that they can be considered optically thin. In order to go beyond the uniform model approximation and study the core space structure, additional information is needed. X-ray monochromatic images based on Ar line emission supply the additional information needed for the analysis of the core spatial structure. In this case, the idea is to search in parameter space for the temperature and density gradients that yield the best fits to both the space-integrated line spectrum and the spatial distribution of monochromatic emissivity [6]. Furthermore, these gradients are also constrained to have emissivity-weighted averages that are consistent with the results of the uniform model analysis.

INDIRECT-DRIVE IMPLOSION EXPERIMENTS

The indirect-drive implosion experiments were performed at the OMEGA laser facility of the Laboratory for Laser Energetics at the University of Rochester. The target consisted of a Au hohlraum with a plastic capsule inside. The Au hohlraum was 2550 μm long, 1600 μm in diameter, and had 1200 μm Laser Entrance Holes (LEH). The plastic capsule had an external diameter of 510 μm with a wall thickness of 35 μm (i.e. initial core diameter was 440 μm), and it was placed at the center of the hohlraum. The capsules were designed to avoid burn through of the hohlraum X-ray radiation. Numerical simulation indicates that at the collapse of the implosion approximately 37% of the shell mass remains unablated and in some state of compression [9]. The core was filled with 50 atm of D_2 and 0.1 atm of Ar. The small (tracer) amount of Ar resulted in a typical optical depth of the Heβ line of 0.2, and even less for the Lyβ line.

These hohlraum targets were irradiated with 30 UV OMEGA beams, split into 15 beams per LEH that were arranged in two cones of 5 and 10 beams each. The beam cones were pointed in such a way that they produced two rings of beams on each end of the hohlraum. The laser energy per beam was 500 J, for a total UV laser energy of 15 kJ. The hohlraum radiation temperature was 210 eV.

Three diagnostic holes placed on the side walls of the hohlraum provided lines-of-sight for a Gated-X-ray-Monochromatic–Imager (GMXI) [10], a new pinhole-array X-ray Multi-Monochromatic-Imager (MMI) [11], and a streaked X-ray crystal spectrometer. GMXI can record gated (Δt=80 ps) Ar Heβ images with a spatial resolution of 10 μm and a spectral resolution of 22 eV. MMI is a pinhole-array instrument that records numerous narrow-band (75 eV) X-ray images in the photon energy range from 3000 eV to 5000 eV with 10 μm spatial resolution. Data from MMI can be used to construct monochromatic images from several lines as well as continuum-based images. The streaked spectrometer (SSCA) uses a flat RbAP crystal, and has time and spectral resolutions of 30 ps and 500 (res. power), respectively.

PROCESSING OF SPECTRAL AND IMAGE DATA

The time-resolved Ar K-shell X-ray line spectra recorded with the SSCA streaked crystal spectrometer was processed using IDL [7]. As an illustration, Fig. 1 shows the spectra recorded in OMEGA shot 24900. Early in time, Au M-band emission from the wall of the hohlraum is observed. Late in time, the implosion collapses and the Ar K-shell emission begins.

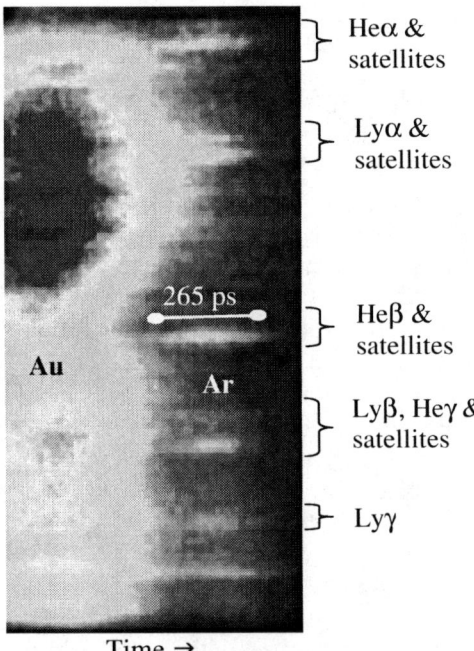

Figure 1. Time-resolved X-ray spectra from OMEGA shot 24900 recorded with the SSCA streaked spectrometer, showing Au emission from the hohlraum early in time and Ar line emission late in time.

Several steps are performed prior to the extraction of spectral lineouts for analysis from the image data. A wedge file correction gives the correlation between film density and intensity, after which corrections for photocathode efficiency, crystal reflectivity, and Be filter transmission are applied. Following these corrections, an approximation to "the tail" of the Au emission late in time is determined using an IDL 2D surface fitting routine. This Au emission is effectively subtracted from the data, leaving behind a close approximation to pure Ar line emission. In order to determine the time histories of the spatially-averaged core temperature and density, a sequence of time-resolved X-ray spectra lineouts are then extracted from the corrected data and analyzed using a uniform model approximation.

The monochromatic images recorded by GMXI and MMI provide information on the spatial structure of the core, in the form of angle-dependent and angle-averaged intensity radial lineouts. For each monochromatic image, an accurate determination of the image center is important. To this end, a center determination technique has been developed which systematically calculates centers and extracts the corresponding radial lineouts from each image. To find the center of a specific image, several contour levels are defined and then fit (in a least-square-minimization fashion) with best-fit circles. An arbitrary contour level is characterized by a set of $\{x_i, y_i\}$ coordinates. Given the location of a circle's center $\{x_c, y_c\}$ and its radius R, the distance from the circle's center to a point i on the contour along the radial direction is defined by Eq. 1.

$$d_i = \sqrt{(x_i - x_c)^2 + (y_i - y_c)^2} \tag{1}$$

The distance along the radial direction from point $\{x_i, y_i\}$ to the circle is then $|d_i - R|$, and the idea is to minimize the sum of the square of these distances, as seen in Eq. 2.

$$Q^2 = \sum_i (R - d_i)^2 = Q^2(R, x_c, y_c) \tag{2}$$

In principle, this is a three-parameter minimization problem. However, for a given $\{x_c, y_c\}$, the problem becomes a one-parameter optimization with the analytical solution $R = \langle d_i \rangle$. Thus, the problem reduces to finding the two parameters $\{x_c, y_c\}$ that minimize the dispersion of the set $\{d_i\}$ (see Eq. 3), which is easily solved numerically within the framework of an IDL code.

$$Q^2 = \sum_i (\langle d_i \rangle - d_i)^2 = Q^2(x_c, y_c) \tag{3}$$

This procedure is implemented one contour level at a time, and automatically executed for a family of contour levels associated with a given image. We find centers for several contour levels and subsequently perform an average of the $\{x_c, y_c\}$ center coordinates. The dispersion about this average value is used to estimate the uncertainty in the center of image location.

Eight equally-spaced radial lineouts are then extracted from the image, starting from the average center point, and they are averaged to determine the angle-averaged radial lineout. The edge of the core is determined using a conservation of mass argument, and the resulting lineout, from core center to edge, is fit with a Gaussian function. This curve is then Abel-inverted in order to retrieve the emissivity profile in the plasma (spherical) volume as a function of the radius [12]. As an illustration of the monochromatic images recorded in the OMEGA indirect-drive implosions, Fig. 2 shows a Heβ monochromatic image of the core recorded with GMXI in OMEGA shot 23686, the collection of contours for image center determination, and eight angle-dependent and the angle-averaged radial lineouts. The quality of the circle fits to the contour levels and the deviation of angle-dependent lineouts from the angle-averaged lineout give an idea of the deviation from a (perfect) spherical implosion. In turn, this information is important in determining the uncertainties in the core gradients. In this case, the image is time-integrated over the period of emission of the Heβ line.

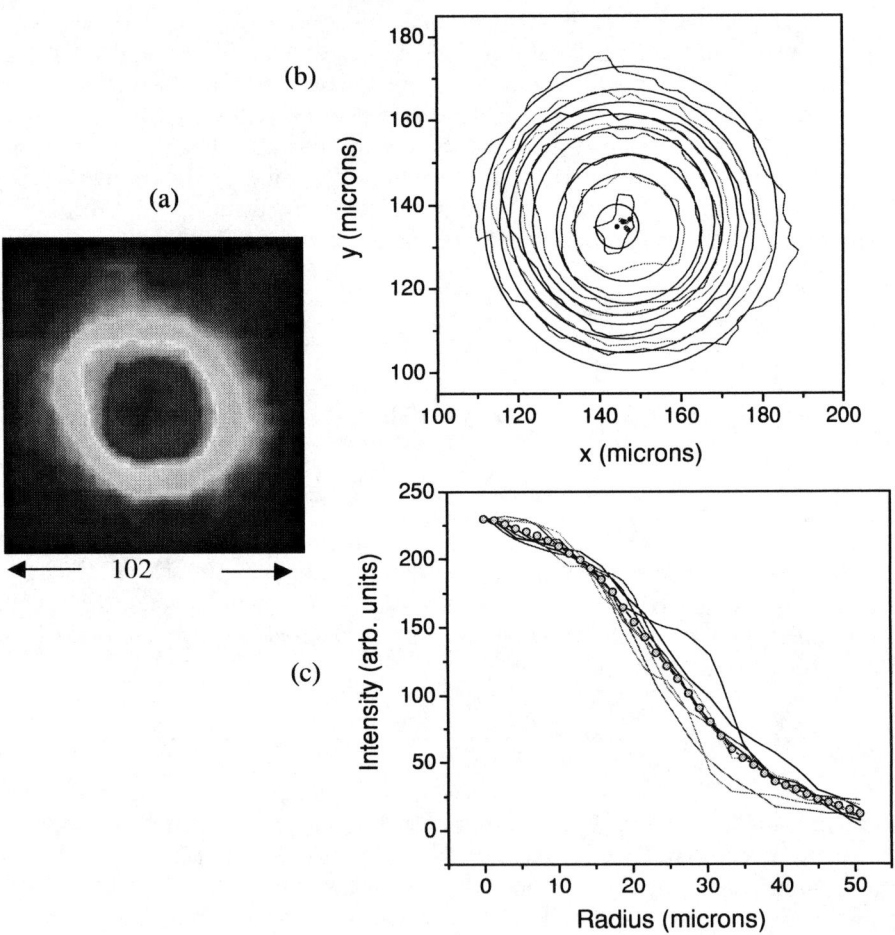

FIGURE 2. Implosion core Ar Heβ monochromatic image from OMEGA shot 23686 (a). Contour plot mapping of image and circle fits (b). Angle-dependent (solid lines) and angle-averaged (solid circles) intensity radial lineouts (c).

CORE GRADIENT ANALYSIS

X-ray line spectra have commonly been used to characterize spatially-averaged core temperature and density [1-5]. To go beyond this core-averaged uniform model analysis, it is necessary to employ spatial information regarding core structure. Analysis of space-integrated X-ray line spectra alone cannot be used to unambiguously determine plasma gradients in an implosion core. The monochromatic images discussed above provide the extra information needed to effectively determine the density and temperature gradients in the implosion core.

The spectroscopic determination of gradients is based on selecting the set of electron temperature and density gradients, $T_e(r)$ and $N_e(r)$, that yield the best self-consistent fits to the spatially resolved monochromatic emissivity and the spatially integrated line spectra. The gradients are also constrained by the fact that emissivity-weighted averages of temperature and density gradients should be consistent with the results obtained from the analysis of the spectrum using a uniform plasma model approximation [6]. Fig. 3 displays the results of the temperature and density core gradient determination for OMEGA shot 23686, time-integrated over the interval of emission of the Heβ line. The synthetic spectrum and emissivity were calculated with the spectral model of Ref. 8. Fig. 3a shows the uniform and gradient model fits to the space-integrated X-ray line spectrum. We note that both fits are of comparable quality. The uniform model analysis results in an electron temperature $T_e = 875$ eV and an electron density $N_e = 8 \times 10^{23}$ cm^{-3}. The self-consistent gradients have emissivity-weighted averages $< T_e >_\varepsilon = 860$ eV and $< N_e >_\varepsilon = 7.8 \times 10^{23}$ cm^{-3}, thus showing good consistency with the results of the uniform model analysis. The spatial emissivity profile in the plasma source extracted via the Abel inversion of the angle-averaged intensity radial lineout from the monochromatic X-ray image in Fig. 2, is displayed in Fig. 3b, which also shows the fit obtained with the gradient model. Finally, Fig. 3c shows the gradients obtained from the analysis of the data from Figs. 3a and 3b.

The algorithmic implementation of the gradient analysis is based on a novel application of Genetic Algorithms (GA) to plasma spectroscopy. First, GA's were tested in the analysis of space-integrated time-resolved X-ray spectral data. This case represents a single criterion problem where the GA searches a two-dimensional parameter space (i.e. T_e and N_e) for the (T_e, N_e) set that yields the best fit to the data based on a least-square-minimization [13,14]. This algorithm-driven analysis technique was then extended to the case of a two-criteria problem where both spectrum and emissivity have to be simultaneously fit. The multi-criteria aspect was handled with a niched-Pareto optimization technique that works with a distribution of solutions obtained by the GA, and extracts from it the optimal solution (if any), i.e. the solution that best satisfies all criteria [13,15]. Before applying it to real data, this algorithmic technique was extensively tested with "synthetic data" from hydrodynamic simulations. These test-studies indicated that the self-consistent analysis of simultaneous X-ray space-integrated line spectra and space-resolved monochromatic images can be used to uniquely determine gradients in implosion cores, and that the niched-Pareto GA (NPGA) is an efficient algorithm to implement it. We emphasize that the idea of using a NPGA algorithm is general and can also be applied to other problems of multi-criteria data analysis.

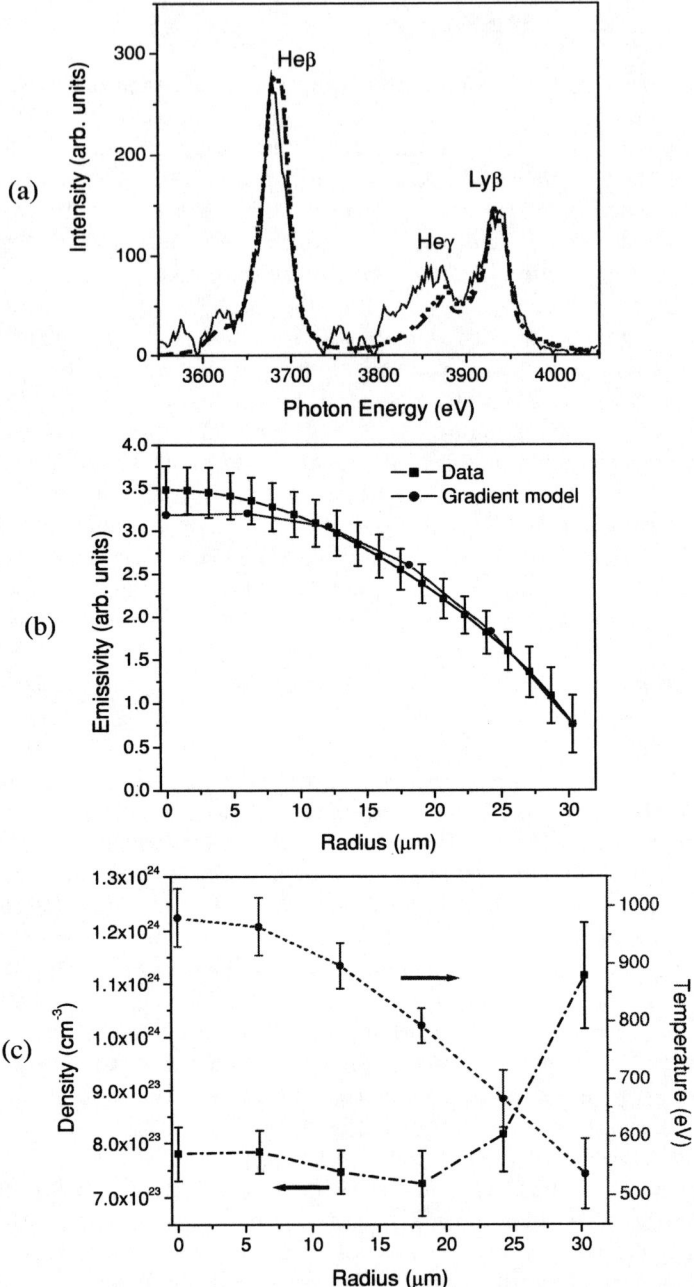

FIGURE 3. OMEGA shot 23686. Experimental x-ray line spectrum (−) with a comparison of spectral fits from the uniform model analysis (···) and the gradient model analysis (− −) (a). Emissivity profile extracted from the Heβ monochromatic image and compared to the emissivity calculated by the gradient model (b). Self-consistent density and temperature core gradients (c).

CONCLUSIONS

A spectroscopic core gradient determination technique has been successfully used to expand the understanding of plasma implosion cores in ICF experiments. We have used a novel method to determine plasma gradients that are time-integrated over the Heβ line emission interval in OMEGA indirect-drive implosion cores. The core gradient determination is based on the self-consistent analysis of data from X-ray line spectra and X-ray monochromatic images. The availability of a multi-objective genetic-algorithm-driven search and optimization procedure is important for the practical implementation of the core gradient analysis. Work is currently in progress to expand our current two-criteria analysis to a three-criteria core gradient determination based on the simultaneous analysis of X-ray line spectra and Heβ and Lyβ emissivity profiles.

ACKNOWLEDGMENTS

This work was supported by DOE NLUF Grant DE-FG03-01SF22225, and LLNL.

REFERENCES

1. H.R. Griem, Phys. Fluids B **4**, 2346 (1992).
2. H.R. Griem, *Principles of Plasma Spectroscopy*, Cambridge University Press (1996).
3. D.A. Haynes, Jr., D.T. Garber, C.F. Hooper, Jr., R.C. Mancini, Y.T. Lee, D.K. Bradley, J. Delettrez, R. Epstein and P.A. Jaanimagi, Phys. Rev. E **53**, 1042 (1996).
4. N.C. Woolsey, B.A. Hammel, C.J. Keane, A. Asfaw, C.A. Back, J.C. Moreno, J.K. Nash, A. Calisti, C. Mosse, R. Stamm, B. Talin, L. Klein and R.W. Lee, Phys. Rev. E **57**, 4650 (1998).
5. S.P. Regan, J.A. Delettrez, R. Epstein, P.A. Jaanimagi, B. Yaakobi, V.A. Smalyuk, F.J. Marshall, D.D. Meyerhofer, W. Seka, D.A. Haynes, Jr., I.E. Golovkin and C.F. Hooper, Jr., Phys. Plasmas **9**, 1357 (2002).
6. I. Golovkin, R. Mancini, S. Louis, Y. Ochi, K. Fujita, H. Nishimura, H. Shirga, N. Miyanaga, H. Azechi, R. Butzbach, I. Uschmann, E. Förster, J. Delettrez, J. Koch, R.W. Lee and L. Klein, Phys. Rev. Letters **88**, 045002-1 (2002).
7. Interactive Data Language (IDL) v5.5 (2001), Research System, Inc., 4990 Pearl E. Cr., Boulder, CO 80301.
8. I.E. Golovkin and R.C. Mancini, J. Quant. Spectrosc. Radiat. Transfer **65**, 273 (2000).
9. H.E. Dalhed, private communication.
10. F.J. Marshall and J.A. Oertel, Rev. Sci. Instrum. **68**, 735 (1997).
11. J.A. Koch, private communication.
12. K. Bockasten, Journal of the Optical Society of America **51**, 943 (1961).
13. I.E. Golovkin, PhD Dissertation, University of Nevada, Reno (2000).
14. I.E. Golovkin, R.C. Mancini, S.J. Louis, R.W. Lee and L. Klein, J. Quant. Spectrosc. Radiat. Transfer **75**, 625 (2002).
15. C.A. Coello-Coello, D.A. Van Veldhuizen and G.B. Lamont, *Evolutionary Algorithms for Solving Multi-Objective Problems*, Kluwer Academic Pub. (New York, New York, 2002).

Hydrogen line broadening in Dense Plasmas: Important physics

S.Alexiou

Abstract.
In view of recent high density experiments on hydrogen lines and highly exaggerated reports of new couplings and effects, this work discusses some *known* physics not normally considered in the analysis of these works that is of much more importance in the dense plasma regime than recently proposed exotic effects.

In the last Spectral Line Shapes Conference[1] convergence of quantal and nonperturbative semiclassical calculations for broadening of isolated ion lines was reported. The convergence was achieved by allowing for penetrating collisions. Further work on a much larger set of isolated ion lines seems to verify this convergence. Note that theory-experiment agreement has always been less than satisfactory for such lines.

In contrast for hydrogen lines theory-experiment agreement has been very good for lines without a central component(such as H_β) ever since accurate quasistatic microfield distributions became available(with the exception of the dip, which needs a good description of the dynamics too) and has been practically perfect for all lines when ion dynamics is taken into account for low and moderate densities. Recently, however, some high density experiments have reported[2, 3, 4] deviations from the standard theory(ST) i.e. perturbative impact theory for electrons and quasistatic ions. In these experiments ions are usually well quasistatic, so the main issue is electron broadening. Although the degree of confidence given to these experiments must be carefully evaluated(at least one seems to have been affected by optical thickness effects[5]), the theoretical calculations should be even more carefully examined, especially as a number of new couplings and effects have been proposed, some of them quite exotic and actually known to be missing important physics and lead to incorrect results. An example is the so-called "Advanced Generalized Theory"(AGT)[6] (p.562):(the present author never saw this paper before it was printed) "However, even for the La line at densities of the order of $10^{17} cm^{-3}$, where the experimental width is by a factor of two greater than the KG width and the entire difference between the two widths was usually "blamed" on the ion dynamics, it turns out the following: The AGT eliminates about one half of this discrepancy just by treating electrons more accurately than in the KG theory, which might mean that the ion-dynamical contribution could be about a factor of two smaller than it was previously thought". This statement is cast in an even stronger form in the abstract of the same source: "...is in reality about a factor of two smaller than it was previously assumed." The truth is that more than a decade before that publication, a number of authors, for instance Hegerfeld and Kesting(HK)[7], had used exact calculations to establish the importance of ion dynamics for the Grützmacher-Wende experiments(i.e. La at the density range mentioned above). As early as 1988, HK had confirmed by including much more physics and an exact treatment(i.e. solving the Schrödinger equation(SE)) that: a) There is no coupling between electrons and ions for the line and parameters in question, that is the joint electron-ion simulation produces an autocorrelation function that is the product of the autocorrelation functions solved with ions and electrons alone and b) the electronic autocorrelation function is in excellent agreement with the ST. In other words, HK treat ion dynamics exactly, treat electron-ion coupling exactly, go beyond the impact approximation and provide an exact semiclassical dipole solution of the dynamics, while the AGT does none of that, yet still claims that HK is in error[1].

In the present work we concentrate on some physics that is *not* included in either standard or exotic calculations and that is important for the conditions of these experiments. Although it would also be very interesting to access the importance of quantal effects too, such issues are not discussed here.

QUALITATIVE CONSIDERATIONS

The ST for hydrogen emitters gives a collision operator with a logarithmic divergence:

$$\Phi \approx ln(\rho_{max}/\rho_{min}) \qquad (1)$$

where ρ_{max} and ρ_{min} are maximum and minimum impact parameters. As long as the argument of the logarithm is large, these cutoffs are not too important. However, for large densities/small temperatures, this ratio becomes small(for instance 1.6 in a case considered below) and cutoffs are critical. In such a case the anwers obtained by different choices of the cutoffs and assumed strong collision contributions can differ significantly. Therefore treating small impact parameters right is important at high densities. One should keep in mind that what physically removes the divergence at small impact parameters(and preserves perturbation theory) is the breakdown of the long range dipole approximation. Therefore taking account of penetration can be critical, i.e. the dipole approximation should be replaced by the full Coulomb interaction:

$$e\mathbf{r} \cdot \mathbf{E}(t) \to e[|\mathbf{r} - \mathbf{R}(t)|^{-1} - R^{-1}(t)] \qquad (2)$$

Although one may remove at least some of the logarithmic divergence for instance by neglecting time ordering or by dressing the electron field(e.g. by the ion+z- component of the electronic field, as in the AGT), this does *not* address the *fundamental* reason for the finite electronic contribution.

---

[1] Theorywise, the AGT, which uses as the unperturbed Hamiltonian the static ion plus electronic field in that direction is an interesting idea; however, the U-matrix elements are connected by the SE and unitarity. The inaccuracy in the computation of one of them will necessarily show in the others and in general it is not possible to compute some parts of the problem inaccurately without affecting all components. This raises the question that if it is important to treat the z-component of the electronic field nonperturbatively, why is it ok to treat the x and y-components perturbatively.

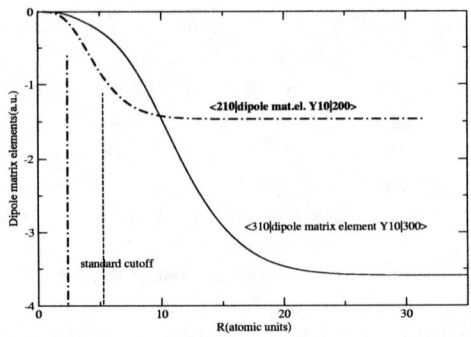

FIGURE 1. Onset of penetration vs. standard estimate(vertical lines).

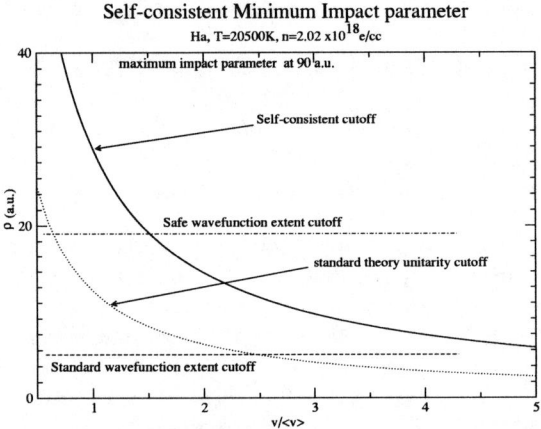

FIGURE 2. Importance of penetration for minimum impact parameter determination.

In [1] it was shown that the standard estimate of the wavefunction extent(WE) $n_u^2 a_0/Z$, with n_u the upper level principal quantum number, a_0 the Bohr radius and Z the spectroscopic charge number, is seriously inaccurate for modern day purposes. Fig.1, which plots dipole matrix elements for L_α and H_α versus perturber position R, shows that things are no different in the case of hydrogen. This means that the contribution from small impact parameters for which penetration is important, is much smaller than the standard predictions and that the WE cutoff employed to preserve the dipole approximation(i.e. $n_u^2 a_0/Z$) should be revised.

To illustrate the importance of these issues, consider the H_α measurement in [4]. In Fig.2 we plot the ST unitarity-based cutoff(dotted line) $(n_u^2 - n_l^2)\hbar/mv$, the self-consistently determined minimum impact parameter(solid line), obtained by solving

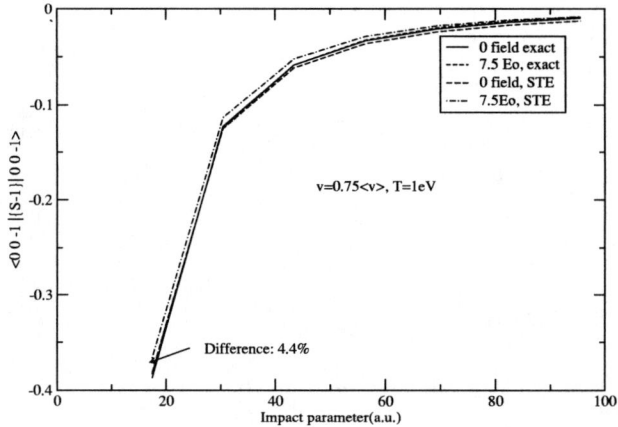

FIGURE 3. Exact vs. STE calculation

$\{I - S_u(\rho,v\ )S_l^{\dagger}(\rho,v)\} = I$ (to second order in perturbation theory for the S-matrix product) and the standard(dashed) $(n_u^2 - n_l^2)a_0$ and safe(based on Fig.1)(dot-dashed) WE minimum impact parameter cutoffs, where the subscripts u and l stand for "upper" and "lower" levels respectively. The regime above the solid line is the region of applicability of the dipole perturbative ST. It is clear that the WE cutoff significantly affects the minimum impact parameter determination and furthermore the perturbation is quite strong(since we are significantly below the unitarity-based $\rho_{min}(v)$) at the onset of penetration. Thus for $v < 1.5\langle v \rangle$ perturbation theory is still not valid at impact parameters $\lesssim 20a_0$, while for higher velocities the weak collision contribution is overestimated. This measurement will be discussed further along with other experimental data later on.

CALCULATIONS

In the calculations shown in Figs. 3-5 the SE has been solved numerically to obtain the S-matrices required for the exact collision operator. This way all nonperturbative aspects are treated exactly and penetration is included, if the relevant code option is enabled. All calculations reported here are dipole only in order to be able to compare with simplified calculations.

It should be mentioned that such calculations are not new[8]. The main difference is that in the present calculations we also allow for a quasistatic ion field in the calculation of impact broadening. Such a coupling was considered in recent works(e.g. "AGT" mentioned above). Common wisdom[9] holds that it is usually permissible to ignore the electric field in the computation of electron broadening, though not in the final profile. However, as the density increases a progressively larger part of electrons becomes nonimpact and is thus correlated with the static ion field[10]. The issue of a possible electron-ion coupling at high densities is an important one for hydrogen, because if this coupling is unimportant the S-matrix may be solved analytically[11, 12, 13] under

305

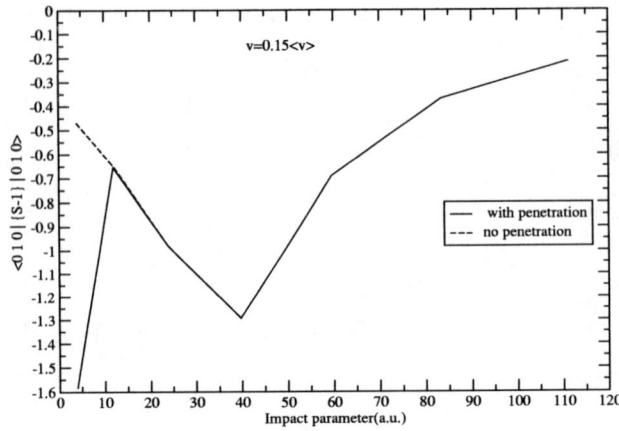

FIGURE 4. Effects of penetration

the assumptions of a pure dipole interaction with no penetration, no-quenching, no fine structure etc. In practice, however, this solution is almost never used in the literature.

All static effects may be treated exactly by using the parabolic basis, i.e. eigenstates of the instantaneous ion field. All calculations refer to $Ly - \alpha$ at 1eV temperature(for which line penetration and nonperturbative effects are the weakest). Within standard assumptions(e.g. electron motion is not affected by the ion field) the only effect of a static field on the broadening operator is to turn hydrogen into a nonhydrogenic species. The corresponding calculation takes into account the finite energy spacings between the states of the upper/lower level and is referred to as STE. The difference from the ST is the imaginary exponentials. Most of the differences between the ST and STE are due to cutoffs. Including these differences, at 1 eV, $10^{18}e/cc$, and even at 5eV and $10^{19}e/cc$, the L_α widths differ by less than 10%. As long as perturbation theory is ok, this is *all* the coupling between electrons and ions. Fig.3 checks STE against exact semiclassical calculations, i.e. the SE is solved and the S-matrix is computed in the parabolic basic for the ion field in the z-direction and the result is integrated over angles describing the orientation of the collision axes with respect to the coordinate frame. The results show that as long as perturbation theory holds, STE is fully satisfactory.

PROBLEMS WITH THE STANDARD THEORY

Problems with the ST arise on two fronts: Neglect of penetration and nonperturbative effects. With respect to the former, the point is that *when* penetration is important, it makes a substantial difference, as Fig.4 illustrates. Fig.5 compares the difference penetration makes against the difference made by the electron-ion coupling.

Even though for L_α penetration is not important, it is still more important than electron-ion coupling. With regard to the later, the best thing ST could do is to compute unitarity-based error bound for the contribution of that part of the phase space(i.e.

306

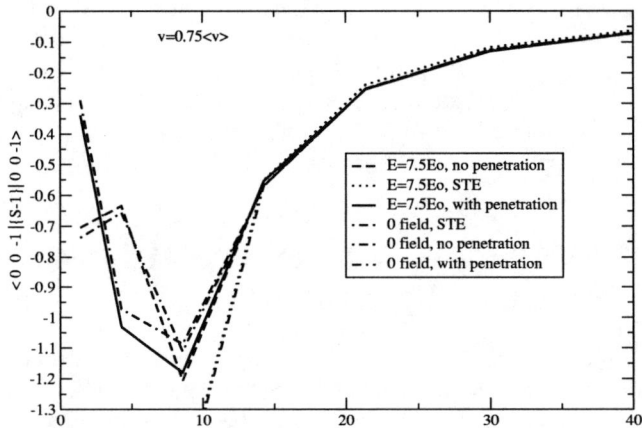

FIGURE 5.   Relative importance of electron-ion coupling and penetration

velocities and impact parameters) which could not be computed nonperturbatively. For high densities and low temperatures this part can be comparable to what can be computed presumably reliably by perturbation theory. What is worse, the standard demarcation line between these two regimes was drawn by neglecting the matrix nature and estimating the part involving the dipole matrix elements

$$\langle \alpha | \mathbf{r} | \alpha'' \rangle \cdot \langle \alpha'' | \mathbf{r} | \alpha' \rangle + \langle \beta' | \mathbf{r} | \beta'' \rangle \cdot \langle \beta'' | \mathbf{r} | \beta' \rangle - 2 \langle \alpha | \mathbf{r} | \alpha' \rangle \cdot \langle \beta' | \mathbf{r} | \beta \rangle = (n_u^2 - n_l^2)^2 a_0^2 \quad (3)$$

In detail this means writing down the expression for the (matrix elements of the) S-matrix as a function of ρ and v (estimating the dipole matrix elements) and equating this to unity. This gives a ρ(v) curve, which is the most sophisticated form of the Weisskopf Radius WR(often the WR is taken to refer to this value at the average velocity). The WR is a fairly naive concept that is often misused in the literature. For instance WR arguments are often used to argue for ions being quasitatic, but the WR is simply a very simple way to check *unitarity, not* time scales. What the WR really tells us is that *if* we were to compute the relevant S-matrix elements by perturbation theory, then this would break down at the WR(which is a function of velocity v, though often the average velocity is used) This in turn means that *if* the WR were correct, then collisions with shorter impact parameters would be strong, and the rest weak. This is used in quasistatic approach justifications, essentially saying that strong collisions are static and hence estimating the degree to which a species(usually ions) is static. As pointed out 22 years ago[14], static collisions are strong, but the converse is not true. Consequently using the WR to argue for time scales is wrong and furthermore leads to the absurd result for lines with a central component that at high enough densities electrons too are quasistatic, in which case there is no mechanism left(except Doppler and natural broadening) to broaden the central component, i.e. if we could turn off Doppler and natural broadening, we'd have δ-functions for the central components at very high densities.

In addition to this erroneous, yet still invoked[15] use of the WR, the very calculation of this cutoff is in principle flawed for two reasons: First, one solves a matrix equation as if it were a scalar as in Eq.(3). Second, one computes the second-order S-matrix expansion and believes that unitarity is satisfied if this is small. However, this will be true if the U-matrix has remained small for *all* times(which is true in the case of hydrogen with no static field), which is not always the case for nonhydrogenic species, in which case this perturbative check is fooled[16]. Safe perturbative formulas have since been given[17], that will correctly account for this effect.

OTHER EXOTIC EFFECTS AND COMPARISON WITH EXPERIMENTS

Electron-ion coupling is not the only exotic effect proposed to explain the high density experiments. Other effects include[18] a narrowing effect allegedly caused by the motion of the perturbing electron in the field of the nearest-neighbor ion(the fact that the plasma is electrically neutral and that a nearest-neighbor electron also exists is ignored) as well as a residual ion impact width(among other issues, the fact that at high densities for every ionic configuration that is "impact" there are many more which are "ion dynamic" -and of course even more that are "quasistatic"- is also ignored. Among other things, by assuming that what is not quasistatic is impact, based on the known to be erroneous Weisskopf radius criterion, the dynamic ionic contribution is grossly overestimated, since at a given set of plasma conditions the impact broadening is an upper bound for the possible contribution). By combining these "effects", in the so-called UAGT("Upgraded" AGT) impressive agreement with most high density experimental data is obtained, for instance Fig.1 in Ref.[18]. It is interesting however, to compare this agreement with ref.[5] which is essentially an *exact* dipole calculation, that essentially confirms the results of the ST(roughly correctly reproduced in that figure, except that error bounds are not given). As proposed in ref.[5] and confirmed by the experimental group[18] the original lowest density points were affected by self-absorption. The exact dipole calculation in [5] *agrees* with the new measurement around 2.4×10^{18}e/cc (the only point common in the original and new measurements) *without* invoking any of these additional effects, as well as with the highest density measurements within the theoretical error bars, which arise from the need to preserve the dipole and semiclassical approximations(not unitarity, which is satisfied automatically in the exact dipole calculations). With regard to the dipole approximation the $n_u^2 a_0$ cutoff was used in [5]. The results in [18] have much lower error bars because they simply *ignore* these considerations and pretend one can integrate down to 0 impact parameters.

We should stress that these additional "effects" are quite important for achieving "agreement" with experiment. For instance in [4] let's look at the point at T=20500K and density 2.02×10^{18}e/cc. Roughly one fourth of the UAGT width is a residual ion impact width. The ST is in agreement with the FFM and MMM results. A self-consistent(i.e. with self-consistent unitarity-preserving $\rho_{min}(v)$ determination) gives an STE width almost twice as large as the ST. However, these values are computed with a minimum impact parameter also larger than the de Broglie wavelength $\hbar/mv$ and WE ($n_u^2 - n_l^2)a_0$, and a strong collision contribution based on the assumption of a constant $\{I - S_u(\rho,v)S_l^\dagger(\rho,v)\} = 1$ for strong collisions. When we replace this WE cutoff by $18a_0$(as obtained from Fig.1), the weak collision width is 49Å in the ST and 32Å in the STE case, i.e. the region for which penetration is important(and which may not be accurately computed by either the ST without penetration or the more exotic approaches which again ignore these issues) is responsible for a large part of the width. In short, all these approaches are insufficient for comparing with experiments in that regime.

CONCLUSIONS

Recent high density experiments with hydrogen lines can be important for checking the contributions of close collisions(in particular for testing the adequacy of a semiclassical treatment of close collisions down to the de Broglie wavelength, which is a theoretically important issue[1]. To do so, however, requires a theoretical analysis that takes into account important physics such as penetration and nonperturbative aspects. Thus far there is *no* evidence against the adequacy of the standard electron impact broadening picture, provided close collisions are treated accurately. Exotic effects proposed to explain the apparent discrepancies between a dipole nonperturbative approach and experiment are clearly incorrect, both in theory and in terms of their results.

REFERENCES

1. S.Alexiou, and R.W.Lee, "Electron Line Broadening in Plasmas: Resolution of the Quantum vs. Semiclassical Calculation Puzzle," in *Spectral Line Shapes*, edited by J.Seidel, AIP Conference Proceedings 559, American Institute of Physics, New York, 2001, pp. 135–143.
2. St.Böddecker, S.Günter, A.Könies, L.Hitzschke, and H.-J.Kunze, *Phys.Rev.E*, **47**, 2785–2791 (1993).
3. A.Escarguel, B.Ferhat, A.Lesage, and J.Richou, *JQSRT*, **64**, 353–361 (2000).
4. S.Flih, and Vitel, Y., "Experimental Profiles of Hydrogen Balmer α Line Emitted in weakly Non-ideal Plasma," in *Spectral Line Shapes*, edited by J.Seidel, AIP Conference Proceedings 559, American Institute of Physics, New York, 2001, pp. 30–32.
5. S.Alexiou, and E.Leboucher-Dalimier, *Phys.Rev.E*, **60**, 3436–3438 (1999).
6. Touma, J., Oks, E., Alexiou, S., and A.Derevianko, *JQSRT*, **65**, 543–571 (2000).
7. G.Hegerfeld, and V.Kesting, *Phys.Rev. A*, **37**, 1488–1496 (1988).
8. K.Bacon, K.Y.Shen, and J.Cooper, *Phys.Rev.*, **188**, 50–56 (1969).
9. Baranger, M., *Atomic and Molecular Processes*, Academic, New York, 1962, p. 523.
10. Alexiou, S., *Phys.Rev.Lett.*, **76**, 1836–1839 (1996).
11. H.Pfennig, *JQSRT*, **12**, 821–837 (1972).
12. H.Pfennig, *Z.Naturforsch.a*, **26**, 1071– (1971).
13. V.S.Lisitsa, and G.V.Sholin, *Sov.Phys.JETP*, **34**, 484–489 (1972).
14. J.Seidel, "Theory of Hydrogen Stark Broadening," in *Spectral Line Shapes*, edited by B.Wende, W. de Gruyter, Berlin, 1981, pp. 3–40.
15. E.Oks, "Dramatic Effect of the accalaration of electrons by the ion field on spectral line shifts in plasmas: advanced theory," in *these proceedings*, 2003.
16. S.Alexiou, *Phys.Rev.Lett.*, **75**, 3406–3409 (1995).
17. A.Poquérusse, S.Alexiou, and E.Leboucher-Dalimier, *JQSRT*, **56**, 797 (1997).
18. E.Oks, "Comparison of the Latest Experimental H-alpha Results at the Bochum's Gas-Liner Pinch with the Upgraded Advanced Generalized Theory," in *Spectral Line Shapes*, edited by J.Seidel, AIP Conference Proceedings 559, American Institute of Physics, New York, 2001, pp. 54–57.

Balmer α and Lyman β
Emitted by a Laser-Produced Plasma

J.F. Kielkopf* and N.F. Allard[†]

*Department of Physics, University of Louisville, Louisville, KY 40292, USA
[†] CNRS Institut d'Astrophysique, 98 bis Boulevard Arago, 75014 Paris
& Observatoire de Paris-Meudon, Département Atomes et Molécules en Astrophysique
92195 Meudon Principal Cedex, France

Abstract.
    A laser-produced plasma in hydrogen is being used to study the shift, shape and width of the Balmer α and Lyman β spectral lines emitted by the $n = 3$ state of atomic hydrogen perturbed by H^+ and electrons. Recent measurements at densities up to 10^{20} cm^{-3} show that the Balmer α center shifts monotonically to the red with increasing density. The largest shifts seen, of the order of 200 Å, occur with shift-to-width ratios of up to 0.25. At densities below 10^{19} cm^{-3} the shift-to-width ratio is an order of magnitude smaller, in agreement with published gas-liner pinch plasma experiments, but the laser-plasma explores an order of magnitude higher density. The broad core profiles are consistent with unified line shape theory calculations based on exact H_2^+ potentials. Balmer α exhibits satellites, some of which are close to the unperturbed line center, and are expected to influence the shift and shape of the line at high H^+ density. The satellites may influence the line shift at the highest densities observed.
    Lyman β, emitted in the extreme ultraviolet, can be seen in a windowless pulsed gas jet target, rather than the static cell used for Balmer α. Preliminary experimental results indicate that with these techniques it will be possible to detect the satellites in its line wing due to H-H$^+$ interactions that were recently predicted [6].

INTRODUCTION

The shift of the Balmer α spectral line due to interactions with electrons and ions in a dense plasma has been a subject of recent theoretical [1, 2] and experimental [3] interest. In well-diagnosed plasmas such as the gas-liner pinch, electron densities up to 10^{19} cm^{-3} have produced shifts as large as 35 Å, in approximate agreement with theoretical models. Recent work in our laboratory on the far wing of Balmer α [4] emitted by a laser-produced plasma [5] in pure hydrogen offered the opportunity to simultaneously measure the shift, shape and line width of the central line region at an order of magnitude higher density.

LASER-PLASMA SOURCE

A plasma is produced by a 300 mJ pulse from a 1064 nm Nd:YAG laser focused into a cell containing 99.999 % pure H_2. Images of typical laser-produced plasmas in H_2 at atmospheric pressure reveal that the laser light is self-focused by the plasma it forms.

CP645, Spectral Line Shapes: Volume 12, 16th ICSLS, edited by C. A. Back
© 2002 American Institute of Physics 0-7354-0100-4/02/$19.00

Bright spots of 1064 nm laser light deviated 90° from the incident direction are seen along the axis in a dentritic pattern as the incident laser propagates through the gas, and through the plasma of its own making.

For the experiments described here, a static cell was filled to initial pressures of 600 kPa, corresponding to atomic densities of 3.2×10^{20} H-atoms/cm^3. The laser produces a channel in the gas and deposits its energy suddenly compared to the dynamical time scale for expansion of the heated gas. The resulting shock wave expands cylindrically from the laser focal region at speeds greater than Mach 10. A central axial channel develops, while much of the atomic emission occurs from an outwardly expanding shock front. The shock compresses, dissociates, and ionizes the molecular hydrogen, leaving a hot atomic hydrogen plasma bubble in its wake.

The dissociation of the molecular gas and the densities of molecules, atoms, ions and electrons may be calculated for cylindrical or spherical shock wave geometries. A typical result of these models for the distribution of the low-lying excited states of H, and for neutral molecular gas, shows that the excited states are present in a shell behind the front, inside of which the atoms are ionized but the density is low, and outside of which the atoms are unexcited or the molecules are not even dissociated. The neutral and electron densities are dependent on the propagation of the shock, and since the emission arises from a thin zone where the population of the initial excited states is high, the densities may be "tuned" by delaying the time of observation from the initiation of the plasma, typically by a few 10's of nanoseconds. After delays of a few μs, the shock front has passed out of the region of observation, and the emission is due to neutrals remaining in the hot bubble. The bubble has a lifetime exceeding 100 μs, during which it evolves from a spheroid with axial flow back toward the laser. The results reported here are based on data from the shock front in the early phase of its expansion, but after the exciting laser pulse has ceased.

SHIFTS AND WIDTHS

For all of the data on Balmer α shown here, the plasma was imaged on the slit of a 0.25 m focal length Czerny-Turner aspheric optic monochrometer with a 1200 g/mm holographic grating. To optimize the signal-to-noise ratio, a relatively wide slit was used. This set the instrumental resolution at approximately 6 Å. In the time delay region of interest, this is much narrower than the width and shift of the line. The grating has a broad smooth spectral response, so corrections were not made for the variation in system response with wavelength. Signals were detected with a fast photodiode and gated amplifier, and recorded digitally under computer controlled acquisition. Typically 30 laser shots were averaged for each sample in the time-spectrum data space. Wavelengths were determined with respect to a low pressure Hg reference source.

Observations were made over a wide range of delays, from the shortest at which the Balmer α line was indistinguishable from the continuum, to the longest when it was too narrow to measure well. The Balmer α line clearly shifts to the red at high density. There is some evidence of reversal at 100 ns, suggesting that at the earliest times the line is not optically thin. This may affect the line shape and width at highest density,

311

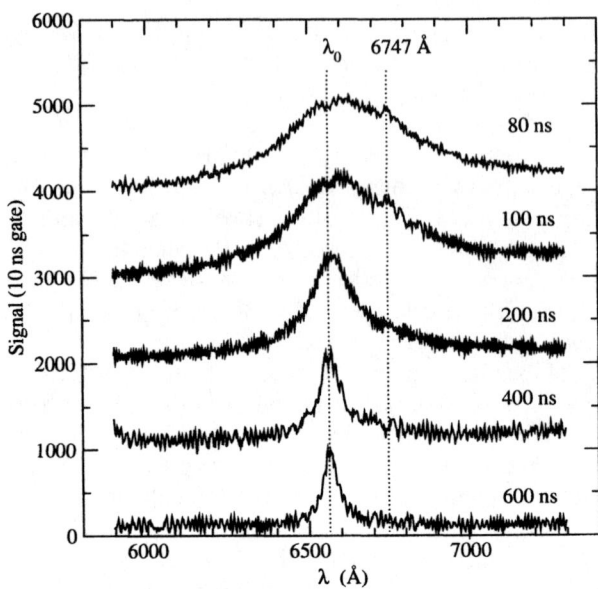

FIGURE 1. Hα with delays from 80 to 600 ns.

probably increasing the apparent line width. We do not model the radiative transport in the plasma, and while the emission results from a shell with narrowly defined conditions, there will be small contributions from a range of densities and temperatures. The filling H_2 gas in the cell is ultra high purity, and impurities which might result from small contamination are not seen. However, a "feature" at 6747 Å remains unidentified. It may be instrumental, but it appears in some spectra and not in others.

Figure 1 shows the profiles of Balmer α with a selection of moderately long delays. At 600 ns the shell is far from the axis, and the density is lowest. At 80 ns the density is high. λ_0 indicated in the figure is the unperturbed line center, and in the 80 and 100 ns data there is a slight hint of reversal which would be due to a cooler zone of atomic hydrogen outside the inner hot dense plasma shell. Such a zone would result from photodissociation and excitation ahead of the shock front.

Figure 2 shows the profiles with shorter delays to illustrate the broadening and shift at some of the highest densities observed. The core line widths and shifts are nominally in agreement with the theory of the profile that includes broadening due to H-H^+ collisions but not neutrals, shown here for three different ion densities. Since ion and electron collision broadening are comparable effects, it is reasonable that a profile including only ion effects may fit the observations. Note that these calculations predict a smooth profile in the line core region at these densities, and a monotonic shift to the red with increasing ion density. The theory is described in a recent paper by Kielkopf, Allard, and Decrette [4] on the wing of H α. The densities shown here are in agreement with the expectations of shock models of the plasma. However, at the highest densities the efficiency of the coupling of the incident laser to the plasma is uncertain because of self-focusing, which

312

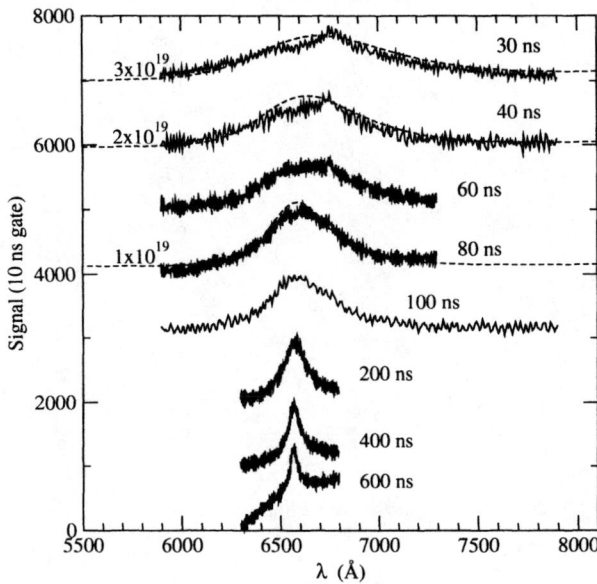

FIGURE 2. Hα with delays down to 30 ns.

increases at the high pressures used in these cells compared to the atmospheric pressure for which the models have been tested. We expect, on the basis of these models, that the density is highest at the shortest observed delays, and that with increasing delay the temperature falls slowly and density falls very rapidly. Thus the sequence shown in Figs. 1-2 represent a monotonic range of densities.

Figure 3 shows the full width at half-maximum intensity (FWHM) and shift of Hα. Both increase monotonically with decreasing delay and increasing density. At very short delays the spectra for this sample were noisy, and the line was nearly lost in the continuum. This may account for the low width in one data point. Excepting this one point, the widths approach 1000 Å before the line disappears, and the shifts exceed 200 Å to the red. There is no evidence of a blue shift, but there is some structure in the line at the highest density. The shift and width are measured by fitting a Lorentzian profile to the data with a least-squares adjustment of the width, shift, and continuum; they do not take into account the shape of the observed line, although they do represent the average behavior of the profile.

The ratio of shift to width increases with decreasing delay or increasing density. Above 100 ns the ratio is nearly constant at about 0.05, approximately the value seen in the arc experiments at densities below 10^{19} cm^{-3} [3]. Below 100 ns the shifts increase dramatically until, at the smallest delays and highest densities sampled, the shift-to-width ratio approaches or exceeds 0.3.

With a 3 ns gate width and a faster response detector than used for the shift measurements shown in Figs. 1-4, the sudden change in the line core at the shortest delays (highest densities) becomes apparent. A clear feature at ≈ 6800 Å appears, while Hα

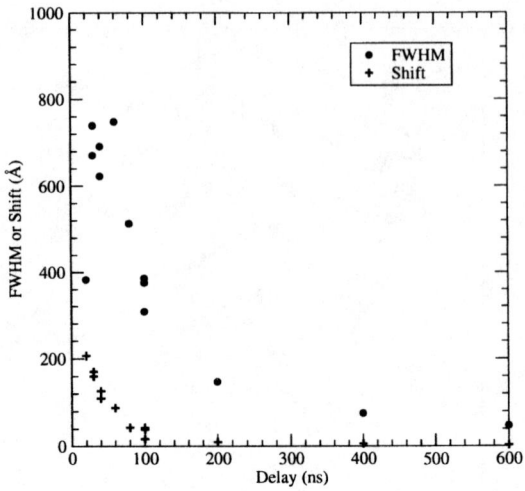

FIGURE 3. The full width at half-maximum intensity (FWHM) and shift of H α.

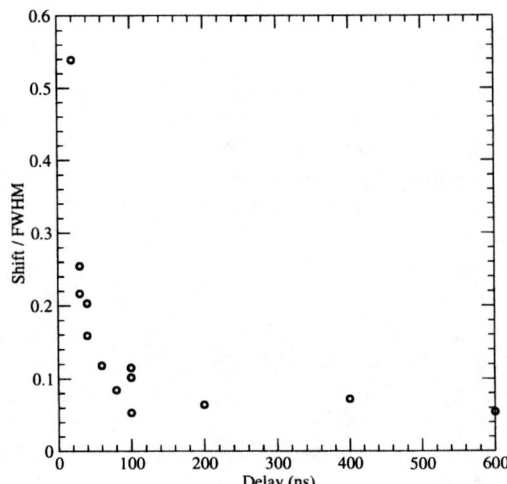

FIGURE 4. The shift-to-width ratio of Balmer α.

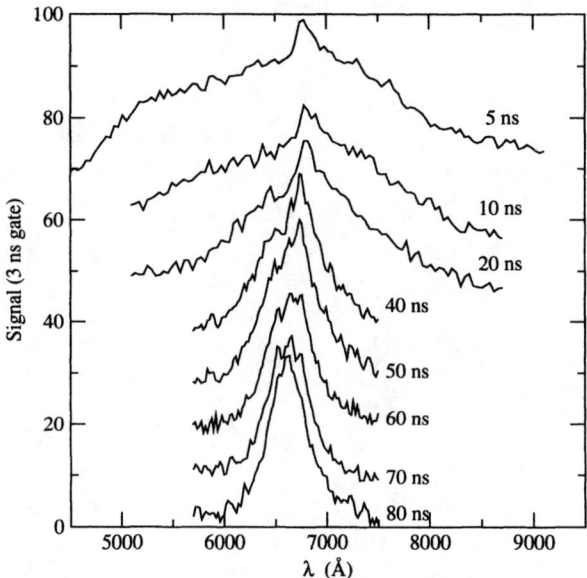

FIGURE 5. Balmer α at densities approaching 10^{20} cm^{-3}.

may extend to the blue where it is lost in Hβ. The asymmetry introduced by the Hβ region gives the impression of a blue-shifted Hα at densities probably of the order of 10^{20} cm^{-3}. Representative spectra at the shortest delays are shown in Fig. 5. Blue shifts arising from the coupling of ion and electron broadening have been discussed recently by Oks [8]. Such shifts are non-linear in electron density, and should be manifested in the high density regime studied here.

Extreme values of the shift-to-width ratio are characteristic of broadening when satellites become a dominate part of the profile. At high density it becomes very probable that radiation from the excited state occurs while the atom is interacting with a perturber, in this case H$^+$, in the vicinity of a potential minimum. As described in Ref. [4], the unified theory predicts satellites due to long range minima in the excited state potentials of H$_2^+$. One of these, at 6830 Å, is close enough to influence the core of the line at ion densities near the upper limit of those studied here, and may be an important factor in determining the shift at high density. A theoretical profile of Balmer α showing the line center is in Fig. 6. A complete theory which includes these effects, as well as broadening by electrons, is needed for an exact comparison.

The same initial n=3 state responsible for Balmer α is responsible for the region between Lyman α and Lyman γ in the vacuum ultraviolet. Of course there is also an overlap of contributions from other states, but the spectrum associated with Lyman β is largely determined by states of H$_2^+$ that asymptotically approach a free atom in the n=3 state in collision with H$^+$. At the lower density required to operate a pulsed gas jet source, there is marginal evidence in the spectrum of the Lyman series of satellites of Lyman β due to H-H$^+$ that have been predicted theoretically by Allard *et al.* [7] and

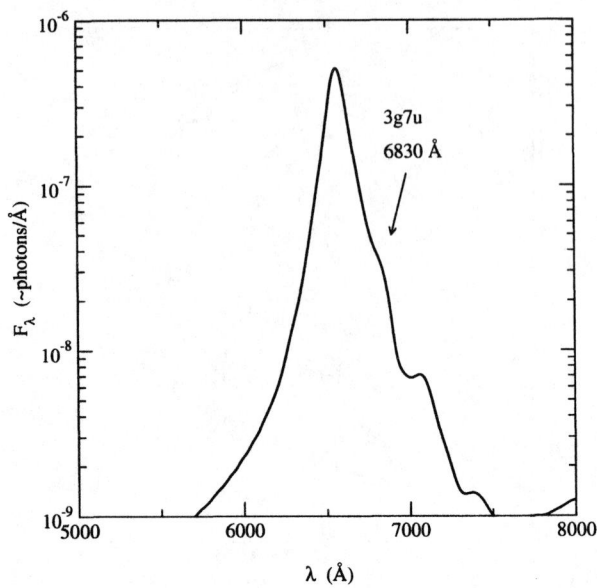

FIGURE 6. The unified line shape theory for H-H$^+$ collisions at 10^{15} H$^+$ cm^{-3}.

have been observed in hot hydrogen-rich white dwarf stars. Preliminary data shown in Fig. 7 show that the two satellites of Lyman β between 1050 Å and 1100 Å may be present in the experiment.

CONCLUSION

At a typical static cell pressure of 600 kPa, the plasma produced by a 300 mJ, 6 ns, 1064 nm laser pulse is sufficiently dense that Balmer α disappears into the continuum in the earliest resolvable emission. Density may be selected by delaying the spectroscopic measurement from the time of the plasma-production, and in these experiments the line was studied with delays up to 600 ns. The determination of the line shift is made difficult by the large width of the line, and by the continuum background. Nevertheless, with allowance for line asymmetry, apparent reversal, and a continuum background, the observed shift-to-width ratio more than 100 ns after the laser pulse agrees with previous lower density observations. At earlier times and higher density, the shifts increase monotonically with density up to a red shift of 200 Å, above which the line disappears in the background. The core profiles also are consistent with far line wing observations and theory [4].

Work on Lyman β emitted by a plasma generated in a pulsed-gas jet target is underway. Preliminary experimental results suggest the detection of satellites in the line wing due to H-H$^+$ interactions predicted theoretically [6, 7].

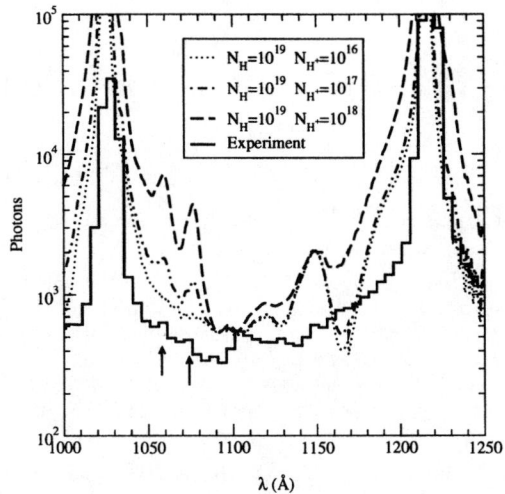

FIGURE 7. Observations and theory for Lyman β. Lyman α at 1215 Å is on the right.

ACKNOWLEDGMENTS

The work at the University of Louisville was supported by grants from the Atomic, Molecular, and Optical Sciences Program of the U.S. Department of Energy, Office of Basic Energy Sciences.

REFERENCES

1. Griem, H. R., *Phys. Rev. E*, **64**, 058401 (2001).
2. Griem, H. R., *Contrib. Plasma Phys.*, **41**, 223–226 (2001).
3. Böddeker, S., Günter, S., Könies, A., Hitzschke, L., and Kunze, H.-J., *Phys. Rev. E*, **47**, 2785–2791 (1993).
4. Kielkopf, J. F., Allard, N. F., and Decrette, A., *Eur. Phys. J. D*, **18**, 51–59 (2002).
5. Kielkopf, J. F., *Phys. Rev. E*, **63**, 016411 (2001).
6. Allard, N. F., Kielkopf, J. F., Drira, I., and Schmelcher, P., *Eur. Phys. J. D*, **12**, 263–268 (2000).
7. Allard, N. F., Royer, A., Kielkopf, J. F., and N., F., *Phys. Rev. A*, **60**, 1021–1033 (1999).
8. Oks, E., *J. Phys. B*, **35**, 2251–2260 (2002).

Asymmetry and Shifts Interdependence in Stark Profiles

Alexander Demura*, Volkmar Helbig† and Dragan Nikolić**

*Hydrogen Energy and Plasma Technology Institute,
Russian Research Center "Kurchatov Institute", Moscow 123182, Russia
†Institut für Experimentelle und Angewandte Physik, Universität Kiel,
Leibnizstr. 19, D-24098 Kiel, Germany
**Department of Physics, Stockholm Centre for Physics, Astronomy and Biotechnology,
Stockholm University, SE - 106 91 Stockholm, Sweden[1]

Abstract. The asymmetry of the Stark profile of the Balmer H_β line of Hydrogen in dense plasmas is analyzed with respect to its dependence on reference points and electronic shifts of the individual Stark components. The strong interdependence of the asymmetry and the shifts is found experimentally and theoretically.

INTRODUCTION.

The observed Stark profiles, emitted from high density plasmas, exhibit noticeable asymmetry varying with density. Besides the asymmetry the profiles are characterized also by the shift, but for the asymmetrical profiles this notion needs in a special definition. Among known definitions are : the shift of maximum (MS), the center of gravity shift (CGS), the estimated line center shift (ELC), defined by the average of three measurements of the line center shift from the half distance between points in the profile corresponding to 1/2, 1/4 and 1/8 of the intensity in maximum [1]. Obviously the task of the shift definition becomes much more complicated for the profiles having several maxima. For example, in the case of the hydrogen Stark profiles without the central Stark components such as H_β one can introduce the shift of the dip (DS) [2], the shift of the blue (MBS) and red (MRS) maxima [3], the shift of the half distance between the maxima or the mean maximum shift (MMS) [4] and ELC that is characterized by the asymmetry of the profile and could be called the mean asymmetry shift (MAS). In the experiment the asymmetry is characterized conventionally by the following ratio called the asymmetry parameter $A(\Delta\lambda)$

$$A(\Delta\lambda) = \frac{I_r(\Delta\lambda) - I_b(|\Delta\lambda|)}{I_r(\Delta\lambda) + I_b(|\Delta\lambda|)} \tag{1}$$

where $\Delta\lambda = \lambda - \lambda_0$, λ_0 being chosen for the reference point, $I_r(\Delta\lambda)$ is the intensity of the red part of the profile $I_b(|\Delta\lambda|)$ is the intensity of the blue part of the profile with

---

[1] e-mail : Alexander.Demura@hepti.kiae.ru, helbig@physik.uni-kiel.de, DNikolic@physto.se

CP645, *Spectral Line Shapes: Volume 12, 16th ICSLS*, edited by C. A. Back
© 2002 American Institute of Physics 0-7354-0100-4/02/$19.00

respect to the reference point. It is easily seen that the definition of the reference point is not unique. However, in the experiment usually the reference point is determined by the signal obtained at very low densities, i.e. by the unperturbed line frequency.

In the present study it was experimentally and theoretically shown that in the case of the hydrogen H_β line *the asymmetry is a very sensitive function of the reference point*. This experimental finding is presented in the Fig. 1a where the asymmetry parameter A has been calculated taking i) the unperturbed line position and ii) the central dip as reference point. It is obvious from the Fig. 1a that this has a dramatic influence on the asymmetry parameter close to the line center.

The experiment was performed in a standard wall stabilized arc discharge running in argon and a second running in argon with a small admixture of hydrogen. The difference spectra from the two discharges gave the pure hydrogen spectrum. Special care was taken to guarantee that the electron density, the number densities and the plasma temperature were the same in both discharges. The experimental procedure and the plasma analysis have been described in detail in [5]. The results communicated here were obtained in [6] for a plasma temperature of $T = 13620$ K and an electron density of $N_e = 1.36 \times 10^{17}$ cm^{-3}.

In the case of H_β it is possible to deduce from the experimental Stark profiles three trends that scale with the increasing electron density : i) in the line center the blue peak gains intensity with respect to the red one; ii) the positions of the two peaks shift to the red with respect to the reference point; iii) for equal distances from the reference point the red wing gains intensity with respect to the blue one.

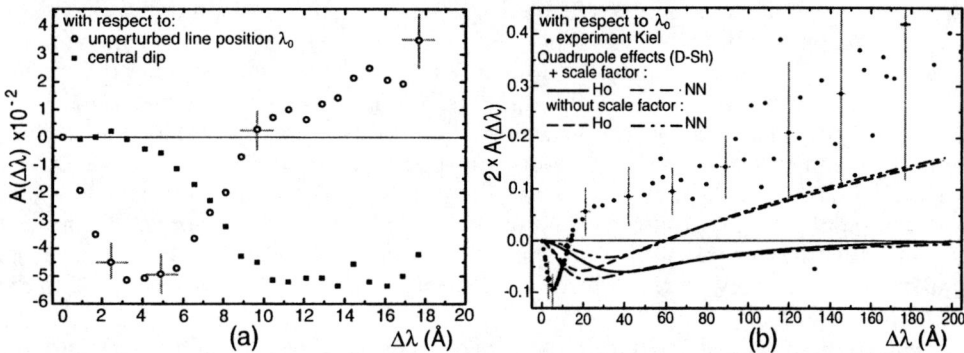

FIGURE 1. (a) Asymmetry parameter of H_β near the line center measured with respect to the unperturbed line position and the dip. (b) Comparison of experimental results with the quadrupole effects contribution.

As is well known from the earlier measurements the H_β dip is shifted with respect to the unperturbed line position and gets deeper with the increasing plasma density [2]. Also in the earlier studies a very close correlation was found between the shift of the dip and some average shifts of the line caused by electronic collisions [4]. That is why the Fig. 1a may be interpreted as the drastic shift influence on asymmetry functional behaviour near the line center. In total this gives an idea of the strong interdependence of asymmetry and the electronic shift values.

Basing on the above mentioned arguments this paper presents the results of a fitting procedure of theoretical profiles to experimental results in detail paying special attention to the functional behavior of the asymmetry parameter and proving the idea of a strong and principal influence of the electronic shift values on the asymmetry parameter. The sensitivity of the functional behavior of the asymmetry parameter on the electronic shift values makes it even possible to use it for their determination. This study exploits parabolic quantization that allows to introduce quite physical notion of the individual widths and shifts of separate Stark components [7]. The diagonal approximation with respect to the electronic impact operator is used throughout this paper.

THE FIRST ORDER QUADRUPOLE EFFECTS.

The spectral lines of Hydrogen and hydrogen-like radiators are strongly influenced by the linear Stark effect due to the plasma microfield. Conventional theories in this case yield symmetrical profiles. The experimentally observed trends in asymmetries can be described theoretically by applying a multipolar expansion to the interaction potential. This allows one to introduce the small parameter that governs the hierarchy of the resulting perturbation expansions $\varepsilon = \frac{n^2 a_0}{R_0} \ll 1$, where n is the principal quantum number, a_0 is the Bohr radius, R_0 is the mean interparticle distance. In this case the leading first order term in ε of the contribution to the asymmetry of the Stark profiles is provided by quadrupole corrections to the energy and wave functions (WF) of the Stark sublevels. Taking in mind that ion dynamics does not influence strongly H_β profile it is possible to describe the line broadening in the electric ion microfield in the quasistatic approximation. On the other hand the electron broadening is in impact regime for the main part of the profile. Namely, the difference between broadening regimes of ions and electrons preserves quadrupole contributions from almost complete cancellation. Indeed, the first order quadrupole effects is averaged out in the impact approximation. In the Fig. 1b a comparison of the pure quadrupole asymmetry with the experimental data is presented. The theoretical results are calculated along with the many-body approach of Demura-Sholin for the Holtsmark microfield distribution function (Ho) and the nearest neighbor distribution (NN) [8]. The significant dramatic influence of the conversion from the circular frequency scale adopted in the theory to the wavelength scale used in the experiments is also shown.

In the present simulations the effects of electron Debye screening of the ion microfield and the ion-ion correlations on the microfield distribution as well as the polarization terms in the interaction potential of the charged plasma particles with the radiator are neglected [9].

THE SECOND ORDER CONTRIBUTION TO ASYMMETRY DUE TO QUADRATIC STARK EFFECT.

It was many times explained that the second order of the asymptotic perturbation expansion in the application to the asymmetry problem includes the second order quadrupole terms, the octupole and the quadratic Stark effect (QSE) corrections to the energy and the wave functions [7]-[8]. However, such consistent consideration was never done for the many-body case due to its complexity. Instead, in many papers only the QSE corrections to the energy were included [12]-[13]. It gives additional red shifts of the Stark components and as a rule was reported to improve coincidence with experimental results.

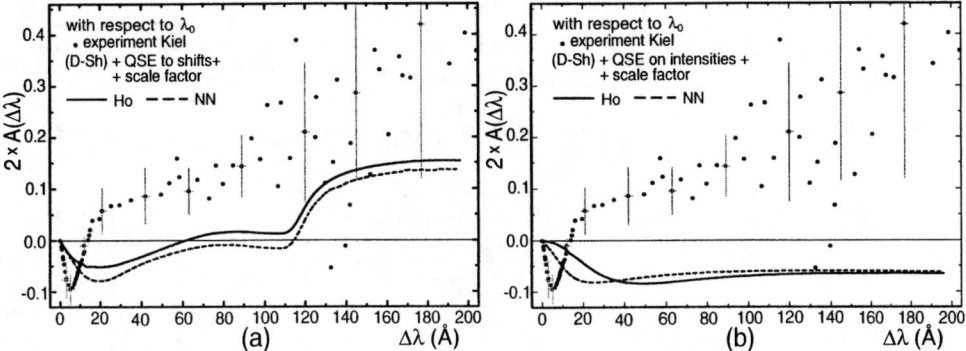

FIGURE 2. (a) Asymmetry behaviour with QSE corrections to energy (shifts). (b) Asymmetry with QSE corrections to intensity.

However, this way is inconsistent not only in the sense of the general theoretical concepts but also with respect to the consistency of only the QSE consideration, that demands the inclusion of the both corrections to the energy and to the wave functions. The latter corrections induce changes in the oscillator strengths of the Stark components proportional to the microfield values.

The separate influence of the QSE corrections on the energy and on the intensities (WF) of the Stark components on the asymmetry behavior is shown in the Fig. 2a and 2b, respectively. The bump located around 110 Å in the calculations taking into account the quadratic Stark effect contribution to the energy (Fig. 2a) as was specially studied is due to the switching from the linear to the quadratic terms predominance in the denominator of the sub-Lorentz profile of the individual Stark component at the efficient field values in the integral corresponding to the average over the field values. For the QSE calculations the data tabulated in [10] were used.

THE DEPENDENCE OF THE ASYMMETRY ON THE ELECTRONIC SHIFTS OF THE STARK COMPONENTS.

The calculated asymmetry of the profiles taking into account of the electronic shifts of the Stark components are compared with experimental results in the Fig.3. The values of electronic shifts contain contributions stemming from $\Delta n = 0$ terms due to the recoil effects as well as contributions from $\Delta n = \pm 1, \pm 2$ terms due to the quenching collisions [12]. These values were provided by [11] running the program of [12] - [13]. However, the shift dependence on the detuning was neglected since the difference in the standard impact theory approach used here and the Green-function formalism containing explicit dependence on plasma dielectric function [12]. The most important feature of those

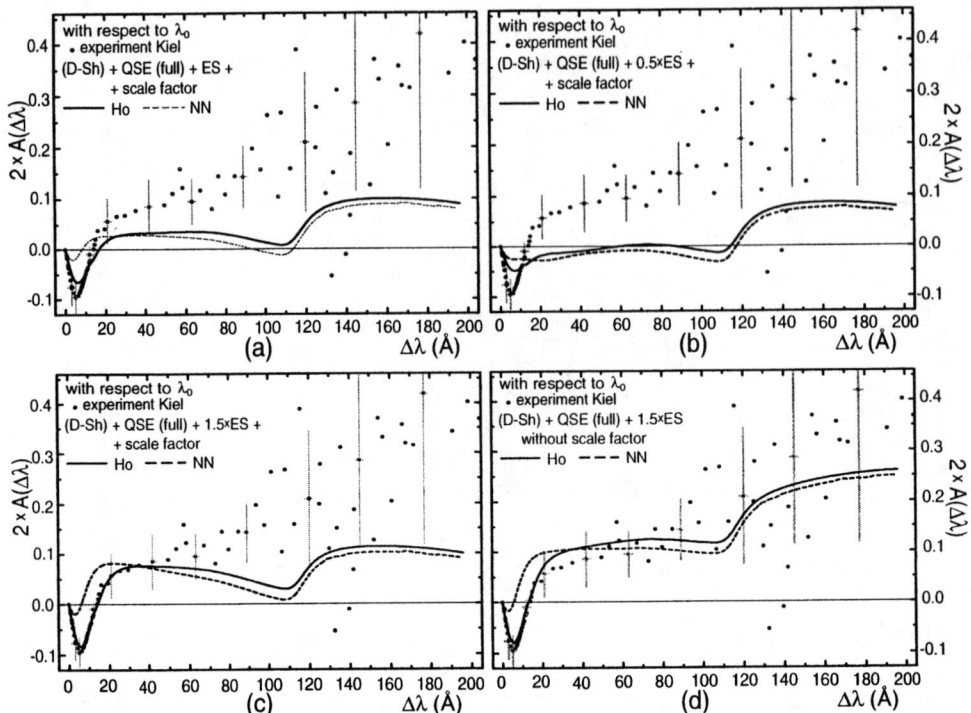

FIGURE 3. Comparison of experimetal asymmetry with calculated one taking account of the electronic shifts (ES) of the individual Stark components

shifts is that they are "red". In the context of continuing discussion on the elaboration of more rigorous approaches of the electronic shifts evaluation it must be stated that for the aims chosen for this paper it has minor importance, since the resulting electronic shifts values are of the same order of magnitude [14]. In the Fig. 3 it is shown the sensitivity of the asymmetry functional behaviour on the shift values by using the shift values multiplied by factors 0.5 and 1.5.

CONCLUSIONS.

- As it is shown experimetally and theoretically in this work the asymmetry is a very sensitive function of the reference point and hence the electronic shift values. It is seen quite evidently that such *fine features of asymmetry as locations of extrema and cross-over points could not be fixed without inclusion of the electronic shift values.*
- Moreover, it is evident from the performed comparison that the asymmetry also is a sensitive function of plasma microfield statistics, depending on what generation function is used for the microfield distribution and the first moments of the non-

uniformity tensor (in our case Holtsmark or nearest neighbor).

- It was also firstly demonstrated that in the quasistatic broadening due to the quadratic Stark effect the corrections to the intensities have the same importance as corrections to the energies and partly cancel each other.

- Examination of other contributions of the same order -octupole shifts and more higher order -the quadratic in the microfield corrections to intensities of Stark components shows sensitivity of the resulting asymmetry values to the inclusion of the higher order terms of asymptotic expansion (see Fig. 4).

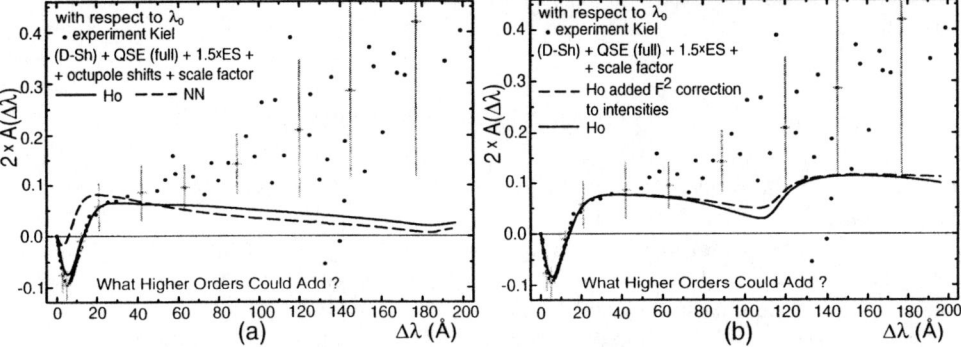

FIGURE 4. Asymmetry behaviour with higher order corrections.

Thus the statement that the asymmetry could be explained only by quadrupole terms does not hold namely due to asymptotic character of the perturbation expansion series. However, the example shown is incomplete from the beginning and the physical sense could be attributed only to the result of summation of all terms of the same order.

- The overall coincidence of the experimental and theoretical results with increasing value of detuning becomes worser as well as accuracy of experimantal points. It could be thus resumed that this difference signifies poor present understanding of the Stark profile formation mechanisms in the wings of spectral lines. Indeed, the H_β profile represents itself a unique case of the very wide Stark broadened line which profile could be observed on detunings making quite a number of half width values and rather small overlapping with profiles of neighboring lines. That is why this discrepancy which might be strongly obscured in the other lines here becomes quite evident.

- It is considered here that the Boltzman factors together with the ω^4 factor in the emission coefficient form the background spectrum for the line shape functions and should be applied to the whole spectra. Evidently the Boltzman factor enhance the red side of the spectrum with respect to the blue one, while the action of the ω^4 factor is opposite. To separate the line shape function it is necessary to subtract the background, that includes Boltzman and ω^4 factors, from the observed spectra. These factors being included directly into the expression of the line shape function both break normalization for the interval of detunings $-\infty, +\infty$, and it necessary then to reconsider the definition of the shape function in the real spectral intervals.

As the present calculations show the inclusion of these factors leads in the case of H_β line to the strong blue (negative) asymmetry at all detunings from the line center (in the nearest and far wings too) that is opposite to experimental observations. The inclusion of Boltzman factors also in fact force to reconsider the impact approach and one can not calculate the rates in the conventional manner, but have to solve the set of coupled equations for populations and line shape formation at once for the transitions between Stark sublevels etc. .

All calculations performed in this work were done with the help of very speedy Mathematica code developed by D. Nicolić on the basis of the approach presented in [8],[15].

ACKNOWLEDGMENTS

This work was initiated during the stay of A.D. as visiting professor along with the DAAD program in Kiel University, whose hospitality is greatly acknowledged.

The authors are grateful to Dr. S. Sorge for providing the electronic shift values, to Dr. M.A. Gigosos and Dr. M.A. Gonzalez for cooperation, numerous instructive discussions on the subject and making available their results on trivial asymmetry.

It is a pleasure to thank to Dr. S. Günter, Dr. E. Oks, Dr. G.V. Sholin for discussions, and Dr. C. Stehlé for cooperation and discussions.

The work of A.D. was partly supported by DAAD and RFBR (00-02-16453), and D. N. acknowledges financial support from the Swedish Science Research Council (VR).

REFERENCES

1. Wiese, W.L., Kelleher, D.E., Paquette, D.R., *Phys. Rev. A* **6**, 1132-1153 (1972).
2. Djurović, S., Mijatović, Z., Kobilarov, R., *Contrib. Plasma Phys.* **28**, 229 - 231 (1988).
3. Halenka, J., *J.Q.S.R.T.* **42**, 571-573 (1989).
4. Mijatović, Z., Pavlov, M., Djurović, S., *Phys. Rev. A* **43**, 6095- 6097 (1991).
5. Helbig, V., Nick, K., *J.Phys. B* **14**, 3573-3583 (1981).
6. Boebel, T., Diploma thesis, University of Kiel (1995).
7. Demura, A., Stehlé, C., " Effects of Microfield Nonuniformity in Dense Plasmas" in *Spectral Line Shapes* **v.8**, editors D. May, J. Drummond, E. Oks, AIP Conference Proceedings 328, N.Y.: American Institute of Physics, 1995, pp. 177-208.
8. Demura, A.V., Sholin, G.V., *J.Q.S.R.T.* **15**, 881-899 (1975).
9. Stehlé, C., Gilles, D., Demura, A.V., *Europhys. J. D* **12**, 355-367 (2000).
10. Hoe-Nguyen, Banarjea, E., Drawin, H.-W., Herman, L., *J.Q.S.R.T.* **5**, 835 -865 (1965).
11. Sorge, S., private communication (2002).
12. Günter, S., Könes, A., *J.Q.S.R.T.* **52**, 819-824 (1994).
13. Günter, S., Könes, A., *Phys. Rev. E* **55**, 907-911 (1997).
14. Oks, E., *J. Phys. B* **35**, 2251-2260 (2002).
15. Demura, A.V., Pleshakov,V.V., Sholin G.V., *Atlas of Detailed Stark Profiles of Hydrogen Spectral Lines in Dense Plasmas*, Preprint IAE-5349/6, Moscow, Kurchatov Institute of Atomic Energy, 1991, pp.1-97.

Evaluation of a Quantum Theory
for the Width and Shift of an Ar-Perturbed K
Emission Line in the Non-Impact Regime
I. Morse Potentials

W.C. Kreye* and J.F. Kielkopf[†]

* RICC/CaTS, Wright State University, Dayton, OH 45435, USA
[†]Department of Physics, University of Louisville, Louisville, KY 40292, USA

Abstract.
     Complete spectral-line shapes were computed for a modeled Ar-perturbed K spectral line in the impact approximation and in the non-impact regime. Morse potentials were used instead of more representative K/Ar pseudo-potentials because of computation-time restrictions. Our model consisted of an upper initial state characterized by a Morse potential with a depth of 100 cm^{-1}, and a lower final state characterized by a zero potential. The reduced mass was that of the K/Ar system. The theory was based on the non-impact derivation of Baranger [1]. We expanded his results for convenient manipulation in the computation. Three important aspects of our work were: the wave functions were expanded in terms of partial waves; initial conditions were devised for the solutions of the Schrödinger equations; a change was made in our programming for the evaluation of an important matrix expression which had a significant effect on the onset pressure of the non-impact regime. We found for these short-range Morse potentials that at pressures above about 300 torr, the non-impact regime should be used.

INTRODUCTION

The pressure region in which the impact approximation is valid is an important factor in analyzing the line shapes from neutral-atom (Ar) broadening of alkali-atom (K) emission spectra. We present here computations which establish the critical pressure, P_c, above which the impact approximation is no longer valid and where the non-impact-regime theory should be used. In general, most treatments of such line shapes invoke the impact approximation. For example see Kreye and Kielkopf [2], Allard [3], and Leo *et al.* [4]. Discussions of semi-classical theory of the non-impact regime are given in Allard and Kielkopf [5] and Kielkopf [6]. The quantum-mechanical theory of the non-impact regime is given by Baranger [1], resulting in his Eq. (29).

At each pressure, P, we compute one full line shape which is based on the impact approximation, and one which is based on the non-impact regime. Our model consists of an upper initial state which is characterized a Morse potential. It should be pointed out that analytical Morse potentials were used instead of numerical K/Ar pseudo-potentials because of computation-time restrictions. The lower final state is characterized by a zero potential. The depth of the well in the Morse potential (-100 cm^{-1}) and the reduced mass are those of the K/Ar system.

CP645, *Spectral Line Shapes: Volume 12, 16th ICSLS*, edited by C. A. Back
© 2002 American Institute of Physics 0-7354-0100-4/02/$19.00

MODEL

The upper state of our model consists of two parts: typically, the first is a very shallow $(-.001 \text{ cm}^{-1})$ square-well potential (SWP), extending from 0 to A. The second part is a Morse-type potential, extending from A to $2A$. The Morse-type potential is a Morse potential which has been slightly modified for computational convenience. It has a depth of 100 cm^{-1}, a representative depth for the low-lying excited states of K/Ar. The reduced mass, μ, is that of the K/Ar system. The non-interacting lower state is characterized by a zero potential which gives rise to a plane wave. The inner SWP is necessary for proper initial conditions (IC) for the wave function, $R_{i,l}(kr')$. Schiff [7] shows that the spherical Bessel function, $j_l(kr')$, is the correct IC *only* for the SWP. For the IC of the wave function, $R_{i,l}(kr')$, at A, we must first solve the Schrödinger equation from 0 to A with the SWP of $-.001 \text{ cm}^{-1}$. We then use the resulting $R_{i,l}(r' = A)$ from this Schödinger equation as the IC for the next solution of the Schrödinger equation in which the Morse potential is used. This solution yields $R_{i,l}$ at $2A$, which is the basis for computing the phase shifts, $\psi_{i,k}$ and $exp(-i\vec{k}\cdot\vec{r}')$. Three Morse potentials are used, ranging from $A = 0.4$ to $A = 1.2 \text{ Å}$.

THEORY AND COMPUTATIONAL/ANALYTICAL DETAILS

Baranger [1] treats both the impact approximation and the non-impact regime. In general, he finds the line shape $F(\omega)$

$$F(\omega) \propto \int ds[\phi(s)]^N \exp(i\omega(s)), \tag{1}$$

where N is the number of perturbers and $n = N/\mathcal{V}$ is the perturber density. By defining $\phi(s) = 1 - g(s)/\mathcal{V}$, where $\mathcal{V}$ is an effective volume, he expresses the Fourier transform, $g(s)$, as a sum of the impact term, $g_1(s)$, and the non-impact term, $g_2(s)$.

From his Eq. (29), the impact term is

$$g_1(s) = is(\exp(-i\vec{k}\cdot\vec{r}'|V_1|\psi_{i,k}), \tag{2}$$

and the non-impact term becomes

$$g_2(s) \propto \int d\phi' \int \sin\theta' d\theta' \int d\varepsilon' \varepsilon'^{1/2} \cdot |\langle \psi*_{f,k'}|V_f - V_i|\psi_{i,k}\rangle|^2$$
$$\{\frac{2\sin^2[(\varepsilon - \varepsilon')s/2]}{(\varepsilon - \varepsilon')^2} + i\frac{\sin[(\varepsilon - \varepsilon')s]}{(\varepsilon - \varepsilon')^2}\}, \tag{3}$$

where $\varepsilon = k^2\hbar^2/2\mu$, $\varepsilon' = k'^2\hbar^2/2\mu$ and $V_f = 0$.

In the standard expansion of the $\psi's$ in terms of partial waves l, typically,

$$\psi_{i,k} = \Sigma_{l=0}^{l_{max}}(2l + 1)i^l \exp(i\delta_{i,l})R_{i,l}(kr')P_l(\cos\theta). \tag{4}$$

326

For the first region from 0 to 1.2 Å, the phase shift $\delta_{i,l}$ is solved for in the standard manner, and $R_{i,l}(kr')$ is the solution of the Schrödinger equation with the SWP. For this study, $l_{max} = 30$ for the longest range Morse potential.

For the SWP, (U_0), the Schrödinger equation for $R_{i,l}(kr')$, is

$$\frac{d^2 R_{i,l}}{dr'^2} + \frac{2dR_{i,l}}{r'dr'} + [k^2 - \frac{U_0 h}{\mu c} - \frac{l(l+1)}{r'^2}]R_{i,l} = 0. \tag{5}$$

This equation yields $R_{i,l}(r' = 1.2)$ Å: this value serves as the IC for the solution of the next Schroedinger equation in which the Morse potential is used.

The following change was made from the presentation in Ref. [8]. An important term within our expansion of Baranger's Eq. (29) is

$$\xi_l \equiv |<exp(-i\vec{k'}\cdot\vec{r}|V_i|\psi_{i,k}>|^2 \equiv |\Sigma_{l=0}^{l_{max}}(2l+1)^2 \cdot G_l|^2, \tag{6}$$

where the symbol G_l represents a double integral over r' and θ. In our previous computations [8], we had assumed, for simplicity, that

$$\xi_l = \Sigma_{l=0}^{l_{max}}|(2l+1)^2 \cdot G_l|^2. \tag{7}$$

In other words, the summation is external to the $|...|^2$ brackets. In the present study, we now employ the rigorous expression in Eq. (6) in which the summation is within the $|...|^2$ brackets. This placing of the summation within the $|...|^2$ brackets would have increased the CPU time drastically. We predicted about 250 CPU hours for the $l_{max} = 250$ which we used for the pseudo-potential in the earlier study [8]. Therefore, we have had to replace the K/Ar pseudo-potentials with the more rapidly evaluated Morse potentials, which require only about 30 CPU hours. In order to relate these Morse potentials to the pseudo-potentials for the K/Ar system, we have assigned a depth of -100 cm^{-1} and a K/Ar reduced mass.

From Eq. (1) and Eq. (2), the non-impact line shape is

$$F_{NI}(\omega) = \frac{2}{2\pi}\int_0^{S_{max}} ds \cdot \exp(-n\{\Re[g_1(s)] + \Re[g_2(s)]\}) \cdot$$
$$\cos(-n\{\Im[g_1(s)] + \Im[g_2(s)]\} + \omega s). \tag{8}$$

RESULTS

Figure 1 presents the Morse-3 potential (which extends from 1.2 to 2.4) Å, combined with the SWP. The SWP of -.001 cm^{-1} has been increased by a factor of 500 to show it on the figure.

Figure 2 shows the line shapes for the Morse-1 potential (0.4 to 0.8 Å). The solid curve is the non-impact curve, and the dotted one is the impact curve. Since the pressure is very low (10 torr), the widths are approximately equal.

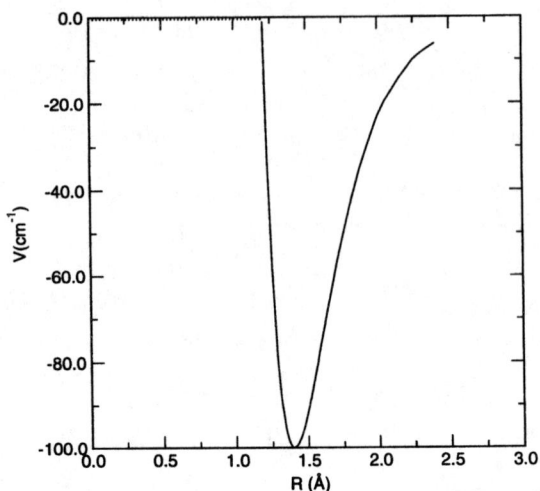

FIGURE 1. The two potentials which comprise Morse-3. The first part is a square well potential (SWP), which has a depth of -.001 cm^{-1} and which extends from 0 to 1.2 Å. The second is a slightly modified Morse potential which connects, continuously, to the SWP and extends from 1.2 to 2.4 Å. In the figure, the SWP has been multiplied by a factor of 500 to make it visible.

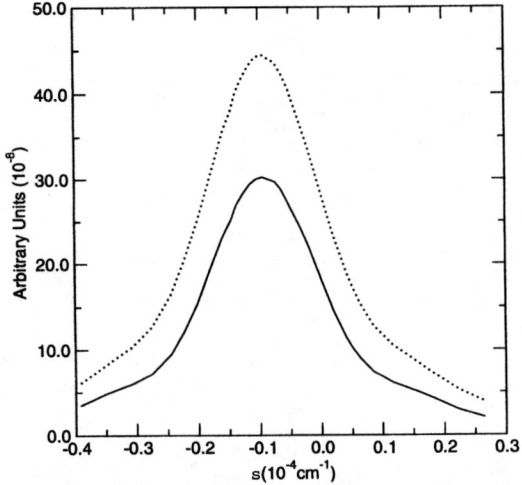

FIGURE 2. Spectral line shapes for the Morse-1 potential (0.4 to 0.8 Å) at a very low pressure of 10 torr. The solid line represents the non-impact regime, and the dotted curve is for the impact approximation.

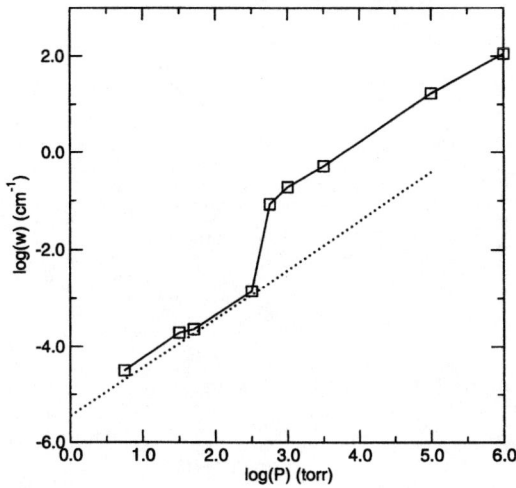

FIGURE 3. Curves showing $\log(w)$ *vs* $\log(P)$ for the Morse-2 potential (0.6 to 1.2 Å) In this figure and in all the remaining ones, the solid curve represents the non-impact regime and the dotted line, the impact approximation. The intersection of the non-impact curve with the impact curve corresponds to the critical pressure, P_c. w is the HWHM.

Figures 3 and 4 show the curves of $\log(w)$ *vs* $\log(P)$ and $\log(-d)$ *vs* $\log(P)$ for the Morse-2 potential. It extends from 0.6 to 1.2 Å. In this figure, and in all others, the solid line represents the non-impact regime and the dotted line represents the impact approximation. The half width at half maximum intensity (HWHM) is w.

Figures 5 and 6 show the curves of $\log(w)$ *vs* $\log(P)$ and $\log(-d)$ *vs* $\log(P)$ for the Morse-3 potential, which extends from 1.2 to 2.4 Å.

DISCUSSIONS

We define the pressure at which the non-impact curve joins the impact curve as P_c, the critical pressure. At higher pressure, the non-impact theory should be used, whereas at lower pressures, the impact approximation is adequate. From the $\log(w)$-*vs*-$\log(P)$ curves, P_c is about 300 torr for the two longest range Morse potentials. On the other hand, the P_c for the corresponding shift curves, ($log(-d)$ *vs* $log(P)$), is about 100 torr.

REFERENCES

1. Baranger, M., *Phys. Rev.*, **111**, 481 (1958).
2. Kreye, W. C., and Kielkopf, J. F., *J. Phys. B*, **30**, 2075 (1997).

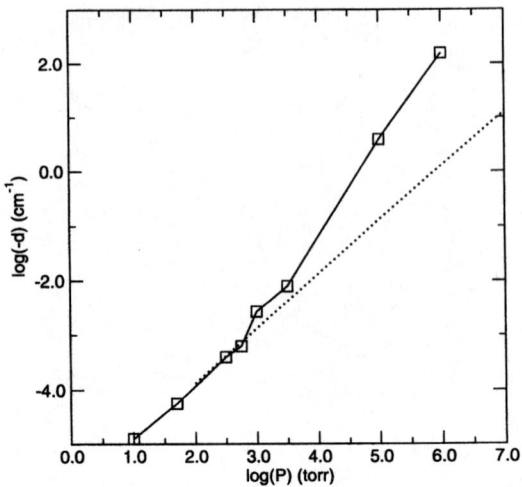

FIGURE 4. Curves showing $log(-d)$ *vs* $\log(P)$ for the Morse-2 potential (0.6 to 1.2 Å). See Fig. 3.

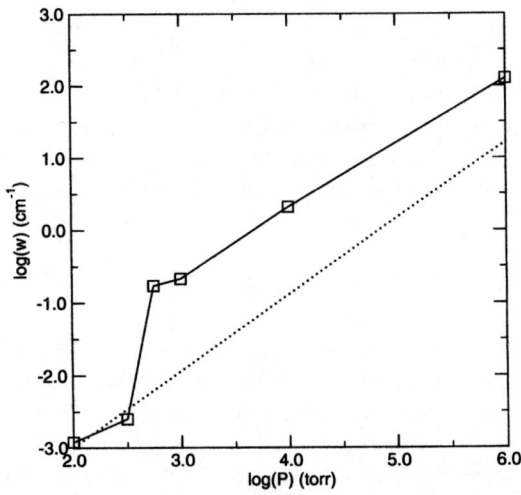

FIGURE 5. Curves showing $log(w)$ <u>vs</u> $log(P)$ for the Morse-3 potential (1.2 to 2.4Å). See Fig. 3.

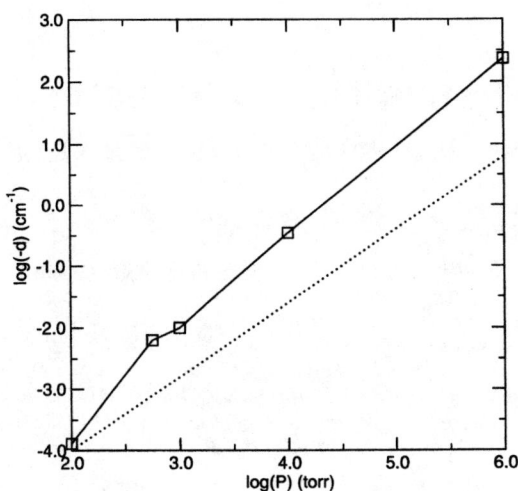

FIGURE 6. Curves showing $\log(-d)$ *vs* $\log(P)$ for the Morse-3 potential (1.2 to 2.4 Å). See Fig. 3.

3. Allard, N. F., *J. Phys. B*, **11**, 1383 (1978).
4. Leo, P. J., Peach, G., and Whittingham, I. B., *J. Phys. B*, **50**, 591 (1995).
5. Allard, N. F., and Kielkopf, J. F., *Rev. Mod. Phys.*, **54**, 1103 (1982).
6. Kielkopf, J. F., *J. Phys. B*, **16**, 3149 (1983).
7. Schiff, L., *Quantum Mechanics*, McGraw-Hill, New York, 1949.
8. Kreye, W. C., and Kielkopf, J. F., "Evaluation of a Quantum Theory for the Width and Shift of Ar-Perturbed K Spectral Lines in the Non-Impact Regime," in *Spectral Line Shapes Volume 11 AIP Conf. Proc. 559*, edited by J. Seidel, AIP, New York, 2000, p. 341.

Advanced Simulations for Signatures of Charge Exchange in Heterogeneous Plasma Emission

E.Leboucher-Dalimier[1], E.Oks[2], P.Angelo[1], P.Sauvan[3], E.Dufour[1], R.Schott[1], F.Philippe[1], A.Poquerusse[1]

[1]*Physique Atomique dans les Plasmas Denses-LULI, CNRS-CEA-Ecole Polytechnique-Université Paris 6, 75252 Paris cedex 05 and 91128 Palaiseau cedex, France*
[2]*Physics Department, 206 Allison Lab., Auburn, AL 36849, USA*
[3]*Instituto de Fusión Nuclear, E.T.S. de Ingenieros Industriales U.P.M., Jose Gutierrez Abascal, 28006 Madrid, Spain*

Abstract. We present an advanced theory of x-dips in spectral lines emitted from laser-produced plasmas. We compare predictions of this theory with our previous experimental results where, in the process of a laser irradiation of targets made out of aluminum carbide, we observed two dips in the Lyγ aluminum line perturbed by fully stripped carbon. Our theory gives a reasonable agreement with our experimental results. The results are of importance for the diagnostics of fundamental processes as it opens up a way to experimentally produce not-yet-available fundamental data on charge exchange between multi-charged ions, virtually inaccessible by other experimental methods. From the theoretical viewpoint, the x-dips are the only one signature of charge exchange in profiles of spectral lines emitted by plasmas and they are the only one quasi-molecular phenomenon that could be observed at relatively "low" densities of laser-produced plasmas, all those aspects emphasize the interest for studying heterogeneous plasma emission.

INTRODUCTION

Charge Exchange (CE) is a fundamental process that occurs and has been observed in many plasma situations: in Tokamaks between multi-charged impurity ions and hydrogen atoms, in axial discharges between multi charged and cold atoms or low charged ions, in solar plasmas...

Signatures of this process are of great importance for getting data on to CE cross sections, for the control of the kinetics of species abundances and of ionic populations (specially in the case of population inversions) and for determining the equations of state.

Up to now only situations where one of the two colliding partners had a low charge Z had been experimentally studied. Recently theoretical developments have shown the importance to find new experimental approaches to measure cross-sections of CE where both colliding partners have charges Z, Z'>>1. In the first experiments, a laser irradiation of solid aluminum carbide targets was used. Here the experimental approach is based on CE-caused dips (called x-dips) in the emitted spectral lines, Al Lyγ in this case. This case was one of the most favorable in the list of the predicted spectral lines of interest emitted by hydrogen-like ions of nuclear charge Z perturbed by some particular fully stripped ions of a nuclear charge $Z' \neq Z$ [1].

The x-dips are a new promising phenomenon that was first observed in the profile of the neutral hydrogen line $H\alpha$ emitted from helium plasma of a gas-liner pinch [2]. Concerning the possibility of observing the x-dips in spectral lines of multi-charged ions in laser-produced plasmas, it might be the only one way to obtain experimental data on CE between multi charged ions.

In a previous experimental paper we presented experimental results where Al $Ly\gamma$ emitted from aluminum carbide plasma exhibits two x-dips [3]. The positions of the dips were compared with predictions of the analytical theory that is asymptotic-valid for relatively large inter-nuclear distances R [4]. In the present work we report simulations that significantly advance the underlying theory. In these simulations we do not use the assumption that values of R are relatively large and we take into account the shielding by electrons, which results in a polarization shift of energy levels. Finally, we compare predictions of this advanced theory with the experimental results.

CHARGE EXCHANGE (CE) IN LASER-PRODUCED PLASMAS

Let us consider electron terms in the field of two quasi-static Coulomb centers of charges Z and Z' ($Z' \neq Z$, i.e. heterogeneous Coulombien system)) separated by a distance R. Specifically, we consider a term of principal quantum number n, which asymptotically ($R \to \infty$) is a radiator term (Z-term), and a term of principal quantum number n', which asymptotically is a perturber term (Z-term). We consider the situation where in the vicinity of R_0 there occurs an avoided crossing of the term n with the term n'. We note that only a tiny minority of crossings in the case of $Z' \neq Z$ are avoided crossings because the selection rules for the electron to tunnel between the two adjacent potential wells are very restrictive.

Due to the avoided crossing, CE appears as an additional channel for decay of the excited state of the Z radiator ion and the lifetime of this state is shortened, so that the collisional width increases. This means that this radiator will emit a broader collisional profile, that is consequently less intense in its central part and more intense in the wings than it should be if the CE process was absent. Thus at the frequency corresponding to the avoided crossing might appear a dip (x-dip) in the line profile.

There are upper and lower limits determining the range of electron densities, where the x-dips can be observed. Both the lower and one of the two competing values for the upper limit come from the requirement that the distance R_0 should not differ significantly from the most probable inter-nuclear distance. The second upper limit comes from the condition that the dynamical electron and ion Stark broadening should not wash out the dip.

Among the best candidates for the present study are the x-dips in Al $Ly\gamma$ (Z=13) perturbed by fully stripped C (Z'=6). Two dips were expected and then observed in laser-produced plasmas of fluorine carbon in the density range [2 10^{20}-10^{22} cm^{-3}].

These dips correspond to the avoided crossing of the (Z, n=4, q=−2) term with the (Z', n'=2, q'=0) term and the avoided crossing of the (Z, n=4, q=−3) term with the (Z', n'=2, q'=−1) term. Here we have introduced the electric quantum numbers q and q'. This case has been selected for comparison between the asymptotical analytical results and the advanced simulations of the x-dips positions as presented below.

ASYMPTOTIC ANALYTICAL THEORY

It is assumed in this theory that the CE-caused avoided crossings occur at relatively large distance, i.e.

$$R_0 >> \text{maximum} (n^2/Z, n'^2/Z') \tag{1}$$

This assumption is more or less justified for the two avoided crossings considered above: the crossing distance R_0 for both is approximately 8au and this distance about 6.5 times greater than maximum $(n^2/Z, n'^2/Z')$.

The condition (1) being met, it allows to treat separately the Z terms (perturbed by the Z' ion) and the Z' terms (perturbed by the Z ion) involved, as well as to use 1/R expansion for the transition energies.

In the following we concentrate the simulation results on the first x-dip corresponding to the avoided crossing of the (Z, n=4, q=−2) term with the (Z', n'=2, q'=0) term. From the asymptotic theory the avoided crossing distance is exactly 8au and the associated dip is localized at $\Delta\lambda \approx 6.7$m   from the center of the Al Lyγ line, i.e. on the red wing.

ADVANCED SIMULATIONS USING THE DICENTER MODEL

It is important to improve the accuracy of the asymptotic analytical results by numerical computations that would not employ the assumption (1) reducing the validity to large R. Moreover it is important to note that the electron shielding that has not been included so far, while it could be important for large distances, resulting in a polarization shift of the energy levels. These two reasons call for advanced simulations that would be valid for all R and would include the electron shielding.

We achieved this goal by using the code IDEFIX that is a quasi-molecular code adapted to these plasma conditions. Two ionic centers, Al^{13+} and C^{6+} in this particular case, are nearest neighbors sharing one bound electron. This dicenter is immersed in a non-uniform free electron gas and the emitting cell is confined such that the total potential is equal to zero at the boundary and the global quasi neutrality is required. By using IDEFIX the energy levels are obtained via an iterative self-consistent approach coupling the Poisson equation for all charges to the Schrödinger equation for the bound electrons.

We performed two different runs of the code. In the first run we neglected the electron shielding (formally $N_e=0$). In the second run, the electron shielding was incorporated for the density $N_e=10^{22}$cm^{-3}, i.e. the upper limit of the density range where the dips should be observed.

334

For the pair of the (Z, 4)-term of $q = -2$ and the (Z', 2)-term of $q' = 0$, at $N_e = 0$ the avoided crossing occurs at $R_0 = 8.00$ au and at the energy $E_1 = -6.12$ au (see Fig. 1). At $R_0 = 8.00$ au, the energy of the ground level of Al XIII is $E_0 = -85.25$ au. Thus the absolute position of the crossing should be 79.13 au in the energy scale or 5.758 Å in the wavelength scale. Since for the center of the Lγ line of Al XIII (i.e., for its unperturbed position) the IDEFIX yields 5.751 Å, the relative position of the expected x-dip becomes $\Delta\lambda = 7$ mÅ (on the red side). This value is close to the corresponding asymptotic result of $\Delta\lambda \approx 6.7$ mÅ given above.

At $N_e = 10^{22}$ cm^{-3}, the allowance for the electron shielding shifts the crossing of the same pair of terms to $R = 7.43$ au and $E_1 = -4.32$ au (see Fig. 2). However, with the allowance for the electron shielding, the energy of the ground level of Al XIII at $R = 7.43$ au is now $E_0 = -83.45$ au. Thus the absolute position of the crossing at $N_e = 10^{22}$ cm^{-3} remains nearly the same as it was at $N_e = 0$: 79.13 au in the energy scale or 5.758 Å in the wavelength scale. At $N_e = 10^{22}$ cm^{-3}, for the center of the Lγ line of Al XIII the IDEFIX yields 5.752 Å. Therefore, with the allowance for the electronic shielding the relative position of the expected x-dip becomes $\Delta\lambda = 6$ mÅ (on the red side).

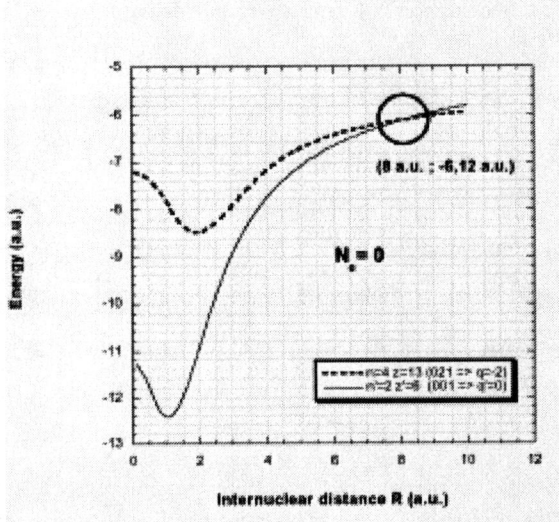

FIGURE 1. Crossing of the term {n = 4, q = -2} of Al XIII perturbed by C^{6+} (dashed line) with the term {n' = 2, q' = 0} of C VI perturbed by Al^{13+} (solid line) *without* the allowance for the electronic shielding, i.e., formally for the electron density $N_e = 0$. The avoidance of the crossing is not shown.

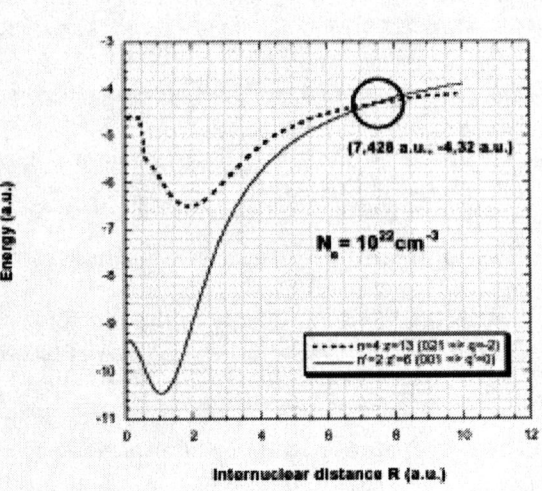

FIGURE 2. Crossing of the term $\{n = 4, q = -2\}$ of Al XIII perturbed by C^{6+} (dashed line) with the term $\{n' = 2, q' = 0\}$ of C VI perturbed by Al^{13+} (solid line) *with* the allowance for the electronic shielding for the electron density $N_e = 10^{22}$ cm^{-3}. The avoidance of the crossing is not shown.

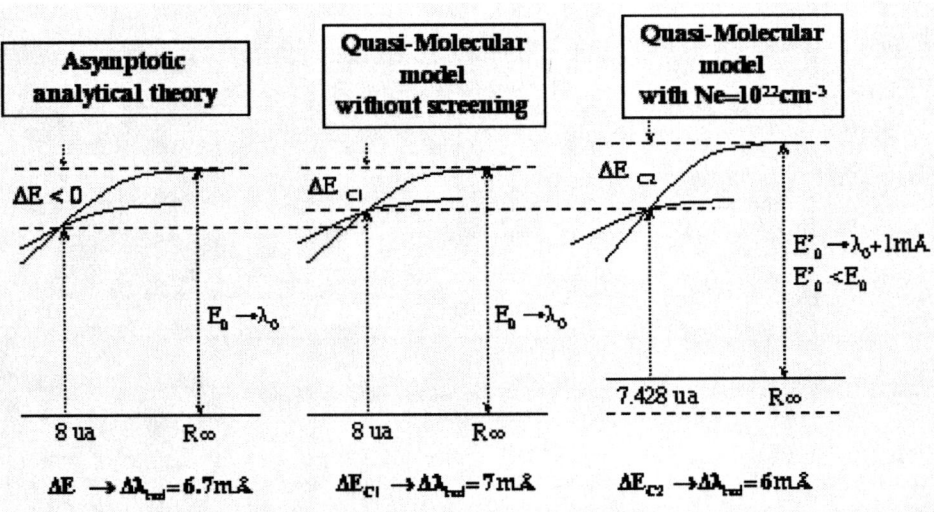

FIGURE 3. Schematics for a comparison between the different computations for the x-dip in Al Lyγ perturbed by C^{6+} corresponding to the avoided crossing of Al-C (n=4, q=-2) term with C-Al (n'=3, q'=0) term.

A comparison between the asymptotic results and the advanced simulations (without and with electron screening) for the positions of the avoided crossing of Al-C (n=4, q=-2) term with C-Al (n'=3, q'=0) term and for the correlated dip positions is given in figure 3. We show schematically the evolution of the levels positions and of the avoided crossings and demonstrate qualitatively the differences between the results. This comparison calls for the following comments:

First, there is not much difference between the asymptotic results and the quasi-molecular model without electron screening: the avoided crossing occurs at the same R_0 and nearly the same energy. This justifies the validity of the assumption (1).

There is more difference between the results from the quasi-molecular model including self-consistently the free electron screening and the other results. This screening perturbs the crossing distance and the levels positions because of the polarization shift, but not the energy for the crossing. Finally the correction is about 14% for the x-dip position.

FIRST OBSERVATION OF X-DIPS IN LASER-PRODUCED PLASMAS

Our experimental discovery of dips in the Lyγ of Al XIII perturbed by fully stripped carbon in laser-produced plasma [3] is described briefly below and is compared to the above simulations made by the advanced theory.

The experiment was performed on the nanosecond laser facility at LULI. A high intensity (4 10^{14}Wcm^{-2}) 4ω laser beam is focused during 500ps onto a structured massive aluminum carbide target. The experimental setup designed to optimize both the generation and the diagnostics of the dips in the Al Lyγ line is shown in fig.4. The new designed targets and spectrograph [3,5] are well suited for the optimization of the emission intensity and for the control of both the transverse inhomogeneity and re-absorption. More precisely a slit located at 2 mm from the plasma allows a transverse magnification of about 100. As for the spectral resolution (R=8000), it is achieved by using a PET crystal in the Johann geometry. Such spectroscopic characteristics are decisive for the observation of very fine structures in spectral lines.

Figure 5 shows the evolution of the experimental profile of Al Lyγ as the spatially-integrated slice Δz increases in thickness from z=0 at the bottom of the densest part of the plasma until z=14 μm (solid line) or z=20 μm (dashed line). For all laser shots corresponding to the same conditions, the observed spectra demonstrate the same qualitative features.

For Δz=14μm the profile exhibits two pronounced dips in the *near* wing. The first dip located at 6 mÅ from the center of the line should correspond to the x-dip that has been simulated up above. The second one located at 9 mÅ from the center of the line is also in excellent agreement with the theoretical calculations by the code IDEFIX including the electron shielding (not presented here). It has been shown by hydro-simulations [6] that the electron densities involved in the spatial interval Δz=14 μm are consistent with a good visibility of the dips.

For Δz=20 μm, only discontinuities of the slope of the near red wing are still visible as remnants of the x-dips. In this case the average density involved in the space integration is too low for observing the dips.

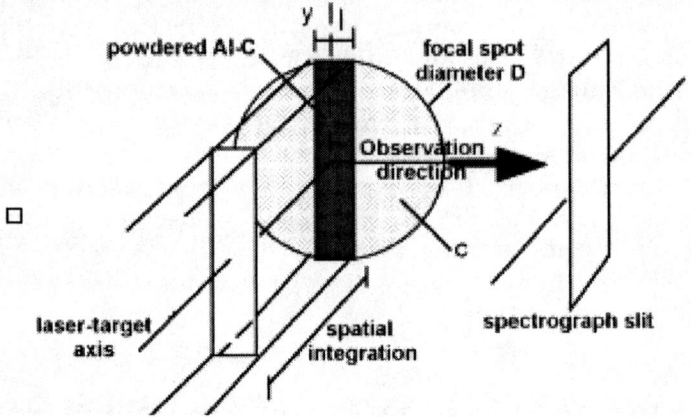

FIGURE 4. Experimental set up, including the design of the structured target. The aluminum carbide plasma is confined in carbon plasma in the direction of the observation, thus allowing the control of the re-absorption. The Al_4C_3 strips, centered on the laser beam, are well suited for the optimization of the emission. The spectrograph slit ensures a spatial resolution along the laser-target axis.

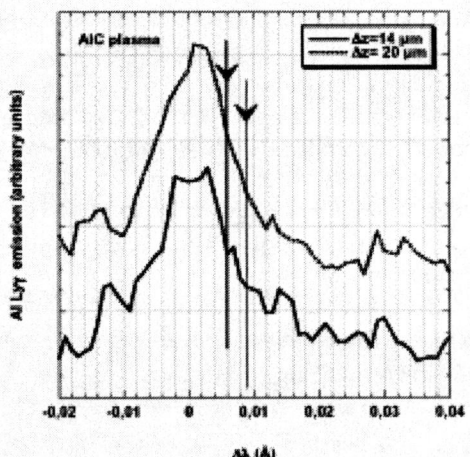

FIGURE 5. Experimental profiles of the Lyγ line of Al XIII emitted from aluminum carbide plasma for two spatially integrated slices Δz, the thickness of which is counted from the bottom of the crater toward the corona plasma. For Δz = 14 μm corresponding to the densities 10^{20}-10^{22}cm^{-3}, the profiles exhibit two pronounced dips in the *near* red wing. For Δz = 20 μm, only discontinuities of the slope of the line profile in the *near* red wing are still visible as remnants of the x-dips.

As a matter of fact the densest part of the plasma is hidden by the slit (15μm). The observed dips, being located in the *near* wing, correspond to the part of the line profile where the signal-to-noise ratio is reasonable. Besides, they are reproducible in all spectra observed for different Δz. These two features distinguish the observed dips from the jitter in the far wings. The latter corresponds to the experimental noise and its spectral positions are not reproducible at the same position in spectra observed for different Δz.

CONCLUSION

We have presented here an advanced theory and simulated x-dips positions in the near red wing of the line Al Lyγ emitted by a aluminum carbide plasma generated by laser irradiation. It is shown that the quasi-molecular approach of the interactions Al-C and C-Al is more accurate than the asymptotic theory for reproducing the experimental results up to now. It has been shown also that the correction due to the free electron environment is important.

The x-dips will serve in the future for producing not-yet-available data on CE between multi-charged ions.

Let us conclude with a comparison between x-dips and Quasi-Molecular Satellites (QS) [3,7], the latter been due to extrema in the transition energies. They have one thing in common: during the collision of the two ions, a quasi-molecule is formed. However, first, x-dips and QS are observed in plasmas of different densities (two or three orders of magnitude). Second the fact that for a short time, the bound electron is shared between two ions does not necessarily mean that CE should occur during the emission of the QS.

REFERENCES

1. E. Oks and E. Leboucher-Dalimier, *J. Phys. B* **33**, 3795 (2000).
2. St. Böddeker, H.-J. Kunze, and E. Oks, *Phys. Rev. Lett.* **75**, 4740 (1995).
3. E. Leboucher-Dalimier, E. Oks, E. Dufour, P. Sauvan, P. Angelo, R. Schott, and A. Poquerusse, *Phys. Rev. E* **64**, R065401 (2001).
4. E. Oks and E. Leboucher-Dalimier, *Phys. Rev. E* **62**, R3067 (2000).
5. E. Leboucher-Dalimier, P. Sauvan, P. Angelo, A. Poquerusse, R. Schott, E. Dufour, E. Minguez, and A. Calisti, *J.Quant.Spectrosc.Radiat.Transfer* **71**, 493 (2001).
6. F. Ogando and P.Velarde, *J.Quant.Spectrosc.Radiat.Transfer* **71**, 541 (2001).
7. E. Leboucher-Dalimier, E. Oks, E. Dufour, P. Angelo, P. Sauvan, R. Schott, and A. Poquerusse, *The European Physical Journal D*, to be published in august 2002

Access to Spectrally Resolved Ultra-Dense Hot Low Z Emissivities and Opacities

R. Schott[1], F. Philippe[1], P. Angelo[1], E. Dufour[1], A. Poquérusse[1], E. Leboucher-Dalimier[1], P. Sauvan[2], P. Velarde[2], F. Ogando[2], E. Minguez[2], J. M. Gil[3], J. G. Rubiano[3], R. Rodriguez[3], P. Martel[3] and R. Mancini[4]

[1]*Physique Atomique dans les Plasmas Denses-LULI, CNRS-CEA-Ecole Polytechnique-Université Paris 6, 75252 Paris cedex 05 and 91128 Palaiseau cedex.*
[2]*Instituto de Fusion Nuclear, E.T.S. de Ingenieros Industriales U.P.M., 28006 Madrid, Spain.*
[3]*Departamento de Fisica, Universidad de Las Palmas de Gran Canaria, Spain.*
[4]*Physics Department University of Nevada, Reno, NV 89557, USA*

Abstract. We present here experimental studies of broadened bound-bound emissivities and opacities of low Z plasmas that are the best candidates for exhibiting ion and electron correlations. First we report on an emission experiment where a new target design is used to access the highest densities. Such targets irradiated by an intense long laser pulse generate plasmas well adapted to model and extract opacities. The measurements are compared to theoretical results obtained from simulations involving new atomic/molecular physics models that take into account detailed line profiles. We end up with the description of an absorption experiment in progress, this experiment using the same targets.

INTRODUCTION

The present work is devoted to measurements of bound-bound opacities of laser-shocked dense aluminum plasmas by emission and absorption spectroscopy. For those plasma conditions corresponding to astrophysical situations, the white dwarf atmosphere, the center of the sun, there is a lack of information, computations and measurements. The study is then decisive for the knowledge of the radiative transfer and the optimization of the X-ray sources. With the access to the new generation of large-scale facilities and the intense lasers, it is now possible to reach partly, in laboratories, the temperature-density domain [10^5 K-1 g.cm^{-3}; 10^6 K-10^2 g.cm^{-3}]. The study of such extreme conditions for low Z plasmas is extremely interesting for the new atomic/molecular physics involved and for the crucial role of the line broadening [1]. Because of the low Z and the high density the strong ionic correlations act through the Stark effect in the spectral line shapes and consequently on opacities. The quasi-molecular model has been proved to be adapted to this plasma regime (strong correlations, moderate coupling) for the simulations of opacities [2].

Two diagnostics, « emission spectroscopy » and « absorption spectroscopy », for the measure of bound-bound opacities are developed in this work and they are both at the limit of the feasibility: as a matter of fact the plasma is too "cold" for emission measurements and too "hot" for absorption measurements. The aim of the work is to

CP645, *Spectral Line Shapes: Volume 12, 16th ICSLS*, edited by C. A. Back
© 2002 American Institute of Physics 0-7354-0100-4/02/$19.00

compare both diagnostics. At first some important improvements devoted to the target design and to the spectrographs are presented.

IMPROVING THE EXPERIMENTAL SETUP FOR THE ACCESS TO DENSE PLASMAS

The experiments have been carried out at the Ecole Polytechnique, Palaiseau. The neodymium laser chain provides a 600 ps gaussian pulse with energy of 80 J and a wavelength of 1,053 μm. In the emission experiment, the initial ω beam is converted into a 4ω beam thanks to KDP crystal (figure 1). The 4 ω beam (25J delivered in 500ps) is focused onto a massive aluminum target, the 100 μm focal spot allowing an intensity as high as 2.10^{14} W.cm^{-2}. The absorption experiment described in the last section requires a second beam at double frequency in order to irradiate a backlighter. There is no need to have a 4ω beam for that because the requirement of high density is less important than the brightness of the source in the spectral window for the aluminum absorption.

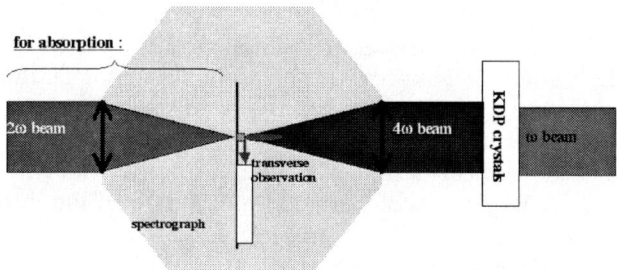

FIGURE 1. Interaction chamber schematic.

The Targets

Structured targets are used in this experiment (see figure 2). A slice of aluminum, the emissive material, is inserted in a carbon substrate. This configuration enables the aluminum plasma to be confined transversally by the carbon plasma. A second interest stems from the control of the transverse optical path through the aluminum thickness. For a thickness lower than 70 μm, the transverse inhomogeneity due to the laser irradiation pattern of the focal spot can be assumed to be negligible. Obviously for such targets directly coated on the spectrograph slit, the densest part of the aluminum plasma is very difficult to access in a transverse direction because it is located in the shocked region inside the crater. The first attempt to overcome this difficulty was to incline the target over the slit, allowing a quasi-transverse observation inside the crater. However, this design generates space integrations over plasma regions where the density is lower than the one we want to probe, and the very dense region that can be accessed transversally is rather limited.

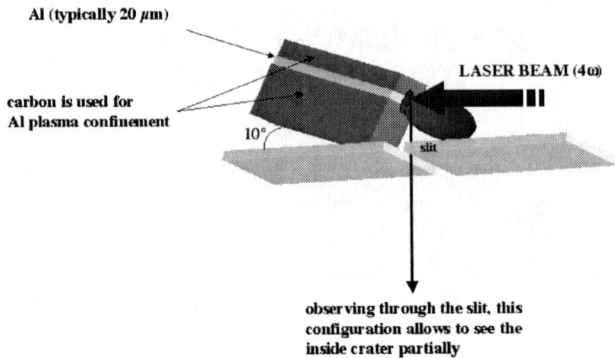

FIGURE 2. Structured target.

Recently, a second attempt to create and access ultra-dense and hot plasma has been successful. The figure 3 shows a schematic of this new design. This time, the laser hits the edge of a three layered-foil (CH/Al/CH) lying on a support so that the foil sticks out the support. The total thickness of the foil must not exceed the focal spot diameter, so that the plastic is entirely converted into plasma during the laser shot. This plasma ensures the aluminum plasma confinement. The laser-target interaction gives rise to a nick instead of a crater, and this innovation is the key point for a rigorous transverse observation of the ultra-dense plasma. Moreover, using this implementation, the space resolution involves a larger part of ultra-dense plasma, and the integration over less dense regions is removed.

This newly designed target is also suitable for absorption spectroscopy. As a matter of fact, it is now possible for the back-lighter emission to go through the densest part of the aluminum plasma.

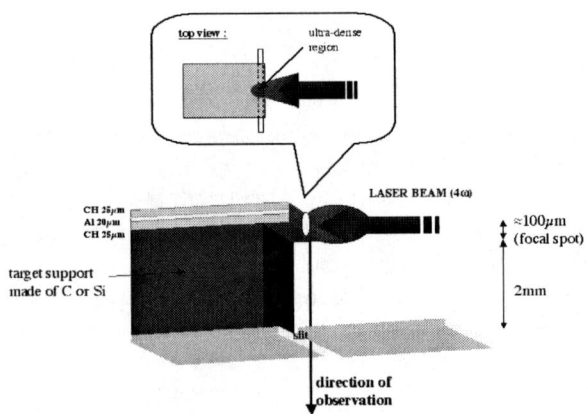

FIGURE 3. New target design.

The Spectrograph

The reabsorbed emission is assessed with an ultra-high resolution ($\Delta\lambda/\lambda$=8000) spectrograph (see figure 4). A PET crystal, curved in order to fit a circle whose radius is twice the Rowland circle radius, is used in Johann geometry. The spectral window includes the aluminum Lyman and helium-like lines from 5.3 to 7.4 Å. A beryllium window protects the DEF film from the visible light.

The aluminum plasma is very close to the slit (2 mm), allowing a very high space resolution with a transverse magnification of 100. On the figure 4, the upper edge of the slit acts as a knife and the film detects the emission with progressive space integration along the laser-target axis. For instance, the emission on the film at 0.2 mm from the bottom of the lines corresponds to the emissive volume of the densest part of the plasma extending on about 2 μm. This volume corresponds to the plasma to be probed in this work by emission or absorption spectroscopy.

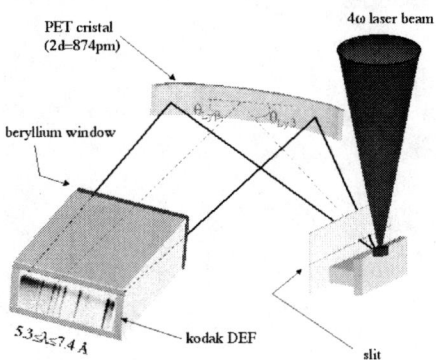

FIGURE 4. The different elements of the spectrograph.

The Improvements

An example of the detected emission from two different targets is given on the left part of figure 5. The upper film has to be related to a previous inclined target while the lower film corresponds to a newly three-layered foil target. The spectral windows and the dispersions are the same for both films. A comparison between the two films reveals that the maximum broadening of the lines is more important with the newly designed target than with the previous target. This broadening is the signature of the Stark effect, i.e. the density effect; thus denser plasma is observed with the new design. The distance, on the film, over which this consequent Stark broadening is observed gives a clue to the extension of the dense plasma region. This distance is increased with the three-layered target because the observation takes place in the nick where most of the dense region is located while the previous inclined target only allowed the probing of a small part of the crater.

The right part of figure 5 displays several scans taking into account all instrumental corrections. These scans are connected to the films on the left hand side. For both sets, space integrations over the first 4, 8 and 12 μm of dense plasma are performed, and the spectra are shifted from one to another for more clarity. The last visible Lyman and Helium lines are respectively Lyε and Heε for the previous target and Lyγ and Heγ for the newly designed target. According to the Inglis-Teller limit, this statement provides a complementary argument for the access to higher densities.

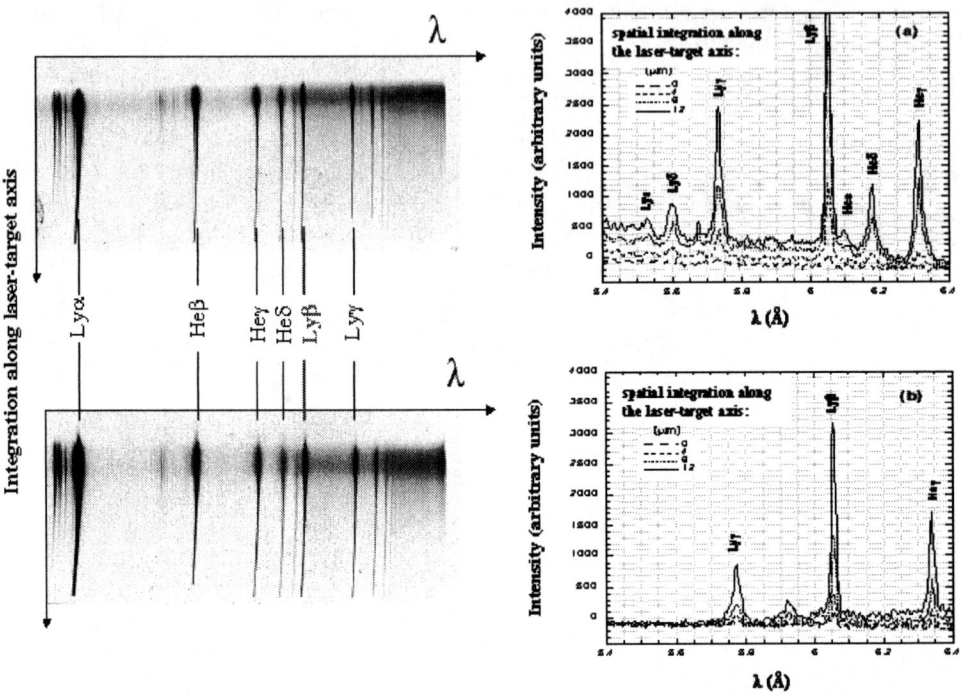

FIGURE 5. Two films and their respective corrected scans: the upper corresponds to a previous target and the lower to a newly designed target.

OPACITY MEASUREMENTS BY EMISSION SPECTROSCOPY

The Experimental Protocol

The measures presented here deal with aluminum bound-bound opacities. First, emission experiments have been modeled to extract opacities. For an aluminum thickness 2l lower than 70 μm, the 2D hydro simulations show that the density and the temperature are more or less homogeneous transversally. The reabsorbed line intensity for a slice dz of plasma can be written as:

$$I(z,\omega,2l)dz = dzDS(z)\left[1 - e^{-\tau(z,\omega,2l)}\right] \qquad (1)$$

where D stand for the focal spot diameter. $S(z)$ and $\tau(z,\omega,2l)$, respectively the source function and the optical depth, can be expressed as follow:

$$S(z) = \frac{\hat{\varepsilon}(z)}{\hat{k}(z)} \quad \text{and} \quad \tau(z,\omega,2l) = \varphi(z,\omega)\hat{k}(z)2l \tag{2}$$

where the opacity $\hat{k}(z)$, the emissivity $\hat{\varepsilon}(z)$ and the normalized line profile $\varphi(z,\omega)$ have been introduced. The aim of this work is to probe the spectrally resolved opacity $k(z,\omega) = \varphi(z,\omega)\hat{k}(z)$. Introducing its value at the maximum ω_0 of the line, we have:

$$k(z,\omega) = k(z,\omega_0)\frac{\varphi(z,\omega)}{\varphi(z,\omega_0)} \tag{3}$$

The opacity at ω_0 can be inferred from the ratio between the two reabsorbed intensities from two different targets having $2l_1$ and $2l_2$ as thickness

$$\frac{I_2(z,\omega_0,2l_2)}{I_1(z,\omega_0,2l_1)} = A\frac{1-e^{-k(z,\omega_0)2l_2}}{1-e^{-k(z,\omega_0)2l_1}} \tag{4}$$

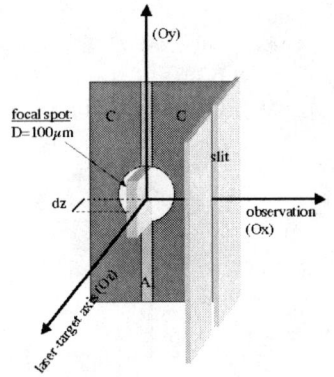

FIGURE 6. Modeling opacities and emissivities.

This ratio is written here with the assumptions that the redistribution is complete (opacity and emissivity have the same shape) and that the ionic populations are decoupled from the radiative transfer and the line profile [3]. This last point is true because the optical depth will never exceed 1 ($\tau \le 1$) for the lines in consideration. A is an adjustable parameter that should be theoretically equal to one. However it turns out to be not true from an experimental point of view. The complication here comes from the control of the position (z) and the extension (dz) of the plasma, the reproducibility of the focal spot diameter (D) and of the laser intensity for the two shots leading to the same source functions $S(z)$. All these experimental uncertainties are gathered in the A parameter, which can be inferred from the same ratio (5) applied to the line wings that are obviously considered as optically thin (i.e. $\tau \ll 1$) even for thick foils:

$$\frac{I_2(z, wings, 2l_2)}{I_1(z, wings, 2l_1)} = A\frac{l_2}{l_1} \qquad (5)$$

The line profile in formula (3) is properly measured from the spectrally resolved intensity in the line wing domain where the plasma is optically thin. For thin targets this domain is enlarged towards the line center.

In conclusion, from emission spectroscopy and using this model, the spectrally resolved opacity should be known accurately at the maximum and in the wings but an extrapolation should be done for the intermediate spectral regions. In fact as we will see, the measures for the maximum are submitted to uncertainties.

Al Lyß Opacity Results and Discussion

The figure 7 a) consists of two different opacity profiles for Al Lyß obtained from the comparison between 20 and 60 μm targets for the upper one, and 20 and 40 μm targets for the lower one. The extrapolations explained above are shown with dotted lines. From the described method, the opacity profiles should be comprised between these two curves. As the transverse inhomogeneity grows progressively when increasing the aluminum thickness from 40 to 60 μm, the error bars drawn on the graph are due to the limits of the validity of the model. We emphasize that all these opacity measurements concern the same plasma zone, i.e. the shocked zone.

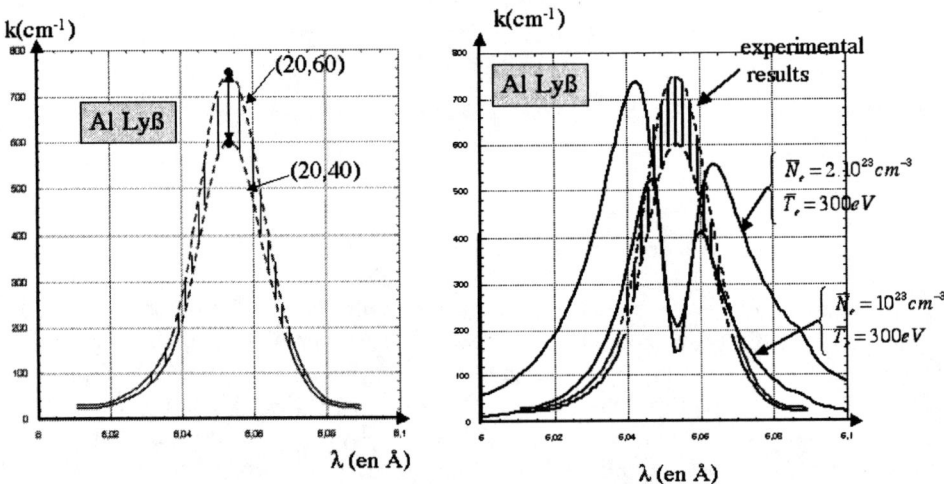

FIGURES 7. a) Al Lyß opacity profile. The upper and lower curves are obtained respectively from the comparisons between 20 and 60 μm targets and 20 and 40 μm targets. b) Opacity simulations and experimental data. The simulations are plotted with solid lines for the temperature 300eV and the densities 10^{23} and $2.10^{23} cm^{-3}$ respectively. The experimental estimate is in the chopped domain.

The opacity measurements have been compared to simulations [2]. The code IDEFIX, based on a quasi-molecular model, has been applied to opacities and emissivities. This code computes the line profile emitted by two nearest neighbor ions

346

separated by R and the oscillator strengths for the same ionic separation. An average over the nearest neighbor probability distribution is performed. The codes RATION [4] or M3R [5] provide the ionic population data.

Figure 7 b) presents two opacity simulations given by solid lines for the plasma conditions [10^{23}cm^{-3}-300ev], and [2.10^{23}cm^{-3}-300eV] respectively. The evolution with the density is drastic. On the same graph are also given the previous experimental opacity profiles (figure 7). From the comparison between the simulations and the experimental profiles, the density corresponding to the experiment is found to be 10^{23}cm^{-3}. This density, which is a discriminating parameter, is coherent with the Inglis-Teller limit and the hydro simulations.

The Hydro Simulations

A code taking into account the radiation transport and using an Adaptative Mesh Refinement (AMR) has been adapted to our target geometry for the hydro simulations [6]. For this, the 3D geometry of the target is converted into a 2D cylindrical geometry as shown on figure 8. For this modeling, the aluminum inner cylinder has a diameter corresponding to the thickness of the aluminum foil and the plastic outer cylinder is fitted to the focal spot.

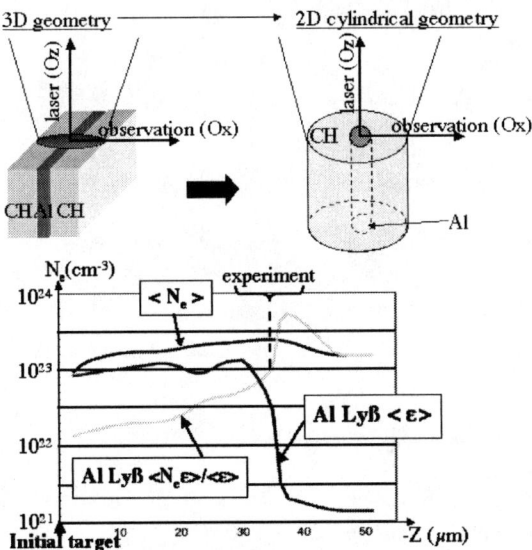

FIGURE 8. Above: cartoon of the modeling operated on the 3D geometry of the target before processing the 2D hydro simulations. Below: results of 2D hydro simulations. The three curves represent respectively the averaged density, the Al Lyß averaged emissivity and the averaged density weighed by the Al Lyß emissivity.

We have plotted on the same figure results that are time averaged and involve a transverse integration along the observation direction. The averaged density, the averaged Al Lyß emissivity, and the averaged density weighted by Al Lyß emissivity are given. The latter is the curve of interest for the purpose. According to the previous paragraph where the re-absorption has been investigated for the density 10^{23}cm^{-3}, the plasma of interest corresponds to the beginning of the shock front region.

OPACITY MEASUREMENTS BY ABSORPTION SPECTROSCOPY

Absorption Experiment

This last section is devoted to the description and the first results of an absorption experiment that is a more natural method for measuring opacities in cold dense plasmas. For moderately hot/cold ultra-dense plasmas characterized by moderate source functions, the natural method based on the most intense measured signal is not obvious. Moreover it is important to underline the long-term goal of our experiment that is to corroborate the previous measurements obtained from emission spectroscopy to the new ones to be obtained from absorption spectroscopy.

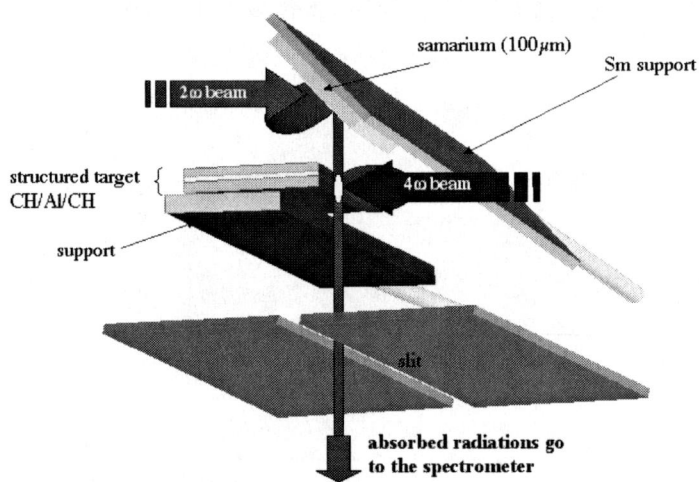

FIGURE 9. Schematic of the absorption experiment

A schematic of the experimental setup is given on figure 9. The samarium has proved to be a suitable back-lighter for the absorption through the three-layered CH/Al/CH foil described previously. The ultra-resolved transition array (UTA) exhibits a very bright quasi-continuum of emission, bright enough to observe entire absorption lines. The experimental issue is the alignment between the most emissive part of the samarium and the densest part of the aluminum plasma confined in the plastic, for the spectral window of absorption. For this purpose, the two supports of samarium and aluminum are able to move along different directions with micrometer steps. Furthermore, the thickness of the plastic on both sides of aluminum is such that the resulting plasma is optically thin. It is also very important to notice that the aluminum emission is more or less negligible with respect to the samarium emission.

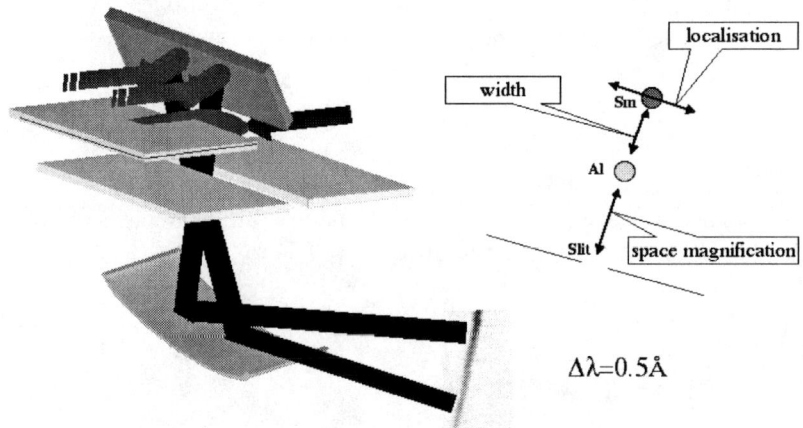

FIGURE 10. Summary of the different possible adjustments driving the magnification and the absorption spectral window.

The design of this experiment engenders typical features in the absorption spectrum. A part of this spectrum, typically a quarter, exhibits a spectral window where the samarium continuum is absorbed by the aluminum plasma. On the rest of the film, samarium and aluminum direct emissions overlap, so that both the aluminum lines and the samarium quasi-continuum can be observed. The figure 11 tries to sum up the spectrum parameters that can be adjusted along by tweaking some distances. First, the lateral position of the samarium plasma with respect to the aluminum plasma controls the localization of the spectral window where absorption takes place. On the figure, two possible focal spots are represented on the samarium with the respective paths for the absorbed radiations, resulting in two different spectral windows for the absorption spectrum. Second, the distance between the two plasmas (samarium and aluminum) is related to the width of this spectral window. The closer the samarium to the aluminum is, the larger the window is. For a separation of 1 mm, this width is about 0.5 Å. Finally the separation between the aluminum plasma and the slit drives the magnification along the space integration axis, i.e. the laser-target axis.

The success of this experiment lies on the abundance of the ionic population of the lower levels involved for the aluminum transition considered. Consequently, it is easier to get absorption in Helium-like lines than Hydrogen-like lines due to the ionic populations of their respective lower levels. So the efforts were first focused on Al Heγ line.

First Al Heγ Absorption Spectrum

Preliminary results for the Al Heγ line are given on figure 12. This figure gives a scan around this line (see the whole spectral window on figure 11) for the emission reabsorbed from the densest part of the aluminum plasma. The spectrum being absorbed through the densest part of the aluminum plasma is shown by the solid line.

The aluminum geometrical depth was about 20 μm. Neglecting the aluminum self-emission, the absorbed samarium intensity is written as follows:

$$I_{absorbed-Sm} \propto \varepsilon_{Sm} e^{-\tau_{Al}}$$ (6)

where ε_{Sm} stands for the emissivity and τ_{Al} for the aluminum optical depth. The aluminum optical depth at central frequency can be inferred from these considerations, and is found to be about 0.5, leading to opacity of 255 cm^{-1}.

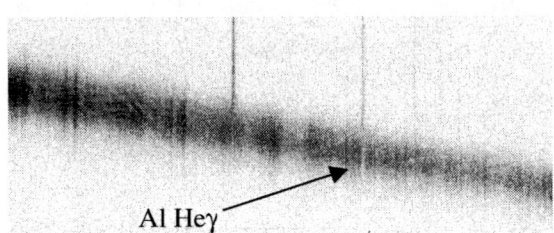

Al Heγ

FIGURE 11. Emission collected on the film during a shot for the absorption experiment. The absorption of the Sm quasi-continuum by the Al Heγ line is pinpointed.

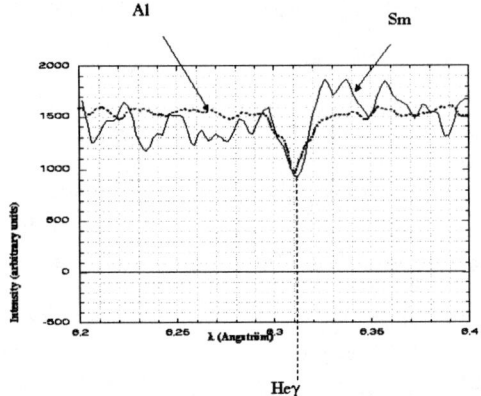

FIGURE 12. Comparison between the absorption and the emission spectra of Al Heγ for the same plasma conditions Solid line gives the spectrum of the samarium reabsorbed by the aluminum ultra-dense plasma. Dotted line corresponds to a flip of the aluminum emission for the same 20 μm thickness. The zero level has been raised arbitrarily up to the samarium quasi-continuum.

On the same graph, the dotted curve reproduces the opposite of the aluminum emission spectrum, i.e. the emission of a 20 μm target, for the same plasma conditions. It appears that both profiles have more or less the same shape and broadening. Actually, this isn't surprising for the optical depth is rather low.

CONCLUSION

First measures of bound-bound opacities have been performed for hot, dense laser shocked aluminum plasmas. For that, new targets and spectrographs have been designed and they have shown their capability for optimizing the diagnostics. We have underlined that the spectral line shape, which is the probe of ionic correlations, is determinant. We have shown the limits of emission spectroscopy that gives not a direct measure of opacity and requires the comparison between different reabsorbed emissions from plasmas resulting from different laser shots. The absorption experiment is in progress. We plan to compare emission and absorption results for selected lines (Lyβ, γ and Heβ, γ) and compare systematically with opacity simulations obtained with different models in order to validate the simulations.

In the next future we plan to quantify the experimental results yet obtained for the bound-free opacities and interpret them within the frame of a model including new recombination processes due to the formation of quasi-molecules in ultra-dense plasmas.

ACKNOWLEDGMENTS

This work has been supported by the "Improve Human Potential and Socio-Economic Base" Program of the European Community, contract n° HPRI CT 1999-0052.

REFERENCES

1. Gauthier, P., et al., *Phys. Rev.* E**58**, 942-950 (1998), Leboucher-Dalimier, E., et al., *J. Quant. Spectrosc. Radiat. Transfer* **71**, 493-504 (2001), Sauvan, P., et al., *J. Quant. Spectrosc. Radiat. Transfer* **71**, 675-687 (2001)
2. Sauvan, P., et al., proceeding in 16th International Conference on Spectral Line Shapes (ICSLS)
3. Scott, H. A., *J. Quant. Spectrosc. Radiat. Transfer* **71**, 689-701 (2001)
4. Lee, R. W., Whitten, B. L., and Strout, J. I., *J. Quant. Spectrosc. Radiat. Transfer,* **32**, 91 (1984)
5. Mancini, R. C. and Minguez, E., First International NLTE Atomic Kinetic Workshop, Gaithersburg (1996)
6. Velarde, P., and Ogando, F., *J. Quant. Spectrosc. Radiat. Transfer,* **71**, 541-550 (2001)

Ultra-dense Hot Low Z Line Transition Opacity Simulations

P.Sauvan*, E.Mínguez*, J.M.Gil#, R.Rodríguez#, J.G.Rubiano#, P.Martel#, P.Angelo‡,
R.Schott ‡, F.Philippe ‡, E. Leboucher-Dalimier ‡, R.Mancini§, A.Calisti¶

*Instituto de Fusión Nuclear, C/ José Gutiérrez Abascal 2, 28006 Madrid, Spain
#Departamento de Física, Universidad Las Palmas de Gran Canaria,
35017 Las Palmas de Gran Canaria, Spain
‡Physique Atomique dans les Plasmas Denses-LULI,CNRS-CEA-Ecole Polythecnique-
Université Paris 6, 75252 Paris cedex 05, France
§Physics Department University of Nevada, Reno, NV 89557, USA
¶PIIM, Université de Provence, Centre St Jérome, 13397 Marseille, France

Abstract. In this work two atomic physics models (the IDEFIX code using the dicenter model and the code based on parametric potentials ANALOP) have been used to calculate the opacities for bound-bound transitions in hot ultra-dense, low Z plasmas. These simulations are in connection with experiments carried out at LULI during the last two years, focused on bound-bound radiation. In this paper H-like opacities for aluminum and fluorine plasmas have been simulated, using both theoretical models, in a wide range of densities and temperatures higher than 200 eV.

INTRODUCTION

The possibility to obtain fast and accurate opacity data for a wide range of plasma conditions is very useful when we have to study the radiation emitted from laser produced plasmas or astrophysical radiation. The use of analytical potentials reflecting the electronic structure of an ion immersed in the plasma is very useful to calculate in a very fast way the energy levels of this ion. Moreover, these potentials are appropriate for a large range of ionic configurations, that allows calculating the ionic energy levels in a lot of different plasmas conditions.

On the other hand, the dicenter model is able to calculate accurate ionic energies for plasmas at very high densities. In this work, we performed the first comparisons between the ANALOP code using the analytical potential and the IDEFIX code using the dicenter model. These comparisons are the preliminary works in order to extend the accuracy of the analytical potential results to the high-density range of plasmas.

An explanation of both models is being done in the next two sections, and a comparison focused on fluorine and aluminum Lyman β lines provides the behavior of both codes.

CP645, *Spectral Line Shapes: Volume 12, 16th ICSLS*, edited by C. A. Back
© 2002 American Institute of Physics 0-7354-0100-4/02/$19.00

THE DICENTER MODEL

The relevance of the dicenter model to reproduce line shapes emitted from highly correlated plasmas has been demonstrated [1]. This model takes the advantage of treating exactly the interaction between the emitter and the first perturbing ion [2]. The numerical code IDEFIX, based on the dicenter approach, is able to calculate hydrogen and helium-like line shapes for high-density plasmas regime [3]. Recently, this code has been improved in order to include the effect of the surrounding plasma on the dicenter cell [4] (Molecular Stark effect). For one bound electron the corresponding Hamiltonian is:

$$H(R,F,\Omega) = \frac{p^2}{2} + \frac{Z_e}{r_1} + \frac{Z_p}{r_2} + V_e - \vec{d}.\vec{F} \tag{1}$$

The first part of the Hamiltonian is the interaction of the bound electron with the quasi-molecule embedded in a free electron gas. The second part ($-\vec{d}.\vec{F}$ term) is the linear Stark interaction of the bound electron with the screened ionic microfield produced by all ions in the plasma except the emitter and perturber. This Hamiltonian depends on three parameters, which are the distance between the two ions of the dicenter R, the magnitude of the field F, and the direction of the field with respect to the molecular axis Ω. These two last parameters are arising from the scalar product $\vec{d}.\vec{F}$.

Absorption Cross Section In The Dicenter Model

Since the Hamiltonian depends on the three parameters R,F,Ω, the resulting energies and oscillator strengths will depend on these parameters. For one set of parameters R, F, Ω, the expression of the absorption cross section is:

$$\sigma_\omega^{abs}(R,F,\Omega) = 2\pi^2\alpha \sum_{lower} \rho_l \frac{2}{3g_l} \sum_{upper} \Delta E_{u-l} \left|\vec{d}_{u-l}\right|^2 \frac{1}{\pi} \frac{\Phi_{u-l}}{(\Delta E_{u-l} - \omega)^2 + \Phi_{u-l}^2} \tag{2}$$

where Φ_{u-l} is the electronic broadening operator, ΔE_{u-l} and $\left|\vec{d}_{u-l}\right|$ the transition energy and the dipole moment respectively are depending on R,F,Ω. ρ_l, g_l and α stand for the population of the lower level l, the statistical weight and fine structure constant respectively.

Each transition involved in the cross section is represented by a normalized Lorentzian profile, multiplied by the dipolar moment of the transition. The final cross section is obtained by summing all the cross sections depending on the parameters R,F,Ω taking into account the probability to find a given set of parameters for given plasma conditions.

$$\sigma_\omega^{abs} = \int \sigma_\omega^{abs}(R, F, \Omega) \, dR \, dF \, d\Omega \tag{3}$$

ANALYTICAL POTENTIAL MODEL

The advantage of the analytical potentials in the calculation of atomic properties for the study of hot plasmas with respect to self-consistent calculations is the considerable reduction in the computing time. This reduction becomes more important when we develop models that handle each configuration into the plasma.

We have developed an opacity code called ANALOP [5] to calculate optical properties for plasmas based on analytical potentials. For an ion having a nuclear charge Z and N bound electrons embedded in the plasma, it was proposed the following effective analytical potential [6,7]

$$U_{eff}(r) = -\frac{1}{r}\left\{(N-1)(\Phi(r) - \eta(r)) + \left[Z - N + (N-1)\eta(0)\right]e^{-ar} + 1\right\} \tag{4}$$

where
$$\eta(r) = \frac{1}{2}a\int_0^\infty e^{-a|s-r|}\Phi(s)\,ds \ . \tag{5}$$

a being the inverse of the Debye radius given by

$$a = \left(\frac{\rho Z^*(Z^*+1)}{kT}\right)^{1/2} \tag{6}$$

Here ρ is the ion density and $\Phi(r)$ is a screening function given by [8]

$$\Phi(a_1, a_2, a_3, r) = \begin{cases} e^{-a_1 r^{a_3}} & , for \ N > 11 \\ (1 - a_2 r)e^{-a_1 r} & , for \ 8 \le N \le 11 \quad or \quad N = 2,3 \\ e^{-a_1 r} & , for \ 4 \le N \le 7 \end{cases} \tag{7}$$

with

$$a_k = c_{1k} Z^4 + c_{2k} Z^3 + c_{3k} Z^2 + c_{4k} Z + c_{5k} \ . \tag{8}$$

The coefficients c_{ik} were obtained for the ground state of isolated ions from He-like to U-like [9].

This non-isolated parametric potential (4) has been obtained by solving the Poisson equation, assuming the linearized Debye-Hückel approximation, and taking into account the reaction of the plasma-charge density to the optical electron [6].

For a given ionic stage and into the context of the Independent Particle Model (IPM), the energy levels and orbital functions for each level (bound and free spectrum) is obtained by solving the Dirac equation using the effective potential given by (4). Because total energies are necessary for obtaining the ionic populations, an expression proposed in the Density Functional Theory [10] is used in ANALOP.

The expression of the bound-bound photo-absorption cross section for a given line is as following (in a_0^2, a_0 is the Bohr radius)

$$\sigma_\omega^{abs} = 2\pi^2 \alpha \overline{\phi(\omega)} \sum_l \rho_l \sum_u f_{u-l}^{abs} \tag{9}$$

Here ρ_l stands for the populations of the lower mono-electronic levels involved in the transition, $\overline{\phi(\omega)}$ is the normalized Stark profile of the considered line. These profiles are calculated with the Pim Pam Poum (PPP) code [11] using as input data the energy transitions and oscillator strengths given by the ANALOP code.

In the equation (9), f_{u-l}^{abs} is the absorption oscillator strength, which is given into the context of IPM (the quantities in the equation are expressed in atomic units)

$$f_{ul}^{abs} = \frac{2\Delta E_{u-l}}{3g_l} \left| \langle \psi_u | r | \psi_l \rangle \right|^2 . \tag{10}$$

Here ΔE_{u-l} is the transition energy, ψ_k is the normalized wavefunction associated to the monoelectronic level k, and g_l is the statistical weight of the initial level.

Finally, the absorption coefficients for the bound-bound transitions have been evaluated, both with ANALOP and IDEFIX codes, through the following expressions

$$\kappa_\omega = N_l \sigma_\omega^{abs} \tag{11}$$

where σ_ω^{abs} stands for the photo-excitation cross section, N_l is the abundance of ions in the initial state of the transition which is calculated with M3R code [12] using ANALOP energies and oscillator strengths.

RESULTS

In figure 1 we have plotted the profiles from the Dicenter Model and PPP for fluorine Lyman β for the macroscopic plasma parameters Ne = 5.10^{22} cm^{-3} and Te = 500 eV. This figure shows a good agreement between both profiles at low density, being the differences between both intensities due to the fine structure effect not included in IDEFIX.

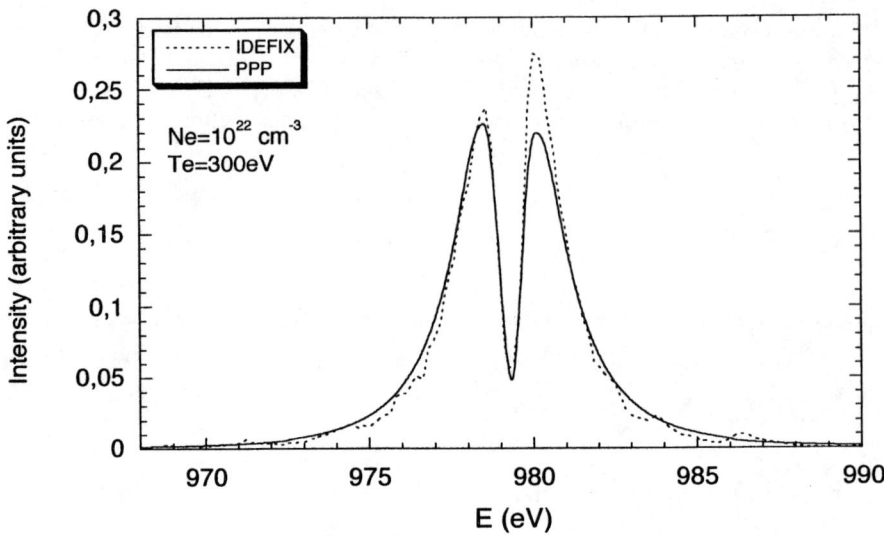

FIGURE 1. Profiles comparison for the fluorine Lyman β between IDEFIX and PPP at "low" density.

However the divergence with the PPP profile increases rapidly with the density. Moreover, the Lyman β line exhibits a stronger asymmetry as the density increases. These behaviors can be observed in figure 2.

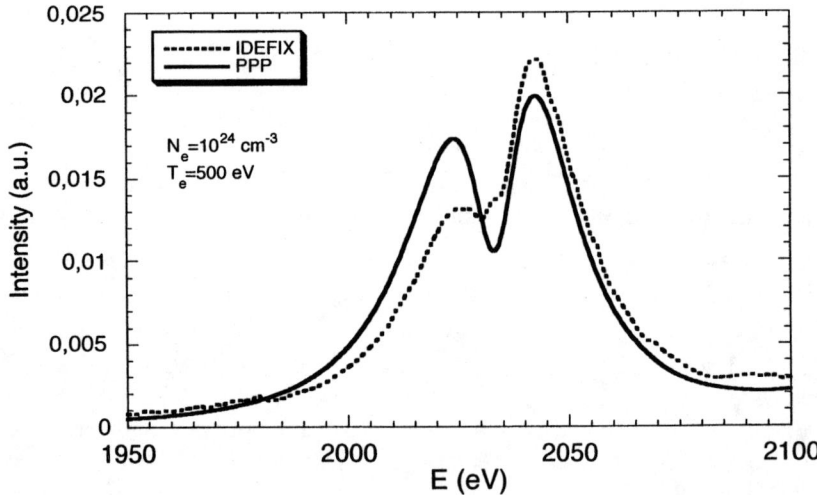

FIGURE 2. Profiles comparison for the aluminum Lyman β between IDEFIX and PPP at high density.

In figure 3 we present the absorption cross section for aluminum Lyman β for the macroscopic parameters $Ne = 5.10^{22}$ cm^{-3} and Te = 500 eV. The plasma coupling parameter Γ, corresponding to these conditions is 1.1. It can be seen from the figure some discrepancies between the non-isolated analytical potential and the dicenter model. These discrepancies were expected, because of the use of the Debye-Hückel approximation in the determination of the non-isolated potential which is not valid at high densities, i.e. Γ>1. On the same figure we give results from the ion sphere potential model. They are close to IDEFIX results.

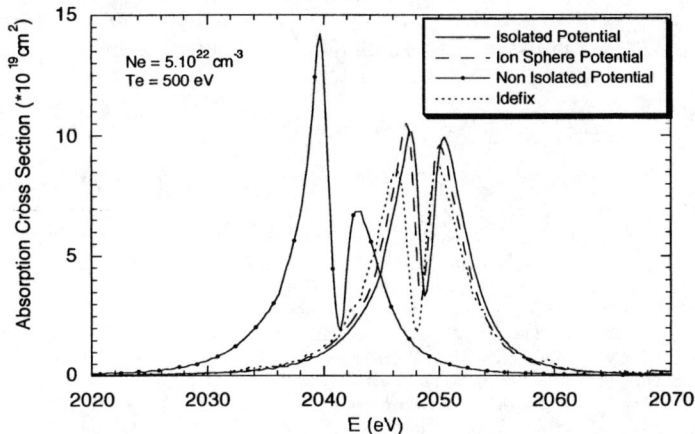

FIGURE 3. Absorption cross section for the aluminum Lyman β. Γ = 1.1.

The analytical potential shows its applicability at low densities, i.e. with a coupling parameter less than one. In this range of density this potential reproduces the main effects of the plasma environment on the atomic magnitudes, such as the continuum lowering and the red shift of the lines with respect to the isolated situations, however, for low density the dicenter model is inapplicable. As an example in figure 4 the fluorine line profile at low density (Γ=0.1) is shown.

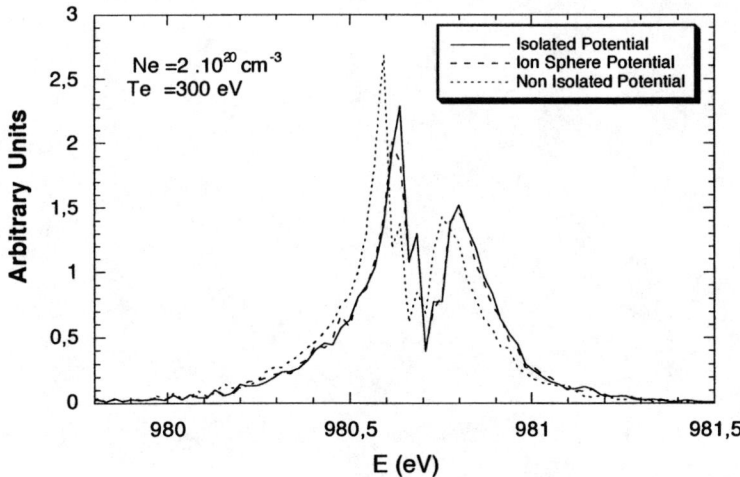

FIGURE 4. Fluorine Lyman β line profile (Γ=0.1).

CONCLUSIONS

We have shown that the introduction of the plasma effect in the analytical potentials assuming the linearized Debye-Hückel approximation is valid at low density and as it was expected it gives very inaccurate results at high-density plasmas. At these high-densities, results with the ion sphere potential are closer to the dicenter model ones. A new non-isolated potential taking into account dense plasma effect is now under development which will give the non-isolated potential presented in this work for low densities, and the ion sphere model for highest densities.

REFERENCES

1. Leboucher-Dalimier, E., et al., *J. Quant. Spectrosc. Radiat. Transfer* **71**, 493-504 (2001).
2. Gauthier, P., et al., *Phys. Rev. E* **58**, 942-950 (1998).
3. Sauvan, P., et al., *J. Quant. Spectrosc. Radiat. Transfer* **65**, 511-525 (2000).
4. Sauvan, P., et al., *J. Quant. Spectrosc. Radiat. Transfer* **71**, 675-687 (2001).
5. Mínguez, E., et al., *Nucl. Instrum. Methods A* **415**, 539 (1997).
6. Gil, J. M., et al., "Analytical Potentials Including Temperature and Density Effects for Calculation of Plasma Optical Properties" in *Advances in Laser Interaction with Matter and Inertial Fusion*, edited by Velarde G., Martínez-Val JM., Mínguez E. and Perlado JM., Pulisher World Scientific, Singapore 1997, pp. 385-388
7. Gil, J. M., et al., Submitted to *J. Quant. Spectrosc. Radiat. Transfer* (2002).
8. Martel, P., et al., *J. Quant. Spectrosc. Radiat. Transfer* **54**, 621 (1995).
9. Martel, P., et al., *J. Quant. Spectrosc. Radiat. Transfer* **4**, 623 (1998).
10. Rajagopal, A. K., *Adv. Chem. Phys.* **41**, 59 (1980).
11. Calisti, A., et al., *Phys. Rev. A* **42**, 5433 (1990).
12. Mancini, R. C. and Minguez, E., First International NLTE Atomic Kinetic Workshop, Gaithersburg (1996)

Calculations and measurements of x-ray Thomson scattering spectra in warm dense matter

G. Gregori[*], S. H. Glenzer[*], R. W. Lee[*], D. G. Hicks[*], J. Pasley[†], G. W. Collins[*], P. Celliers[*], M. Bastea[*], J. Eggert[*], S. M. Pollaine[*] and O. L. Landen[*]

[*]*Lawrence Livermore National Laboratory, PO Box 5508, Livermore, CA 94551*
[†]*Imperial College, University of London, Prince Consort Rd., London, England*

Abstract. We present analytical expressions for the dynamic structure factor, or form factor $S(k, \omega)$, which is the quantity describing the inelastic x-ray cross section from a dense plasma or a simple liquid. Our results, based on the random phase approximation (RPA) for the treatment on the charged particle coupling, can be applied to describe scattering from either weakly coupled classical plasmas or degenerate electron liquids. Our form factor correctly reproduces the Compton energy downshift and the usual Fermi-Dirac electron velocity distribution for $S(k, \omega)$ in the case of a cold degenerate plasma. The results shown in this work can be applied to interpreting x-ray scattering in warm dense plasmas occurring in inertial confinement fusion experiments. We show that electron density, electron temperature and ionization state can be directly inferred from such measurements. Specifically, we present as an example, use the results of experiments performed at the Vulcan laser facility at the Rutherford Appleton Laboratories (UK) on a LiH target.

1. INTRODUCTION

Diagnostics of dense plasmas poses several difficulties as currently adopted experimental techniques are rather limited in probing particle densities, temperatures and charge states of warm dense matter. Optical techniques, for example, can only provide information on surface layers of dense plasmas since they are opaque to visible or UV light. On the other hand, the emerging interest in understanding the properties of matter under extreme conditions, as the ones achieved in inertial confinement fusion (ICF) experiments [1], necessitates the developing of finite temperature dense matter probes. In ICF implosion experiments a variety of plasma regimes are created, and of particular interest are Fermi degenerate (or quantum) plasmas, characterized by a Fermi temperature greater than the electron kinetic temperature. Moreover, equation of state (EOS) predictions for various degenerate plasmas can only be resolved by accurate measurements of the chemical state of the materials. However, uncertainties in the present data and the lack of reliable independent measurements of temperature and density have made the validation of current models and calculations difficult.

We investigate the possibility of extending spectrally resolved Thomson scattering [2] in the x-ray regime for the diagnostics of solid density plasmas. This method was first discussed by Landen *et al.* [3] as a viable diagnostics alternative in ICF experiments. In

CP645, *Spectral Line Shapes: Volume 12, 16th ICSLS*, edited by C. A. Back
© 2002 American Institute of Physics 0-7354-0100-4/02/$19.00

Ref. [3], calculations were presented for scattering parameters $\alpha = 1/k\lambda_D \ll 1$, where λ_D is the Debye length and $\mathbf{k} = \mathbf{k}_0 - \mathbf{k}_1$ is the difference between the wave-number of the scattered and the incident probe radiation. In the present work, we provide a theoretical expression for the scattering form factor to represent x-ray Thomson scattering for arbitrary α parameter. In addition, our treatment can be applied in the description of scattering from degenerate to weakly coupled plasmas. For plasmas obeying the classical statistics, the electron-electron coupling constant is defined as (see, *e.g.,* Ichimaru [4]) $\Gamma = e^2/4\pi\varepsilon_0 k_B T_e d$, where T_e is the electron temperature and $d = (3/4\pi n_e)^{1/3}$ the ion-sphere radius, with n_e the electron density. In other words, Γ is the ratio between the potential and the kinetic energy of the electrons. For coupling between different charged particles, we also need to account for the ionization state of the material.

In an ideal plasma, $\Gamma \ll 1$ and the kinetic energy dominates the particle motion with negligible inter-particle coupling, while in a strongly coupled plasma, $\Gamma \gg 1$, the electrostatic (Coulomb) forces determine the nature of the particle motion. Weakly coupled plasmas lie in the range $\Gamma \lesssim 1$. The extension of definition of the coupling constant Γ to the quantum domain (*i.e.,* a degenerate plasma) is discussed by Liboff [5]. In this case, quantum diffraction prevents the electrons to get arbitrarily close to each other and Γ is now the ratio between the potential and the Fermi energy, E_F, of the electrons. Having $E_F = \hbar^2(3\pi^2 n_e)^{2/3}/2m_e$, as electron density increases, in contrast to a classical plasma, the coupling constant decreases, since $\Gamma \equiv \Gamma_q = e^2/4\pi\varepsilon_0 E_F d \sim n_e^{-1/3}$.

2. THEORY

2.1. Basic definitions

We are interested in describing the scattering from a uniform plasma containing N ions per unit volume. If Z_A is the nuclear charge of the ion, the total number of electrons per unit volume in the system, including free and bound ones, is $Z_A N$. Let us now assume we probe such a system with x-rays of frequency ω_0 such that $\hbar\omega_0 \gg E_I$, with E_I the ionization energy of any bound electron, *i.e.,* the incident frequency must be large compared to any natural absorption frequency of the scattering atom, which allows us to neglect photoabsorption. During the scattering process, the incident photon transfers momentum $\hbar\mathbf{k}$ and energy $\hbar\omega = \hbar^2 k^2/2m_e = \hbar\omega_0 - \hbar\omega_1$ to the electron, where ω_1 is the frequency of the scattered radiation. Under these conditions we can distinguish between electrons that are *kinematically* free with respect to the scattering process and *core* electrons that are tightly bound to the atom. If a_n is the orbital radius of the electron with principal quantum number n, kinematically free electrons satisfy the relation [6, 7] $ka_n \gtrsim 1$ (in the hydrogenic approximation, $a_n \sim a_B n^2/Z_A$ with $a_B = 4\pi\varepsilon_0 \hbar^2/m_e e^2$ the Bohr radius), while the opposite inequality applies for core electrons. This condition is equivalent to assuming that $\hbar\omega$, the energy transferred to the electron by Compton scattering, is larger than its binding energy. In the non-relativistic limit $(\hbar\omega \ll \hbar\omega_0)$

$$k = |\mathbf{k}| = \frac{4\pi}{\lambda_0}\sin(\theta/2),\qquad(1)$$

with λ_0 the probe wavelength and θ the scattering angle. We denote with Z_f and Z_c the number of kinematically free and core electrons, respectively. Clearly, $Z_A = Z_f + Z_c$. To avoid possible confusions, we should stress that Z_f is conceptually different from the *true* ionization state of the atom. It includes both the truly free (removed from the atom by ionization) and the valence (weakly bound) electrons; thus $Z_f = Z + Z_v$, where Z is the number of electrons removed from the atom, and Z_v is the number of valence electrons. In the limiting case of a liquid metal, $Z = 0$, and only the valence (or conduction) electrons need to be considered.

2.2. Scattering cross section

Following the approach of Chihara [8, 9] the scattering cross section is described in terms of the dynamic structure factor of all the electrons in the plasma

$$\frac{d^2\sigma}{d\Omega d\omega} = \sigma_T \frac{k_1}{k_0} S(k,\omega), \tag{2}$$

where σ_T is the usual Thomson cross section and $S(k,\omega)$ is the total dynamic structure factor defined as

$$S(k,\omega) = \frac{1}{2\pi} \int e^{i\omega t} \langle \rho_e(\mathbf{k},t)\rho_e(-\mathbf{k},0)\rangle dt, \tag{3}$$

with $\langle ... \rangle$ denoting a thermal average and

$$\rho_e(\mathbf{k},t) = \sum_{s=1}^{Z_A N} \exp\left[i\mathbf{k}\cdot\mathbf{r}_s(t)\right], \tag{4}$$

is the Fourier transform of the total electron density distribution, with $\mathbf{r}_s(t)$ the time dependent position vector of the s-th electron. Assuming the system is isotropic, as in the case of interest here (liquid metals or plasmas), the dynamic structure factor depends only on the magnitude of k, not on its direction. The next step consists in separating the total density fluctuation, Eq. (4), between the free (Z_f) and core (Z_c) electron contributions, and separating the motion of the electrons from the motion of the ions. The details of procedure are given by Chihara [8, 9], thus obtaining for the dynamic structure:

$$S(k,\omega) = |f_I(k) + q(k)|^2 S_{ii}(k,\omega) + Z_f S_{ee}^0(k,\omega) + Z_c \int \tilde{S}_{ce}(k,\omega-\omega')S_s(k,\omega')d\omega'. \tag{5}$$

The first term in Eq. (5) accounts for the density correlations of electrons that dynamically follow the ion motion. This includes both the core electrons, represented by the ion form factor $f_I(k)$, and the screening cloud of free (and valence) electrons that surround the ion, represented by $q(k)$ [10]. $S_{ii}(k,\omega)$ is the ion-ion density correlation function. The second term in Eq. (5) gives the contribution in the scattering from the free electrons that do not follow the ion motion. Here, $S_{ee}^0(k,\omega)$ is the high frequency part of the electron-electron correlation function [11] and it reduces to the usual electron feature

[12, 13] in the case of an optical probe. Inelastic scattering by core electrons is included in the last term of Eq. (5), which arises from Raman transitions to the continuum of core electrons within an ion, $\tilde{S}_{ce}(k, \omega)$, modulated by the self-motion of the ions, represented by $S_s(k, \omega)$. We point out that in Eq. (5) electron-ion correlations are implicitly accounted in the first term, since, as shown by Chihara [8], the electron-ion response function can be written in terms of the ion-ion response function. We observe that the total density correlation function must obey the relation [14]

$$S(k, -\omega) = \exp(-\hbar\omega/k_B T_e)S(k, \omega), \qquad (6)$$

which is a consequence of detail balance. This gives rise to asymmetry in the spectrum as we will discuss further in the next sections.

The ion-ion correlations reflect the thermal motion of the ions and/or the ion plasma frequency, and since we cannot currently experimentally access this low frequency part of the spectrum, we can approximate $S_{ii}(k, \omega) = S_{ii}(k)\delta(\omega)$. We thus only need to calculate the static structure factor for ion-ion correlations. We shall also observe that for typical conditions in dense plasmas for ICF experiments, the ions are always non-degenerate, since their thermal de Broglie wavelength is much smaller than the average interparticle distance. On the other hand, the electrons can exhibit some degree of degeneracy, and in the case of very cold and dense plasmas, they will obey the Fermi-Dirac distribution. Under these conditions, and within the framework of the random phase approximation (RPA), we can calculate $S_{ii}(k)$ using the semi-classical approach suggested by Arkhipov and Davletov [15], which is based on a pseudo-potential model for the interaction between charged particles to account for quantum diffraction effects (*i.e.*, the Pauli exclusion principle) and symmetry [16]. The correlation function is then calculated at the effective temperature $T_{cf} \sim (T_e^2 + T_q^2)^{1/2}$, where $T_q = T_F/(1.3251 - 0.1779\sqrt{r_s})$, with $r_s = d/a_B$. From quantum Monte Carlo calculations this corrected temperature was shown [17] to reproduce the exact quantum statistics at kinetic temperatures well below the Fermi temperature ($T_e \ll T_F$).

The free electron density-density correlation function that appears in the second term of Eq. (5) can be formally obtained through the fluctuation-dissipation theorem [18]:

$$S_{ee}^0(k, \omega) = -\frac{\hbar}{1 - \exp(-\hbar\omega/k_B T_e)} \frac{\varepsilon_0 k^2}{\pi e^2 n_e} \text{Im}\left[\frac{1}{\varepsilon(k, \omega)}\right], \qquad (7)$$

where $\varepsilon(k, \omega)$ is the electron dielectric response function. In the case of an ideal classical plasma, the plasma dielectric response is evaluated from a perturbation expansion of the Vlasov equation [19]. The resultant form for the density correlation function is then known as the Salpeter electron feature [12]. This approach, however, fails when the electrons become degenerate or nearly degenerate as quantum effects begin to dominate. Under the assumption that interparticle interactions are weak, so that the nonlinear interaction between different density fluctuations is negligible, the dielectric function can be derived in the random phase approximation (RPA) [20, 21]. In the classical limit, it reduces to the usual Vlasov equation.

We shall stress the point that in the limit of the RPA, strong coupling effects are not accounted for, thus limiting the model validity to plasma conditions in the range $\Gamma \sim 1$.

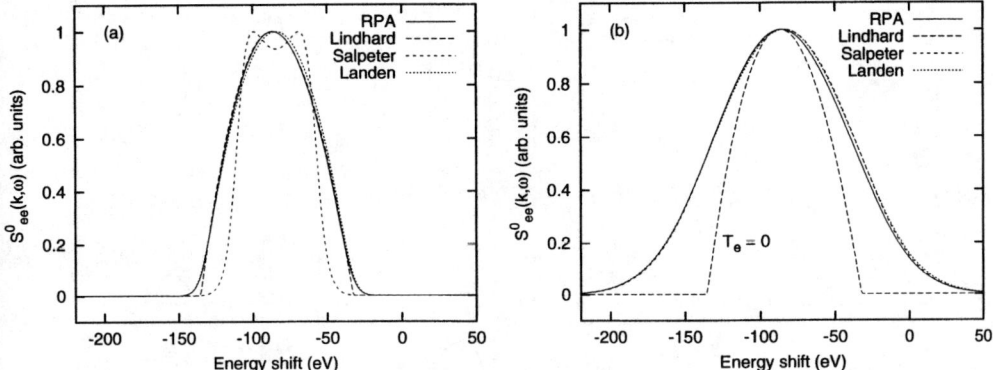

FIGURE 1. Free electron dynamic structure $S_{ee}^0(k,\omega)$ for $n_e = 1.0 \times 10^{23}$ cm^{-3} at $T_e = 1$ eV (a) and $T_e = 10$ eV (b). The probe radiation is $\lambda_0 = 0.26$ nm and the scattering angle is $\theta = 160°$, and $\alpha = 0.40$ (a) or $\alpha = 0.29$ (b).

Use of the RPA at larger couplings may still provide fairly accurate results if $kd \gtrsim 1$ [22, 23]. In the cases studied here, the plasma are within the range of validity. However, extensions to strong coupling are possible in terms of a local field correction [24] of the dielectric response functions, but they are significantly more complex and can be obtained only through the solution of the hypernetted chain (HNC) equation [25] or molecular dynamics simulations [26].

In Fig. 1 we have plotted normalized line profiles of $S_{ee}^0(k, \omega)$ calculated assuming incident x-rays with $\lambda_0 = 0.26$ nm, corresponding to the Ti He-α 4.75 keV emission line, and a scattering angle of $\theta = 160°$. The various models compared with the RPA in Fig. 1 are the analytical Lindhard-Sommerfeld theory [21], which is exact for $T_e = 0$ eV, the classical Salpeter form factor, and the calculations of Landen *et al.* [3] which is a direct representation of the electron distribution function. We observe that the RPA calculation automatically includes the effect of the Compton energy downshift in the scattered spectrum. This is not true, for example, in the Salpeter and Landen approximations since momentum transfer from the photons to the electrons is neglected there. Thus, in order to compare with the RPA, we need to translate the entire line profile an amount that corresponds to a shift of $\hbar^2 k^2/2m_e$ in energy. At a density of $n_e = 1.0 \times 10^{23}$ cm^{-3}, the Fermi temperature is $T_F = 7.85$ eV. We indeed see that, at temperatures lower than T_F, when quantum effects are important, the Salpeter result deviates from the RPA one. On the other hand, at $T_e = 10$ eV ($T_e > T_F$), the Salpeter formula agrees very well with the RPA since now the kinetic temperature is comparable with T_F. From Fig. 1 we also see that at $T_e = 1$ eV the calculated profile of $S_{ee}^0(k, \omega)$ is parabolic, while at $T_e = 10$ eV the profile is Gaussian. The transition from a parabolic to a Gaussian profile, as the electron temperature is raised, corresponds to the transition from Fermi to Boltzmann statistics in the electron velocity distribution.

It is customary to describe scattering processes in terms of the parameter $\alpha = 1/ks$, where s is the characteristic screening length of the electrostatic interactions. For a

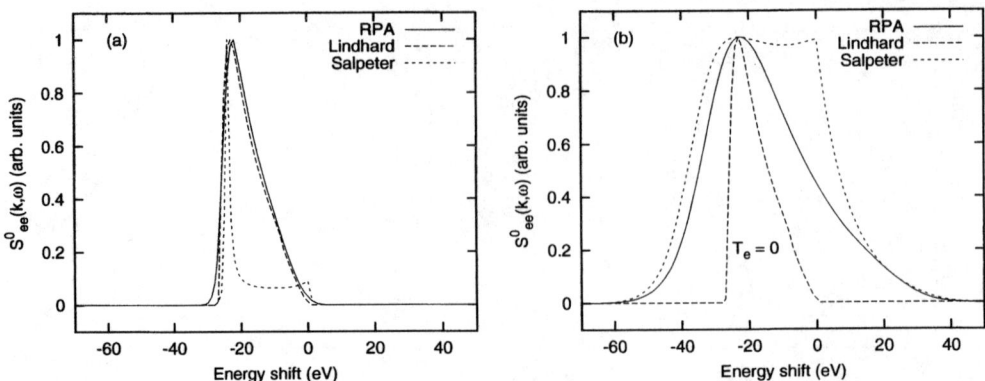

FIGURE 2. Free electron dynamic structure $S_{ee}^0(k, \omega)$ for $n_e = 1.0 \times 10^{23}$ cm^{-3} at $T_e = 1$ eV (a) and $T_e = 10$ eV (b). The probe radiation is $\lambda_0 = 0.78$ nm and the scattering angle is $\theta = 160°$, and $\alpha = 1.17$ (a) or $\alpha = 0.85$ (b).

classical plasma s coincides with the Debye length λ_D. If $\alpha < 1$ the electrons behave as uncorrelated scatters, while for large α parameters the scattering reflects their collective motion. In a classical plasma $\alpha \sim (T_e/n_e)^{1/2}$, and the nature of the scattering depends on both the electron temperature and the electron density. As the plasma becomes degenerate, the Debye length does not represent anymore the screening of the Coulomb forces. However, the classical results are still valid if, instead of using the kinetic temperature, they are evaluated at the effective temperature T_{cf} [17]. Fig. 4 shows $\alpha = const$ contours in the T_e-n_e plane for typical experimental conditions. We see that in the case of an ideal ($r_s \to 0$) degenerate electron liquid this approximation yields $s \sim \lambda_{TF}$, where $\lambda_{TF} = \sqrt{2\varepsilon_0 E_F/3n_e e^2}$ is the Thomas-Fermi screening length. Thus, $\alpha \sim n_e^{-1/6}$ and the type of scattering (uncorrelated or collective) is independent of T_e and weakly dependent on the electron density. The collective nature of the scattering can thus be investigated by only changing the wavelength of the probe x-ray.

Dynamic structures for collective scattering (*i.e.*, large α parameters) are shown in Fig. 2, which correspond to a longer probe radiation of wavelength $\lambda_0 = 0.78$ nm (Al He-α 1.6 keV emission line), all the other conditions being the same as in Fig. 1. In both Figs. 1 and 2 we see the strong asymmetry in the line profiles resulting from the detail balance relation (6).

The last term in Eq. (5) corresponds to the density correlations of the tightly bound electrons within each single ion, and it arises from electron-hole and bound excitations of the inner core electrons. The Fermi *golden rule* in the first order perturbation theory can be used to calculate the spectrum resulting from electron-hole excitations [27, 28]. As discussed by Mizuno and Omura [7] inner core electrons can be excited by the probe radiation to continuum states and the corresponding spectrum of the scattered radiation is that of a Raman-type band. Since the Raman band has width comparable or larger than the Compton band [29], we can regard this type of contribution as yielding only a small background [27]. This seems consistent with the results presented by Glenzer [30] on x-ray scattering from moderately heated beryllium targets.

3. THOMSON SCATTERING PROFILES: COMPARISON WITH EXPERIMENTS

Based on the theory outlined in the previous sections, we are now able to calculate the full Thomson scattering profile for x-ray probes at arbitrary scattering angle, for either classical or quantum plasmas. The only limitation is that the degree of coupling must not be too large to invalidate the limits of the RPA. We have obtained synthetic line profiles for the Ti He-α 4.75 keV radiation probe at $\theta = 160°$ scattering angle. In addition, we have assumed that the probe material consists of LiH ($Z_A = 4$) at $T_e = 1$ eV and at various compressed densities. To simulate actual experimental data, the theoretical line profile from Eq. (5) has been convoluted with a Gaussian instrument function with 40 eV FWHM. From Fig. 3 we can see that synthetic line profiles tend to be fairly similar since the broadening of the Compton profile goes as $\sqrt{T_F} \sim n_e^{1/3}$. The effect of the ionization state on the line profiles can also be seen in Fig. 3. Here, we have plotted synthetic lineshapes for different values of Z_f (or Z_c) with $n_e = 1.0 \times 10^{23}$ cm^{-3} ($T_F = 7.85$ eV) and $T_e = 1$ eV or $T_e = 10$ eV. We see dramatic differences in the simulated lineshapes for the various Z_f. This effect then suggests that x-ray Thomson scattering can also be implemented as a diagnostics tool for the ionization state of solid density plasmas based on the difference in the intensity between the unshifted and the Compton shifted peaks. This possibility was suggested by Landen *et al.* [3] since current optical techniques cannot directly measure the number of free electrons in solid density plasmas, and it extends the ionization state measurements based on visible light Thomson scattering [31]. On the other hand, the ratio of the scattered intensities between the shifted and the unshifted peaks is only sensitive to Z_f which is *not* the same as Z, the true ionization state of the material. The measure of Z_f will thus only provide an upper bound to Z, unless the number of valence electrons can be calculated or determined by other techniques.

Preliminary data obtained from a LiH target, with initial density of 0.77 g/cc, probed at the Vulcan laser facility at the Rutherford Appleton Laboratory with Ti He-α 4.75 keV x-rays have been compared with our theoretical model. The initial ion density is thus $n_i^0 = 5.8 \times 10^{22}$ cm^{-3}. The LiH target has been shocked and heated with two beams with approximate energy of 50 J/beam at 2ω, and 1 ns pulsewidth. X-ray line radiation has been generated by the interaction of a 110 J, 0.6 ns pulsewidth, 2ω laser beam on a Ti foil. The probe beam has been delayed 0.8 ns with respect to the heater beams. By shielding the view of the LiH target with 50 μm thick Au foils, a scattering angle of 160° can be selected. The scattering data have been collected using a mosaic HOPG (graphite) crystal used in focus mode in order to achieve high resolution and efficiency. The crystal was positioned 7.5 cm from the LiH target and the data were collected on an x-ray CCD camera 15 cm from the target. Comparison between the experimental line profile and the theoretical ones, convoluted with an instrument function of 30 eV FWHM, are shown in Fig. 4. We have $n_e \simeq 1.8 \pm 0.6 \times 10^{23}$ cm^{-3} ($T_F = 12.5$ eV) and $T_e \simeq 0\text{-}5$ eV, with $Z_f = 3.2 \pm 0.1$. From Fig. 4, we see that typical error in the density fit is of the order of 30%, which is to be expected for preliminary quality data. These values corresponds to

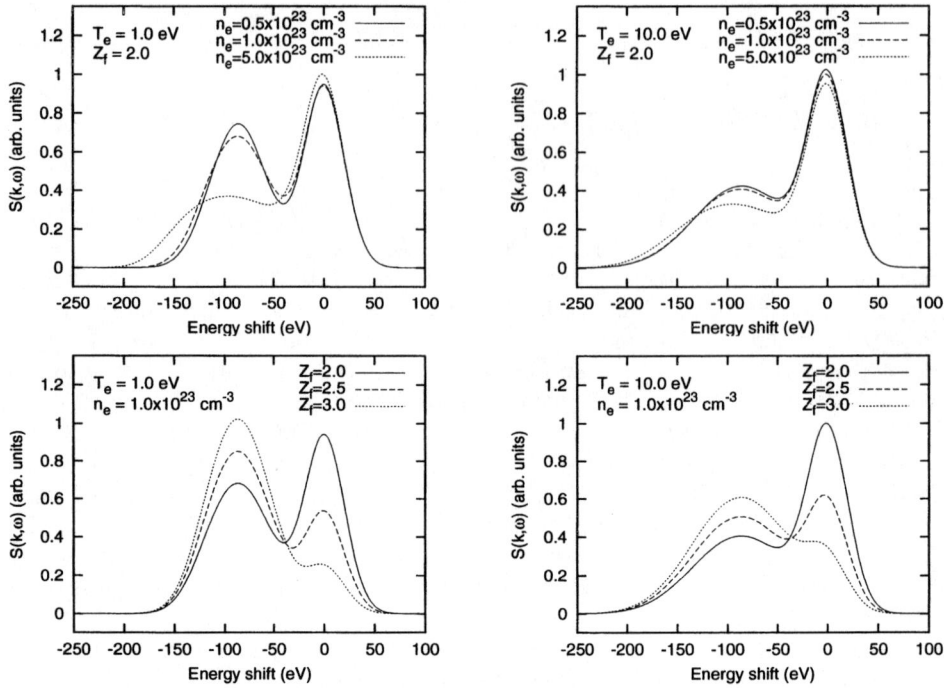

FIGURE 3. Synthetic dynamic structure $S(k, \omega)$ calculated for LiH target ($Z_A = 4$) at $T_e = 1$ eV and $T_e = 10$ eV. The probe radiation is $\lambda_0 = 0.26$ nm and scattering angle $\theta = 160°$.

a compression ratio $\chi = n_e/Z_f n_i^0 \sim 0.6$-$1.4$. The theory reproduces sufficiently well the measured profile, suggesting that the analysis reported here may be fairly adequate in describing the scattering process. Also, measurement of the number of free (and bound) electrons can be performed directly from the experimental data. The temperature-density domain of interest for typical experimental conditions is also shown in Fig. 4. From the fitting results, we see that the Vulcan experiment lies on the line where the α parameter changes its slope, which corresponds, as we have previously discussed, to a change from the classical Debye screening to the Thomas-Fermi screening.

4. SUMMARY AND CONCLUDING REMARKS

In this paper we have presented analytical expressions for the inelastic x-ray form factor that can be easily applied to interpreting scattering experiments in solid and super-solid density degenerate-to-hot plasmas. We have shown that x-ray Thomson scattering can be used as an effective diagnostic technique in plasmas produced under extreme conditions as the ones occurring in ICF experiments or to simulate scattering conditions found in the interiors of planets. This new technique will be useful, for example, to directly

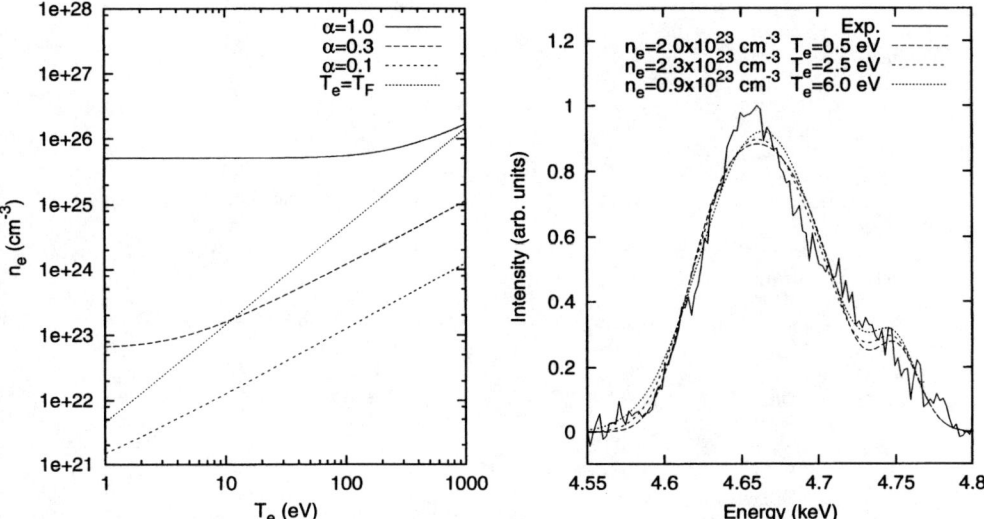

FIGURE 4. Left: calculated $\alpha = const$ contours for $\lambda_0 = 0.26$ nm and $\theta = 160°$. The line $T_e = T_F$ is also plotted in figure. Right: preliminary experimental scattering profile from LiH target obtained with a probe radiation $\lambda_0 = 0.26$ nm and at a scattering angle of $\theta = 160°$. Fitting parameters are given in the figure.

measure the electron temperature, ionization state or electron conductivity for EOS model validation.

Preliminary comparison with experiments conducted at the Vulcan laser facility on LiH targets have showns fair agreement between the model discussed in this paper and the experimental data, suggesting that x-ray Thomson scattering can be implemented as a viable diagnostics for low Z solid density plasmas.

ACKNOWLEDGMENTS

This work was performed under the auspices of the U.S. Department of Energy by the University of California Lawrence Livermore National Laboratory under Contract No. W-7405-ENG-48. We also acknowledge support from Laboratory Directed Research and Development grant No. 02-ERD-13.

REFERENCES

1. Lindl, J. D., *Inertial Confinement Fusion*, Springer-Verlag, New York, 1998.
2. Glenzer, S. H., Alley, W. E., Estabrook, K. G., de Groot, J. S., Haines, M. G., Hammer, J. H., Jadaud, J.-P., MacGowan, B. J., Moody, J. D., Rozmus, W., Suter, L. J., Weiland, T. L., and Williams, E. A., *Phys. Rev. Lett.*, **82**, 97 (1999).

3. Landen, O. L., Glenzer, S. H., Edwards, M. J., Lee, R. W., Collins, G. W., Cauble, R. C., Hsing, W. W., and Hammel, B. A., *J. Quant. Spectrosc. Radiat. Transfer*, **71**, 465 (2001).
4. Ichimaru, S., *Rev. Mod. Phys.*, **54**, 1017 (1982).
5. Liboff, R. L., *J. Appl. Phys.*, **56**, 2530 (1984).
6. Bushuev, V. A., and Kuz'min, R. N., *Usp. Fiz. Nauk*, **122**, 81 (1977).
7. Mizuno, Y., and Ohmura, Y., *J. Phys. Soc. Japan.*, **22**, 445 (1967).
8. Chihara, J., *J. Phys. F: Met. Phys.*, **17**, 295 (1987).
9. Chihara, J., *J. Phys.: Condens. Matter*, **12**, 231 (2000).
10. Riley, D., Woolsey, N. C., McSherry, D., Weaver, I., Djaoui, A., and Nardi, E., *Phys. Rev. Lett.*, **84**, 1704 (2000).
11. Ichimaru, S., *Basic Principles of Plasma Physics*, Addison, Reading, MA, 1973.
12. Salpeter, E. E., *Phys. Rev.*, **120**, 1528 (1960).
13. Evans, D. E., and Katzenstein, J., *Rep. Prog. Phys.*, **32**, 207 (1969).
14. Hansen, J.-P., and McDonald, I. R., *Theory of Simple Liquids*, Academic, London, 2000.
15. Arkhipov, Y. V., and Davletov, A. E., *Phys. Lett. A*, **247**, 339 (1998).
16. Baus, M., and Hansen, J.-P., *Phys. Rep.*, **59**, 1 (1980).
17. Perrot, F., and Dharma-Wardana, M. W. C., *Phys. Rev. B*, **62**, 16536 (2000).
18. Kubo, R., *J. Phys. Soc. Japan*, **12**, 570 (1957).
19. Landau, L. D., Lifshitz, E. M., and Pitaevskii, L. P., *Physical Kinetics*, Pergamon, Oxford, 1995.
20. Pines, D., and Bohm, D., *Phys. Rev.*, **85**, 338 (1952).
21. Pines, D., and Nozieres, P., *The Theory of Quantum Fluids*, Addison-Wesley, Redwood, CA, 1990.
22. Cauble, R., and Boercker, D. B., *Phys. Rev. A*, **28**, 944 (1983).
23. Boercker, D. B., Lee, R. W., and Rogers, F. J., *J. Phys. B*, **16**, 3279 (1983).
24. Ichimaru, S., Mitake, S., Tanaka, S., and Yan, X.-Z., *Phys. Rev. A*, **32**, 1768 (1985).
25. Carley, D. D., *Phys. Rev.*, **131**, 1406 (1963).
26. Hansen, J.-P., "Molecular Dynamics Simulation of Coulomb Systems in two and three Dimensions," in *Molecular Dynamics Simulation of Statistical Mechanical Systems*, edited by G. Ciccotti and W. G. Hoover, North-Holland, Amsterdam, 1986, vol. 97.
27. Platzman, P. M., and Tzoar, N., *Phys. Rev.*, **139**, A410 (1965).
28. Eisenberger, P., and Platzman, P. M., *Phys. Rev. A*, **2**, 415 (1970).
29. Issolah, A., Garreau, Y., Levi, B., and Loupias, G., *Phys. Rev. B*, **44**, 11029 (1991).
30. Glenzer, S. H., *Bull. American Phys. Soc.*, **46**, 325 (2001).
31. Glenzer, S. H., Alley, W. E., Estabrook, K. G., de Groot, J. S., Haines, M. G., Hammer, J. H., Jadaud, J.-P., MacGowan, B. J., Moody, J. D., Rozmus, W., Suter, L. J., Weiland, T. L., and Williams, E. A., *Phys. Plasmas*, **6**, 2117 (1999).

INNOVATIVE THEORY
AND EXPERIMENTAL TECHNIQUES

The Latest Advances and New Physics Based on the Generalized Theory of Stark Broadening

Eugene Oks

Physics Department, Auburn University, Auburn, AL 36849, USA

Abstract. The following topics are presented: 1) new non-model analytical results for multiparticle ion-dynamical broadening; 2) dramatic effect of the acceleration of electrons by the ion field on spectral line shifts in plasmas – advanced theory.

NEW NON-MODEL ANALYTICAL RESULTS FOR MULTIPARTICLE ION-DYNAMICAL BROADENING

The Generalized Theory (GT) of Stark broadening was based on the following three corner-stones.

1. It incorporated a strong *indirect coupling* between the electron and ion broadenings, the coupling being carried out via the radiating atom acting as an intermediary [1].

2. It allowed also for a strong *direct coupling* between the electron and ion broadenings, represented by the acceleration of the perturbing electrons by the ion nearest to the radiator [2, 3].

3. It provided a significantly *improved treatment of the ion dynamics*, based on the analytical solution for this problem [4].

The code based on the GT eliminated significant discrepancies between previous theories and benchmark experiments by various experimental groups at different plasma sources [5-9]. The reliability of the temperature and density measurements based on this code was proven over broad ranges of T and N_e [9-11]. Here we present an important *advance* in one of the three corner-stones of the GT: in the *non-model, analytical theory of the ion-dynamical broadening*.

Analytical results of the GT depend on the number of ions $\tilde{N}_w$ in the sphere of the ionic Weisskopf radius R_w:

$$\tilde{N}_w = (4\pi/3)NR_w^3, \qquad R_w = 3\,X_{\alpha\beta}\hbar/(m_e v). \qquad (1)$$

CP645, *Spectral Line Shapes: Volume 12, 16th ICSLS*, edited by C. A. Back

Here N and v are the perturbers density and velocity, respectively; $X_{\alpha\beta} = n_\alpha(n_1-n_2)_\alpha - n_\beta(n_1-n_2)_\beta$] is the usual combination of the parabolic quantum numbers of the Stark sublevels α and β involved in the radiative transition. For the ion broadening usually $\tilde{N}_w \gg 1$, i.e., ions are "mostly quasistatic".

For the case of $\tilde{N}_w \gg 1$, we applied the GT formalism to the ion broadening in [4] using the following features of the GT: 1) for $\tilde{N}_w \gg 1$, the nonadiabatic term is much smaller than the adiabatic term and can be neglected; 2) the remaining adiabatic contribution is obtained NONPERTURBATIVELY. As a result we obtained in [4] the ion-dynamical part of the *correlation function* $C(\tau) = \exp(-\gamma_{ad}\tau)$ for $\tau \gg 1/\Omega_w$, where $\Omega_w = v/R_w$ is the Weisskopf frequency. Physically this was a residual "impact" contribution from "mostly quasistatic" ions.

That ion-dynamical theory from [4] employed an assumption of *completed collisions* ($\Omega_w\tau \gg 1$), which led to the divergence of the integral over ion impact parameters ρ at large ρ. Therefore the validity condition of the *impact approximation* had to be used for finding a cutoff ρ_{max}.

Here we briefly present an advanced *non-model, analytical theory of the MULTIPARTICLE ion-dynamical broadening*, that is free from these deficiencies: it does NOT assume completed collisions and it does NOT employ the impact approximation.

For a given τ, the ion-dynamical part of the correlation function originates from the collisions, for which the instants of the closest approach t_0 fall within the interval $(-\tau/2, \tau/2)$. The rest of the perturbing ions are quasistatic. If $\tilde{N}_w \gg 1$, the majority of ions are quasistatic. If $\tilde{N}_w \ll 1$, only a minority of ions are quasistatic. The electric field $\mathbf{F}$, produced by quasistatic ions at the location of the radiator, reduces the symmetry of the problem to the spherical symmetry (characteristic for the unperturbed radiator) to the axial symmetry. The component of the ion-dynamical field $E_{\shortparallel}(t)$, which is parallel to $\mathbf{F}$, is responsible for the *adiabatic* contribution to the broadening. In the GT, this component is taken into account *exactly*, while the two other components – by the perturbation theory. Again, for $\tilde{N}_w \gg 1$, the adiabatic contribution predominates, thus making the entire result nonperturbative.

The primary difference in the starting point here from the starting point in [4] is that here we do NOT assume that $\tau \gg \rho/v$, i.e., we do NOT assume that collisions are complete. Instead, we consider a general case of the arbitrary ratio $v\tau/\rho$ and thus allow, in particular, for *incomplete collisions*.

Based on the above, the starting formula for the ion-dynamical part of the correlation function $C(\tau)$ can be written as follows:

$$C(\tau) \cong \exp[-NV(\tau)], \qquad V(\tau) \cong 2\pi v \int_0^\infty d\rho\, \rho \, <\{1 - \exp[iI(v,\rho,\tau)]\}> \int_{-\tau/2}^{\tau/2} dt_0, \qquad (2)$$

$$I(v,\rho,\tau) = (vR_w/2)\int_{-\tau/2}^{\tau/2} dt\, E_{\shortparallel}(v,\rho,t) = (\cos\theta\, R_w/\rho) \int_0^{v\tau/(2\rho)} dx\, (1+x^2)^{-3/2} = G(v,\rho,\tau)\cos\theta. \qquad (3)$$

372

Here <...> stands for the angular average, θ is the angle between vectors ρ and $\mathbf{F}$, and

$$G(v,\rho,\tau) = v\tau R_w/[2\rho(\rho^2 + v^2\tau^2/4)^{1/2}].\tag{4}$$

After performing the angular averaging, the expression <...> in (2) becomes

$$<...> = 1 - [\sin G(v,\rho,\tau)]/G(v,\rho,\tau).\tag{5}$$

Expressing ρ via G (using (3)) and changing the variable of the integration in (2) from ρ to G, we obtain the following final form of the ion-dynamical part of the correlation function:

$$C(\tau) = \exp[- 6\tilde{N}_w \, f(\Omega_w\tau/4)], \qquad f(x) = x \int\limits_0^\infty dy \, [1 - (\sin y)/y]/[y^2 \, (y^2 +1/x^2)^{1/2}].\tag{6}$$

Eq. (6) is valid for *any* value of $\Omega_w\tau$, *including* $\Omega_w\tau \ll 1$ (i.e., *"totally" incomplete* collisions). In particular, for $\Omega_w\tau \gg 1$, Eq. (6) yields

$$C(\tau) = \exp[- \tilde{N}_w \, (\Omega_w\tau/4) \, \ln(\Omega_w\tau/4)],\tag{7}$$

what corresponds to a *quasi-Lorentzian* lineshape and practically coincides with our result from [4]. However, for $\Omega_w\tau \ll 1$ the ion-dynamical part of the correlation function takes a different shape:

$$C(\tau) = \exp[- (3\pi/32) \, \tilde{N}_w \, (\Omega_w\tau)^2],\tag{8}$$

what corresponds to a *Gaussian* lineshape. This might be *the first time* when a HOMOGENEOUS Stark broadening leads to the Gaussian lineshape.

Thus, for the most interesting case of $\tilde{N}_w \gg 1$, the *total* correlation function for the ion broadening (and consequently, the lineshape) has the following *three physically-different regions* shown in Fig. 1:
- *static* region: $\Omega_w\tau \ll 1/\tilde{N}_w^{1/2}$
- dynamic *Gaussian* region: $1/\tilde{N}_w^{1/2} \ll \Omega_w\tau \ll 1$
- dynamic quasi-*Lorentzian* region: $1 \ll \Omega_w\tau$

In the quasi-Lorentzian region, the correlation function contains both adiabatic and nonadiabatic contributions, but the former predominates over the latter for $\tilde{N}_w \gg 1$.

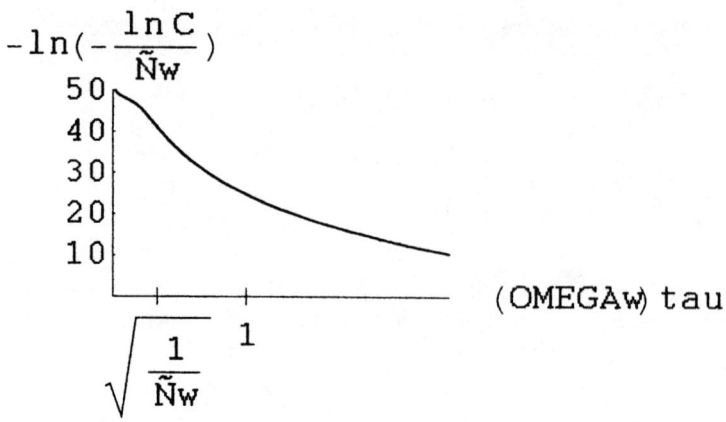

Figure 1. Double logarithm of the correlation function for the ion broadening vs. the product of the ion Weisskopf frequency Ω_w and the correlation time τ - for the case where the number of ions in the ionic Weisskopf sphere $\tilde{N}_w \gg 1$: 1) static region $\Omega_w\tau \ll 1/\tilde{N}_w^{1/2}$; 2) dynamic Gaussian region $1/\tilde{N}_w^{1/2} \ll \Omega_w\tau \ll 1$; 3) dynamic quasi-Lorentzian region $1 \ll \Omega_w\tau$.

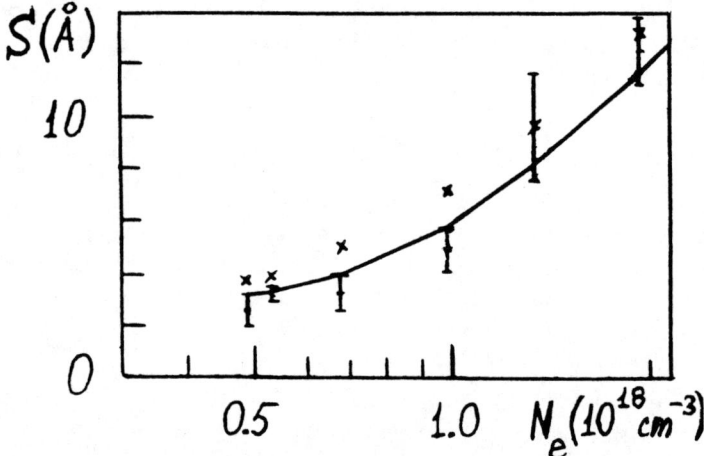

Figure 2. Stark *width* (FWHM) of the hydrogen H_α line measured in the latest benchmark experiment in the gas-liner pinch by the Kunze's group [6] (vertical bars). The crosses represent the results of the Griem's theory [12]. The dashed band shows the output of our code (the widths of the band represents an estimated theoretical error).

An application example is given in Fig. 2. It shows the *width* (FWHM) of the hydrogen H_α line measured in the latest benchmark experiment in the gas-liner pinch by the Kunze's group [6]. The experimental width is given by vertical bars. The crosses represent the results of the Griem's theory [12]. The dashed band shows the output of our code (the widths of the band represents an estimated theoretical error). It is seen that *our code is in a good agreement with the benchmark experiment* [6], while the *Griem's theory yields a discrepancy of a factor of two.*

We emphasize that our analytical results describe the ION-DYNAMICAL WIDTH for an ARBITRARY value of $\tilde{N}_w$. This is especially important, e.g., for Balmer and Paschen lines of hydrogen or deuterium used for *diagnostics of tokamak plasmas*, covering *low and high values of the principal quantum number n*: $\tilde{N}_w < 1$ for low-n lines, but $\tilde{N}_w > 1$ for high-n lines. For these experiments, our findings allow to *describe the entire range of n by a single analytical result.*

Finally we reiterate that our results:
- are obtained from FIRST PRINCIPLES (no model assumptions)
- use MULTIPARTICLE approach (no binary approximation)
- go BEYOND THE IMPACT approximation
- do NOT involve any *diverging integrals*
- do NOT involve any *cutoffs*

DRAMATIC EFFECT OF THE ACCELERATION OF ELECTRONS BY THE ION FIELD ON SPECTRAL LINE SHIFTS IN PLASMAS: ADVANCED THEORY

A significant effect of the Acceleration of Electrons by the Ion Field (AEIF) on widths of spectral lines in high-density plasmas was first studied in one of our previous papers [2]. This narrowing phenomenon is the realization of a *direct* coupling between electrons and ions. That theory was successfully applied for explaining H_α widths measured in the underwater laser-produced plasma [8].

In [2] we focused on the situation where impact parameters of perturbing electrons ρ $\ll R$ (R is the distance from the nearest perturbing ion to the atom). We showed that in this case the electron broadening operator acquires the following *reducing factor*: $g_0(v)$ $= 1/(1 + 2\rho_c/R)$, where $\rho_c = Z_p e^2/(mv^2)$ is the so-called Coulomb radius. For $\rho \gg R$, the *reducing factor* was set back to *unity* – as an *approximation*.

Here we briefly present an *advanced theory* of the effect of the AEIF on *shifts* of spectral lines, where we rigorously treat the situation where $\rho \gg R$ – without the above approximation. For $\rho \gg R$, the *reducing factor is the same as the factor distinguishing* the electron broadening operator for *H-like ions* from electron broadening operator for *hydrogen atoms*: $g_1(v, \rho) = 1/(1 + \rho_c^2/\rho^2)$. The above expressions for g_0 and g_1 can be combined in the following formula: $g(v, \rho) =$ $1 - 1/[1 + R/(2\rho_c) + \rho^2/(2\rho_c^2)]$. This formula coincides with g_0 for $\rho \ll R$ and with g_1 for $\rho \gg R$. In the intermediate range of ρ, the relative inaccuracy of (4) is < 5%. Further details can be found in our recent paper [3].

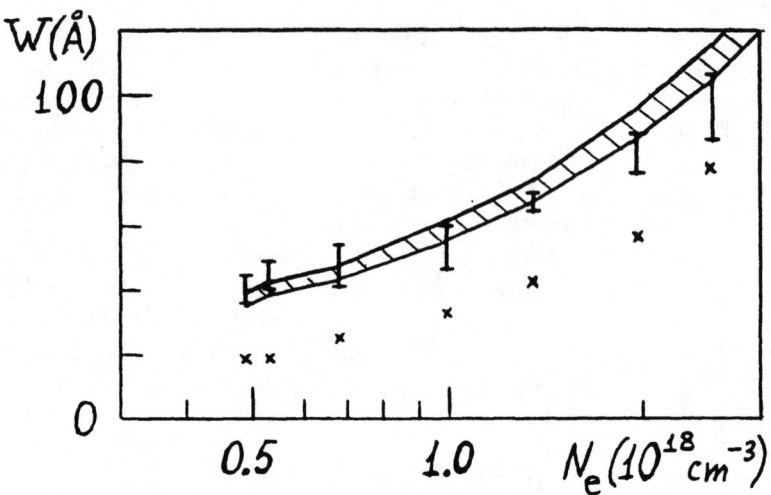

Figure 3. H_α shifts vs. the electron density N_e for the conditions of the experiment [6] in the gas-liner pinch. The measured shifts are shown by vertical bars. The solid line corresponds to our theory. The crosses represent the Griem's theory of the shifts [13].

We applied the results to two sets of experiments where H_α shifts were measured in high-density plasmas. Fig. 3 shows experimental and theoretical H_α shifts as a function of the electron density N_e for the conditions of the latest benchmark experiment at the gas-liner pinch [6], where the temperature was up to 10 eV. The experimental shift is given by vertical bars. The crosses represent the results of the Griem's theory [13]. The solid line corresponds to our theory. It is seen that *our theory is in a good agreement with the benchmark experiment* [6], while the *Griem's theory yields a discrepancy up to 50%*. For the conditions of this experiment, practically the entire difference between the results of the present theory and the theoretical results by Griem is due to the AEIF.

For the experiment in the underwater laser-produced plasma [9], where the temperature was 0.77 eV, i.e., by one order of magnitude lower than in the experiments at the gas-liner pinch, the Griem's theory [13] overestimated the experimental shifts even more dramatically – by a factor of two. In distinction to this, our theory resulted in the excellent agreement with the experimental shifts from [9]. The figure comparing the theories with these experiments can be found in our recent paper [3].

CONCLUSIONS

We obtained analytical results describing the ION-DYNAMICAL WIDTH for an ARBITRARY number of perturbers $\tilde{N}_w$ in the sphere of the Weisskopf radius. This is especially important, e.g., for Balmer and Paschen lines of hydrogen or deuterium used for *diagnostics of tokamak plasmas*, covering *low and high values of the principal quantum number n*. For these experiments, our findings allow to *describe the entire range of n by a single analytical result*.

It should be emphasized that our results on the ion-dynamical width:
- are obtained from FIRST PRINCIPLES (no model assumptions)
- use MULTIPARTICLE approach (no binary approximation)
- go BEYOND THE IMPACT approximation
-do NOT involve any *diverging integrals*
- do NOT involve any *cutoffs*

We also showed that the development of *the advanced theory of the Acceleration of Electrons by the Ion Field* is important for explaining the experimental *shifts* in high-density plasmas. We *eliminated a discrepancy up to a factor of two* between the shifts measured in the benchmark experiments and the Griem's theory of the shifts.

REFERENCES

1. Ispolatov, Ya., and Oks, E., *JQSRT* **51**, 129 (1994); Touma, J.E., Oks, E., et al, *JQSRT* **65**, 543 (2000).
2. Oks, E., *JQSRT* **65**, 405 (2000).
3. Oks, E., *J. Phys. B* **35**, 2251 (2002).
4. Oks, E., *Phys. Rev. E*, Rapid Communications **60**, R2480 (1999).
5. Böddeker, St., Günter, S., Könies, A., Hitzschke, L., and Kunze, H.-J., *Phys. Rev. E* **47**, 2785 (1993).
6. Büscher, S., Wrubel, T., Ferri, S., and Kunze, H.-J., *J. Phys. B* **35**, 2889 (2002).
7. Flih, S.A., and Vitel, Y., AIP Conf. Proc. 559, New York, 2001, p. 30.
8. Escarguel, A., Ferhat, B., Lesage, A., and Richou, J., *JQSRT* **64**, 353 (2000).
9. Escarguel, A., Oks, E., Richou, J., and Volodko, D., *Phys. Rev. E* **62**, 2667 (2000).
10. Oks, E., Bengtson, R.D., and Touma, J., *Contrib. Plasma Phys.* **40**, 158 (2000).
11. Oks, E., AIP Conf. Proc. 559, New York, 2001, p. 54.
12. Griem, H.R., *Principles of Plasma Spectroscopy*, Cambridge Univ. Press, Cambridge, 1997; *Spectral Line Broadening by Plasmas*, Academic, New York, 1974.
13. Griem, H.R., *Phys. Rev.* A **28**, 1596 (1983); **38**, 2943 (1988).

Hydrogen-like Spectral Line Intensities: Quasiclassical Representation in Parabolic Variables

L.A. Bureyeva[1], V.S. Lisitsa[2] and D.A. Shuvaev[2]

[1]*Institute of Spectroscopy of the RAS, Troitsk, Moscow Region, 142190, Russia*
E-mail: bureyeva@sci.lebedev.ru
[2]*RRC "Kurchatov Institute", Kurchatov Sq., 46, Moscow, 123182, Russia*

Abstract. A quasiclassical approach to the calculations of radiative transition probabilities in parabolic variables is under consideration. Kramers formulas for total radiative transition intensities are obtained in parabolic coordinates. The quasiclassical limit of Gordon formulas for radiative transition probabilities between highly excited atomic states is investigated. The transition from quantum Gorgon results to classical M.Born formulas for spectral line intensities in parabolic variables is obtained. The simple universal quasiclassical formulas for Stark component intensities are applied for calculations of statistical and dynamical spectral line shapes in plasmas.

INTRODUCTION

Spectral line intensities of hydrogen and hydrogen-like ions are of permanent interest for numerous applications especially for spectral line shape calculations. The problem is solved by well known Gordon formulas for dipole matrix elements both in spherical and parabolic coordinates [1]. However in specific applications of these formulas one is faced with very tedious calculations, especially for highly excited (Rydberg) atomic states. Really for values of principle quantum numbers $n=10^2$ the total quantity of matrix elements is of order of 10^{13}.

The solution of the problem is a transition to a quasiclasssilcal or pure classical description of atomic transition probabilities or spectral line intensities. The most famous formulas for such a description is well known Kramers formulas [1, 2].

It is demonstrated many times that Kramers probabilities result in a good accuracy even in the case of low values of principle quantum numbers, see, for example [3, 4]. Note that classical Kramers results are applicable for radiation transitions not only in the case of small change of the initial electron energy (or the principle quantum number) but also in the case of its strong change that is for strongly inelastic electron radiative transitions, see for details [3].

The Kramers formulas are presented in spherical variables related to the electron energy E (or principle quantum number n), orbital momentum l and its projection m conservation in the central field. It is well known however that an electron motion in a Coulomb field may be considered in parabolic variables as well corresponding to the

CP645, *Spectral Line Shapes: Volume 12, 16th ICSLS*, edited by C. A. Back
© 2002 American Institute of Physics 0-7354-0100-4/02/$19.00

conservation of the energy E (or n), an electric quantum number $k=n_1-n_2$ (n_1, n_2 are standard parabolic quantum numbers), connected with the eccentricity of the electron orbit and the same magnetic quantum number m. The case of the parabolic coordinate system is of especial interest because it corresponds to the electron motion in an external electric field. The case is of importance for numerous applications connected with observations of hydrogen-like atomic (or ions) spectra in plasmas [5, 6].

In reality, the observations of atomic spectra are connected with simultaneous self-consistent determination of both energy Stark shifts and correspondent intensities of Stark components. The self-consistensy is possible only in ξ, η parabolic representation of the electron motion in a Coulomb field. So one can consider the radiation emitting electron motion along a Keppler ellipse in two variables: spherical and parabolic. The corresponding variables are presented on Fig.1.

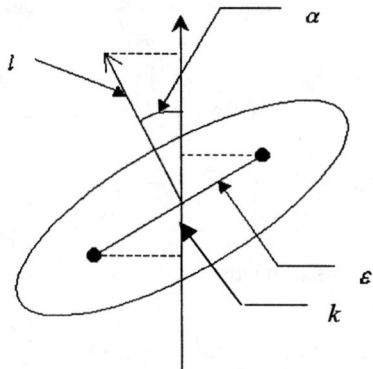

FIGURE 1. Spherical and parabolic variables for electron motion along a Keppler ellipse.

One can see a relationship between orbital momentum l and its projection m (spherical variables) versus eccentricity ε with its projection k ("electric" quantum number) and the same orbital momentum projection m (parabolic variables). Just former ones are connected with an energy shift dE in an electric field $\mathbf{F}$ according the relation:

$$dE = Const\,|\,F\,|\,nk, \qquad Const = \frac{3}{2}\frac{ea_0}{\hbar}. \tag{1}$$

If one is interested in intensities of Stark components as functions of their shift in the electric field then the correct consideration of corresponding intensities must be done just in parabolic variables.

It is well known that the electron motion in parabolic coordinates corresponds to two independent motions along ξ and η - directions of a parabolic coordinate frame. So to find the intensities of radiation emission in parabolic variables it is necessary to find double Fourier series corresponding to these two motions in the parabolic coordinates. It is in contrast with well known one dimensional Fourier series in a spherical basis well known in the literature.

The situation is of the same nature both in classical and quantum mechanics. Really the spectral line intensities in spherical basis are expressed in terms of integrals of electron radial wave functions whereas in a parabolic basis the intensities of radiative transitions are expressed in terms of double integrals of wave functions corresponding to the electron motion along ξ-, η-directions. The former ones are expressed in term of hypergeometric function found by Gordon, see [1]. The quasiclassical limits of the general quantum formulas are well investigated in a spherical variables whereas the analogous results in a parabolic basis are absent up to present time.

The first investigations of a classical electron motion in parabolic coordinates were performed by M. Born [7], but his results were also not investigated for a long time. It is a goal of the present paper to investigate a transition from quantum Gordon formulas to the classical Born results. The corresponding results provide simple quasiclassical formulas for radiative transition probabilities to make it possible simple estimations of radiative emission intensities in parabolic variables.

TOTAL RADIATION EMISSION INTENSITIES. LIFE TIME OF AN ATOMIC ENERGY STATE IN PARABOLIC VARIABLES

We begin with a consideration of the total intensity of a radiation emission in a Coulomb field integrated over all frequencies of transitions. This quantity determines a probability $A(n,l)$ of a radiative transition per unit time or the inverse life time of an energy level in atomic physics. The quantity (in a. u.) is well known in a spherical bases (c is the speed of light):

$$A(n,l) = 4Z^4 \left[\pi \sqrt{3} c^3 n^3 (2l+1)^2 \right]^{-1}. \tag{2}$$

Let us make a transformation of eq. (1) to the parabolic basis. It is well known that the parabolic and spherical basis are connected by Clebsh-Gordan coefficients determining a transformation from spherical wave functions to parabolic ones [8]. From classical point of view the squared of Clebsh-Gordan coefficient is a joint probability of an appearance of specific values of spherical and parabolic quantum numbers l, k, m. One can obtain the probability by two ways: 1) by pure classical calculations determining a portion of the electron phase space where conservation conditions for the electron motion are fulfilled [9] or 2) by a transition to the quasiclassical limit in general generation formulas for Clebsh-Gordan coefficients [8].

The result takes the form:

$$[C(n,k \mid l,m]^2 = \frac{2l}{\pi \left[\left(l^2 - l_{min}^2 \right) \left(l_{max}^2 - l^2 \right) \right]^{\frac{1}{2}}}, \tag{3}$$

where

$$l_{min}^2 = \frac{1}{2} \left\{ \left[(n-1)^2 + m^2 - k^2 \right] - \left[\left[(n-1)^2 + m^2 - k^2 \right]^2 - 4(n-1)^2 m^2 \right]^{\frac{1}{2}} \right\},$$

$$l_{max}^2 = \frac{1}{2}\left\{ \left[(n-1)^2 + m^2 - k^2\right] + \left[\left[(n-1)^2 + m^2 - k^2\right]^2 - 4(n-1)^2 m^2\right]^{\frac{1}{2}}\right\},$$

where the electric quantum numbers k are used.

Values l_{max}, l_{min} determines a relationship between parabolic and spherical variables. The physical sense of the interrelation between both types of variables is clear from the electron trajectory in a Coulomb field presented on Fig. 1. It follows directly from Fig. 1 a following relation between spherical (orbital momentum) and parabolic (a projection of the orbit eccentricity) variables:

$$\frac{k}{n} = \varepsilon \cdot \sin(\alpha) = \varepsilon \cdot \sqrt{1 - \frac{m^2}{l^2}}. \qquad (4)$$

To determine Kramers formulas for radiative transition probabilities in parabolic coordinates let us multiply eq. (2) in a spherical basis by the squared of Clebsh-Gordan coefficients (3) and integrate over l:

$$B(k,m) = \int_{l_{min}}^{l_{max}} A(n,l)\left[C(n,k\,|\,l,m)\right]^2 dl. \qquad (5)$$

Performing the integration with account of eqs. (3) one obtains

$$B(k,m=0) = \frac{4Z^4}{\pi^2 \sqrt{3}c^3}\frac{2}{n^3}\frac{1}{\left[(n-1)^2 - k^2\right]^{\frac{1}{2}}}, \qquad (6)$$

$$B(k,m) = \frac{4Z^4}{\pi\sqrt{3}c^3}\frac{1}{n^3(n-1)|m|}. \qquad (7)$$

Eqs. (6,7) are just "parabolic analogs" of Kramers formulas in the spherical basis. It is of interest to note that the probabilities $B(k,m)$ doesn't depend on the electric quantum number k in the case $m>0$.

Let us compare the simple results (6,7) with the precise calculations [10] of the transition probabilities in the parabolic basis. Such a comparison is presented in Table 1 for $m=0$ and in Table 2 for $m>0$.

TABLE 1. The comparison of Kramers radiation emission probabilities $B^{Kr}(k,m=0)$ given by eq. (6) from the energy level $n=10$ described by quantum numbers $k,m=0$ with the results of numerical calculations [10].

k	$B^{Kr}(k,m=0)$ Eq. (6)	[10]
1	0.73	0.48
3	0.77	0.57
5	0.88	0.77
7	1.15	1.17
9	2.63	2.15

TABLE 2. The same as in Table 1 but for $m>0$. $B^0(m)$ is the quasiclassical formula from [4], $B(m)$ is the quantum numerical calculations, $B^{Kr}(m)$ is calculation according eq. (7) averaged over k-numbers

m	$B^{Kr}(m)$ Eq. (7)	$B^0(m)$	$B(m)$
1	1.129	1.045	1.0871
2	0.556	0.518	0.552
3	0.363	0.341	0.366
4	0.265	0.254	0.272

It follows from the tables that Kramers probabilities in parabolic variables are of a reasonable precision especially for $m>0$. It must be noted that the data presented belongs to the moderate values of principle quantum numbers $n=10$.

CLASSICAL INTENSITIES OF RADIATIVE TRANSITIONS BETWEEN TWO HIGHLY EXCITED ATOMIC STATES

Intensities of radiative transitions between two atomic energy levels in parabolic variables are expressed in the frame of quantum theory in terms of matrix elements of dipole momentum being a rather complicated combination of hypergeometric functions depending on parabolic quantum numbers of initial n, n_1, n_2, m and final n', n_1', n_2', m', see [1].

Let us look at its classical analog. It can be obtained by a Fourier transformation of an electron trajectory in parabolic coordinates. Corresponding calculations was performed by M. Born [7] and they expressed in terms of a double Fourier series which presents "relative intensities" of radiation emission by Z- and X- components of the electron trajectory:

$$Z_{\Delta n_1 \Delta n_2}^{\Delta m=0} \approx \frac{1}{\Delta n} \left\{ \sigma_2 J_{\Delta n_1}(\Delta n \sigma_1) J_{\Delta n_2}'(\Delta n \sigma_2) - \sigma_1 J_{\Delta n_1}'(\Delta n \sigma_1) J_{\Delta n_2}(\Delta n \sigma_2) \right\} =$$

$$= \frac{1}{\Delta n} \left\{ -\frac{K}{\Delta n} J_{\Delta n_1}(\Delta n \sigma_1) J_{\Delta n_2}(\Delta n \sigma_2) - \sigma_2 J_{\Delta n_1}(\Delta n \sigma_1) J_{\Delta n_2 + 1}(\Delta n \sigma_2) + \right.$$

$$\left. + \sigma_1 J_{\Delta n_1 + 1}(\Delta n \sigma_1) J_{\Delta n_2}(\Delta n \sigma_2) \right\}, \tag{8}$$

$$X_{\Delta n_1 \Delta n_2}^{\Delta m=\pm 1} \approx \frac{1}{\Delta n} \left\{ \frac{n \sigma_1 \sigma_2}{\sqrt{n_1 n_2}} J_{\Delta n_1}(\Delta n \sigma_1) J_{\Delta n_2}(\Delta n \sigma_2) - \frac{\sqrt{n_1 n_2}}{n} J_{\Delta n_1 \pm 1}(\Delta n \sigma_1) J_{\Delta n_2 \pm 1}(\Delta n \sigma_2) \right\},$$

where $J_p(g)$ are Bessel functions with integer indexes p,

$$\sigma_1 = \frac{1}{2\sqrt{nn'}} \sqrt{(n'+k'-1)^2 - m^2}, \qquad \sigma_2 = \frac{1}{2\sqrt{nn'}} \sqrt{(n'-k'-1)^2 - m^2}$$

$$\Delta n_1 = \frac{1}{2}(\Delta n + K \mp \Delta m), \quad \Delta n_2 = \frac{1}{2}(\Delta n - K \mp \Delta m)$$

$$K = (n_1-n_1')-(n_2-n_2')=\Delta n_1-\Delta n_2 \text{ и } k = n_1-n_2, \quad k' = n_1'-n_2'. \tag{9}$$

Here differences of quantum numbers $\Delta n = n-n'$, $\Delta n_1=n_1-n_1'$, $\Delta n_2=n_2-n_2'$ as well as the difference K of electric quantum numbers of upper and lower levels are introduced and separately the electric quantum number k' of the lower level is labeled. The differences play an important role in an analytical presentation of the intensities below. One must take into account that the frequency shift accordance of a spectral line components in an external electric field $\mathbf{F}$ can be expressed in terms of the quantum numbers K and k as follows

$$d\omega = Const \, | \, F \, |(\, Kn + \Delta n k' \,), \qquad Const = \frac{3}{2} \frac{ea_0}{\hbar}. \tag{10}$$

The same classical results (8,9) we obtained from quantum mechanical general Gordon formulas [1] using formulas of transformations and approximations for hypergeometric functions in the case of small changes in principle quantum numbers Δn, namely when $\Delta n^2/4nn' \ll 1$ [11].

The general results (8, 9) may be simplified essentially in the case of small changes in principle quantum numbers Δn. Expanding relative amplitudes (8,9) into a series of the parameter $\Delta n^2/4nn'$ one obtains the following expressions for the intensities of π- and σ-components:

$$I^{\pi}_{K=0}(m) \approx Z^2_{K=0} \cong \left(\frac{1}{2}\right)^{2(\Delta n+1)} \left(\frac{\Delta n^2}{4nn'}\right)^{\Delta n} \frac{\left[(n'^2-k'^2)^2 - 2m^2(n'^2+k'^2) + m^4\right]^{\Delta n/2}}{\left(\frac{\Delta n}{2}\right)!^2 \left(\frac{\Delta n+1}{2}\right)!^2} \left(\frac{k'}{n}\right)^2$$

$$I^{\pi}_{K\neq0}(m) \approx Z^2_{K\neq0} \cong \left(\frac{1}{2}\right)^{2\Delta n} \left(\frac{\Delta n^2}{4nn'}\right)^{\Delta n} \left(\frac{K}{\Delta n^2}\right)^2 \frac{\left[(n'^2+k'^2)^2 - m^2\right]^{\frac{1}{2}|\Delta n+K|} \left[(n'^2-k'^2)^2 - m^2\right]^{\frac{1}{2}|\Delta n-K|}}{\left(\frac{|\Delta n+K|}{2}\right)!^2 \left(\frac{|\Delta n-K|}{2}\right)!^2}$$

(11)

$$I^{\sigma}_K(m) \approx \left(X^{\Delta m=+1}\right)^2 + \left(X^{\Delta m=-1}\right)^2 \cong \left(\frac{1}{2}\right)^{2(\Delta n-3)} \left(\frac{\Delta n^2}{4nn'}\right)^{\Delta n} \frac{1}{\Delta n^4} \frac{\left[(n'+k')^2 - m^2\right]^{\frac{1}{2}|\Delta n-1+K|} \left[(n'-k')^2 - m^2\right]^{\frac{1}{2}|\Delta n-1-K|}}{\left(\frac{|\Delta n-1+K|}{2}\right)!^2 \left(\frac{|\Delta n-1-K|}{2}\right)!^2}$$

(12)

It is seen that he intensities are separated into two blocks in accordance with the combinations of quantum numbers K and k, namely: large scale block describing by the index K and small scale block describing by k'.

Integrating eqs. (11) and (12) over m one obtains total relative intensities of Stark π- and σ-components. Thus for the transitions with $\Delta n=1$ $(n \sim n')$ one obtains

$$I^{\pi}_k \sim \sum_m \left(Z^m_m\right)^2 \cong \frac{1}{6n^2}\left(n^3 + 3n^2 k' - 3nk'|k'| - k'^2|k'|\right)$$

$$I^{\sigma}_k \sim \sum_m \left[\left(X^{m-1}_m\right)^2 + \left(X^{m+1}_m\right)^2\right] \cong \frac{1}{3n^2}\left(2n^3 - 3n^2|k'| + k'^2|k'|\right)$$

(13)

This limiting case was discovered by Gulyaev [12] in his analyses of quantum Gordon formulas. The simple results (13) make it possible to present the intensities of such transitions in a universal form that is as a composition of two blocks in the large scale K and analytical description in the small scale k'.

APPLICATIONS. STATISTICAL AND DYNAMIC INTENSITIES OF SPECTRAL LINES

Let us consider very briefly one application of the obtained results connected with dynamical and statistical intensities of spectral lines. The problem is well known [1] and it relates to the calculation of spectral line shapes in dense or rarified media. In the first case the statistical intensities of Stark components corresponds to a population of components proportional to their statistical weighs provided by collisions. In this case the total spectral line intensity is equal to the sum of intensities of specific Stark components according their statistical weighs. Another dynamical intensities correspond to the case when collisions are not strong as compared with radiative transitions so the populations of atomic states don't correspond to their statistical weighs but they are determined by a distribution function $f(n,k,m)$ over parabolic variables. The distribution function (or relative population of atomic states) is found from corresponding kinetic equation for Stark sublevels populations.

It is of general interest to investigate the n,k,m-distribution of atomic energy levels with the dominant dielectronic recombination (DR) population source $q(n,k,m)$ [13] resulting in a radiative three dimensional cascade in rarefied plasmas. The problem was considered in details in paper [11].

The process under consideration appears as follows: a free electron with an energy E less than the ion core excitation energy $\hbar\omega_c$ (without change of core quantum number n_c) excites the core and is captured at a highly excited (Rydberg) state with large values of the principle (n) quantum number, then the core is stabilized by the radiative transition at the core frequency ω_c and we look for a radiative cascade of the captured electron over the Rydberg atomic state n,k,m .

As it was shown in [11] the classical radiative cascade equation in parabolic variables may be written in the form:

$$\dot{n}\frac{\partial}{\partial n}f + \dot{k}\frac{\partial}{\partial k}f + \dot{m}\frac{\partial}{\partial m}f = q(n,k,m) \tag{14}$$

where $f(n,k,m)$ is a distribution function in three-dimensional phase space $\{n,k,m\}$, $\dot{n},\dot{k},\dot{m}$ are corresponding radiative losses of quantum numbers. All derivatives are expressed in the parabolic variables. The solution of eq. (14) is

$$f(n,k,m) = \varphi(n,k,m) + \int_{n+1}^{\infty}\frac{q(n',k[n',m,C_1],m[n',k,C_2])}{|\dot{n}(n',k[n',m,C_1],m[n',k,C_2])|}dn' \tag{15}$$

where $\varphi(n,k,m)$ is the initial condition and characteristics C_1 and C_2 have the forms [11]:

$$m^2 = k^2(\ln\frac{n^2}{n^2-k^2}-k^2C_2)^{-1}, \qquad k^2 = n^2\left(1-C_1^{2/3}m^2\right)+m^2 \tag{16}$$

Using expression for $q(n,k,m)$ in parabolic variables obtained in [13] it follows from (15) the general expressions for the distribution function $f(n,k,m)$

$$f(n,k,m,Z) = \frac{B(Z,T)}{Z^4}\left(\frac{\pi\sqrt{3}n^4|m|}{4}+\frac{\sqrt{2}}{4}\pi\frac{l_{min}^5}{3}\left(\frac{c^3l_{min}^3}{\omega_c^2\pi m}\int_1^{\infty}\frac{t^2e^{-2\left(\frac{l_{min}}{l_{ef}}\right)^3 t^3}}{\sqrt{t^2-1}}dt\right)^{1/4}\right), n<<n^* \tag{17}$$

$$f(n,k,m,Z) = \frac{B(Z,T)}{Z^4}\frac{c^3}{\pi\omega_c^2}\left(\frac{\pi\sqrt{3}|m|^3}{4}+\frac{1}{9}\frac{|m|^7}{(n+1)^3}\right)\int_1^{\infty}\frac{t^2e^{-2\left(\frac{l_{min}}{l_{ef}}\right)^3 t^3}}{\sqrt{t^2-1}}dt , n>>n^* \tag{18}$$

where

$$n^{*3} = \frac{c^3l_{ef}^2}{\omega_c^2\pi l_{max}}\left(\frac{l_{min}}{l_{ef}}\right)^2\int_1^{\infty}\frac{t^2e^{-2\left(\frac{l_{min}}{l_{ef}}\right)^3 t^3}}{\sqrt{t^2-1}}dt , \qquad l_{ef} = \left(\frac{3Z^2}{\omega_c}\right)^{1/3}$$

Let us calculate the statisstical I^{stat} and dynamic I^{dyn} intensities for hydrogen-like ions using the distribution function above. The statistical iintensity I^{stat} is determined by the direct sum over parabolic quantum numbers:

$$I^{stat}(n,\Delta n) = \sum_{K,k,m} I(K,n,\Delta n,k,m)\delta\left(\frac{\Delta\omega}{\omega_F} - Kn - \Delta nk\right), \quad \omega_F = \frac{3ea_0 F}{2\hbar} \quad (19)$$

The dynamical intensity I^{dyn} is obtained by weighing of Stark components intensities with the normalized distribution function mentioned above:

$$I^{dyn}(n,\Delta n) = \sum_{K,k,m} I(K,n,\Delta n,k,m)\frac{f(n,k,m)}{\sum_{k,m} f(n,k,m)}\delta\left(\frac{\Delta\omega}{\omega_F} - Kn - \Delta nk\right) \quad (20)$$

The results of the calculations of the statistical and dynamic intensities for $H_{n\alpha}$ are shown in Fig. 2. One can see an essential difference between both types of intensities.

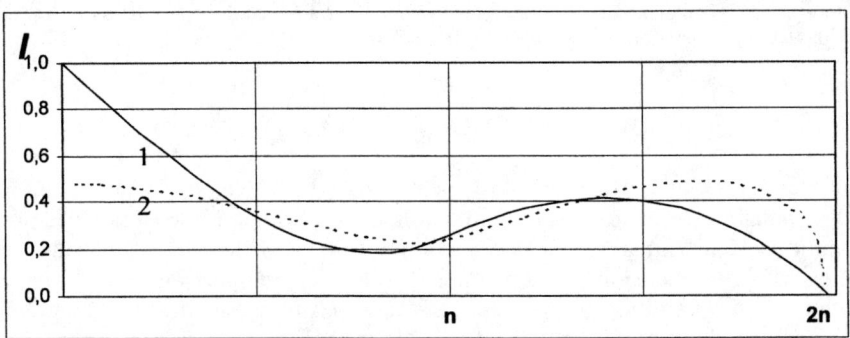

FIGURE 2. Total (π- and σ-components) statistical (1) and dynamic (2) intensities (in relative units) for $H_{n\alpha}$ as functions of $\Delta\omega$ (in units of ω_F).

ACKNOWLEDGMENTS

This work was supported in part by a grant from the Russian Ministry of Industry, Science and Technology and in part by the grant 01-02-16305 from the RFBR.

REFERENCES

1. Bethe, H. A., and Salpeter, E.E., *Quantum mechanics of One- and Two-electron Atoms*, Berlin, Heidelberg: Springer, 1958.
2. Jackson, J., *Classical Electrodynamics,* New York: Wiley, 1975.
3. Kogan, V.I., Kukushkin, A.B., Lisitsa, V.S., *Phys. Rep.* **213**, 1 (1992).
4. Bureyeva, L. A., and Lisitsa, V. S., *A Perturbed Atom,* Gordon & Breach, 2000.
5. Gream, H., Spectral line broadening by plasmas, New York: Academic Press,1974
6. Hanh, Y., *Rep. Prog. Phys.*, **60**, 691 (1997).
7. Born, M., *The mechanics of the Atom*, New York: Frederick Ungar Publishing CO, 1934.
8. Varshalovich, D.A., Moskalev, A.N., Khersonskii, V.K., *Quantum theory of angular momentum,*. Singapure-New Jersy –Hong Kong: Word Scientific, 1980.
9. Bureyeva, L.A., Kato, T., Lisitsa, V.S., Namba, C., J.Phys. B, **34**, 3909 (2001)
10. Herrick, D.R., *Phys. Rev. A*, , **12**, 1949 (1975).
11. Bureyeva, L.A., Lisitsa V.S., Shuvaev, D.A., *Sov. Phys.-JETP*, (in press), (2002).
12. Gulyaev, S. A., *Soviet Astron. J.*, **53**, 1010 (1976).
13. Bureyeva, L.A., Kato, T., Lisitsa, V.S., Namba, C., Phys. Rev. A, **65**, 032702 (2002)

Universal Scaling of Linewidth for Keilson—Storer Model

D.A. Shapiro

Institute of Automation and Electrometry,
Siberian Branch Russian Academy of Sciences, Novosibirsk 630090, Russia

Abstract. The spectral line shape is obtained for gas, where collisions are described by the Keilson — Storer model. The approximate scaling law is found: the line width as a function of transport collision frequency demonstrates almost universal behavior.

INTRODUCTION

Dicke effect [1], i.e., the narrowing of a spectral line due to velocity changing collisions without dephasing (or analogous effect of molecular Q-branch line mixing) is observed with confidence in Raman spectra on vibrational levels of hydrogen and other simple molecules. Measurement of line shape and width after comparison with theory allows one to find transport collision frequencies. Calculations for gas with large Doppler broadening were carried out within models of soft [2] or hard [3] collisions. For soft collisions the change of velocity vector after a collision is assumed to be small. For hard collisions the velocity change is considered as large; the Maxwellian distribution is restored after several collisions. More precise experiments [4] have found the width somewhere between the predictions of limiting models. General semiclassical analytical theory [5] requires the full set of eigenvalues and eigenvectors of collision operator, unknown even for rigid spheres. Therefore the numerical calculation is necessary with some realistic kernel approximation, see [6], or molecular dynamic simulation [7].

Although the Keilson — Storer (KS) model [8] is more complicated than soft and hard collisions, it does allow analytical solution in the form of continued fraction [9]. Then it is possible to study the qualitative features of line broadening or narrowing without long numerical calculations. KS is the simplest kernel that takes into account the correlation of velocity after and before the collision act. The single parameter p (velocity persistence) is equal to 0 for hard collision limit and 1 for soft one. The line shape and width can be calculated for $0 \leqslant p \leqslant 1$. The KS model allows tracing the smooth transition between the limiting cases. The alternative is Rautian — Sobelman model [10], which involves one parameter, "hardness" of collisions, and gives the interpolation between soft and hard collisions, too. The continued fraction was exploited for calculations of the light-induced drift [11], rotational relaxation of spherical molecules [12]. The approximate scaling law proposed for rigid spheres [13] is based on the assumption of fixed diffusion coefficient. It is tested in the present paper for KS model.

CP645, *Spectral Line Shapes: Volume 12, 16th ICSLS*, edited by C. A. Back
© 2002 American Institute of Physics 0-7354-0100-4/02/$19.00

DIFFERENCE KINETIC EQUATION

Profile of absorption line $I(\Omega)$ can be expressed in terms of the off-diagonal element of density matrix $\rho(\vec{v})$ averaged over velocities $\vec{v}$ of radiator

$$I(\Omega) = \frac{1}{\pi}\text{Re}\langle\rho\rangle, \quad \langle\rho\rangle = \int \rho(\vec{v})\,d\vec{v}. \tag{1}$$

The off-diagonal element obeys the kinetic equation [14]

$$(\Gamma + v - i\Omega + i\vec{k}\vec{v})\rho(\vec{v}) - \int A(\vec{v}|\vec{v}')\rho(\vec{v}')\,d\vec{v}' = W(\vec{v}). \tag{2}$$

Here ω and ω_{mn} are frequencies of the field and transition, $\Omega = \omega - \omega_{mn}$ is the detuning, Γ, v, and $A(\vec{v}|\vec{v}')$ are the relaxation constant, collision frequency and the kernel of collision integral, $\vec{k}$ is the wavevector of radiation. Dephasing is taken into account by constant Γ, which is assumed below to be equal to zero. The distribution over velocities $W(\vec{v})$ is equilibrium

$$W(\vec{v}) = \frac{1}{\left(\sqrt{\pi}v_T\right)^3}\exp\left(-\frac{v^2}{v_T^2}\right), \quad v_T = \sqrt{\frac{2T}{m}},$$

where T, m are the temperature in energy units and the mass of the radiator.
    The KS model of the kernel is

$$A(\vec{v}|\vec{v}') = v\frac{\exp\left[-\dfrac{(\vec{v}-p\vec{v}')^2}{(1-p^2)v_T^2}\right]}{\left(\sqrt{\pi(1-p^2)}v_T\right)^3}, \tag{3}$$

where $0 \leqslant p \leqslant 1$ is the velocity persistence. Parameter p means the correlation coefficient between the velocity $\vec{v}$ after a collision act and velocity $\vec{v}'$ before it. Limiting values $p \to 0$ or $p \to 1$ correspond to statistically independent or strongly coupled initial and final velocities of the active particle, i.e., significant or weak velocity changing. We consider elastic collisions, then the rate v is the same constant as in Eq.(2), while the dephasing can be included into the relaxation constant Γ.
    Simplifying Eq. (2) we integrate the distribution $\rho(\vec{v}) = W(v_x)W(v_y)\rho(v_z)$ ($v_z = \vec{k}\vec{v}/k$) over transverse velocities with respect to the wavevector $\vec{k}$, and go to one-dimensional kernel

$$(\Gamma + v - i\Omega + ikv_z)\rho(v_z) - \int\limits_{-\infty}^{\infty} A(v_z|v_z')\rho(v_z')\,dv_z' = W(v_z), \tag{4}$$

$$W(v_z) = \frac{e^{-v_z^2/v_T^2}}{\sqrt{\pi}v_T}.$$

Denote the dimensionless longitudinal velocity $\xi = v_z/v_T$, then the normalized eigenvectors and eigenvalues of KS kernel can be written as

$$\int_{-\infty}^{\infty} A(\xi|\xi')\psi_n(\xi')\,d\xi' = \lambda_n \psi_n(\xi), \tag{5}$$

$$\psi_n(\xi) = \frac{H_n(\xi)}{(\sqrt{\pi}2^n n!)^{1/2}}e^{-\xi^2}, \quad \lambda_n = v(1-p^n),$$

where H_n are the Hermitian polynomials:

$$H_n(\xi) = (-1)^n e^{\xi^2}\frac{d^n}{d\xi^n}e^{-\xi^2},$$

see. e.g., [12, 15]. Kinetic equation (2) involves term $i\vec{k}\vec{v}$, therefore (5) are its asymptotic eigenvectors at $k \to 0$.

Expanding distribution $\rho(\xi)$ over the basis we write

$$\rho(\xi) = e^{-\xi^2}\sum_{n=0}^{\infty}\left(\frac{iv}{kv_T}\right)^n c_n H_n(\xi), \tag{6}$$

where c_n are the coefficients of decomposition. Then substitute (6) into Eq. (4), multiply by $H_m(\xi)$, and integrate over ξ from $-\infty$ to ∞. Using the orthogonality, recurrence relations [16]

$$\int_{-\infty}^{\infty} e^{-\xi^2}H_m(\xi)H_n(\xi)\,d\xi = \sqrt{\pi}2^n n!\,\delta_{mn}, \quad 2\xi H_n = H_{n+1} + 2nH_{n-1},$$

and integral

$$\int_{-\infty}^{\infty} e^{-ax^2}H_n(x+b)\,dx = \sqrt{\frac{\pi}{a}}\left(a-\frac{1}{a}\right)^{n/2}H_n\left(b\sqrt{\frac{a}{a-1}}\right), \quad a > 1,$$

we obtain the infinite system of algebraic equations with three-diagonal matrix

$$r_0 c_0 - c_1 = \frac{1}{q\sqrt{\pi}}, \tag{7}$$

$$a c_{m-1} + r_m c_m - (m+1)c_{m+1} = 0, \quad m = 1, 2, \ldots,$$

$$r_m = g + 1 - p^m, \quad g = \frac{\Gamma - i\Omega}{v}, \quad q = \frac{v}{kv_T}, \quad a = \frac{1}{2q^2}.$$

To reduce the only inhomogeneous equation (7) to the same form as other equations we introduce the auxiliary coefficient $c_{-1} = 0$. Then the system takes the form

$$a c_{m-1} + r_m c_m - (m+1)c_{m+1} = \frac{\delta_{m0}}{q\sqrt{\pi}}, \quad m = 0, 1, 2, \ldots. \tag{8}$$

The second boundary condition $c_m \to 0, m \to \infty$ is necessary for convergence of expansion (6). We obtain the second-order difference equation that may be considered as other representation of integral kinetic equation.

388

SOLUTION OF DIFFERENCE EQUATION

To solve the difference kinetic equation let us divide each equation of the chain (8) at $m = 1, 2, \ldots$ by c_m:

$$a\frac{c_0}{c_1} + r_1 - 2\frac{c_2}{c_1} = 0, \quad a\frac{c_1}{c_2} + r_2 - 3\frac{c_3}{c_2} = 0, \quad a\frac{c_2}{c_3} + r_1 - 4\frac{c_4}{c_3} = 0, \quad \ldots, \tag{9}$$

hence

$$\frac{c_1}{c_0} = \frac{a}{-r_1 + 2\frac{c_2}{c_1}}, \quad \frac{c_2}{c_1} = \frac{a}{-r_2 + 3\frac{c_3}{c_2}}, \quad \frac{c_3}{c_2} = \frac{a}{-r_3 + 4\frac{c_4}{c_3}}, \quad \ldots \; .$$

We can find c_0 from Eq. (7):

$$J \equiv q\sqrt{\pi}\,c_0 = \frac{1}{r_0 - \frac{c_1}{c_0}} = \frac{1}{r_0 + \frac{a}{r_1 - 2\frac{c_2}{c_1}}} = \frac{1}{r_0 + \frac{a}{r_1 + \frac{2a}{r_2 - 3\frac{c_3}{c_2}}}} = \ldots \; .$$

J tends to the infinite continued fraction

$$J = \cfrac{1}{r_0 + \cfrac{a}{r_1 + \cfrac{2a}{r_2 + \cfrac{3a}{r_3 + \ldots}}}}. \tag{10}$$

The other approach is to denote the ratio of neighbor components as Q_n and get the recurrence relation or first-order difference equation

$$Q_n = \frac{c_{n-1}}{c_n}, \quad n = 1, 2, 3, \ldots, \quad Q_n = \frac{1}{a}\left(\frac{n+1}{Q_{n+1}} - r_n\right). \tag{11}$$

From Eq. (7) we find

$$J = \frac{1}{r_0 - \frac{1}{Q_1}}. \tag{12}$$

This form is more compact, but after iterations (11) it reduces to continued fraction (10) again.

In KS model the diffusion coefficient $D = v_T^2/2\nu_{tr}$ is defined by the transport frequency. To fix the transport frequency in the difference equation (8) we use new coefficients d_n: $c_n = (1-p)^n d_n$, i.e., rewrite the series (6) with powers of transport frequency. Only parameters α, β and p enter to the difference equation for d_n

$$\alpha d_{n-1} + \left(\beta + \sum_{m=0}^{n-1} p^m\right) d_n - (n+1)d_{n+1} = 0, \quad n = 1, 2, 3, \ldots;$$

$$\alpha = \frac{a}{(1-p)^2} = \frac{k^2 v_T^2}{2v_{tr}^2}, \quad \beta = \frac{g}{1-p} = \frac{\Gamma - i\Omega}{v_{tr}}.$$

389

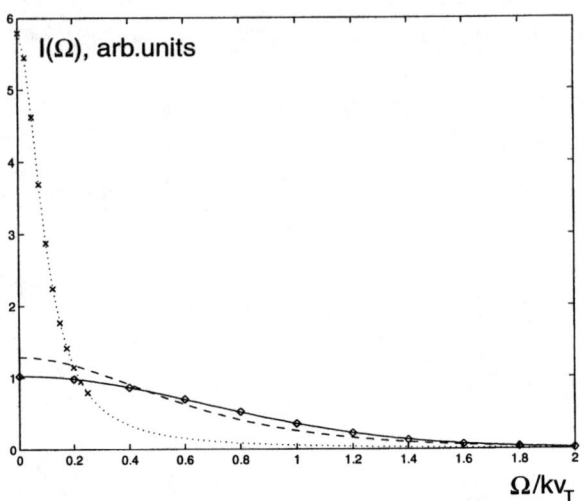

FIGURE 1. Spectral line profile at different parameter $q = v/kv_T = 0.1$ (solid line); 1 (dashed); 10 (dotted): fitting by the Doppler contour (circles), Lorentzian contour (crosses).

Find $c_1 = -1/q_0\sqrt{\pi}$, $q_0 = q(1-p) = v_{tr}/kv_T$, then $d_0 = c_0$ and the fraction can be obtained from the first equation of the chain:

$$I(\Omega) = \frac{1}{\pi v_{tr}} \mathrm{Re} \cfrac{1}{\beta + \cfrac{\alpha}{\beta + 1 + \cfrac{2\alpha}{\beta + 1 + p + \cfrac{3\alpha}{\beta + 1 + p + p^2 + \dots}}}}, \tag{13}$$

The absorption depends on the pressure only via the transport frequency.

LINE SHAPE AND WIDTH

Numerical calculations are carried out by recurrence relations (11). Fig.1 shows typical line profiles at $p = 0.5$. Contours are fitted by Doppler at $q = 0.1$ or Lorentzian at $q = 10$. The transition occurs from Doppler contour at $q \ll 1$ to Lorentzian one at $q \gg 1$ due to Dicke effect. The agreement is satisfactory for both limits.

Half width $\Omega_{1/2}$ at half maximum was determined numerically. In general case, the HWHM normalized by the Doppler width is dimensionless function of two dimensionless parameters p, q.

$$\frac{\Omega_{1/2}}{kv_T} = G(q, p). \tag{14}$$

The width $\Omega_{1/2}/kv_T$ is shown in Fig.2 as a function of normalized transport frequency $q_0 = q(1-p)$. In these variables all the curves run closely and the width is nearly

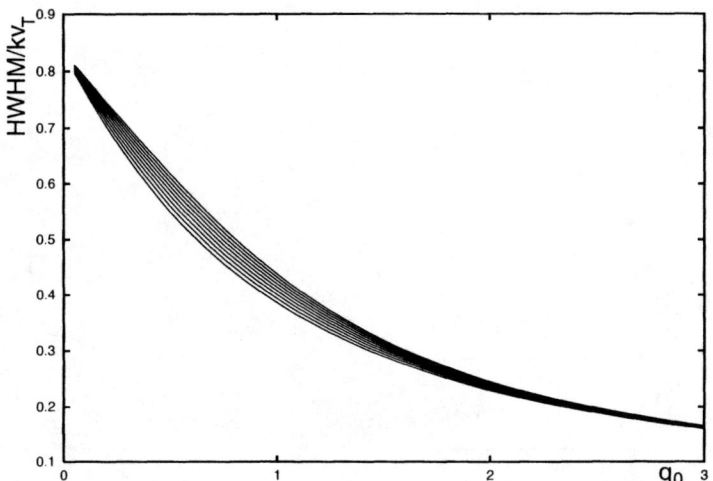

FIGURE 2. Half width at half maximum in kv_T units as a function of normalized transport frequency at $\Gamma = 0$. Curves from the bottom upwards correspond to persistence $p = 0.1 \div 0.9$ (step 0.1).

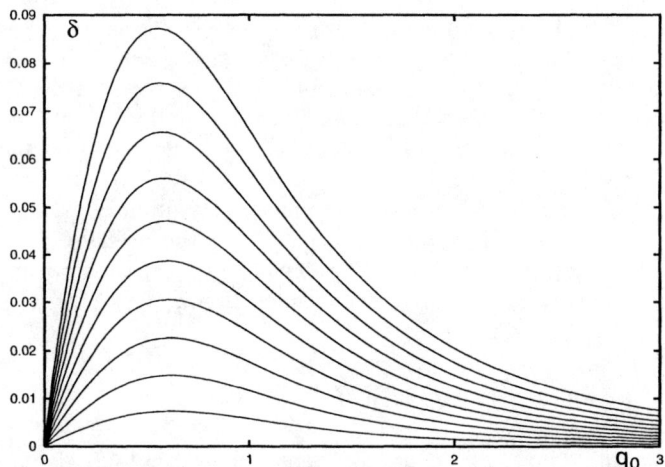

FIGURE 3. Deviation $\delta = \left(\Omega_{1/2}(0.9999) - \Omega_{1/2}(p) \right) / kv_T$ of KS model from soft collisions as a function of parameter transport frequency at $\Gamma = 0$. Solid curves from the top down correspond to the persistence $p = 0 \div 0.9$ (step 0.1). Hard collisions are displayed by the topmost curve, soft collisions are shown by the q_0-axis.

independent of the persistence. Function G becomes approximately a function of one variable: $G(q, p) \approx G(q(1 - p))$. An analogous approximate scaling law was observed in numerical calculation within rigid spheres approximation with one dimensional kernel [13]. It was based on the assumption of fixed diffusion coefficient. Note that the similar scaling was observed in the kangaroo model for different mass ratio and interaction potentials [17].

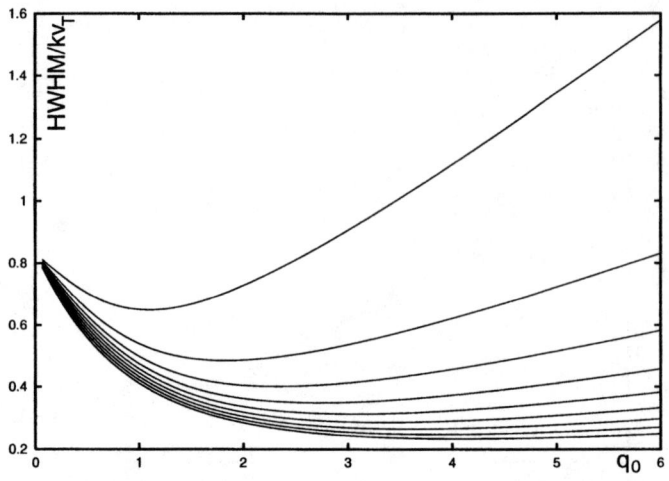

FIGURE 4. The same as in Fig. 2, but at $\gamma = 0.025$.

Curves in Fig.2 are almost confluent. To resolve them better and determine the degree of non-universality let us plot the difference between the neighbor curves. It is more convenient to compare each curve not with neighbor, but with soft collisions, choosing, eg $p = 0.9999$. Fig.3 shows the relative deviation in the line half width at $\Gamma = 0$. It is plotted as functions of normalized transport frequency $q_0 = v_{tr}/kv_T$. The maximal difference is about 0.09 at $q_0 \simeq 0.5$.

Inelastic collisions can be taken into account by relaxation constant Γ that is to be proportional to the pressure: $\Gamma/kv_T = \gamma q$. Fig.4 shows the pressure dependence of the line width, like in Fig.2, but at $\gamma \neq 0$. The dephasing breaks the scaling. Function in Eq.(14) depends on not only on transport frequency $G(p,q,\gamma) = G_1(q_0, \gamma)$.

LORENTZ LIMIT

The KS model permits one to compare different formulas in hydrodynamic limit $v_{tr} \gg kv_T$. As Fig.2 shows, all the curves tend to the universal asymptotics at $q_0 \gg 1$. It is enough to keep only two terms of fraction (13) and get the "universal formula" [18]

$$I(\Omega) = \frac{c_0}{\sqrt{\pi}kv_T} = \frac{1}{\pi}\text{Re}\,\cfrac{1}{\Gamma - i\Omega + \cfrac{k^2v_T^2}{2(\Gamma + v_{tr} - i\Omega)}}. \tag{15}$$

Other approach [5] requires the set of eigenfunctions φ_n and eigenvalues $\lambda_n, n = 0, 1, 2, \ldots$, of the integral operator $\hat{K}$:

$$\hat{K}\varphi_n = v\varphi_n - \int \tilde{A}(v_z|v_z')\varphi_n(v_z')\,dv_z' \quad \tilde{A}(v_z|v_z') = A(v_z|v_z')\sqrt{\frac{W(v_z')}{W(v_z)}}.$$

For KS the set (5) is known. The correction to the Lorentzian width is given by the second order of the perturbation theory

$$\delta\Gamma = k^2 \sum_{n\neq 0} \frac{\langle 0|v_z|n\rangle\langle n|v_z|0\rangle}{\lambda_n}. \tag{16}$$

The only nonzero is the "dipole" matrix element

$$\langle 0|v_z|1\rangle = \int_{-\infty}^{\infty} \varphi_0(v_z)v_z\varphi_1(v_z)\,dv_z,$$

then we retain one term in sum (16) and obtain $\delta\Gamma = (kv_T)^2/2v_{\text{tr}}$ in agreement with (15).

ACKNOWLEDGEMENTS

Author is grateful to A.D. May, A.M. Shalagin, and E.V. Podivilov for fruitful discussions. The work was supported by Russian Ministry of Industry, Science and Technology (program "Physics of quantum and nonlinear processes") and Russian Foundation for Basic Research (grant # 00-02-17973).

REFERENCES

1. Dicke, R. H., *Phys. Rev.*, **89**, 472–473 (1953).
2. Galatry, L., *Phys. Rev.*, **122**, 1218 (1961).
3. Nelkin, M., and Ghatak, A., *Phys. Rev. A*, **135**, A4–A9 (1964).
4. Forsman, J. W., Sinclair, P. M., May, A. D., Duggan, P., and Drummond, J. R., *J. Chem. Phys.*, **97**, 5355–5362 (1992).
5. Alekseev, V. A., and Malyugin, A. V., *Zh. Eks. Teor. Fiz*, **80**, 897–915 (1981), [Sov. Phys. JETP, **53** (3) 456 (1981)].
6. Ciurylo, R., Bielski, A., Drummond, J. R., Lisak, D., May, A. D., Pine, A. S., Shapiro, D. A., Szudy, J., and Trawinski, R. S., "High-resolution studies on the influence of velocity-changing collisions on atomic and molecular line shapes," in *16th International Conference on Spectral Line Shapes*, Berkeley, CA, 2002.
7. Hoang, P. N. M., Joubert, P., and Robert, D., *Phys. Rev. A*, **65**, 012507 (2001).
8. Keilson, J., and Storer, J. E., *Quart. J. Appl. Math.*, **10**, 243–253 (1952).
9. Shapiro, D. A., *Opt. Comm.*, **206**, 347–354 (2002).
10. Rautian, S. G., and Sobelman, I. I., *Usp. Fiz. Nauk*, **90**, 701–715 (1966), [Sov. Phys. Usp. **9** (5) 701–715 (1967)].
11. Kryszewski, S., and Nienhuis, G., *J. Phys. B: At. Mol. Opt. Phys.*, **20**, 3027–3045 (1987).
12. Temkin, S. I., Suvernev, A. A., and Burshtein, A. I., *Opt. Spektr.*, **66**, 69–76 (1989), [Opt. Spectr. **66** (1) 39–43 (1989)].
13. Shapiro, D. A., and May, A. D., *Phys. Rev. A*, **63**, 012701 (2001).
14. Rautian, S. G., and Shalagin, A. M., *Kinetic Problems of Non-linear Spectroscopy*, North-Holland, Amsterdam, Oxford, 1991.
15. Snider, R. F., *Phys Rev. A*, **33**, 178–181 (1986).
16. Abramowitz, M., and Stegun, A., editors, *Handbook of mathematical functions*, NBS, Washington, 1964.
17. Parkhomenko, A. I., and Shalagin, A. M., *Zh. Eksp. i Teor. Fiz.*, **120**, 830–845 (2001), [JETP, **93** (4) 723-736 (2001)].
18. Rautian, S. G., *Opt. Spektr.*, **86**, 385–387 (1999), [Opt. Spectr. **86** (3) 334–336 (1999)].

A Study of Collision–Induced Absorption in Lennard–Jonesium by Molecular Dynamics Simulations

John Courtenay Lewis

*Department of Physics and Physical Oceanography,
Memorial University
St. John's NF
Canada A1B 3X7*

Abstract. We have carried out systematic studies on mixtures of spherically symmetric molecules with the masses of H_2 and Ar, interacting through pairwise additive Lennard–Jones potentials. The induced dipole moments were of the form of the Lennard–Jones force, but with variable range parameter. Simulations were carried out for densities ranging from 65 Amagat to 900 Amagat, with number density ratios of approximately 0.1/0.9, 0.5/0.5 and 0.9/0.1 H_2/Ar, and with the induced dipole moment range running from 90% to 130% of that of the $H_2 - $ Ar force. The results for low densities, and induced dipole moment range greater than the range of the $H_2 - $ Ar force, have been compared qualitatively with the theory of Lewis and van Kranendonk [1, 2]. For induced dipole moment ranges less than the range of the force a peculiar double–peaked binary collisional line shape is found, which interacts in a complicated way with intercollisional interference.

INTRODUCTION

The first molecular dynamics study of collision–induced absorption was carried out by Lewis and Tjon [3] on a two–dimensional Lorentz gas with exponential induced dipole moment. Birnbaum et al. [4, 5, 6] later carried out molecular dynamics simulations using realistic interactions, including irreducible three–body potentials and induced dipole moments. Recently Lewis [7] has continued the study of simple models with work on lattice gases. Simple model systems have not been widely used in line shape studies but they have an important place in statistical mechanics more generally, particularly in systems where many–body interactions play a significant role. Many–body effects are, however, important in collision–induced spectroscopy, in particular in collision–sequence interference [2] and in pressure narrowing, and it is likely that simple models will also be as illuminating in that context as they have been in other problems in statistical mechanics.

In the present work we have simulated fluids of two components, the "molecules" of which are spherically symmetric with the masses of H_2 and Ar. The intermolecular forces were pairwise additive and the pair potentials were of Lennard–Jones 12–6 form. The thermodynamic properties of "Lennard–Jonesium" are well known from Monte Carlo and molecular dynamics studies. Single–component Lennard–Jonesium is known to have a solid phase and to show critical behavior, though the simulations described

CP645, *Spectral Line Shapes: Volume 12, 16th ICSLS*, edited by C. A. Back
© 2002 American Institute of Physics 0-7354-0100-4/02/$19.00

here have not looked at either of those regions.

The dipole moments induced in $H_2 - Ar$ collisions were also taken to be of Lennard–Jones form, with range parameter σ_μ which was varied systematically. The present work centers on the vector collision–sequence interference effect ("intercollisional interference"), and in particular on its dependence on σ_μ. To the extent that realistic induced dipole moments can be represented in Lennard–Jones form, σ_μ would lie between 1.10σ and 1.15σ. Molecular dynamics thus affords a means of exploring certain many–body effects such as collision–sequence interference for ranges of induced dipole moment which are difficult of access experimentally.

THE INTERACTIONS

The interactions, as stated above, were taken to be pairwise additive sums of central pair forces and induced dipole moments. The pair potentials were Lennard–Jones 12–6 potentials with smooth cutoff at $R = R_c$, where R is the center–of–mass separation of the two molecules:

$$U_{\text{mod}}(R) = 4\varepsilon \left(\frac{\sigma^{12}}{R^{12}} - \frac{\sigma^6}{R^6} \right) + a + bR + cR^2 \text{ for } 0 < R \leq R_c$$

$$U_{\text{mod}}(R) = U'_{\text{mod}}(R) = U''_{\text{mod}}(R) = 0 \text{ for } R \geq R_c$$

The component of the pair force along the line of centers was, therefore,

$$F_{\text{mag}} = -U'_{\text{mod}}(R) = \begin{cases} 48\varepsilon \left(\frac{\sigma^{12}}{R^{13}} - \frac{\sigma^6}{R^7} \right) + b + 2cR & \text{for } 0 < R \leq R_c \\ 0 & \text{for } R \geq R_c \end{cases}$$

and the induced dipole moment was

$$\mu_{\text{mag}} = \begin{cases} 48\varepsilon \left(\frac{\sigma_\mu^{12}}{R^{13}} - \frac{\sigma_\mu^6}{R^7} \right) + b_\mu + 2c_\mu R & \text{for } 0 < R \leq R_c \\ 0 & \text{for } R \geq R_c \end{cases}$$

The present work describes Lennard–Jonesium mixtures which are rough models for the $H_2 - Ar$ system. Three compositions were used: dilute (5%–10%) in H_2, for which the persistence of velocity is small; equal concentration (50%–50%) in H_2 and Ar; and dilute (5%–10%) in Ar, for which the persistence of velocity is close to maximal.

The Lennard–Jones parameters used are shown in Table 1.

In the present work, collision–induced spectra were obtained for induced dipole moment ranges σ_μ varying from 0.90σ to 1.5σ. For $\sigma_\mu = \sigma$ the induced dipole moment is exactly proportional to the intermolecular force; in the present work the parameter ε in the induced dipole moment was chosen equal to that of the corresponding intermolecular force; hence when $\sigma_\mu = \sigma$ for $H_2 - Ar$ the induced dipole moment is exactly equal to the $H_2 - Ar$ part of the intermolecular force; and the collision–induced spectrum is, therefore, identical to the power spectrum of that part of the intermolecular force.

Three temperatures were investigated: 225 K, 300 K, and 400 K. However, all spectra discussed in this paper are for systems at 300 K.

TABLE 1. Values of the Lennard–Jones parameters used in the simulations

	ε K	σ nm
$H_2 - H_2$	33.3	0.2968
$H_2 - Ar$	63.7	0.3199
$Ar - Ar$	121.85	0.3429
$\mu, H_2 - Ar$	63.7	$0.3199 \left(\sigma_\mu / \sigma \right)$

SIMULATION METHOD AND SPECTRAL ANALYSIS

The simulations were carried out in the microcanonical (fixed energy) ensemble, using the Verlet method with a step size of 1 fs, the total induced dipole moment being calculated every 5 fs. To obtain good statistics for the interference dips many thousands of collisions must be followed. At 100 amagat, with a total of 200 particles, each run was of duration 0.7 ns and each spectrum was usually accumulated over 40 such runs. For 900 amagat 600 particles were used with 6 or 8 runs each of about 0.2 ns duration.

A significant difficulty in obtaining accurate estimates of the spectral intensities in the interference dips arises from the fact that periodogram estimates of spectra do not converge in probability as the length of the time series increases, but instead show fluctuations of approximately constant magnitude over smaller and smaller frequency ranges. Two usual prescriptions to obtain a smooth spectral estimate are to smooth the periodogram, or to break long runs up into several shorter ones [8] and average the spectra for each short run. However, even at 900 amagat the dips in spectra of mixtures dilute in Ar are narrow when $\sigma_\mu \simeq \sigma$, and will be partially smoothed away by any conventional smoothing technique. Breaking a long run into several shorter ones is also inapplicable in the analysis of $H_2 - Ar$ simulation spectra because, as stated above, long runs are necessary for good statistics when the persistence of velocity is nearly maximal, as is the case for dilute–Ar spectra at all densities considered. A third prescription [9] is to transform the spectrum to its autocorrelation function, chop the autocorrelation function or force it smoothly to zero for lags greater than about 15% of the total signal length, and transform back to the spectrum. However we found no consistent procedure for doing this which did not partially fill in some interference dips.

The prescription that has been followed in this work has been to average spectra from a fairly large number of long runs, or from a smaller number of long runs with larger number of particles: in essence, a brute force approach.

It was shown in ref.[7] that the use of the Hanning window improves estimation of the spectral density in an interference dip, and the fitting procedure described below uses "Hanned" periodograms. However, for display purposes, in Figs. 2 to 5 of this paper, Savitsky–Golay (32, 32, 4) smoothing [8] of the "Hanned" periodograms has been used.

In a few cases the integrated spectral intensities were checked against the standard Monte Carlo method.

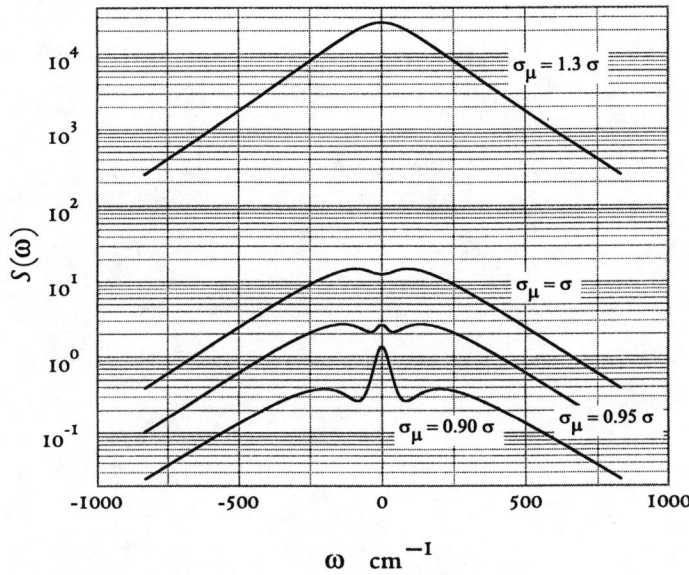

$S(\omega)$

$\sigma_\mu = 1.3\,\sigma$

$\sigma_\mu = \sigma$

$\sigma_\mu = 0.90\,\sigma$

$\sigma_\mu = 0.95\,\sigma$

ω cm^{-1}

FIGURE 1. Binary collisional line shapes for four different values of σ_μ

SIMULATION RESULTS

Fig. 1 shows spectra which would result from isolated binary or bimolecular collisions. The central peak with dip is found for $\sigma_\mu \lesssim 0.96\sigma$ and is due to collisions with small impact parameter and low relative velocity. Hence it becomes weaker as the temperature is increased.

"Normalized" spectra have been divided by the number densities of H_2 and Ar. Fig. 2 shows normalized spectra at 65 amagat for the three different mixtures for $\sigma_\mu = 0.90\sigma$; lines for the different mixtures are in fact not readily distinguishable in this plot. As expected the spectra are very similar, and closely resemble the binary collisional line shapes, with some diminution of the intensity of the central peak.

Fig. 3 shows spectra for mixtures dilute in H_2 at 900 amagat with $\sigma_\mu/\sigma = 1.3$, 1.1, 1.0, and 0.90. The very deep dip for $\sigma_\mu/\sigma = 1.0$ is noteworthy; in this case the induced dipole moment is exactly equal to the intermolecular force, and the dip will, for a run of infinite length, go exactly to zero. For $\sigma_\mu/\sigma = 0.90$ a residue of the central peak is still visible, even at this high density.

At 200 amagat, for given σ_μ/σ, the normalized spectra are nearly superimposable except at low frequencies. Fig. 4 shows the central parts of spectra for 200 amagat and $\sigma_\mu/\sigma = 1.1$ one of which is dilute in H_2 (broad dip) and the other which is dilute in Ar (narrow dip). The sharp difference in width is as predicted by the theory of Lewis and van Kranendonk [1, 2] and is due to the much greater persistence of velocity of Ar relative to H_2 in H_2 – Ar mixtures.

Fig. 5 shows spectra for mixtures with $\sigma_\mu/\sigma = 1.3$ which are dilute in H_2 and with densities increasing from 400 amagat to 900 amagat. It is apparent that the interference

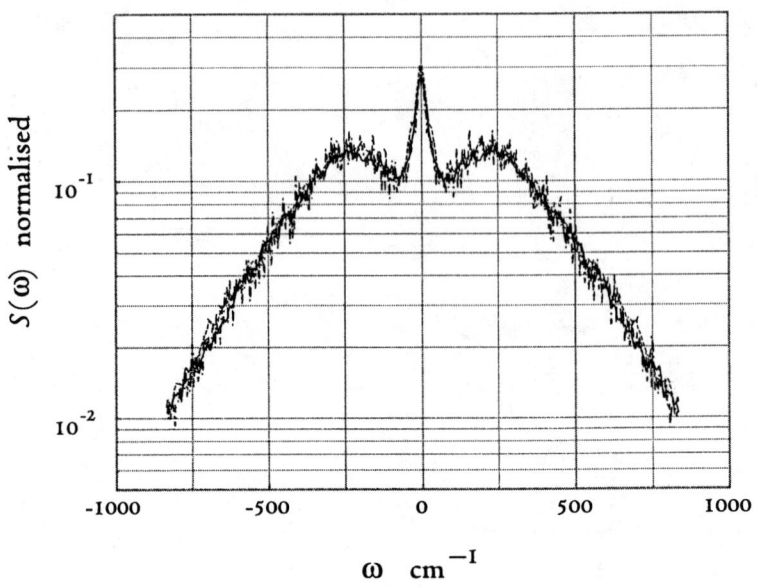

FIGURE 2. Simulation line shapes for the three different mixtures, each with total density 65 amagat and $\sigma_\mu/\sigma = 0.90$

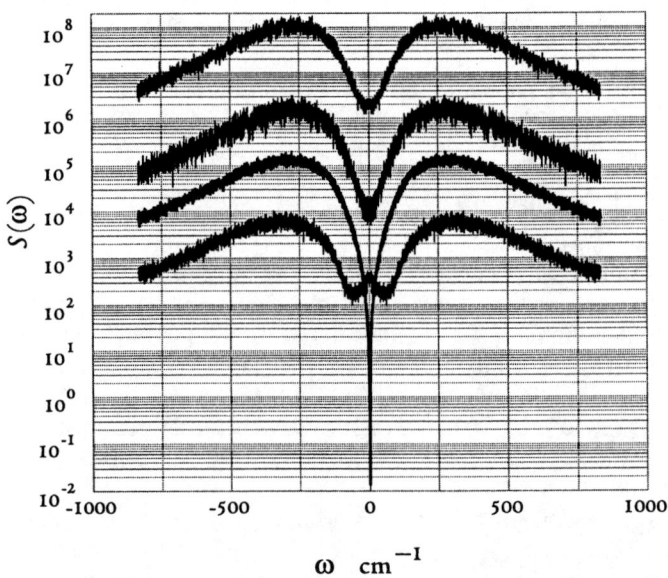

FIGURE 3. Simulation spectra: total density 900 amagat, 5% H_2, for (from top to bottom) $\sigma_\mu/\sigma = 1.3$, 1.1, 1.0, and 0.90.

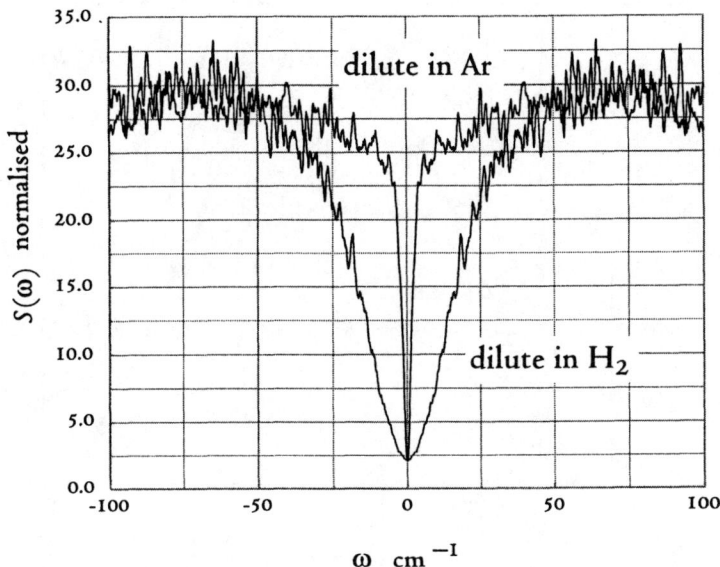

FIGURE 4. Spectra for 200 amagat, $\sigma_\mu/\sigma = 1.1$. Note that the ordinate in this figure is linear rather than logarithmic.

dip broadens with increasing total density, as is expected from the theory of Lewis and van Kranendonk and from experiment, but also deepens, which is in contradiction to prediction. The kinetic theory of Lewis and van Kranendonk predicts that the depth of the dip is independent of density for given σ_μ/σ. The deepening of the dip with increasing density is presumably a high–density effect which cannot be described by low–density kinetic theory.

FITTING THE SPECTRA

The low–density spectra, and most of the high–density ones, can be fit satisfactorily by a four–parameter spectral density of the form

$$S_{\text{fit}}(\omega) = p_1 S_{\text{binary}}(\omega) \left\{ 1 - \frac{p_2}{1 + |p_3 \omega|^{p_4}} \right\}$$

where p_1, p_2, p_3, p_4 are the fitting parameters, and S_{binary} is the spectrum for isolated binary collisions as shown in Fig. 1. Unweighted least mean squares fitting ignores sharp features such as the interference dips; hence the fit is weighted with the reciprocal of a spectral estimate. The "Hanned" periodograms themselves show so much fluctuation at higher frequencies that their reciprocals overweight the downward fluctuations in the spectra which are being fitted; however, the reciprocals of the smoothed "Hanned"

399

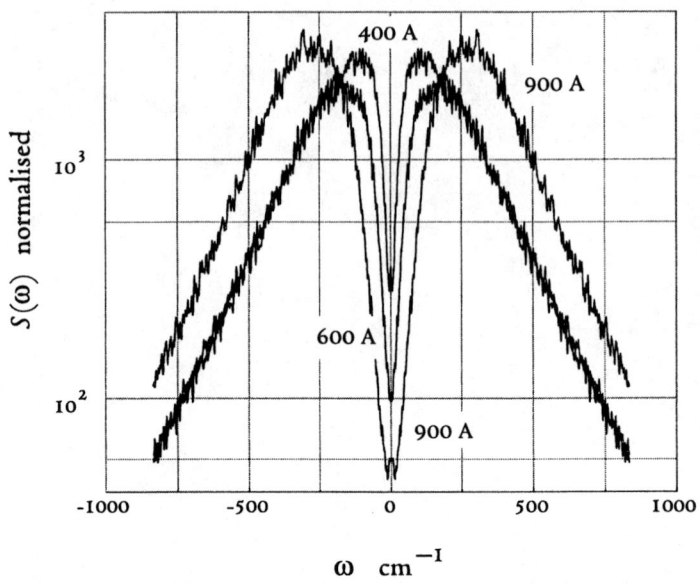

FIGURE 5. Simulation spectra: $\sigma_\mu/\sigma = 1.3$, dilute in H_2, for 400 amagat, 600 amagat, and 900 amagat.

periodograms shown in Figs. 2–5 have proved to be suitable. Results from these fits will be published elsewhere.

In the present work only qualitative comparison with the theory of Lewis and van Kranendonk has been undertaken. However, we note that as part of this work we have simplified the expressions for the interference dip which form part of the detailed low–density kinetic theory and which are given in refs. [1] and [2] so that they can readily be calculated from binary collision dynamics. These calculations will also be published elsewhere.

REFERENCES

1. Lewis, J. C., *Can. J. Phys.*, **50**, 2881–2901 (1972).
2. Lewis, J. C., "Intercollisional Interference – Theory and Experiment," in *Phenomena Induced by Intermolecular Interactions*, edited by G. Birnbaum, Plenum Press, New York, 1985, pp. 215–257.
3. Lewis, J. C., and Tjon, J. A., *Physica A*, **91**, 161–201 (1978).
4. Birnbaum, G., and Mountain, R., *J. Chem. Phys.*, **81**, 2347–2351 (1984).
5. Mountain, R., and Birnbaum, G., *J. Chem. Soc, Faraday Trans. 2*, **83**, 1791–1799 (1987).
6. Guillot, B., Birnbaum, G., and Mountain, R., *J. Chem. Phys.*, **90**, 650–662 (1989).
7. Lewis, J. C., "Lattice Gas Models for Intercollisional Interference," in *Spectral Line Shapes Volume 11*, edited by J. Seidel, AIP Conference Proceedings 559, American Institute of Physics, 2001, pp. 394–396, proceedings of the 15th ICSLS.
8. Press, W. H., Teukolsky, S. A., Vetterling, W. T., and Flannery, B. P., *Numerical Recipes in Fortran*, Cambridge University Press, 1992, second edn.
9. Koopmans, L. H., *The Spectral Analysis of Time Series*, Academic Press, Inc., 1995.

On the Application of Cavity Ringdown Spectroscopy to Measurements of Line Shapes and Continuum Absorption

John G. Cormier[*], Joseph T. Hodges[*], and James R. Drummond[#]

[*] *Chemical Science and Technology Laboratory, National Institute of Standards and Technology,*
100 Bureau Drive, Gaithersburg, MD 20899, USA
[#] *Department of Physics, University of Toronto, 60 St. George Street, Toronto ON, M5S 1A7, Canada*

Abstract. Cavity ringdown spectroscopy (CRDS) is a highly sensitive spectroscopic technique that has been successfully applied to problems such as trace gas detection and the observation of weak spectra. Despite possessing several intrinsic advantages over other techniques, CRDS has not yet been widely used to study spectral line shapes. Therefore, we begin with an introduction to CRDS, followed by a discussion of practical considerations and quantitative data analysis for an important variation of CRDS: high-resolution CRDS. We then briefly discuss the features and objectives of two high-resolution CRDS experiments in our laboratory. The first experiment uses a continuous wave CO_2 laser operating in the 920 cm^{-1} to 1090 cm^{-1} region. The principal objective of this experiment is to make accurate measurements of water vapor continuum absorption. The second experiment uses an external cavity diode laser operating in the 10500 cm^{-1} to 10860 cm^{-1} region. This experiment utilizes the frequency comb of a length-stabilized ringdown cavity to provide a precise measure of frequency intervals in line shape measurements. High resolution spectra of a pressure-broadened water vapor transition are presented.

INTRODUCTION

Cavity ringdown spectroscopy (CRDS) is a relatively new technique in which the absorption coefficient of a gas sample is accurately obtained from measurements of the loss rate of optical energy stored inside a high-finesse, stable optical resonator known as the "ringdown cavity" [1-11]. In the past decade, there has been an explosion of growth in the number of CRDS-related journal articles. Our survey of these articles suggests that interest in CRDS is roughly evenly split between the following two applications: 1. Measurement of absorber amount (including trace gas detection), and 2. Observation of weak absorption spectra.

Despite possessing numerous advantages, there are few accounts of CRDS being used to measure spectral line shapes or continuum absorption [8, 10-12]. One of the most important (and oft-discussed) advantages of CRDS is the high sensitivity to weak

CP645, *Spectral Line Shapes: Volume 12, 16th ICSLS*, edited by C. A. Back
2002 American Institute of Physics 0-7354-0100-4

absorption that is possible in the small volume of a typical ringdown cavity. For example, in a CRDS experiment operating near 760 nm, a noise-equivalent absorption coefficient of 5×10^{-10} cm^{-1} Hz$^{-1/2}$ was measured using a 30 ml volume ringdown cavity [8]. Other important advantages include the inherent accuracy of direct absorption measurements, a relatively simple experimental design, a large dynamic range, and impressive signal-to-noise ratio.

CRDS possesses a further advantage that is especially relevant for line shape studies: the possibility of high spectral resolution (≤ 1 MHz), even with pulsed laser sources. This possibility, realizable in high-resolution CRDS [7, 8, 10, 11], exploits the extremely narrow instrumental line widths inherent to high-finesse optical cavities. High-resolution CRDS was also used to achieve the lowest reported relative uncertainties in decay time constants, 3×10^{-4}, corresponding to a signal-to-noise ratio > 3000:1 [8]. Thus, the main purpose of this paper is to encourage the development of high-resolution CRDS experiments for applications requiring high accuracy, sensitivity and spectral resolution. Line shape phenomena that should be interesting to study with high-resolution CRDS include Dicke narrowing, collision-time asymmetry, speed-dependent effects, line mixing, line shifting and far-wing absorption.

We begin with a discussion of practical considerations and quantitative analysis for high-resolution CRDS experiments. Rather than present a comprehensive treatment of the physics of ringdown cavities, we refer the reader to the more detailed discussions given in Refs [4, 5, 7, 11]. Then, we briefly describe the features and objectives of two high-resolution CRDS experiments in our laboratory. One experiment, dubbed MIR-CRDS, uses a continuous-wave (cw) CO_2 laser and operates in the mid-infrared region of 920 cm^{-1} to 1090 cm^{-1}. The other experiment, dubbed NIR-CRDS, uses a cw external cavity diode laser and operates in the near-infrared region of 10500 cm^{-1} to 10860 cm^{-1}.

PRACTICALITIES OF HIGH-RESOLUTION CRDS

In CRDS, laser energy is injected into the ringdown cavity via the transmission of the front mirror (Fig. 1). Following the abrupt termination of the laser excitation, the time evolution of the energy stored inside the ringdown cavity is measured with a fast detector through the transmission loss of the back mirror. The ringdown cavity is usually designed no differently from a spherical mirror Fabry-Pérot interferometer (useful reviews of Fabry-Pérot theory and design may be found in many books and articles, *e.g.* [13]). In general, the ringdown cavity length will have to be precisely controlled, and so a piezoelectric transducer (PZT) is usually incorporated into the design in order to translate one of the mirrors. The best mirrors for CRDS are high-reflectivity dielectric coatings on transparent super-polished substrates having negligible bulk absorption. Dielectric mirror reflectivities vary substantially depending on the spectral region. Mirror reflectivities as high as R = 0.99999 have been employed in a CRDS experiment near 12900 cm^{-1} [14]. However, CRDS has also been successfully implemented with mirror reflectivities as low as $R = 0.994$ in the 1000 cm^{-1} region [10].

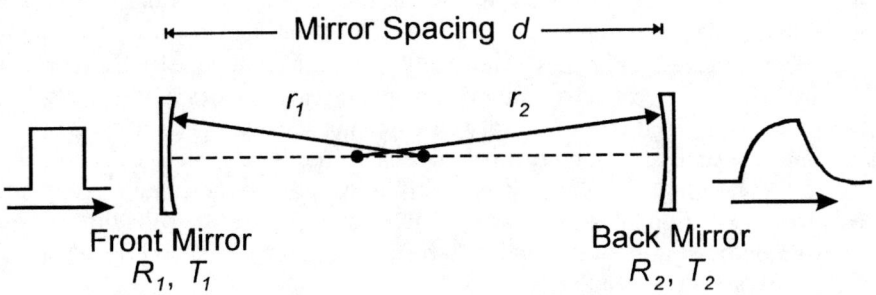

FIGURE 1. Conceptual schematic of a CRDS experiment. The ringdown cavity is formed by two concave, high-reflectivity mirrors spaced a distance d. The time evolution of the optical energy stored in the ringdown cavity is sensed with a fast detector via the transmission loss of the back mirror.

It is well known that self-reproducing stable electromagnetic field patterns, or "optical modes", exist inside optical cavities. The cavity eigenmodes are transverse electromagnetic (TEM) modes, which are well approximated by the product of Hermite and Gaussian functions when Cartesian coordinates are used [15, 16]. These modes are designated TEM$_{mnq}$ where m and n are positive integers identifying the transverse mode orders in the x- and y-directions, respectively, and q is the longitudinal mode order. The lowest order transverse mode of a cavity is the axisymmetric TEM$_{00q}$ mode, corresponding to a Gaussian spatial profile. The eigenfrequencies ν_{mnq} (Hz) for the case of a symmetric cavity are:

$$\nu_{mnq} = \frac{c}{2n_0 d}\left[q + \frac{2}{\pi}(m + n + 1)\arctan\sqrt{\frac{d}{2r - d}} + \frac{\theta}{\pi}\right] \qquad (1)$$

where n_0 is the real part of the refractive index of the cavity medium, r is the mirror radius of curvature, and d is the cavity length. The term θ represents the phase shift per reflection from the mirrors. The quantity $c/(2n_0 d)$ is an important parameter known as the free spectral range (FSR) of the cavity. The full width at half-maximum (FWHM) of the cavity transmission peaks is given by [17]:

$$\Delta\nu_{FWHM} = \frac{\Delta\nu_{FSR}}{finesse} = \frac{c\sum_i L_i}{2\pi n_0 d\sqrt{1 - \sum_i L_i}} \qquad (2)$$

where $\sum L_i$ is the sum of all per-pass losses in the cavity, including mirror losses, absorption of radiation by the gas, *etc.* For example, a 1-m long ringdown cavity with $\sum L_i = 10^{-3}$ will have $\Delta\nu_{FSR} = 150$ MHz and transmission peak widths $\Delta\nu_{FWHM} = 50$ kHz. Thus, the instrument functions of ringdown cavities are comb-like patterns of narrow transverse modes that are repeated over frequency intervals equal to one cavity FSR.

In conventional CRDS, several transverse and/or longitudinal cavity modes may be excited depending on the spatial and spectral overlap of excitation laser energy with cavity modal energies. However, this is unsuitable for high-resolution applications such as line shape measurements, because a multi-peaked instrument function will smear the recorded line shapes. The analysis of multi-mode CRDS signals is also problematic because we have found different optical modes possess significantly different decay rates that do not vary in a predictable way. The modal variation of decay time constants is thought to occur because the mirror surfaces sampled by the different transverse modes have different effective reflectivities, but this hypothesis is difficult to verify. Finally, multi-mode CRDS signals are often distorted by effects such as mode-beating between optical modes, which introduce additional uncertainty to the decay rate measurement [5].

High-resolution CRDS avoids these complications by probing only the TEM_{00q} mode of the ringdown cavity, through a combination of mode-matching and careful selection of cavity geometry (*i.e.* mirror curvature r and separation d). First, the cavity geometry is chosen in such a way that spacing between the lowest-order transverse modes of the cavity is much greater than the bandwidth of the probe laser. This will also prevent the excitation of multiple longitudinal cavity modes. For example, if the laser bandwidth is 60 MHz, then it is necessary to space the TEM_{01q} mode at least 60 MHz away from the TEM_{00q} mode. This could be achieved with a ringdown cavity of length $d = 50$ cm and $r = 80$ cm mirrors, as the mode separation would be 110 MHz. In fact, countless combinations of r and d would be suitable. However, exactly confocal cavities ($r = d$) should be avoided, because optical mode degeneracy greatly complicates the task of establishing single-mode excitation.

The laser beam is then mode-matched to the ringdown cavity. Mode matching is a standard technique employing lenses to modify the spot size and phase-front curvature of a propagating Gaussian beam to exactly overlap the TEM_{00q} mode volume of an optical cavity [18, 19]. An ideal Gaussian laser beam may be mode-matched to the TEM_{00q} mode of the ringdown cavity using a single mode-matching lens. However, mode-matching a real laser beam often requires a combination of lenses and/or spatial filter. Fortunately, it is not necessary to achieve perfect mode-matching in high-resolution CRDS. Rather, the goal is to reduce as much as possible the coupling of laser energy into higher order cavity modes, thereby ensuring that the TEM_{00q} mode can be unambiguously identified in the comb-like cavity spectrum. From the spectra of excited cavity modes in our experiments, we typically find that the TEM_{00q} mode has a relative coupling efficiency of $\geq 90\%$, the TEM_{01q} mode has a relative coupling efficiency of $\leq 10\%$, and higher order modes have negligible relative coupling efficiencies. The TEM_{00q} mode is therefore always easy to identify in cavity spectra.

The general principle we have developed to achieve high-resolution in our CRDS experiments involves locking the narrow TEM_{00q} mode of a high-finesse ringdown cavity to the output of a cw gas laser. The high intrinsic frequency stability and narrow bandwidth of cw gas lasers provide sufficient precision for high-resolution work. In one of the experiments described below, the MIR-CRDS experiment, the cw gas laser is a stable CO_2 laser (measured drift rate: ≤ 5 MHz/day) that also serves as the ringdown excitation laser. Although the MIR-CRDS experiment does not presently have the fine-tuning range needed to study near-center line shapes, the CO_2 laser is line-tunable over a broad range and so the experiment is very useful for measuring absorption spectra that vary slowly with frequency, such as the infrared water vapor continuum [10, 11]. Of course, gas lasers are unsuitable source lasers for most line shape investigations, and this problem is addressed in the second experiment described in this paper, the NIR-CRDS experiment. Here the ringdown cavity is actively length-stabilized to a cw frequency-stabilized HeNe laser (measured drift rate: ≤ 1 MHz/day) and a single-mode external cavity diode laser is used for ringdown excitation and probing of the absorption line shape. In a novel approach, this system uses the stabilized frequency comb of the ringdown cavity as a set of frequency markers for determining frequency intervals of single-mode CRDS spectra. The NIR-CRDS experiment has a fine-tuning scan range large enough to permit near-center line shape investigations. More details on both these experiments are provided following our discussion of data analysis.

QUANTITATIVE HIGH-RESOLUTION CRDS

In a high-resolution CRDS experiment with an unsaturated absorber obeying the Beer-Lambert law, one observes a simple exponential decay of intra-cavity intensity I:

$$I(V_{mnq}, t) = I_0(V_{mnq}) \exp(-t / \tau(V_{mnq})) + I_b(V_{mnq}) \tag{3}$$

where I_0 is the signal at time $t = 0$, I_b is the zero-signal background and τ is the decay time constant. Hereafter the mode index, mnq, associated with the cavity eigenfrequency is omitted for greater clarity. The decay time constant is inversely proportional to the sum of the losses per unit length within the cavity, and is given by:

$$\frac{1}{c\tau(V)} = \frac{1}{d}\sum L_i = \frac{1 - R(V)}{d} + k_{bg}(V) + k(V) \tag{4}$$

where $R(V)$ is the effective reflectivity of the mirror pair and $k_{bg}(V)$ encompasses all losses due to the presence of a buffer gas. Here $k(V) \equiv N\sigma(V)$ is the absorption coefficient of the absorbing gas, where N is the absorber number density and $\sigma(V)$ is the absorption cross section.

When the absorber gas is removed from the ringdown cell, Eq. (4) reduces to:

$$\frac{1}{c\tau_0(V)} = \frac{1 - R(V)}{d} + k_{bg}(V) \tag{5}$$

where τ_0 is the absorptionless decay time constant. Then, Eq. (4) and Eq. (5) may be rearranged to solve for the absorption coefficient:

$$k(V) = \frac{\tau_0(V) - \tau(V)}{c\tau_0(V)\tau(V)} \tag{6}$$

Thus, the absorption coefficient is directly obtained from measurements of the decay time constant alone, and in particular does not require separate measurements of both the ringdown cavity length and the mirror reflectivity.

We conclude this section by sharing some of our experience on CRDS data analysis. We have generally found that the best results are obtained when individual decay events are analyzed separately. We do not recommend the common practice of averaging (or "stacking") decay events before solving for the decay time constant. Signal averaging leads to an underreporting of decay time constant uncertainty, and may also mask important diagnostic clues in individual decay events pointing to problems such as multi-mode excitation and nonlinear detection system response. Such clues may include structured residuals due to mode beating, and correlation of decay time constant with peak signal intensity. We also do not recommend taking the logarithm of the background-corrected data (in order to do a linear fit) where the zero-signal background is estimated from pre- (or post-) trigger quiescent signals. This extrapolation of background intensity may significantly bias the fitted decay time constant [11]. Instead, we recommend using the Levenberg-Marquardt least-squares minimization algorithm to fit a fixed portion of each decay curve to Eq. (3). Thus, three free fit parameters are associated with each decay event: the initial intensity, I_0, the decay time constant, τ, and the constant zero-signal background I_b.

MIR-CRDS EXPERIMENT

The layout for the mid-infrared MIR-CRDS experiment is illustrated in Fig. 2. The laser source is a stable cw CO_2 laser that is line-tunable from 920 cm^{-1} to 1090 cm^{-1}.

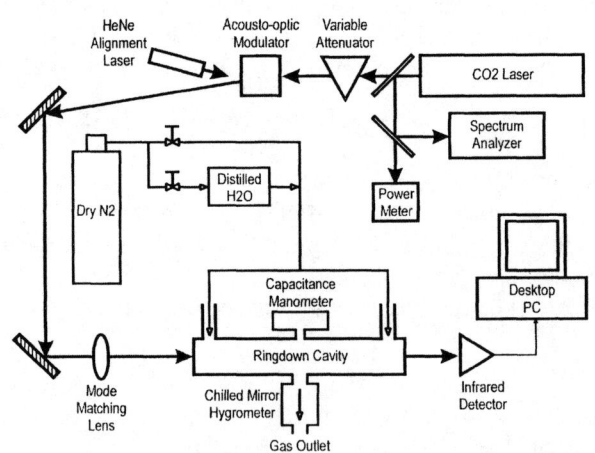

FIGURE 2. Schematic layout for the MIR-CRDS experiment. The main optical beam moves anticlockwise through the apparatus from the CO_2 laser at top right, to the detector at bottom right [10].

At $\tilde{\nu} = 944.195$ cm^{-1}, the 10P(20) transition of CO_2, the laser has a bandwidth

< 120 kHz (1-second measurement) and an output power of 6 W. The laser beam is attenuated with stacked Brewster plates so that only 1 mW of power is incident on the ringdown cavity. Optical pulses (5 μs long) are shaped from the first-order deflected beam of an acousto-optic modulator (AOM) at a repetition rate of 2.5 kHz. The pulses are mode-matched to the ringdown cavity using a single lens (focal length = 75 cm).

The ringdown cavity is formed from two $r = 100$ cm mirrors that are kinematically mounted onto a spacer assembly ($d = 102.5$ cm) designed around three zero-expansion ceramic glass rods (linear expansion coefficient $\alpha = 0.00 \pm 0.02 \times 10^{-6}$ $°C^{-1}$ over the temperature range 0 °C to 50 °C). The 2.54 cm diameter mirrors are made from dielectric coatings layered on ZnSe substrates, and have reflectivity $R = 0.994$ over the spectral range of the CO_2 laser. Horizontal and vertical tilt adjustments of the mirror mounts are made with precision alignment screws. A PZT mounted at one end of the ringdown assembly permits the apparatus to be fine-tuned over 170 MHz, which is slightly greater than the cavity FSR of 146 MHz.

The entire ringdown cavity assembly rests on a three-point kinematic mount inside an electropolished stainless steel gas cell. Infrared radiation enters and exits the gas cell via anti-reflection coated ZnSe windows on the end flanges. Infrared energy exiting the gas cell is sensed with a liquid-nitrogen cooled HgCdTe photovoltaic detector (50 MHz electrical bandwidth). The signal from the dc-coupled output of the detector is digitized by a 12-bit flash converter at a sampling rate of 10^7 s^{-1} and stored on a lab computer for later analysis.

The main objective of the MIR-CRDS experiment is to make accurate measurements of an absorption phenomenon known as the water vapor continuum. The water vapor continuum spans the infrared spectrum and plays a significant role in atmospheric radiative transfer. Despite its importance, we have found large uncertainties in water vapor continuum absorption coefficients, which we discuss separately in an upcoming article [20]. Such uncertainties are certainly forgivable, given the formidable experimental and theoretical challenges that the water vapor continuum has posed since its existence was predicted in 1938 [21]. However, present-day concerns about the possibility of global climate change are adding a sense of urgency to our effort to accurately characterize water vapor continuum absorption so that we may better understand the role it plays in atmospheric radiative transfer.

The chief experimental difficulty arises from the combination of small continuum absorption cross sections and the low saturation vapor pressure of water vapor, resulting in absorption that is too weak for reliable detection in the generally restricted path lengths of the laboratory. In order to measure continuum absorption, investigators have employed a variety of signal-enhancing techniques, most notably large multi-pass cells (*i.e.* White cells) and photoacoustic spectroscopy. However, with uncertainties in water vapor continuum absorption coefficients ranging from ± 20 % for the self-broadened continuum to ± 70 % for the foreign-broadened continuum [11, 20], improved experimental techniques are clearly necessary.

Because it is not possible to detune the experiment from the absorption feature in the case of continuum absorption, we have developed a novel gas flow methodology. Nitrogen gas continuously flows through the ringdown cavity and is exhausted into the laboratory. By diverting a portion of the inlet gas stream over a body of distilled water prior to entering the ringdown cavity, we are able to generate humidities in the ringdown cavity ranging from zero to near-saturation. The gas flow methodology permits the most non-perturbative continuum measurement possible, in that the only change to the ringdown cell is the amount of water vapor in the gas flow. Another benefit of this approach is that it is possible to make water vapor absorption measurements for several different cell humidities in a short amount of time.

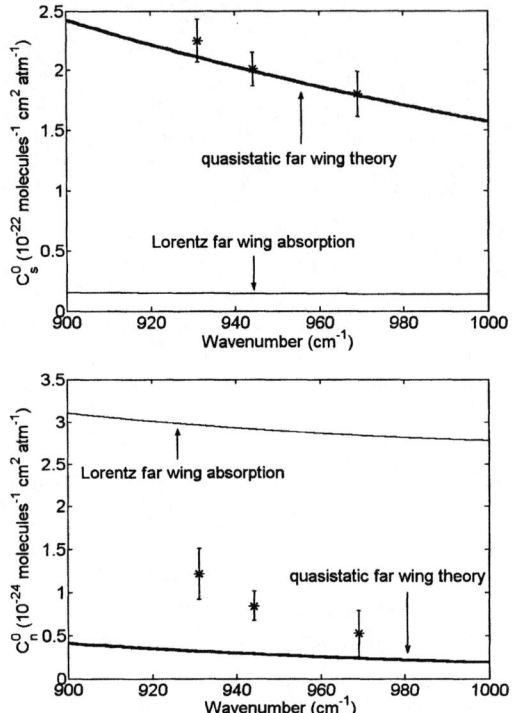

The MIR-CRDS measurements of continuum absorption coefficients in the 900 cm^{-1} to 1000 cm^{-1} region for $T = 296$ K are presented in Fig. 3. Presently, our experimental results have uncertainties of ± 7 % and ± 20 % in the self- and nitrogen-broadened continuum coefficients, respectively. The generally good agreement between our results and theoretical calculations of far wing absorption provides compelling evidence that the water vapor continuum results from the accumulated far wings of strong water vapor absorption lines [22-24]. This is important, as it has been widely misunderstood that the water vapor continuum results from water vapor dimers since this mechanism was first proposed in 1967 [25, 26]. Our measurements of the water vapor continuum are also significant in that they illustrate the degree to which high-resolution CRDS is free of the baseline drift problems that plague most other experimental techniques.

FIGURE 3. Water vapor continuum absorption coefficients in the 900 cm^{-1} to 1000 cm^{-1} region. The self-broadened coefficients are in the top panel and the nitrogen-broadened coefficients are in the bottom panel. Our data are shown as '*' with standard uncertainty error bars. Theoretical calculations of far wing absorption based on the quasistatic approximation are shown as a thick curve [22-24]. For comparison, we also show the far wing absorption calculated using the familiar Lorentz line shape (thin curve).

Ongoing work with the MIR-CRDS experiment is aimed at extending the capability of the apparatus to include studies of temperature dependence. Few low-temperature measurements of the water vapor continuum have ever been attempted due to the extreme sensitivity required; yet accurate continuum coefficients for temperatures ranging from 250 K to 300 K are necessary for atmospheric radiative transfer applications. In addition, we are also preparing to study the foreign-broadened continuum in greater detail, by measuring the effect of other atmospheric gases such as oxygen and argon. Finally, we also plan to extend the spectral coverage of our data.

NIR-CRDS EXPERIMENT

The layout for the external cavity diode laser NIR-CRDS experiment is illustrated in Fig. 4. The system comprises a symmetric ringdown cavity (r = 100 cm, nominal d = 75 cm) whose length is locked to a frequency stabilized HeNe laser at λ = 633 nm (frequency drift < 1 MHz day^{-1}). A cw external cavity diode laser (ECDL) emitting in the spectral range 920 nm to 940 nm is used as probe laser. The ECDL gives a single-

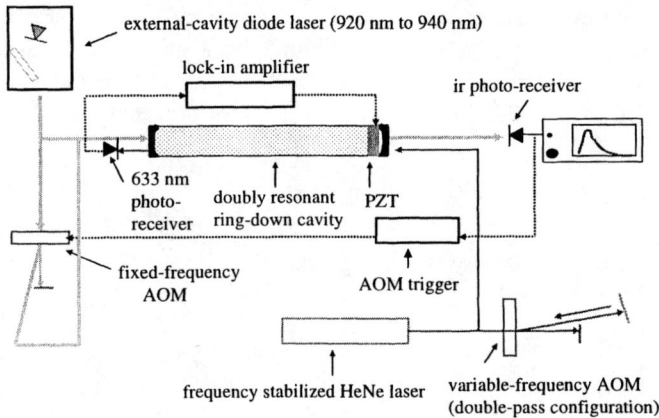

FIGURE 4. Schematic layout for the NIR-CRDS experiment. Numerous optical components have been omitted for clarity. Dashed lines represent electronic connections and solid lines represent visible and infrared laser beams.

mode output (1 s time-averaged linewidth of ≈ 1 MHz) with a mode-hop-free tuning range of 30 GHz. The ringdown mirrors have nominal reflectivities of 0.95 at 633 nm, and 0.99998 near the probe wavelength. A variable frequency AOM is used in a double-pass configuration, thus providing both a variable frequency shift (range of 250 MHz) and frequency modulation (up to 10 MHz) of the 633-nm light without inducing any change in the HeNe beam propagation vector. The instantaneous cavity length is interrogated by measuring the intensity of transmitted HeNe beam with a photodiode. Phase sensitive detection of this transmitted beam at the modulation frequency provides a discriminant signal, which is fed back to actuate the PZT and maintain resonance between the cavity mode frequency and the HeNe laser. This servoing of the cavity length gives time-averaged variations in the FSR ≤ 1 MHz.

At each frequency step, the probe laser frequency is tuned to excite the TEM_{00q} mode of the ring-down cavity. When this signal exceeds a preset value, the probe laser beam is switched off using a fixed frequency AOM. This initiates the passive decay of light from the cavity and gives rise to a single-mode ringdown signal. The process is repeated by increasing the probe laser frequency through one free spectral range of the ring-down cavity to TEM_{00q+1}. In this manner, the frequency steps are independent of nonlinearity, hysteresis and drift in the frequency tuning of the probe laser.

Note that the cavity free spectral range does not limit the minimum frequency step within a spectral scan. Since there is a one-to-one correspondence between the frequency shift of the AOM and the cavity mode spectrum at 633 nm, and since the cavity servo can track changes in the AOM frequency, then for a unit frequency shift in the AOM, the cavity mode frequency spectrum shifts by the amount $\lambda_{\text{HeNe}}/\lambda_p$ where λ_{HeNe} and λ_p are the wavelengths of the HeNe and probe beams, respectively. In this manner, frequency steps as small as 1 MHz can be realized with the current setup.

Three room-temperature spectra of an isolated transition of water vapor at $\tilde{v} = 10689.00 \text{ cm}^{-1}$ are given in Fig. 5. N_2 was the buffer gas for all cases. The HITRAN database reports a line strength of $1.41 \times 10^{-23} \text{ cm}^2 \text{ cm}^{-1} \text{ molecule}^{-1}$ and air-broadening parameter equal to $0.0844 \text{ cm}^{-1}/\text{atm}$ (HWHM) for this transition [27]. The

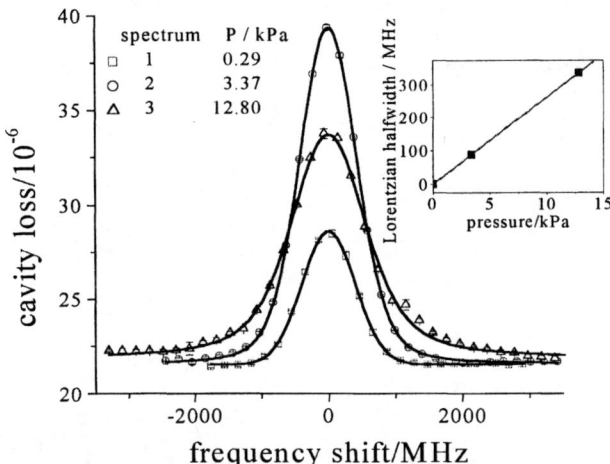

FIGURE 5. Measured single-mode CRDS spectra at 293 K for the 10689.00 cm^{-1} H$_2$O transition (symbols) and corresponding Voigt fits to the data (solid lines). Spectra 2 and 3 correspond to a fixed partial pressure of H$_2$O approximately equal to 2.7 Pa (0.02 Torr). The inset shows the Lorentzian halfwidths (squares) derived from the respective fitted data sets and a linear regression to these data (solid line).

first spectrum corresponds to a total pressure of 0.29 kPa (2.2 Torr), and is therefore near the Doppler limit. The expected linewidth based on a Voigt profile and the broadening parameter given above yield a FWHM of 934.4 MHz. The measured

FWHM, based on a fit to the data, gives 937.7 MHz and is within 0.4 % of the predicted value. Spectra 2 and 3 correspond to measurements at a fixed partial pressure of H_2O and total pressure of 3.37 kPa (25.3 Torr) and 12.8 kPa (96.2 Torr) respectively. Voigt fits to the data give best-fit peak areas agreeing to within 0.2 %. Finally, the Lorentzian half width at each pressure was determined from the respective Voigt fits. The results (squares) and a linear least-squares fit of the pressure-dependent data (solid line), are shown in the inset of Fig. 5 and yield an air-broadening parameter 5.2 % greater than the above-reported HITRAN value. While our early results are not yet definitive, they are presented here in order to illustrate the tremendous potential of high-resolution CRDS.

CONCLUSIONS

The main purpose of this paper is to discuss the recent development of high-resolution CRDS experiments and to encourage the application of this powerful technique to line shape investigations. High-resolution CRDS is distinguished from conventional CRDS by the requirements of a length-stabilized ringdown cavity and single-mode excitation. Although instrumental line widths as low as 1 kHz are readily achievable in high-finesse ringdown cavities, this property has rarely been exploited in order to achieve high resolution. One reason is that most CRDS measurements have been performed using non-length-stabilized ringdown cavities, and consequently over the time interval of a spectral scan the frequency comb of the resonator moves about in frequency space. Another reason is that single-mode excitation has not been achieved in most CRDS experiments, which has the effect of limiting the frequency resolution.

In both experiments described herein, length stabilization was achieved by locking the ringdown cavity to the output of a cw gas laser. The first experiment, dubbed MIR-CRDS, uses a cw CO_2 laser as the ringdown excitation laser, and operates in the 920 cm^{-1} to 1090 cm^{-1} region. In this case, the ringdown cavity is passively locked to the CO_2 laser, which is possible due to the impressive stability of both the ringdown cavity and the laser. The primary objective of this experiment is a full and accurate characterization of water vapor continuum absorption. The second experiment, dubbed NIR-CRDS, uses an external cavity diode laser, and operates in the 10500 cm^{-1} to 10860 cm^{-1} region. This experiment actively locks the ringdown cavity to the output of a frequency-stabilized HeNe laser, permitting the frequency comb of the ringdown cavity to be used as precise frequency markers. The NIR-CRDS experiment has a fine-tuning range of 30 GHz, large enough to permit near-center line shape measurements. For both experiments, we have presented early results, and we hope to report on further progress at a later date.

Finally, we note that high-resolution CRDS offers a further tantalizing possibility: high accuracy (*i.e.* sub-MHz uncertainties) determination of the frequency axis in absorption spectra. Although we do not presently make a high-precision measurement of the cavity FSR, one could adopt the method of Bay *et al.* to measure the cavity FSR [28]. Our preliminary analysis suggests that this technique could permit determination of the cavity FSR to within 10 kHz. It is also interesting to note that this method, which involves frequency modulation techniques, does not require a measurement of the cavity length in order to obtain the FSR of the ring-down cavity. Furthermore, the absolute measurement of laser frequency using a stabilized high-finesse interferometer has already been demonstrated [29], and we believe a similar methodology could be implemented in high-resolution CRDS. In conclusion, high-resolution CRDS experiments based on the general methodologies discussed here could be developed that permit the acquisition of ultra-low uncertainty spectroscopic data. These data could be applied to problems such as atmospheric radiative transfer, atmospheric remote sounding, and the testing of *ab initio* line shape theories.

REFERENCES

1. O'Keefe, A., and Deacon, D. A. G., *Rev. Sci. Instrum.* **59**, 2544-2551 (1988).
2. Romanini, D., and Lehmann, K. K., *J. Chem. Phys.* **99**, 6287-6301 (1993).
3. Zalicki, P., and Zare, R. N., *J. Chem. Phys.* **102**, 2708-2717 (1995).
4. Lehmann, K. K., and Romanini, D., *J. Chem. Phys.* **105**, 10263-10277 (1996).
5. Hodges, J. T., Looney, J. P., and van Zee, R. D., *J. Chem. Phys.* **105**, 10278-10288 (1996).
6. Hodges, J. T., Looney, J. P., and van Zee, R. D., *Appl. Opt.* **35**, 4112-4116 (1996).
7. Looney, J. P., Hodges, J. T., and van Zee, R. D., "Quantitative Absorption Measurements Using Cavity-Ringdown Spectroscopy With Pulsed Lasers," in *Cavity-Ringdown Spectroscopy: An Ultratrace-Absorption Measurement Technique*, edited by K. W. Busch and M. A. Busch, American Chemical Society, New York, 1999, pp. 93-105.
8. van Zee, R. D., Hodges, J. T., and Looney, J. P., *Appl. Opt.* **38**, 3951-3960 (1999).
9. Berden, G., Peeters, R., and Meijer, G., *Int. Rev. Phys. Chem.* **19**, 565-607 (2000).
10. Cormier, J. G., Ciurylo, R., and Drummond, J. R., *J. Chem. Phys.* **116** 1030-1034 (2002).
11. Cormier, J. G., *Development of an Infrared Cavity Ringdown Spectroscopy Experiment and Measurements of Water Vapor Continuum Absorption*, PhD thesis, University of Toronto, 2002.
12. Romanini, D., and Lehmann, K. K., *J. Chem. Phys.* **105**, 81-88 (1996).
13. Vaughan, J. M., *The Fabry-Perot Interferometer: History, Theory, Practice and Applications*, Adam Hilger, Philadelphia, 1989.
14. Romanini, D., Kachanov, A. A., and Stoeckel, F., *Chem. Phys. Lett.* **270**, 538-545 (1997).
15. Boyd, G. D., and Gordon, J. P., *Bell Sys. Tech. J.* **40**, 489-508 (1961).
16. Yariv, A., *Quantum Electronics*, Wiley, New York, 1989.
17. Born, M., and Wolf, E., *Principles of Optics*, 6th ed., Cambridge University Press, New York, 1980.
18. Kogelnik, H., *Bell Sys. Tech. J.* **43**, 334-337 (1964).
19. Kogelnik, H., and Li, T., *Appl. Opt.* **5**, 1550-1567 (1966).
20. Cormier, J. G., and Drummond, J. R., to be published in the forthcoming NATO Science Series volume entitled *Weakly Interacting Molecular Pairs: Unconventional Absorbers of Radiation in the Atmosphere*.
21. Elsasser, W. M., *Phys. Rev.* **53**, 768 (1938).
22. Tipping, R. H., and Ma, Q., *Atmos. Res.* **36**, 69-94 (1995).
23. Ma, Q., and Tipping, R. H., *J. Chem. Phys.* **111**, 5909-5921 (1999).
24. Ma, Q., and Tipping, R. H., *J. Chem. Phys.* **112**, 574-584 (2000).
25. Penner, S. S., and Varanasi, P., *J. Quant. Spectrosc. Radiat. Transfer* **7**, 687-690 (1967).
26. Gebbie, H. A., *et al.*, *Nature* **221**, 143-145 (1969).
27. Rothman, L. S., *et al.*, *J. Quant. Spectrosc. Radiat. Transfer* **60**, 665-710 (1998).
28. Bay, Z., Luther, G. G., and White, J. A., *Phys. Rev. Lett.* **29**, 189-192 (1972).
29. Layer, H. P., Deslattes, R. D., and Schweitzer, W. G. Jr., *Appl. Opt.* **15**, 734-743 (1976).

New Possibilities for Electric Field Measurements in a Plasma with the Use of Laser and Stark Spectroscopy

E.K. Cherkasova[a], V.P. Gavrilenko[b,c], and A.I. Zhuzhunashvili[d]

[a]*Moscow Institute of Physics and Technology, Dolgoprudnyi 141700, Russia*
[b]*General Physics Institute, Russian Academy of Sciences, Department of Plasma Physics, Vavilov street 38, 119991, GSP-1, Moscow, Russia*
[c]*Center for Surface and Vacuum Research, Russian State Committee for Standards, Vavilov street 38, 117334 Moscow, Russia*
[d]*Russian Research Center "Kurchatov Institute", Kurchatov square, 123182 Moscow, Russia*

Abstract. Two laser techniques for the electric field measurements in a plasma have been experimentally tested. The first technique is based on the use of Stark mixing of the Λ - doublet energy sublevels of BH molecules, whereas the second technique is based on the use of Stark splitting of highly lying levels of atomic hydrogen. Both diagnostic techniques were applied to the measurements of electric fields in a hollow cathode discharge. The results of the electric field measurements obtained by both techniques are in good agreement.

INTRODUCTION

Electric field (EF) is one of the main parameters in physics of discharges. The EF in a discharge determines the charged particles energy distribution functions and the charged particle fluxes. Knowledge of these characteristics is in turn of great importance for improving modeling and control of plasma discharges. One of the advantages of using laser spectroscopy for the EF measurements in a plasma is the possibility to perform measurements with high temporal and spatial resolution, making it possible to reduce the minimum detectable EFs as compared with the use of conventional emission spectroscopy (see, e.g., Ref. [1]). In the present work two techniques for the laser-aided EF measurements are experimentally realized. The first technique is based on the effect of appearance of dipole-forbidden lines in a fluorescent spectrum of polar diatomic molecules BH. We note that until recently only three diatomic molecules, BCl, NaK, and CS, have been used for the EF measurements in plasmas [2-7]. The second technique is based on the Stark splitting of highly lying energy levels of hydrogen. In this technique a double resonance in hydrogen atoms is realized. The same hollow cathode discharge is used as a plasma source for both techniques. The BH molecules and the hydrogen atoms were produced in the discharge due to the decomposition of pentaborane, B_5H_9. The spectroscopic measurements were performed at a pressure of 13.3 Pa.

STARK SPECTROSCOPY OF BORON MONOHYDRIDE MOLECULES

A method for laser-induced fluorescence (LIF) measurements of EFs in plasmas and gases based on Stark effect of polar diatomic molecules was suggested and tested in Ref. [2]. Since then it was implemented in a number of experiments [3-7]. The primary characteristic feature of the method is its high sensitivity with respect to EFs that can be explained as follows. Laser is tuned to an allowed $X\,^1\Sigma - ^1\Pi$ transition in a polar diatomic molecule (see Fig. 1). Each rotational level of the molecule in the

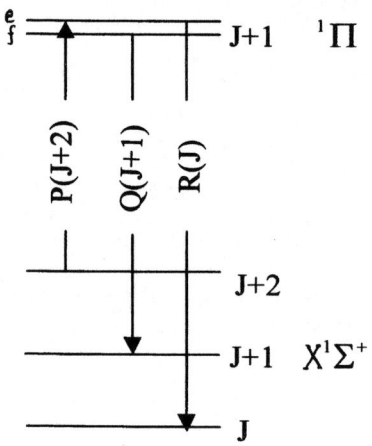

$^1\Pi$ state is split into two closely spaced sublevels (Λ- doublet splitting). Therefore, even moderate EFs can significantly intermix the wave functions of the Λ- doublet sublevels, and hence, even moderate EFs can produce considerable modification of the LIF spectrum of such a molecule. Although there exists a large number of polar diatomic molecules and radicals [8], until recently only three molecules, BCl, NaK, and CS, have been used for LIF measurements of EFs. However, it is desirable to have a larger choice of molecules that can be used for LIF measurements of EFs. This is due to the fact that for a specific discharge it is desirable to utilize for diagnostics the molecules that are either present in the discharge, or could be added to the discharge without changing significantly its properties. From this point of view, the boron monohydride (BH) molecule is of considerable interest. First, its ground

FIGURE 1. Structure of energy levels of a polar diatomic molecule. P-branch laser excitation followed by Q- and R-branch fluorescence are shown by vertical arrows.

electronic state is $X\,^1\Sigma^+$, and as the upper electronic state the $A\,^1\Pi$ state can be chosen. Second, there are the reference data on the dipole moment d_Π and the Λ- doublet splitting Δv_{ef} for the $A\,^1\Pi$ state of the BH molecule (see [8]). It follows from [8] that the values of Δv_{ef} and d_Π for the $A\,^1\Pi$ state of BH are

$$\Delta v_{ef} = 0.0389 - 0.0027(v + 1/2)J'(J' + 1) \text{ cm}^{-1}, \qquad d_\Pi(v = 0) = 0.58\,\text{D} \qquad (1)$$

where v and J' are the vibrational and rotational quantum numbers, respectively. Third, the BH molecules may be present in many discharges in hydrogen that are used for the boron nitride deposition.

As the plasma source, we used the hollow cathode discharge in a discharge chamber with a cylindrical cathode. The laser source was a tunable dye laser pumped by a nitrogen laser. The dye laser was operated at $\lambda \approx 395 - 430$ nm. The output of the dye laser had pulse duration of 20 ns, a spectral width of 0.01 nm, and an output energy of about 1.5 mJ. The hollow cathode was of inner diameter of 8 mm and of the length of 40 mm. The laser beam was directed to the hollow cathode plasma parallel to the axis of the cylindrical cathode. The majority of the fluorescent measurements were performed in the cathode dark space when the radial distance r between the cathode axis and the laser beam was 3 mm. A few of the fluorescent measurements were also performed in the negative glow region when the laser beam was directed along the cathode axis (this corresponds to $r = 0$ mm). The LIF spectra were recorded orthogonal to the direction of the laser beam. Let the axis z of the coordinate system $Oxyz$ be the direction of the laser beam, and the axis y be the direction of the LIF observation; the axis y has the radial direction. In our experiments, the polarization of the laser radiation was random in the plane xy with the electric vector of the laser radiation having the isotropic angular distribution in this plane.

In our work, we employed the laser excitation of the transition belonging to the P-branch, and we used the ratio of intensities of the LIF signal spectral lines belonging to the Q- and R-branches for the EF measurements. For such a laser excitation, we obtained the following formulas for the intensities of the Q- and R-components in the LIF spectrum for recording the spectrum along the axis y

$$I_Q = [(J+2)^2(2J+3)^2(2J+5)]^{-1} \sum_{M=-(J+1)}^{J+1} \sum_{M_1=-(J+1)}^{J+1} (1 - \cos\beta_M \cos\beta_{M_1})$$
$$\times \{2[(J+2)(J+3) + M_1^2] + \sin^2\alpha[(J+1)(J+2) - 3M_1^2]\} \qquad (2)$$
$$\times \{2[(J+1)(J+2) - M^2] + \cos^2\alpha[3M^2 - (J+1)(J+2)]\},$$

$$I_R = [(2J+1)(2J+3)^3(2J+5)]^{-1} \sum_{M=-(J+1)}^{J+1} \sum_{M_1=-(J+1)}^{J+1} (1 + \cos\beta_M \cos\beta_{M_1})$$
$$\times \{2[(J+2)(J+3) + M_1^2] + \sin^2\alpha[(J+1)(J+2) - 3M_1^2]\} \qquad (3)$$
$$\times \{2[J(J+1) + M^2] - \cos^2\alpha[3M^2 - (J+1)(J+2)]\},$$

where $\cos\beta_M = (1 + \Phi_M^2)^{-1/2}$, the value of Φ_M characterises the Stark mixing of the Λ-doublet sublevels,

415

$$\Phi_M(J') = 2Md_\Pi F /[\Delta\nu_{ef}J'(J'+1)], \qquad J' = J+1, \tag{4}$$

M is the magnetic quantum number, and α is the angle between the EF vector F and the axis z. In formulas (2)-(4), J is the rotational quantum number of the lower level belonging to the $X^1\Sigma^+$ state, and $J' = J+1$ is the rotational quantum number of the upper level belonging to the $A^1\Pi$ state. We should note that formulas (2)-(4) were obtained for the case when laser radiation has random polarization isotropically distributed in the plane xy. The observation of the fluorescent radiation was assumed to be along the y axis, the intensities I_Q and I_R in Eqs. (2), (3) being "unpolarized", i. e. each of these intensities is, in fact, the sum of two intensities – with x- and z-polarizations. In our work, we present the EF vector F in the form: $F = F\sin\alpha\, e_y + F\cos\alpha\, e_z$, the e_x, e_y, and e_z vectors being the unit vectors along the axes x, y, and z. When deriving Eqs. (2), (3), similarly to [6] we assumed that populations of the states with different M within each of the Λ-doublet sublevels are equal. Figure 2 shows, as an example, the experimental intensities of the Q- and R-lines in the fluorescence spectra of BH recorded in the case when the laser beam was directed at the radial distance of $r = 3$ mm from the cathode axis, and the discharge current was 0.1 A. These intensities were obtained when the laser was tuned to the P(7) transition.

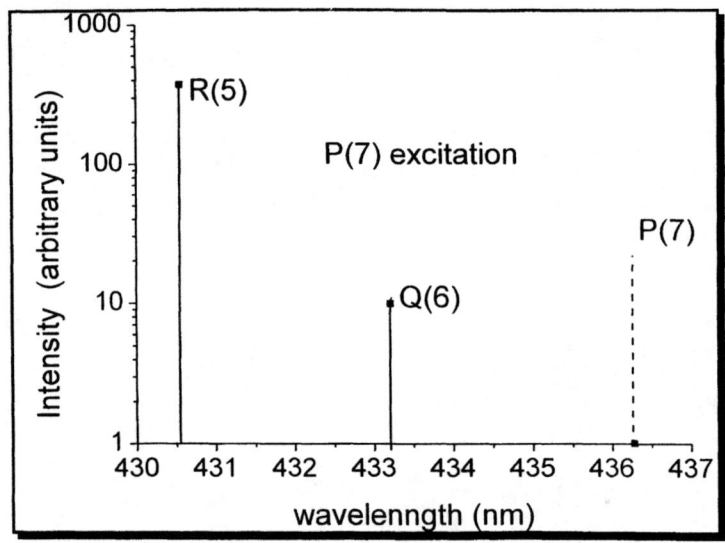

FIGURE 2. Spectrally resolved laser-induced fluorescence from excitation of P(7) transitions in the (0,0) band of the BH $A^1\Pi - X^1\Sigma^+$ system. The discharge current I=0.1 A.

We used the dependence of I_Q/I_R on F for the EF measurements in the hollow cathode discharge. Equations (2) and (3) correspond to the case when the strength and the direction of the EF F are fixed in a plasma. In our experiment, the LIF radiation was collected from a plasma volume having the linear size of about $(0.5-1.0)$ mm. Different BH molecules being in this volume "see" different EF vectors F. Therefore, the employment of Eqs. (2), (3) for the analysis of the experimental intensities of Q- and R-lines of the LIF spectrum of BH gives the EF mean strength F_m. In order to deduce such a mean strength F_m, we used the two dependencies of the I_Q/I_R ratio on F obtained for $\alpha = 0$ and $\pi/2$. After obtaining by this means the two values of F, $F = F^{(\alpha=0)}$ and $F = F^{(\alpha=\pi/2)}$, as the mean value of F we used the following one: $F_m = (F^{(\alpha=0)} + F^{(\alpha=\pi/2)})/2$. Such determination of the EF strength F_m introduces the uncertainty $\Delta F = \pm \left| F^{(\alpha=\pi/2)} - F^{(\alpha=0)} \right|/2$. The other uncertainty in determining the field strength F_m is introduced by the experimental uncertainties for the I_Q and I_R intensities.

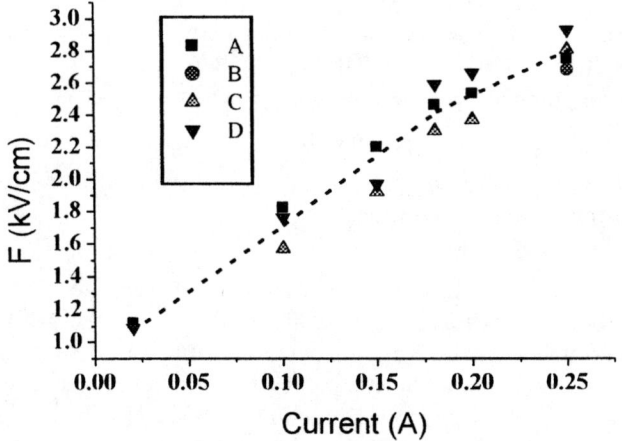

FIGURE 3. The measured electric field strength F versus the current through the hollow cathode discharge. Laser is tuned to the transition $P(N+2)$ between the $X^1\Sigma^+, v''= 0$ and $A^1\Pi, v'= 0$ states of BH. The field strength F was deduced from the $I_{Q(N+1)}/I_{R(N)}$ ratios. where N=5 for the (A) and (B) results, N=8 for the (C) and (D) results. For the (A) and (C) results. the I_Q and I_R intensities were recorded using a photomultiplier; for the (B) and (D) results. the I_Q and I_R intensities were recorded using a photon counting technique. The relative errors for the values of the field strength F are within the (10-25) % range.

Figure 3 demonstrates the results of the measurements of the mean EF F_m versus the discharge current I in the cathode dark space (for $r = 3$ mm). It can be seen from Fig. 3 that the values of F_m obtained by using excitation of two different transitions of BH as well as two different recording techniques (a photomultiplier and a photon counting technique) are in a reasonable agreement. We have also performed the fluorescent measurements of the EF in the negative glow region (the case of $r = 0$ mm) when the applied voltage between the cathode and anode was from 290 V to 340 V. For such measurements, we used the dependence of the I_Q / I_R ratio on F under the $P(6)$ transition excitation by laser radiation. We obtained that the field strength in the negative glow region is about (280 ± 65) V/cm. This EF can be the electric microfield produced by the ions.

STARK SPECTROSCOPY OF ATOMIC HYDROGEN

In the technique based on the Stark spectroscopy of atomic hydrogen, a double resonance in hydrogen atoms was realized. Two counter-propagating laser beams were used, the first laser being in the resonance with the transition $n = 2 \leftrightarrow n = 3$ ($\lambda \approx 656.3$ nm) in a hydrogen atom, whereas the second laser was tuned to the Stark components of the P_ε transition ($n = 3 \leftrightarrow n = 8$, $\lambda \approx 955$ nm), and a signal of fluorescence at the P_ε transition ($n = 8 \rightarrow n = 3$) was detected. Here n is the principal quantum number. The laser excitation scheme has some common features with the schemes proposed in Refs. [9,10]. In our work, the first laser is a tunable dye laser ($\tau_{pulse} \approx 20$ ns) pumped by a Cu vapour laser. The second laser is a grating-tuned colour center laser ($\tau_{pulse} \approx 12$ ns) pumped by a ruby laser. The use of two counter-propagating laser beams enables us to substantially compensate the Doppler shifts that occur at the two transitions in hydrogen ($n = 2 \leftrightarrow n = 3$ and $n = 3 \leftrightarrow n = 8$). Such compensation considerably enhances the fluorescence signal at the P_ε transition. As a plasma source, a hollow cathode discharge was used (see also the previous section). The directions of both laser beams were parallel to the axis of the cylindrical cathode (the z-axis). The fluorescent measurements were performed in the cathode dark space when the radial distance r between the axis of the cylindrical cathode and the laser beams was 3 mm. The LIF spectra were recorded orthogonal to the direction of the laser beams (along the y-axis). Figure 4 demonstrates typical experimental LIF spectra in hydrogen when the voltage between the cathode and anode was 500 V, and the discharge current was 0.2 A. The experimental points are shown by squares; they are connected by a polygonal line. Figure 4,a shows the P_ε fluorescent spectrum of x-polarization, whereas Figure 4,b shows the P_ε fluorescent spectrum of z-polarization. In Fig. 4, the vertical lines show the Stark components for P_ε spectrum calculated with the assumption that the electric field vector is parallel to the z-axis, and the field strength is equal to 2530 V/cm. We can see that the theoretical spectra calculated for x- and z-polarizations well fit the experimental fluorescent spectra. We note that under the same discharge conditions, the field strength deduced from the LIF spectra of BH is approximately 2.5 kV/cm (cf. Fig. 3). The latter value is close to the field strength obtained by using the Stark splitting of the P_ε line in hydrogen.

418

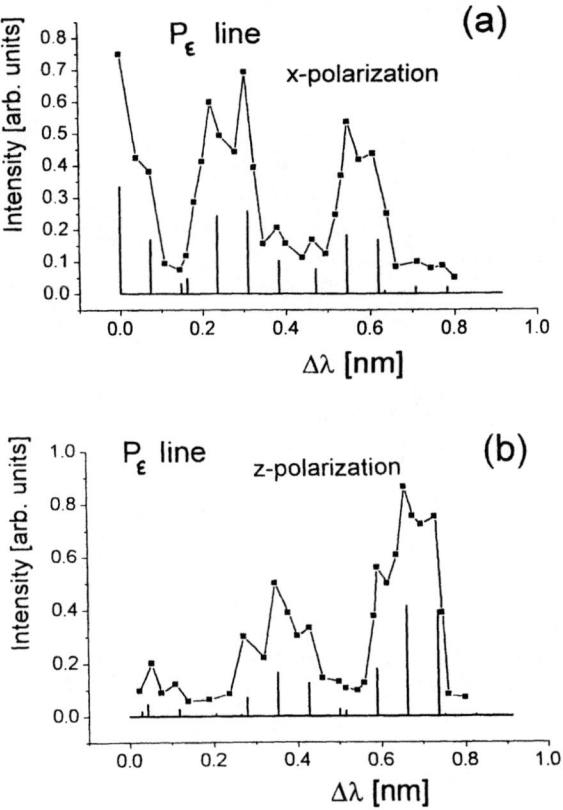

FIGURE 4. Stark spectra (the "red" halves) of the P_ε spectral line ($\lambda \approx 954.6$ nm) of atomic hydrogen. A zigzag line shows spectra of the P_ε line recorded from a hollow cathode discharge. Vertical lines show theoretical Stark spectra of the P_ε line of hydrogen. (a) x-polarization of the photon emitted; (b) z-polarization of the photon emitted. Theoretical Stark spectra of the P_ε line were calculated for the case when the strength of the electric field was equal to 2530 V/cm, and the electric field vector was directed along the z-axis.

CONCLUSION

We have experimentally tested two laser techniques for the measurements of electric fields in a plasma. The first technique is based on the use of Stark mixing of the Λ-doublet energy sublevels of BH molecules, whereas the second technique is based on the use of Stark splitting of highly lying levels of atomic hydrogen. Both diagnostic techniques were applied to the measurements of electric fields in a hollow cathode discharge. The results of the electric field measurements obtained by both techniques are in good agreement.

ACKNOWLEDGEMENTS

This work was supported by the Russian Foundation for Basic Research (project No 01-02-17810).

REFERENCES

1. Gavrilenko, V.P., Ochkin, V.N., and Tskhai, S.N., "Progress in plasma spectroscopic diagnostics based on Stark effect in atoms and molecules", in *Selected Research Papers on Spectroscopy of Nonequilibrium Plasma at Elevated Pressures*, edited by V.N. Ochkin, Proceedings of the International Society for Optical Engineering (SPIE), Vol. 4460, SPIE, Bellingham, WA, 2002, pp. 207-229.
2. Moore, C.A., Davis, G.P., and Gottscho, R.A., *Phys. Rev. Lett.* **52**, 538- 541 (1984).
3. Gottscho, R.A., and Mandich, M.L., *J. Vac. Sci. Technol. A* **3**, 617-624 (1985).
4. Derouard, J., and Sadeghi, N. *Optics Commun.* **57**, 239- 243 (1986).
5. Bowden, M.D., Nakamura, T., Muraoka, K., Yamagata, Y., James, B.W., and Maeda, M., *J. Appl. Phys.* **73**, 3664-3667 (1993).
6. Maurmann, S., Kunze, H.-J., Gavrilenko, V.P., and Oks, E., *J. Phys. B: At. Mol. and Opt. Phys.* **29**, 25-34 (1996).
7. Maurmann, S., Gavrilenko, V.P., Kunze, H.-J., and Oks, E., *J. Phys. D: Appl. Phys.* **29**, 1525-1531 (1996).
8. Huber, K.P., and Herzberg, G. *Molecular Spectra and Molecular Structure. IV. Constants of Diatomic Molecules*, Van Nostrand, New York, 1979.
9. Booth, J.P., Fadlallah, M., Derouard, J., and Sadeghi, N., *Appl. Phys. Lett.* **65**, 819-821 (1994).
10. Czarnetzki, U., Luggenhölsher, D., and Döbele, H.F., *Phys. Rev. Lett.* **81**, 4592-4595 (1998).

Electron Density Measurements in the Peripheral Regions of a Current Sheet with the Use of Helium Spectral Lines

A.G. Frank[a], V.P. Gavrilenko[a,b], and N.P. Kyrie[a]

[a] *General Physics Institute, Russian Academy of Sciences, Vavilov street 38, 119991, GSP-1, Moscow, Russia*

[b] *Center for Surface and Vacuum Research, Russian State Committee for Standards, Vavilov street 38, 117334 Moscow, Russia*

Abstract. Profiles of the HeI 447.1 nm and HeI 492.2 nm spectral lines of neutral helium have been measured in a current sheet plasma of a "Current Sheet 3D" device. We performed theoretical calculations of these profiles using the model microfield method. Analyzing profiles of the HeI 447.1 nm and HeI 492.2 nm spectral lines, we found out that the electron density in the peripheral regions of a current sheet is in the range $(1.0-1.6) \times 10^{15}$ cm^{-3} for different instants of time in the course of the current sheet evolution.

INTRODUCTION

The objective of this paper is the spectroscopic diagnostics of a current sheet plasma by using spectral lines of neutral helium at 447.1 nm (the $2\,^3P - 4\,^3D$ transition) and 492.2 nm (the $2\,^1P - 4\,^1D$ transition) on the "Current Sheet 3D" (CS-3D) device. This device is constructed to study the evolution of current sheets under various conditions and in different two- and three-dimensional magnetic configurations. Current sheets have been attracting a lot of research for many years [1,2] that is due to the fact that such sheets are ideally suited for studying the fundamental problem of plasma physics – magnetic reconnection. In our previous papers [3, 4], we reported on spectroscopic measurements of electron density N_e in the central region of the current sheet. These measurements were based on using the profiles of the HeII 468.6 nm and HeII 656.0 nm spectral lines of the hydrogen-like helium ion. In the present study, the use of spectral lines of neutral helium made it possible to determine N_e in the peripheral regions of the current sheet.

CP645, *Spectral Line Shapes: Volume 12, 16th ICSLS*, edited by C. A. Back
© 2002 American Institute of Physics 0-7354-0100-4/02/$19.00

EXPERIMENTAL DEVICE AND SPECTROSCOPIC MEASUREMENT PROCEDURE

Experiments were carried out on the CS-3D device [3,4]. In the case of electric current excited in a plasma along the null-line of a two-dimensional (2D) magnetic field, a plane current sheet was formed in it [1,2]. This null-line was situated at the axis of the cylindrical vacuum chamber. In the experiments described below the radial magnetic field gradient was 600 G/cm. The preliminarily evacuated chamber was then filled with helium at a pressure of 300 mTorr. The initial plasma was produced in the vacuum chamber by an auxiliary theta-pinch discharge with a strong preliminary ionization. Then, a current aligned with the null-line was excited in the plasma. The current initiated two-dimensional plasma flows which in turn, resulted in the formation of a current sheet. The plasma-current half-period was equal to $T/2 \approx 5$ µs, and its maximum value varied in these experiments from 50 to 80 kA.

The optical scheme used for spectroscopic measurements is similar to that described in Refs. [3,4]. Plasma radiation from the central region of the plasma chamber was collected by an achromatic objective. This region was 1.2 cm in diameter and 60 cm in length. The radiation was then transmitted through the quartz fibers onto the entrance slit of the MDR-3 monochromator and recorded by the MORS-3 multi-channel optical recording system. This system consists of an image tube with a microchannel plate amplifier and a multichannel CCD detector connected to a computer through an adapter, see Ref. [3] for details.

In Refs. [3, 6—8], it was shown that within the region of thickness 1.2—1.5 cm, the current-sheet plasma is strongly inhomogeneous both in the electron temperature and density. Both have maximum in the middle of the sheet, and decrease several times towards the peripheral regions. As a consequence, the helium spectral lines of interest are emitted from the peripheral regions of the current sheet and from the surrounding space. For the current sheets formed under "high-pressure" conditions, this conclusion was also confirmed by the results obtained from the analysis o 2D distributions of plasma emission in different spectral lines [7—9]. Such distributions were obtained with the image tube in combination with the narrow-band ($\Delta\lambda_{1/2} = 1.1$ nm) interference filters and gave a qualitative picture of the plasma current sheet's structure.

Typical experimental profiles of the HeI 447.1 nm and HeI 492.2 nm spectral lines are shown in Figs. 1 and 2 by squares. These experimental profiles contain dipole-forbidden components, which are indicated by the vertical arrows. The presence of the dipole-forbidden components in the spectrum testifies that the plasma electric microfield plays a significant role in the formation of the spectral lines under consideration.

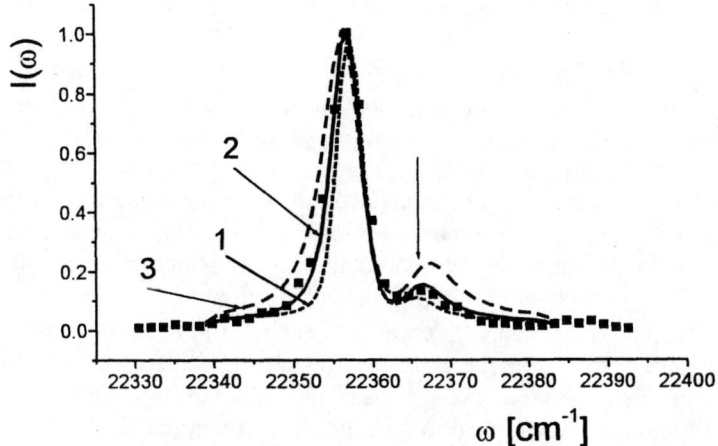

FIGURE 1. Experimental and theoretical profiles of the HeI 447.1 nm spectral line. The experimental profile (shown by squares) refers to the radiation emitted from the peripheral regions of the current sheet. The theoretical profiles are calculated using the model microfield method for three values of the plasma density $N_e = 6.0 \times 10^{14}$ cm^{-3} (curve 1), $N_e = 1.6 \times 10^{15}$ cm^{-3} (curve 2), and $N_e = 3.0 \times 10^{15}$ cm^{-3} (curve 3). The vertical arrow shows the position of the dipole-forbidden spectral component ($2\,^3P - 4\,^3F$).

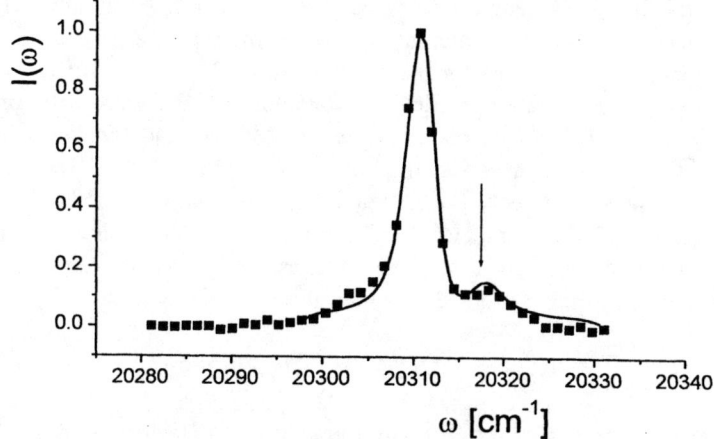

FIGURE 2. Experimental and theoretical profiles of the HeI 492.2 nm spectral line. The experimental profile (shown by squares) refers to the radiation emitted from the peripheral regions of the current sheet. The theoretical profile is calculated using the model microfield method for the plasma density $N_e = 1.02 \times 10^{15}$ cm^{-3}. The vertical arrow shows the position of the dipole-forbidden spectral component ($2\,^1P - 4\,^1F$).

RESULTS AND DISCUSSION

To analyze the experimental profiles of the HeI 447.1 nm and HeI 492.2 nm spectral lines and to obtain information about plasma density in the peripheral regions of the current sheet, we performed calculations of the profiles of these lines based on the model microfield method (MMM) proposed in Ref. [5]. The basic idea of the MMM is that the actual electric microfield produced by charged plasma particles is modeled by a simple stochastic process for which an exact analytical expression for the evolution operator can be derived. In addition, it is assumed in the MMM that the electric microfield $F(t)$ changes abruptly at random times t_j distributed according to the Poisson law, and the probability density of such jumping-times is the function of the field strength: $v=v(F)$. In the time interval $t_j<t<t_{j+1}$, the electric field F is the constant vector. In our calculations we assumed that the temperature of charged particles and helium atoms is equal to 2 eV [3, 7], and that the singly charged helium ions He^+ play the role of perturbing particles.

Figure 1 demonstrates three theoretical profiles of the HeI 447.1 nm line calculated for three values of the plasma density: $N_e=6.0\times10^{14}$, $N_e=1.6\times10^{15}$, and $N_e=3.0\times10^{15}$ cm^{-3}. As one can see, the HeI 447.1 nm line profile is modified noticeably as the density N_e varies from 6.0×10^{14} to 3.0×10^{15} cm^{-3}. Such modification manifests itself as the increase of the main line's FWHM (2^3P-4^3D) and the intensity of the dipole-forbidden line (2^3P-4^3F) when the density N_e is increasing. From Fig. 1, it follows that the theoretical profile calculated for $N_e=1.6\times10^{15}$ cm^{-3} gives the best fit to the experimental profile. The theoretical profile of the HeI 492.2 nm spectral line that provides the best fit to the experimental one is shown in Fig. 2 by a solid line. This theoretical profile was calculated for the plasma density $N_e=1.02\times10^{15}$ cm^{-3}. We note that each of the theoretical profiles is obtained by the convolution of the corresponding Stark profile (calculated using the MMM) and the Gaussian profile (which allows for the Doppler and instrumental broadening). The difference between the values of the plasma density N_e obtained from the analysis of the experimental profiles of the HeI 447.1 nm and HeI 492.2 nm lines (see Figs. 1 and 2, respectively) could be explained by the fact that these profiles were measured at different time instants in the course of the current sheet evolution.

ACKNOWLEDGEMENTS

This work was supported by the Russian Foundation for Basic Research (project No 01-02-17810).

REFERENCES

1. Syrovatskii, S.I., *Annu. Rev. Astron. Astrophys.* **19**, 163-229 (1981).
2. Frank, A.G., *Plasma Phys. Contr. Fusion* **41**, Suppl. 3A, A687-A697 (1999).
3. Büscher, S., Kyrie, N.P., Kunze, H.-J., and Frank, A.G., *Plasma Phys. Reports* **25**, 164-169 (1999).
4. Frank, A.G., Gavrilenko, V.P., Kyrie, N.P., and Markov, V.S., *Contrib. Plasma Phys.* **41**, 85-90 (2001).
5. Brissaud, A., and Frisch, U., *J. Quant. Spectrosc. Radiat. Transfer* **11**, 1767-1783 (1971).
6. Kirii, N.P., Markov, V.S., and Frank, A.G., *JETP Lett.* **56**, 82-87 (1992).
7. Bogdanov, S.Yu., Burilina, V.B., and Frank, A.G., *JETP* **87**, 655-662 (1998).
8. Bogdanov, S.Yu., Kyrie, N.P., Markov, V.S., and Frank, A.G., *JETP Lett.* **71**, 53- 57 (2000).
9. Frank, A.G., Bogdanov, S.Yu., Burilina, V.B., Kyrie, N.P., and Markov, V.S., *Contrib. Plasma Phys.* **40**, 106-112 (2000).

Three-Phase Plasma Arc Atomic Emission Spectrometric Analysis of Environmental Samples Using Ultrasonic Nebulizer

Z. F. Ghatass*, G. D. Roston** and M. M. Mohamed***

* Institute of Graduate Studies and Research, Alexandria University, Egypt
** Department of Physics, Faculty of Science, Alexandria University, Egypt
** gamal_daniel@yahoo.com
*** Medical Research Institute, Alexandria University, Egypt

Abstract. Combination of ultrasonic nebulizer and plasma excitation sources for spectrochemical analysis offers desirable features of low detection limits, high sample through out, wide dynamic range of operation, acceptable precision and accuracy, and simultaneous quantitative analytic capabilities. Moreover, the ultrasonic nebulizer does not require sample preconcentration. Recently we have developed a three phase plasma arc (TPPA) for atomic emission spectrochemical analysis. In the present work, to increase the analytical utility of the three-phase plasma system, ultrasonic nebulizer was used for sample introduction. The effects of the argon gas flow rate, current, excitation temperature have been studied. The analytical calibration curves are obtained for Ca, Cr, Fe, Mg and Mn and detection limits have been calculated. The present technique is used to determine the concentration of elements the Ca, Cr, Fe, Mg and Mn in airborne samples.

1. INTRODUCTION

Atomic emission plasma spectroscopy techniques have enjoyed long and rich heritages as methods which are often uniquely suited to rapid, simultaneous multi-element analysis of species in a wide variety of sample matrices. The success of atomic spectrometric plasma techniques has drawn considerable attention and remains the area of experimental [1] difficulty encountered with plasma sources. Selection of the best sample introduction procedure for an analysis requires consideration of a number of points: (1) the type of sample (solid, liquid or gas), (2) the level and also the range of levels for elements to be determined, (3) the accuracy, and precision, required, and (4) the amount of material available. The measurement technique available will also have a major effect on the choice of the selected procedure. The selection of a suitable sample introduction technique can depend heavily on an available and effective sample preparation procedure where sample introduction and sample preparation are intimately linked [2].

The most common sample introduction is nebulization of liquid by pneumatic nebulizer. These devices are convenient to operate; however, less than 10% of sample is nebulized into useful aerosol with the bulk of the sample being wasted [3]. They also present difficulties with viscous high salt content and microvolume solution. Ultrasonic nebulizer is another means of introducing liquid samples. This gives, lower detection limits when compared with the pneumatic nebulizer [4]. Recently we have developed a three phase plasma arc (TPPA) for atomic emission spectrochemical analysis [5]. To increase the analytical utility of the three-phase plasma system and to come over problems related to the use of pneumatic nebulizer in sample introduction, such as distinguish of the plasma discharge due to large size of the aerosol drops ultrasonic nebulizer have been used. This is presented in this work.

2. EXPERIMENTAL WORK

The major constituents of the three-phase plasma arc system are: (1) the three phase power supply, (2) plasma torch, (3) sample introduction, (4) spectrometer observation system. This is shown in Fig. (1).

A sample of an element, whose concentration to be determined, is introduced into the discharge plasma structure in a suitable form. The light emitted by atoms or ions in the plasma is resolved by means of a spectrometer and the output radiation from the spectrometer is detected by a photomultiplier tube.

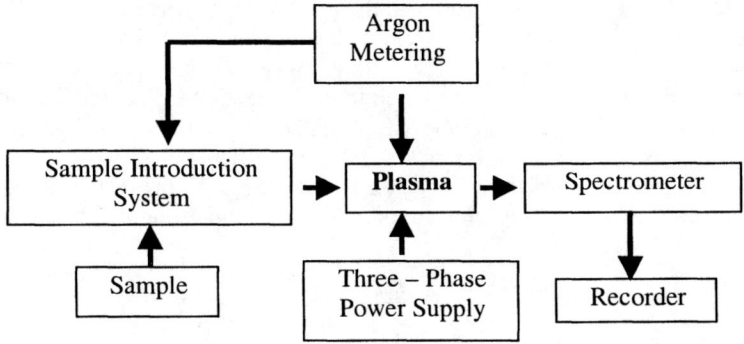

FIGURE 1. Schematic diagram of three phase plasma system.

2. 1. Three Phase Plasma Power Supply

The power used to sustain the argon plasma arc is derived from three-phase ac line voltage. Three-phase ac line voltage consists of three sine waves 120° out of phase from each other. It is transmitted through four wires with one wire corresponding to each phase and the fourth wire serving as a common (natural). The voltage between any two of the three phases was 380 V. This voltage is reduced with three-phase way connected transformer (7.5 kW) to 95 V. Standard 32 A circuit breakers are placed in each leg. To drop the voltage across the plasma and to regulate the plasma current, a 0 - 1.55 Ohm rheostats (consisting of two 16 A, 0 - 1.5 Ohm rheostats in series with other two in parallel) is connected in series between each transformer phase and its corresponding electrode [5-6].

2. 2. Plasma Torch

The plasma torch shown in Fig. (2), consists of three coaxial quartz tubes similar in size to those of a typical ICP torch. The innermost tube containing the aerosol carrier gas is thereby injected into the center of the Plasma. The outer and middle tubes are fused to each other to form an annular space of 3.5 mm wide. The diameter of the outer, middle and inner tubes are 22 mm, 15 mm and 5 mm, respectively. The 5 mm diameter sample introduction tube in the center has a tip with an inner diameter of 1 mm. The sample introduction tube and three 2% thoriated tungsten electrodes (1 mm) diameters and spaced 8 mm from each other are pass through a marble plate that has been machined to fit over the bottom of the middle quartz tube. The electrodes and sample introduction tube slide through their holes to allow their height to be adjusted. However, the holes are small enough for the loss of argon to be negligible.

The argon flowing along the electrodes provides adequate cooling. An undetermined amount of heat is transferred to the sample introduction tube from the warm electrodes. The middle quartz tube stabilizes the plasma by preventing the argon flowing through the outer tube from pushing the plasma away from the electrodes. The sample introduction tube and electrodes are passed through another marble plate with small adjustable brass blocks to attach the electrodes to the plate and the power supply wire. This marble plate is attached to a micrometer stage, so that the electrodes can be adjusted simultaneously and precisely to a particular height in the torch. The torch is mounted on the three direction holders. The electrode tips are usually adjusted to be 2 mm above the middle quartz tube, and the tip of the sample introduction tube is 3-5 mm below the top of the electrodes. The best signal intensities and detection limits are obtained at sample stream flows of 3, 1.5 and 1 L/min respectively. The plasma has a steady audible low buzz when it is operated at the previous set of flow rates. The schematic diagram of the plasma torch is presented in Fig. (2).

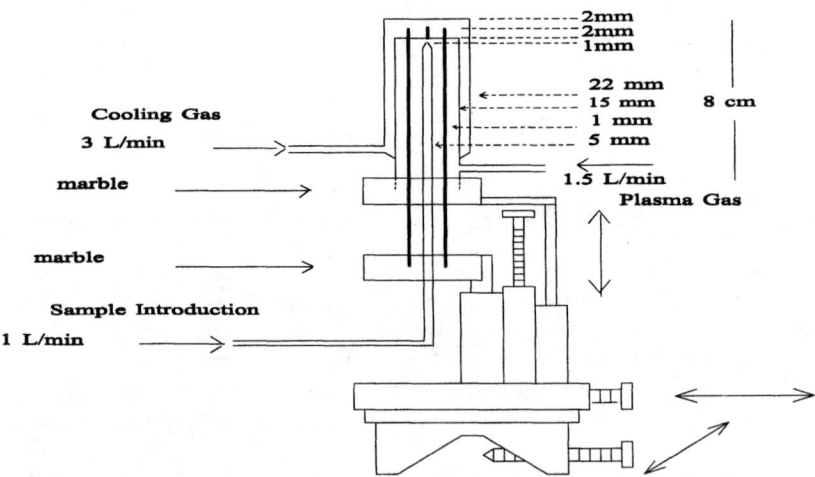

Figure 2. Schematic diagram of the three phase plasma torch.

2. 3. Sampling and Sample Introduction

Airborne sample are collected on Whatman fiber glass filter using a portable dust sampler. The sampling and digestion method were described in details in a previous paper [5]. An ultrasonic nebulizer (Fig.3), (Silentium, manufactured by Norditalla S.N.C) operated at frequency of 1.8 MHz, sample uptake of 2.25 ml/min and a drop size in the range from 0.5 to 3 micron. In this nebulizer, the aerosol product inside the nebulizer impact the front glass wall of the spray chamber, which acts as an impact wall, some of the large size aerosol particles collide and condense and fall back into the reservoir. The remaining fine aerosol passes through the side way tube. This tube helps to trap only fine aerosol particles to the outlet of the nebulizer. The outlet of nebulizer is connected to the plasma with an aerosol transport tube. A homogeneous fine aerosol can be obtained in the front the nebulizer outlet to the plasma.

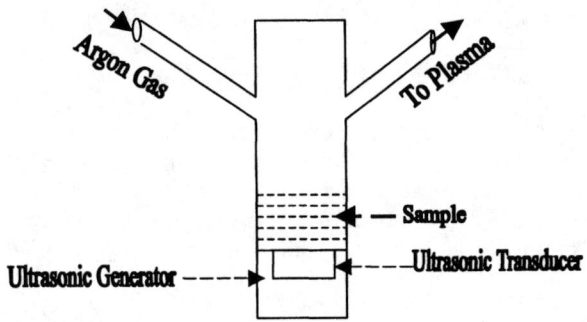

Figure 3. Ultrasonic nebulizer.

2. 4. Spectrometer Observation System

For convenient spatial profiling the discharge structure is mounted on a Czerny-Turner spectrometer model 1704 Spex with 1200 groves/mm and 500 blazed diffraction grating. The wavelength slowing and scanning is achieved by the rotation of the grating, which moves the spectrum across the exit slit. An 8 cm focus lens is used to focus an image (3:1) of the horizontal slit on to the entrance of the spectrometer. . The output signal is detected by a photomultiplier tube (PMT, R 955 Products for Research, INC, USA) operated at 1200 V by a suitable power supply, and recorded by driving chart recorder pen meter.

3. RESULTS AND DISCUSSION

3. 1. Effect of the Nebulizer Gas Flow Rate

The vertical emission profiles along the central vertical axis of the plasma for Mg I 2852.1 Å and Mg II 2802.7 Å are carried out as a function of nebulizer gas flow rate from 0.8 L/min to 1.2 L/min as shown in Fig. (4). The observation height at which the maximum emission is observed is shifted to positions higher in the plasma as the nebulizer gas flow rate is increased. This downward spatial shift and decrease in maximum emission intensity at higher nebulizer gas flow rates reflects the shorter time period that the sample spends in the plasma. At the higher sample gas flow rates, less time is available for complete vaporization, atomization, and excitation of the sample [7].

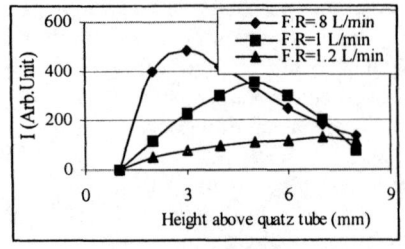 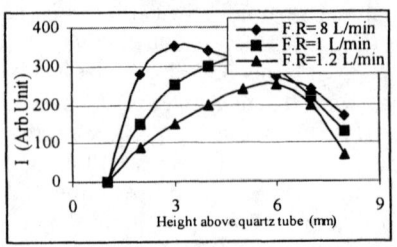

(a)   (b)

FIGURE 4. The vertical emission profiles along the central vertical axis of the plasma for Mg I 2852.1 Å (a) and Mg II 2802.7 Å (b) as a function of nebulizer gas flow rate from 0.8 to 1.2 L/min.

3. 2. Effect of the Electric Current

Current is the most effective factor that can affect the emission intensity of the analyte. Increasing the current from 15 A to 45 A causes increasing of the emission line intensity. The horizontal emission profiles along the central vertical axis of the plasma for Mg I 2852.1 Å and Mg II 2802.7 Å are shown in Fig. (5) for different currents (15, 27 and 45 A). The increase of emission intensity with the current may be attributed to high temperature arising from greater current and electron density.

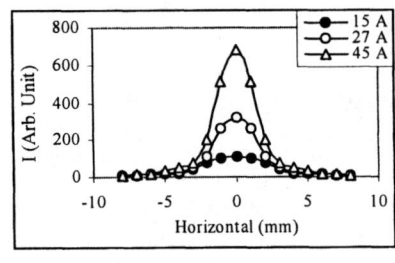 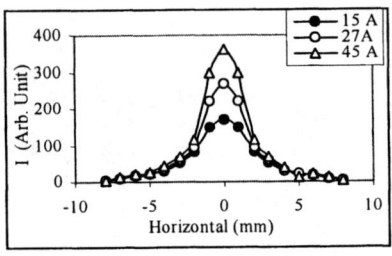

(a)   (b)

FIGURE 5. The horizontal emission profiles along the central vertical axis of the plasma for Mg I 2852.1 Å (a) and Mg II 2802.7 Å (b) for different currents (15, 27 and 45 A).

As current is increased, the increasing diameter of each of the three plasmas causes the plasmas to pinch more strongly on the sample stream. At high currents, the plasma becomes larger enough to restrict sample introduction and cause the emission signal to level off. Although this may be a good current region to use for analytical work, since a well-regulated power supply would be needed, the present torch and rheostats could not sustain these currents for extended time periods without overheating.

3. 3. Determination of the Excitation Temperature (T_{exc})

Theoretical discussions related to temperature determination and plasma diagnostics at atmospheric pressure are found in a number of publications [8-13]. Local thermodynamic equilibrium (LTE) is often assumed to exist in plasma at/or near atmospheric pressure [8].The following equation is derived from the expression for integrated radiance of an atomic line [9]:

$$\ln\left[\frac{I_{ul}\lambda_{ul}}{g_u A_{ul}}\right] = \ln\left[\frac{Lhcn_a}{4\pi Z}\right] - \frac{E_u}{kT_{exc}}$$

(1)

where u, l are subscripts that refer to the upper and lower energy states, g_u is the statistical weight of upper state, h is Planck's constant, I is the integral spectral irradiance, T_{exc} is the excitation temperature, λ is the wavelength, n_a is the number density of atoms, Z is the partition function, c is the speed of light, L is the path length, k is Boltzmann constant, E_u is the energy of the upper state), and A_{ul} is the transition probability. Expressing E_u in units of 10^3 cm^{-1} for convenience in computation, Eq. (1) leads to

$$\log\left[\frac{I_{ul}\lambda_{ul}}{g_u A_{ul}}\right] = C - \frac{625 E_u}{kT_{exc}}$$

where

$$C = \ln\left[\frac{Lhcn_a}{4\pi Z}\right]$$

In this study Ar was chosen as thermometric species [8]. The excitation temperature of the plasma is calculated from argon lines in the range between 4251.18 Å to 4345.16 Å. The wavelengths of the individual lines, their excitation energies and statistical weights of their upper levels and relative transition probabilities are taken from [1] and [8]. The excitation temperature, T_{ex} is obtained as the reciprocal of the slope of a graph of Log($I\lambda$/gA) vs E_u where I is taken as the normalized spectral line irradiance (intensity), as shown in Fig. (6). The average excitation temperature of the plasma is 5250 K.

3. 4. Analytical Calibration Curves

Analytical calibration curves are obtained for several elements as shown in Fig (7). Each data point shown in the figure is the average of three replicate determinations. For all analytical systems, the net analyst measure is determined by subtracting the average of the blank measures from the sample measure. These data has good linearity over two order of magnitude for most elements examined.

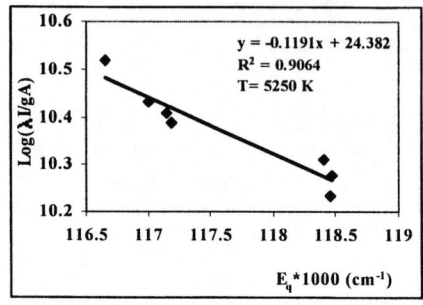

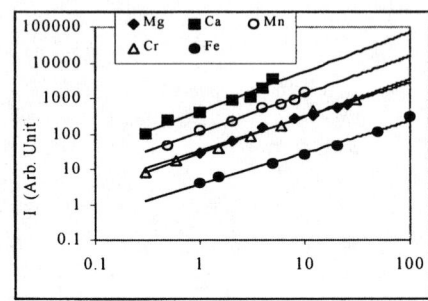

Figure 6. Boltzmann plot.

Figure 7. Analytical calibration curves

3. 5. Detection Limits

The detection limit value represent the concentration which yields an emission intensity two times the standard deviation of the blank signal [14]. The detection limit deduced from this work for several elements with the corresponding limits reported for the ICP and DCP are given in table (1).

Table 1. Detection limits ($\mu g/mL$)

Element	Wavelength Å	Present work	DCP	ICP
Mg	2852.1	0.0350	0.0002	0.0002
Mn	2576.1	0.0240	0.0020	0.0006
Fe	3688.8	0.2420	0.0030	0.0030
Ca	4226.7	0.0200	0.0002	0.0001
Cr	4254.3	0.0950	0.0020	0.0040

3. 6. The Application of the System in the Environment

To assess the practical utility of the present system, the determination of several elements (Mg, Ca, Mn, Cr and Fe) in environmental samples (airborne samples) is undertaken. The environmental samples are collected and digested as previous work [1]. The samples are split into two parts; one part is analyzed by the present technique while the other analyzed by atomic absorption spectroscopy (Perkin Elmer Model 2380) as standard technique.

Table 2. Comparison between the results obtained by the three – phase plasma and the atomic absorption spectrometry for determination of trace metal concentration ($\mu g/m^3$) in airborne samples.

Three – phase plasma spectrometry					Atomic absorption spectrometry				
Mg	Mn	Cr	Ca	Fe	Mg	Mn	Cr	Ca	Fe
27±3	68±5	31±4	3241±21	2456±18	25±3	67±6	31±5	3231±35	2431±25
29±4	73±4	35.±4	2099±18	2600±27	31±3	72±5	34±4	2109±20	2591±20
38±4	94±7	39±3	5021±21	3696±16	36±5	94±9	39±5	5126±17	3721±35
31±2	52±4	35±5	4201±23	3168±14	32±3	51±6	35±5	4301±20	3151±18
39±3	78±3	28±3	3912±18	1954±23	38±4	75±4	28±4	3891±20	1921±27
23±2	57±3	25±3	2916±18	1490±21	23±4	56±4	23±3	2921±15	1510±18
18±2	44±4	24±4	5211±18	2340±16	18±3	42±3	23±3	5232±23	2290±45
21±3	51±6	29±3	3976±25	1335±17	21±4	50±6	29±4	3980±18	1329±19
34±3	91±8	25±5	4201±30	1844±20	32±4	93±8	25±6	4199±35	1799±24
33±4	107±8	30±3	5216±19	3034±34	32±3	97±8	30±5	5209±13	3121±22

The average and standard deviation are calculated as shown in Table (2). It is seen from the table that the results show no significant difference between the two techniques.

ACKNOWLEDGEMENTS

The authors thank Prof. M. S. Helmi for reviewing the manuscript.

REFERENCES

1. Mohamed M. M., Indian J. Pure and Appl. Phys., **35**, 624-635, (1997)
2. Browner R .F. and Buorn A. N., Anal. Chem., **56**, 786, (1984)
3. Ng. C. K. and Caruse J. A., Appl. Spectros., **39**, 719, (1985)
4. Browner R. F and Buorn A. N., Anal. Chem. **56**, 875, (1984)
5. Mohamed M. M, Shalaby E. A. , Ghatass Z. F., Naim M. A., El-Raey M. , Indian J Pure and Appl Phys., **36**, 719 - 728, (1998).
6. Mohamed M.M., Ghatass Z. F., Shalaby EA, Kotb M. M., El-Raey M. , Fresenius, J anal chem., **368**, 809 - 815, (2000).
7. Shields J. P., Lee G. H., Piepmeier H. E. , Appl. spectrosc **42**, 693 - 704, (1988).
8. Kal nicky D. J., Fassel V. A. , Kniseley R. N., Appl. spectrosc. **31**, 137 - 150, (1977).
9. Walters P.E, Chester T.L., and Winefordner J.D., Appl. Spectrosc. **31**, 1, (1977).
10. Boumans PWJM Excitation spectra in analytical emission spectroscopy, El Grove, Ed. Dekker, New York, (1972).
11. Thome A. P., Spectrophysics, Chapman & Hall, London, (1974).
12. Jarosz J., Mermet J. M. , Robin J. P, Spectrochim Acta **33B**: 55 - 78, (1978).
13. Furuta N., Spectrochim Acta **40B**: 1013 - 1022, (1985).
14. Montaser A., Golightly D. W., Inductively coupled plasma in analytical atomic spectrometry, VCH publications, Inc., New York, (1992).

433

Determination of Nickel, Vanadium and Iron in Crude Oil by Three-Phase Plasma Arc Spectrometry

Zekry F. Ghatass

Institute of Graduate Studies and Research, Alexandria University, Egypt
163 Horrya Avenue, P.O. 832 Shatby, Alexandria, EGYPT (z_ghatass@yahoo.com)

Abstract. Three-phase plasma arc (TPPA) with ultrasonic nebulizer is developed for simultaneous determination of trace elements in crude oil samples. Ultrasonic nebulizer is used instead of pneumatic nebulizer in order to minimize the problems caused by the oil viscosity during the operation. This system was used for determination of some trace elements (V, Ni, and Fe) in a crude oil samples. Methyl isobutyl ketone (MIBK) was used to dilute the oil samples. The TPPA instrument offers several advantages including a low cost power supply with no radio frequency, linear dynamic ranges from 4 to 5 of orders of magnitude, and detection limits (0.121, 0.313 and 0.242 (μg/ml) for Ni, V and Fe respectively. The average concentrations were 31 ± 0.45 (μg/ml) for Ni, 40 ± 0.88 (μg/ml) for V and 8 ± 0.74 (μg/ml) for Fe at Balaaiem fields and 2 ± 0.05 (μg/ml) for Ni, 4.8 ± 0.25 (μg/ml) for V and 2 ± 0.10 (μg/ml) for Fe at Western Desert fields.

INTRODUCTION

Determination of heavy metals in the crude oil is important from several points of views. From the environmental point of view the knowledge of the heavy metal content in the crude oil gives an idea of the amount of heavy metal fall-out into the environment as result of the burning of the oil fuel. Crude oil contains many metals, some of which may be found in high concentrations, such as nickel and vanadium [1]. Many of these metals are toxic to humans if present at relatively high levels and exposure to the contaminated air is prolonged. Some of the metals that can be found in crude oils are known to be phytotoxic and zootoxic [1].

Selection of the best sample introduction procedure for an analysis requires consideration of a number of points: (1) the type of sample (solid, liquid or gas), (2) the level and also the range of levels for elements to be determined, (3) the accuracy, and precision, required, and (4) the amount of material available. The high viscosity and flammability of many oils and gasolines cause difficulties when these samples are analyzed by plasma atomic emission spectrometry with pneumatic nebulizer. Nevertheless, many schemes have been developed to overcome these problems. Boumans and Lux-Steiner [2] modified an inductively coupled plasma spectrometry-atomic emission spectrometer (ICP-AES) to analyze both aqueous and organic liquids. Detection limits for various metals in methyl isobutyl ketone (MIBK) and in oil diluted with MIBK were reported. Broekaert et al [3] used a 4 kW argon-nitrogen ICP for multi-element

CP645, *Spectral Line Shapes: Volume 12, 16ᵗʰ ICSLS*, edited by C. A. Back

determinations in organic solution. Oil samples were diluted 1:10 with xylene, and detection limits for Al, Cu, Fe, Mg, Mn, Si, and Zn were reported to be in the sub microgram per gram range. King et al [4] compared flame atomic absorption spectroscopy (AAS) and ICP-AES results for the determination of wear metals in used lubricating oils. The data obtained by ICP-AES compared very favourably with the results from AAS. Recently, we have developed a three-phase plasma arc system for simulations determination of trace metals in environmental and biological samples. In the present work, ultrasonic nebulizer was used to introduce fine aerosol of samples into the plasma. The application possibility of three-phase plasma arc with ultrasonic nebulizer for determination of some trace elements (V, Ni, and Fe) in a crude oil samples was carried out.

EXPERIMENTAL WORK

Three-phase plasma arc system has been previously described [5] and will be only briefly reviewed here. The major constituents of this plasma system (Fig.1) are (i) three-phase power supply, (ii) plasma torch, (iii) sample introduction and (iv) detection system. Three-phase power supply of the plasma system has been previously described in details [5]. The plasma torch consists of three concentric quartz tubes similar in size to those of a typical inductively coupled plasma (ICP) torch [5]. The samples were introduced to the plasma via ultrasonic nebulizer. The combination of ultrasonic nebulizer and three-phase plasma arc offers desirable features of low detection limits, high sample through out, wide dynamic range of operation, acceptable precision and accuracy, and simultaneous quantitative analytic capabilities. Moreover, the ultrasonic nebulizer does not require sample preconcentration. The spectrometer (Czerny-Turner spectrometer model 1704 Spex) was set on a vertically adjustable table and the optical axis was adjusted with the He-Ne laser which was installed behind the three-phase plasma source. The output from the spectrometer is detected by photomultiplier tube (PMT). The output of the PMT is fitted to X-T recorder and optionally a PC.

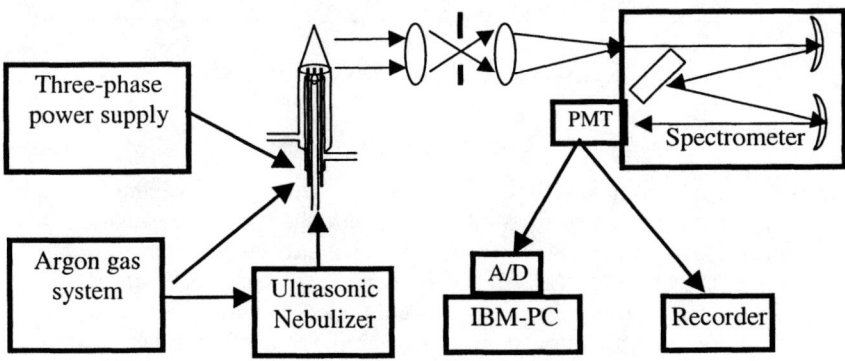

FIGURE 1. Schematic diagram of three-phase plasma system

The effects of the gas flow rate, current; excitation temperatures have been carried out in details. The optimum conditions are shown in Table (1).

Sampling and Dilution Procedures

Alexandria Petroleum Company provided crude oil samples that have been used to conduct a trace metal analysis. The trace metal analysis was carried out for different types of Egyptian crude oil from Western Desert and Balaaiem fields. The sample and blank oils were diluted 2:20 w/v with methyl isobutyl ketone (MIBK) i.e. 2 g of oil was diluted to 20 ml with MIBK [6]. Each sample was divided into two parts, one part was analyzed by the present technique (TPPA) while the other analyzed by atomic absorption spectroscopy as a standard technique at Alexandria Petroleum Company.

TABLE 1. Experimental conditions.

Three phase power supply		Detection	
-Operating current	27 A	system	
-Operating voltage	55 V	-Spectrometer	1 m focal length, Spex,
-Resistance series	0.3 Ω		model 1704, 1200
-Y transformer	7.5 KW		groves/mm and 500 nm
-Electrodes (thoriated tungsten)	1 mm		blazed diffraction grating
Plasma torch assembly			
(all of fused quartz construction)			
-Diameter of outer quartz tube	22 mm	-Enter slit	50 μm
-Diameter of inner quartz tube	15 mm	-Outer slit	50 μm
-Diameter of sample introduction tube	5 mm	-PMT	PRA (R-955), operated at 1200 V
Ultrasonic nebulizer		-Lens	Quartz (12 cm fl)
-Frequency	1.8 MHz		
-Particle size	0.5-8 micron	-Plasma	(3:1) on the enter slit
-Sample uptake	2.25 ml/min	image	
Argon flow rate			
-Outer plasma argon flow rate	3 L/mim	-Chart	10-V strip chart recorder
-Inner plasma argon flow rate	1.5 L/min	recorder	
-Nebulizer gas flow	1 L/min		

RESULTS AND DISCUSSIONS

Background Emission

One of the most steps in developing an emission spectroscopic technique is studying the background spectrum. A spectrum of the background emission was obtained at an observation height of 2 mm relative to the top of the touch over the wavelength region 200-600 nm, as shown in Fig. (2). All scans are not at the same scale sensitivity due to the fact that the optical efficiency of the spectrometer varied over the 200 – 600 nm wavelength region. The background signal increased as wavelength increased from 350 to 600 nm. In addition to an actual change in the emission intensity, this signal change may be due in part to the fact that the grating of the monochromator was blazed at 500 nm, and as the scan moves away from the wavelength at which the grating was blazed, the stray light increases [7]. Important features should be taken into account with respect to

436

background spectrum. First, there are no Ar transitions between 200 to 330 nm. Second, the most intense emission are in the region from 350-450 nm. The peaks in the background emission spectrum corresponding to tungsten emission liens are very sharp. The tungsten emission results from the use of 98% tungsten electrodes while the other 2% is thorium.

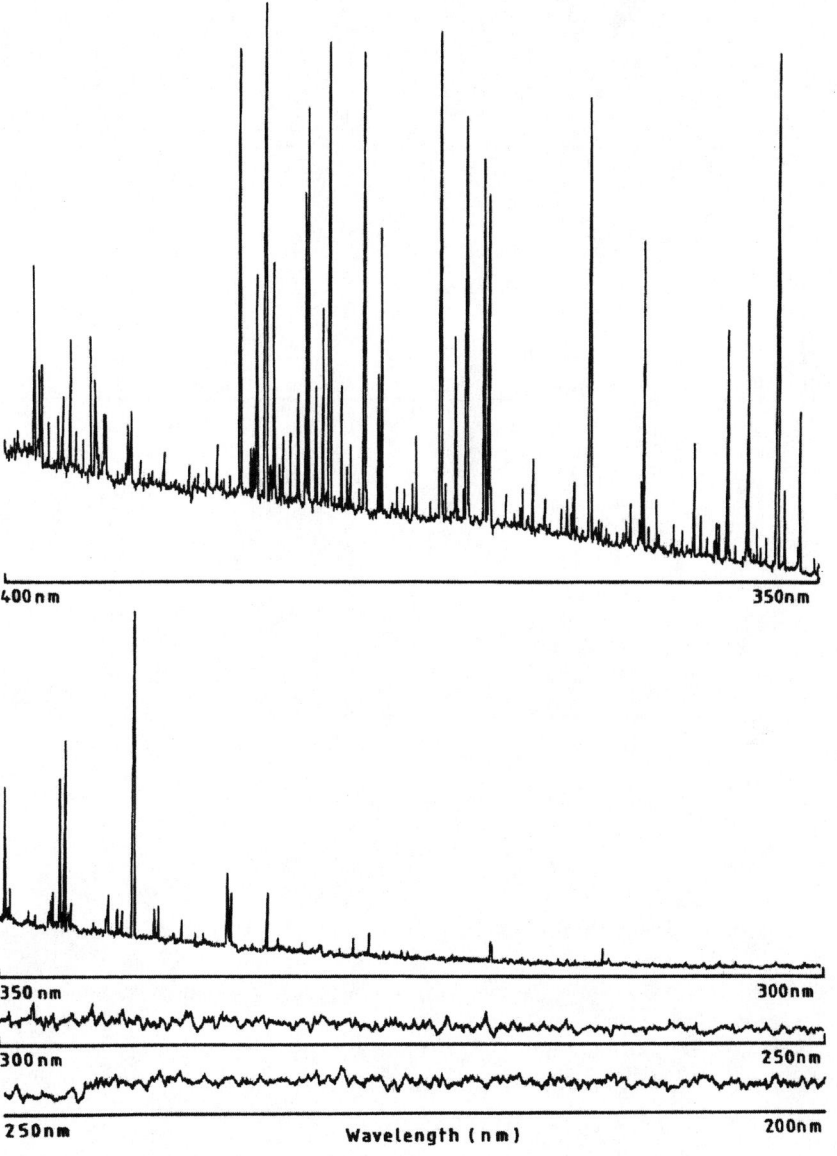

FIGURE 2. Background Spectrum.

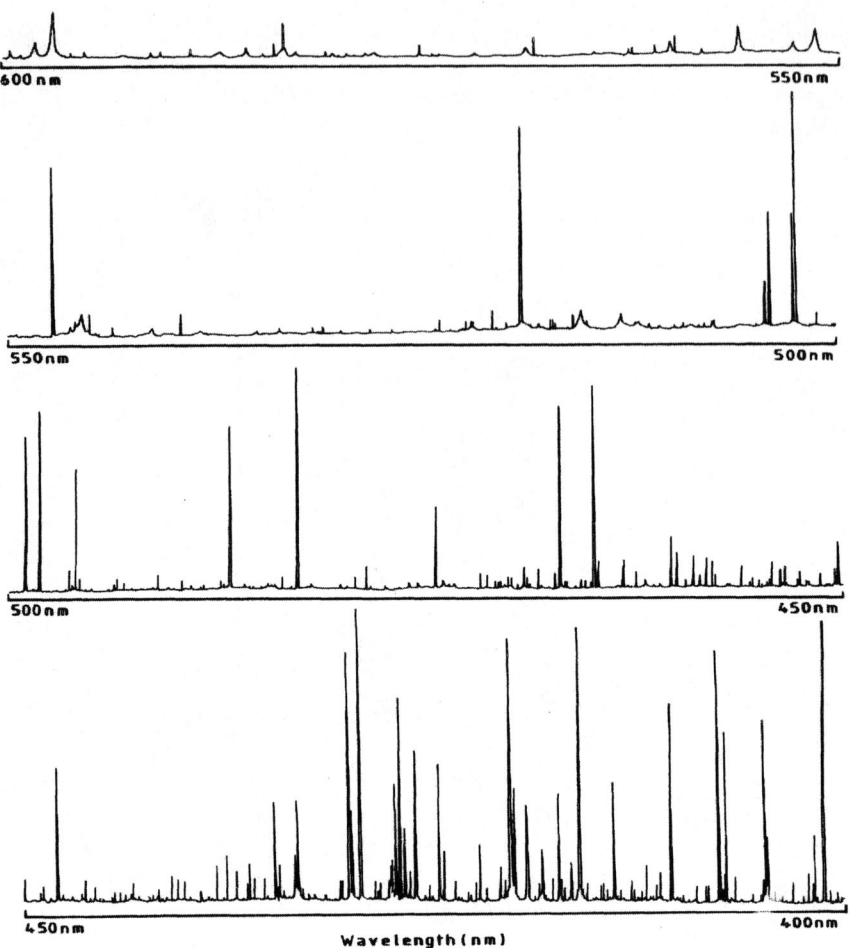

FIGURE 2. Cont.

Excitation Temperature

Since the atom reservoir is used to thermally excite the analyte in atomic emission spectrometry, the excitation temperature is of considerable importance. The excitation temperature is one of the most parameters, which govern the excitation mechanisms of a three-phase plasma arc. In the present work the excitation temperature, T_{exc} is determined by fitting a thermal distribution to the appropriately weighted intensities of a set of atomic transitions for a specific atomic species in the plasma. The atomic species used is argon.

The argon lines of interest fall in the range between 425.118 nm to 434.516 nm as shown in Fig. (3). Each observed Ar wavelength has an associated energy level, E_q, a statistical weight, g_q, and transition probability, A_{qp}, T_{exc} can be calculated from the slope of a plot log $(I_{qp}\lambda/ g_q A_{qp})$ *vs* E_q [8]. The excitation temperature was carried as function of current. There is an increase in excitation temperature with arc current. This increase in excitation temperature in plasma plume with current occurs because of the increase in power dissipation for a constant flow rate of inert gas. The excitation temperatures in this study were varying between 4500 and 6500 K.

Choice of Wavelengths for Crude Oil Analysis

Wavelength scans of a typical group of analysis lines and the concentrations are shown in Fig. (4). The wavelengths were chosen for crude oil analysis after consideration of number of factors, the expected concentration of each element in the sample solutions, prominence of the line, freedom from spectral interferences, and sufficient range of linearity.

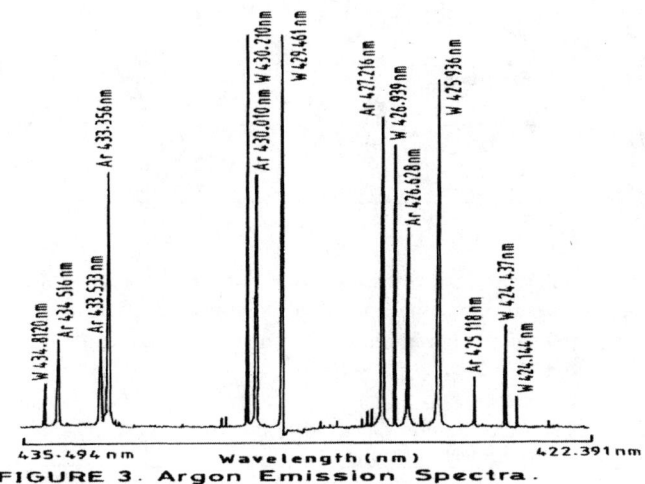

FIGURE 3. Argon Emission Spectra.

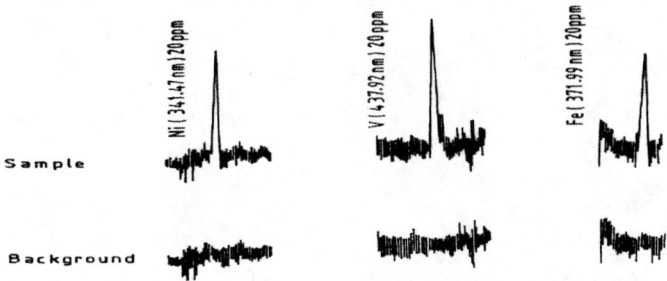

FIGURE 4. Wavelength Scans of Typical Analysis Lines.

Detection Limits and Analytical Curves

Detection limits for three elements Ni (341.47 nm), V (437.92 nm) and Fe (371.99 nm) are given in Table (2). The detection limit values represent the concentration, which yields emission intensity two times the standard deviation of the blank signal [9]. Table (3) shows the slopes of Log-Log emission concentration curves and the trace elements concentration that determined by our system (TPPA) and atomic absorption spectroscopy (AAS). Five to eight concentrations were used for each curve. The slope of the three elements close to one these indicate that there are linear relationship between the emission signal and concentration. In each case the point for the highest concentration clearly fell below a straight line and used omitted for least squares fit. Table (4) demonstrates a comparison between the obtained results in the present work with those reported by other investigators for crude oil from different parts of the world. These results show that the values of Ni, V and Fe of the investigated samples lies within the range reported for other crude oil.

TABLE 2. Detection Limits.

Element	Wavelength λ (nm)	Excitation potential (eV)	Detection Limits (μg/ml)			
			Present Work	[10]	[11]	[12]
Fe	371.99	3.3	0.242	0.03	0.02	0.52
Ni	341.47	3.6	0.121	0.12	---	0.11
V	437.92	3.1	0.313	---	0.01	---

From the results obtained in the present work, it can be seen that the combination between ultrasonic nebulizer and three-phase plasma system offer sensitive technique for the determination of trace elements in crude oil and similar environmental samples.

TABLE 3. Comparison Between Results Obtained from Three-Phase Plasma Arc Spectrometry (TPPA) and Atomic Absorption Spectrometry (AAS).

Element	Wavelength λ (nm)	Log-Log Slope of calibration curves	Concentration (μg/ml)			
			Balaaiem		Western Desert	
			TPPA	AAS	TPPA	AAS
Fe	371.99	0.862	8 ± 0.74	8 ± 0.52	2 ± 0.10	2 ± 0.013
Ni	341.47	1.025	31 ± 0.45	30 ± 0.81	2 ± 0.05	2 ± 0.024
V	437.92	0.902	40 ± 0.88	40 ± 1.02	4.8 ± 0.25	5 ± 0.017

TABLE 4. Comparison Between the Obtained Results in the Present Work With Other Reported for Crude Oils.

Place	Concentration (μg/ml)		
	Fe	Ni	V
Present work			
Balaaiem	**8 ± 0.74**	**31 ± 0.45**	**40 ± 0.88**
Western Desert	**2 ± 0.10**	**2 ± 0.05**	**4.8 ± 0.25**
Gulf of Suez [12]			
Well No. 1	256.4	20.85	31.80
Well No. 2	142.4	33.8	38.82
Well No. 3	243	24.54	38.82
Well No. 4	135.5	22.24	27.2
Soudi Arabia [12]	---	22	68
Venezuela [12]	---	83	377
Red River [12]	---	2.95	0.13
Nigerian [12]	---	4.0	0.4
Iraq [12]	---	---	24.7

CONCLUSION

The combination of ultrasonic nebulizer and three-phase plasma arc illustrated a practical technique that has been used to measure trace elements in crude oil samples. It is expected that the same technique could be applied in measuring heavy metals in environmental samples. The developed technique could overcome several problems associated with pneumatic nebulizer used in previous work.

ACKNOWLEDGEMENT

The author would like to thanks Prof. M. El-Raey for his suggestions and comments, Prof. M.M. Mohamed for his interest, helpful discussions, and reviewing the manuscript, and Prof. S.K. Guirguis for his assistance in preparation and reviewing of the manuscript.

REFERENCES

1. Sadiq, M., and Mian, A. A., Atomos. Environ. **28**, pp 2249-2253 (1994).
2. Boumans, P. W. J. M., and Lux-Steiner, M. C., Spectrochim. Acta, **37B**, pp 3-7, (1984).
3. Broekaert, J. A. C., Keis, F., and Laqua, K., Talanta, **28**, pp 745-752 (1981).
4. King, A. D., Hilligoss, D. R., and Wallace, G. G., At. Spectrosc. **5**, pp 189-191, (1984).
5. Mohamed, M. M., Shalaby, E. A., Ghatass, Z. F., Naim, M. A., El-Raey, M., Indian J Pure and Appl. Phys **36**; pp 719-728 (1998).
6. Fassel, V. A., Peterson, C. A., Abercrombie, F. N., and Kniselsy, R. N., Anal Chem ,**48**, pp 516-519, (1976).
7. Mattoon, T. R., and Piepmeier, E. H., Anal chem. **55**, pp1045-1050 (1983).
8. Kalinicky D. J, Fassel V. A, Kniseley R. N, Appl spectrosc. **31**: pp137-150 (1977)
9. Montaser, A., and Golightly, D.W., Inductively Coupled Plasma in Analytical Atomic Spectrometry, VCH, (1992) pp. 318-323
10. Mohamed, M. M., and Ghatass, Z. F., Fresenius J Anal Chem. **368**: pp 449-455, (2000).
11. Rippetoe, W.E., Johnson, E. R., and Vickers, T. J., Anal. Chem, **47**, pp 436-440 (1975).
12. Mahdy, A. A., Metawe, F., Fakhry, A. A., Eid, M. A., and Ashkar, E. A., Regional Symp. Environ. Stud. (UNARC), Alexandria, Egypt, Ed. M. El-Raey, 1990, pp. 160-176.

X-RAY LASERS

Importance of line shape calculations in x-ray laser modelling

D. Benredjem [*], C. Mossé [†], S. Ferri [†], B. Talin [†] and C. Möller [*]

[*] Laboratoire d'Interaction du Rayonnement X avec la Matière, UMR 8624
Université Paris-Sud, Centre d'Orsay, Bât. 350
91405 Orsay Cedex, France

[†] Laboratoire de Physique des Interactions Ioniques et Moléculaires, UMR 6633
case 232, Université de Provence, Centre de Saint-Jérôme
13397 Marseille Cedex 20, France

Abstract. Modelling of x-ray lasers shows the need for accurate line shape calculations.
We consider two pumping schemes, namely the recombination and the collisional excitation schemes. In the first one, inversion between the populations of the lasing levels occurs during the cooling of the plasma, i.e., after the irradiation of the plasma by an intense laser. We focus our attention on the C^{5+} x-ray laser line at 18.2 nm. This $3d$-$2p$ line shows an important Stark broadening and a negligible Doppler broadening. In the collisional scheme, the inversion occurs during the heating of the plasma. In this case, we concentrate on the Ne-like (10 bound electrons) germanium $3d$-$3p$ lasing line at 23.6 nm. In contrast to the carbon line, this transition shows a negligible Stark broadening and an important Doppler broadening.
We have calculated the emissivity of the amplifying medium taking photon scattering (2nd order process) into account. This process involves the redistribution function which in certain circumstances can be factorized into the product of a frequency redistribution function and a phase function. In fact, in the laboratory frame, the factorization occurs when one neglects the elastic electron-ion collisions which destroy the alignment. We have used the absorption line shapes given by the frequency fluctuation model and the frequency redistribution functions in the laboratory frame obtained by a recent code.
We have also calculated the emission profile. It depends on the x-ray laser intensity and on the photon scattering angle.

INTRODUCTION

In laser-produced plasmas of interest for x-ray lasers, spectral lines are generally broadened by Doppler effect, but in some cases, fine structure and ionic Stark effect also play a role in line width. Many properties of x-ray beams propagating in these media depend strongly on spectral line shapes. An accurate knowledge of the line profile, in particular its width, is thus necessary. The development of x-ray laser (XRL) applications [1] necessitates the improvement of the optical properties of the beam, e.g. coherence [2] and polarization [3]. For example, the temporal coherence of a quasi-monochromatic radiation of wavelength λ is maintained during the coherence

CP645, *Spectral Line Shapes: Volume 12, 16th ICSLS*, edited by C. A. Back
© 2002 American Institute of Physics 0-7354-0100-4/02/$19.00

time τ, $\tau = \gamma\lambda^2/(c\Delta\lambda)$, where γ (of the order unity) is a factor depending on the line profile and $\Delta\lambda$ is the linewidth. τ gives a useful estimate of the characteristic time during which a given polarization state of the macroscopic field is maintained, especially when collisions play an important role in the broadening. In the temporal coherence domain, collisional processes influence the line shape, in conjunction with other effects such as ion motion.

The frequency detuning due to the thermal motion of the ions is responsible for an inhomogeneous broadening, while finite lifetime and collisions give rise to a homogeneous broadening. In contrast to a recent work on spatial coherence [4], it can be shown for neon-like germanium lasing lines that the inhomogeneous broadening overcomes the homogeneous broadening.

Intensity calculations generally involve first-order processes only, i.e., emission or absorption. Higher-order processes such as the coherent scattering of photons are generally ignored, despite the fact that they can play an important role in radiative transfer in dense laboratory plasmas. Second-order processes where the absorption of a photon of frequency ν' is followed by the emission of a photon of frequency ν, occur with a probability involving the frequency redistribution function (FRF). An important advance has been achieved by Omont and co-workers [5] who investigated the FRF for a 3-level atom with the help of a quantum-mechanical description of matter and radiation.

A model for the calculation of the redistribution function in the laboratory frame, based on the Frequency Fluctuation Model (FFM) [6], has been developed [7]. The second-order process mentioned above gives rise to an additional term in the emissivity, whose importance increases with XRL intensity, and which becomes negligible when the collisional deexcitation rate overcomes the spontaneous emission rate.

The objectives of this work are (i) the calculation of the emission line shape and the emissivity in the laboratory frame, and (ii) the investigation of the transport of XRL lines in neon-like germanium (lasing line at 23.6 nm). This line is amplified in the collisional excitation scheme, and saturation has been reported [8]. Our approach of radiation transfer combines the Maxwell wave equation and the Bloch relations, yielding a set of population equations where the interaction of the x-ray beam with the plasma is explicitly taken into account [9].

ABSORPTION PROFILE, FREQUENCY REDISTRIBUTION FUNCTION

The absorption profile is calculated in the FFM. In addition to a streaming motion with a velocity profile, the amplifying medium presents a random thermal motion in the laboratory frame; thus the absorption profile of a moving atom differs from that of the same atom at rest, due to the Doppler frequency shift. Moreover, in the study of second-order processes, the absorbed and the emitted photons travel in different directions; the projection of the atom velocity vector along the wave vectors will then be different for the two photons, and a differential Doppler shift occurs.

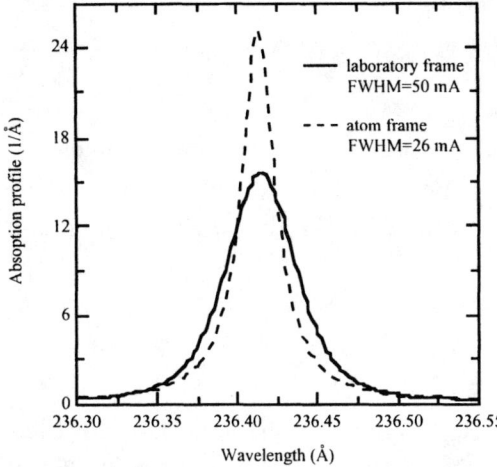

Figure 1 shows the absorption profile calculated in the atom frame and in the laboratory frame for the germanium 2-1 line. The electron density, electron and ion temperatures are fixed at $N_e = 7 \times 10^{20}$ cm^{-3}, $T_e = 700$ eV and $T_i = 350$ eV, respectively. The Doppler shift is responsible for a large inhomogeneous broadening. In fact, the FWHM calculated in the laboratory frame is larger than the FWHM obtained in the atom frame.

FIGURE 1. Absorption profile of the 2-1 x-ray laser line.

Let us consider an atom moving with velocity **v** and let a photon, propagating in direction **n'** within an element of solid angle $d\Omega'$, have its frequency in the range $(\nu', \nu'+d\nu')$ in the laboratory frame. Our purpose is to study its absorption and re-emission by the moving atom. The joint probability that this photon is absorbed and that a photon is emitted in the frequency range $(\nu, \nu+d\nu)$, with a direction of propagation given by **n** within the solid angle $d\Omega$ is $R_{\mathbf{v}}(\nu', \mathbf{n}'; \nu, \mathbf{n})d\nu'd\nu$ $(d\Omega'/4\pi)(d\Omega/4\pi)$, where $R_{\mathbf{v}}$ is the redistribution function [10]. The joint probability is normalized to unity, i. e.,

$$\iint \iint R_{\mathbf{v}}(\nu' \, \mathbf{n}'; \nu, \mathbf{n})d\nu' \, d\nu \frac{d\Omega'}{4\pi} \frac{d\Omega}{4\pi} = 1. \tag{1}$$

It is clear that integration over ν and Ω gives the absorption profile $\Phi_{\mathbf{v}}$ in the laboratory laboratory frame, whose velocity in the atom frame is -**v**:

$$d\nu' \frac{d\Omega'}{4\pi} \iint R_{\mathbf{v}}(\nu', \mathbf{n}'; \nu, \mathbf{n})d\nu \frac{d\Omega}{4\pi} = \Phi_{\mathbf{v}}(\nu')d\nu' \frac{d\Omega'}{4\pi}, \tag{2}$$

with the normalization condition $\iint \Phi_{\mathbf{v}}(\nu')d\nu' \, d\Omega' = 4\pi$.

Let v_0 be the central frequency of the line and let us assume that the velocity of the atom remains fixed during the scattering process. We then have the following simple relations between the frequencies:

$$v' = \xi' + \frac{v_0}{c}\mathbf{n'} \cdot \mathbf{v},\tag{3}$$

$$v = \xi + \frac{v_0}{c}\mathbf{n} \cdot \mathbf{v},\tag{4}$$

where ξ and ξ' are the frequencies in the atom frame. In going from one frame to the other, $\mathbf{n'}$ and $\mathbf{n}$ are unchanged, to the same order of magnitude. Using the above relations, we can easily express R_v in terms of r and ϕ (FRF and absorption profile in the atom frame). We have:

$$R_v(v',\mathbf{n'};v,\mathbf{n}) = \phi(v' - \frac{v_0}{c}\mathbf{n'} \cdot \mathbf{v}) r(v' - \frac{v_0}{c}\mathbf{n'} \cdot \mathbf{v}, v - \frac{v_0}{c}\mathbf{n} \cdot \mathbf{v}) P(\mathbf{n'},\mathbf{n}),\tag{5}$$

where P is the phase function. The subscript $\mathbf{v}$ recalls that the frequency redistribution is produced by an atom of velocity $\mathbf{v}$ in the laboratory frame. To find the net result over all atoms, we must average over the velocity distribution, which is assumed to be Maxwellian, and obtain the line profile $\Phi(v)$ and the redistribution function $R(v',\mathbf{n'};v,\mathbf{n})$.

The calculation of the frequency redistribution function (FRF) is a very complicated problem, due to the atomic structure of multi-electron emitters and to the inhomogeneous and homogeneous line-broadening mechanisms that result from the combination of rapid and slow fluctuations of the microfield perturbation.

The most common assumption, strong mixing limit or complete redistribution, is equivalent to postulating that the scattered resonance radiation has the same spectral shape as the absorption line. More generally, the partial redistribution hypothesis results in fluorescence with a spectral shape consisting of a sharp coherent (Rayleigh) peak centered on the frequency of the absorbed radiation and a redistribution line emitted near the resonance frequency.

Let us concentrate on the angular variation of the FRF. We assume that the elastic electron-ion collisions destroying the alignment are negligible. In this case, the FRF takes the form (see for example Ref. 11, with $c = 0$):

$$R(v',\mathbf{n'}; v,\mathbf{n}) = R(v', v)P(\mathbf{n'},\mathbf{n}),\tag{6}$$

where $P(\mathbf{n'},\mathbf{n})$ is represented by the following 4x4 matrix:

$$P(\mathbf{n'},\mathbf{n}) = \begin{pmatrix} 1 - \frac{1}{4}W_2(1-3\cos^2\Theta) & -\frac{3}{4}W_2\sin^2\Theta & 0 & 0 \\ -\frac{3}{4}W_2\sin^2\Theta & \frac{3}{4}W_2(1+\cos^2\Theta) & 0 & 0 \\ 0 & 0 & \frac{3}{2}W_2\cos\Theta & 0 \\ 0 & 0 & 0 & \frac{3}{2}W_1\cos\Theta \end{pmatrix} \quad (7)$$

where W_i ($i = 1, 2$) are coefficients depending on $6j$-symbols [12] and Θ is the angle formed by the two vectors $\mathbf{n'}$ and $\mathbf{n}$. The 3×3 matrix whose entries depend on W_2 involve linear polarization while the 1×1 matrix depending on W_1 involve circular polarization. In the general case, intensity is a vector formed of four Stokes parameters, $I = (I^{(0)}, I^{(1)}, I^{(2)}, I^{(3)})$. As we will see below, the emission profile and the emissivity involve the product of the FRF with intensity. The result is a 4 components vector. However, when one deals with non-polarized radiation, only $I^{(0)}$ is different from zero and then only the P_{11} entry need be considered.

The W_2 values are tabulated in Ref. 12. $W_2 = 0.35$ for the 2-1 line and $W_2 = 0$ for the 0-1 line (19.6 nm in germanium), which means that the second line does not show angular variation.

EMISSION LINE PROFILE

The emission line shape is written as a function of the absorption profile and of the FRF. We have [13]

$$\Phi_{emiss}(\nu,\mathbf{n},z) = \frac{\dfrac{B_{lu}}{4\pi}\iint R(\nu,\mathbf{n'};\nu,\mathbf{n})I(\nu',\mathbf{n'},z)d\nu'\,d\Omega' \quad + \quad C_{lu}\Phi(\nu)}{\dfrac{B_{lu}}{4\pi}\int\Phi(\nu')I(\nu',\mathbf{n'},z)d\nu'\,d\Omega' \quad + \quad C_{lu}}, \quad (8)$$

where C_{lu} is the collisional excitation rate from the lower to the upper lasing level. The emission profile is normalized in the sense that

$$\frac{1}{4\pi}\iint\Phi_{emiss}(\nu,\mathbf{n},z)d\nu d\Omega = 1. \quad (9)$$

It depends on the amplification length z. We assume that the XRL beam propagates in the z direction of a right-handed orthonormal (laboratory) frame $Oxyz$. Then, $\mathbf{n'}\equiv\mathbf{u_z}$ and

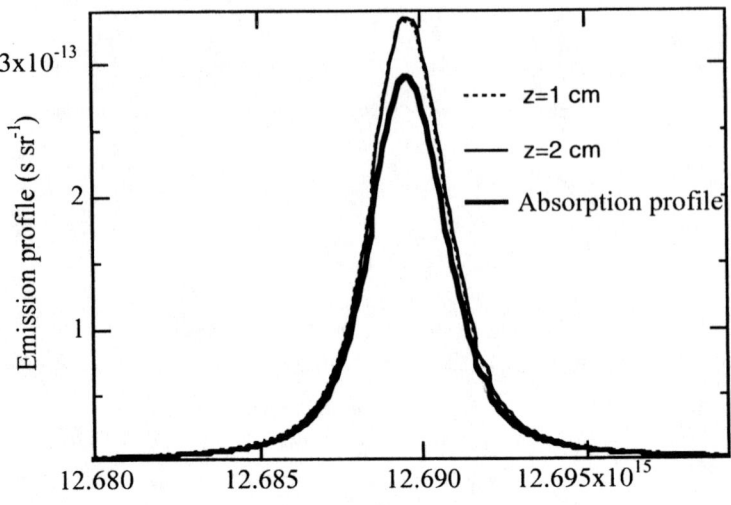

FIGURE 2. Emission and absorption profile of the 2-1 lasing line in Ne-like germanium (23.6 nm) for two amplification lengths and a scattering angle fixed to 0°.

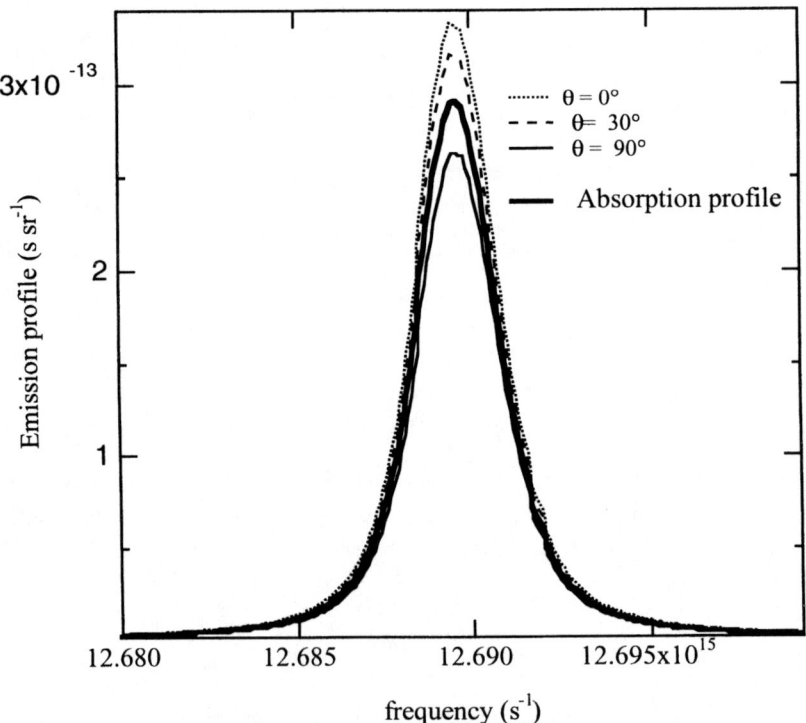

FIGURE 3. Same as Fig. 2 for $z = 1$ cm and a varying scattering angle.

Θ is the angle between the direction of the scattered photon **n** and the z axis.

In Figure 2, we show the emission and absorption profiles of the 2-1 line in neon-like germanium, calculated in the conditions of Figure 1. These conditions are obtained by modelling an experiment performed at the Rutherford Appleton Laboratory [3]. The scattering angle is fixed to 0 and we vary the amplification length z. We can see that the emission profile does not vary much with z. In Fig. 3, we show the angular dependence of the emission profile. The intensity of the X-ray laser is determined by an amplification length of 1 cm. The emission profile varies appreciably with the scattering angle, with a maximum for $\Theta = 0°$, which means that there is a preferential direction of emission, i.e. the direction of amplification.

We have also modeled the experiment of Zhang et al [14]. The Hα line of H-like carbon shows a strong enhancement of the output for $N_e = 1.4 \times 10^{19}\,\text{cm}^{-3}$, $T_e = 9$ eV and $T_i = 5$ eV. Calculations of the absorption profile in the frequency fluctuation model shows an important Stark broadening.

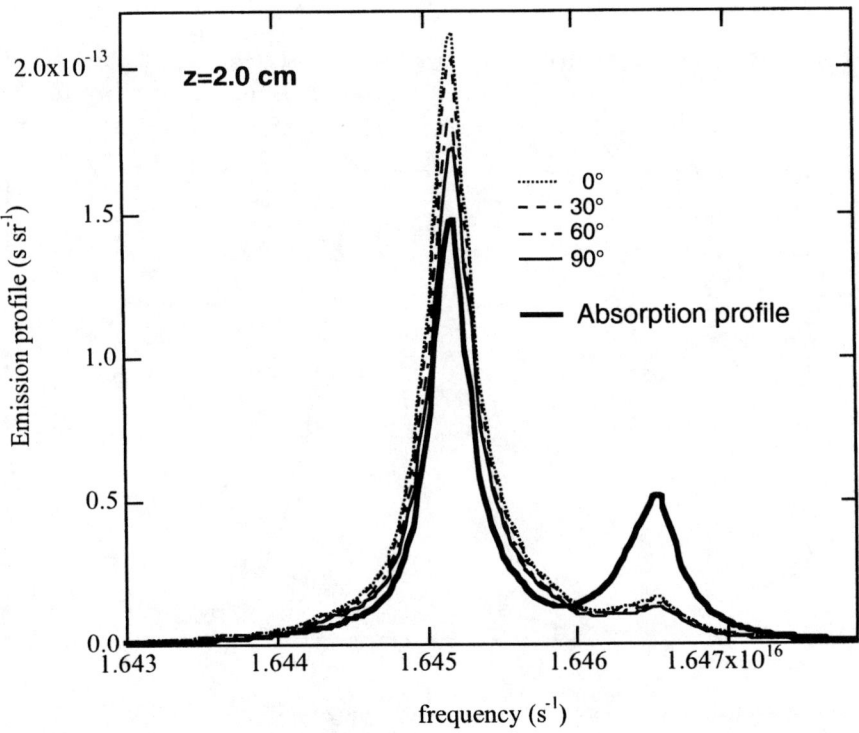

FIGURE 4. Emission and absorption profile of the Hα lasing line in H-like carbon (18.2 nm) for $z = 2$ cm.

Figure 4 represents the emission profile for four scattering angles. As clearly seen, the shape of the emission line differs from that of the absorption line. In particular, the blue component is nearly absent. Owing to the gain narrowing, the XRL photons are absorbed at the frequency of the red component.

EMISSIVITY

When one accounts for frequency redistribution, the emissivity if formed of two contributions, the first one involves spontaneous emission and the second one accounts for second-order processes. The emissivity can be written as:

$$\sigma_{ul}(\nu, \mathbf{n}, z) = \frac{h\nu}{4\pi} N_u(z) A_{ul} \Phi(\nu)$$
$$+ (1 - \varepsilon) N_l(z) \frac{B_{lu}}{c} \frac{h\nu}{4\pi} \int d\Omega' \int d\nu' \, I_{ul}(\nu', z) R(\nu', \mathbf{n}'; \nu, \mathbf{n}), \tag{10}$$

where A_{ul} is the rate of spontaneous emission. The second contribution in the rhs involves the x-ray laser (pump) intensity, and depends on the number of photons

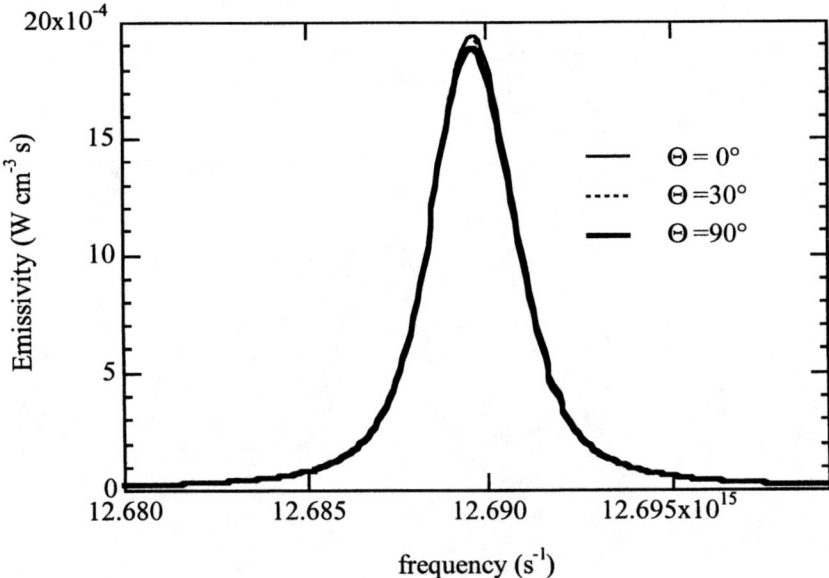

FIGURE 5. Emissivity of the 2-1 lasing line in Ne-like germanium (23.6 nm) for $z = 2$ cm and three values of the scattering angle.

452

absorbed at frequency v', given by $N_l B_{lu} I(v',z)$. The value of the multiplying factor $(1-\varepsilon)$ is determined by the relative rates of collisional deexcitation C_{ul} and spontaneous emission between u and l. In fact, we have, $\varepsilon = C_{ul}/[A_{ul}+C_{ul}]$. When collisional deexcitation overcomes spontaneous emission, we have $\varepsilon \approx 1$, and thus the contribution of redistribution to σ is negligible.

In Fig. 5, we show the emissivity at 23.6 nm in neon-like germanium, for various scattering angles and a propagation length fixed to 2 cm. The correction to the emissivity due to frequency redistribution does not vary with the scattering angle. The Hα line in H-like carbon is more sensitive to the photon scattering angle (see Fig. 6).

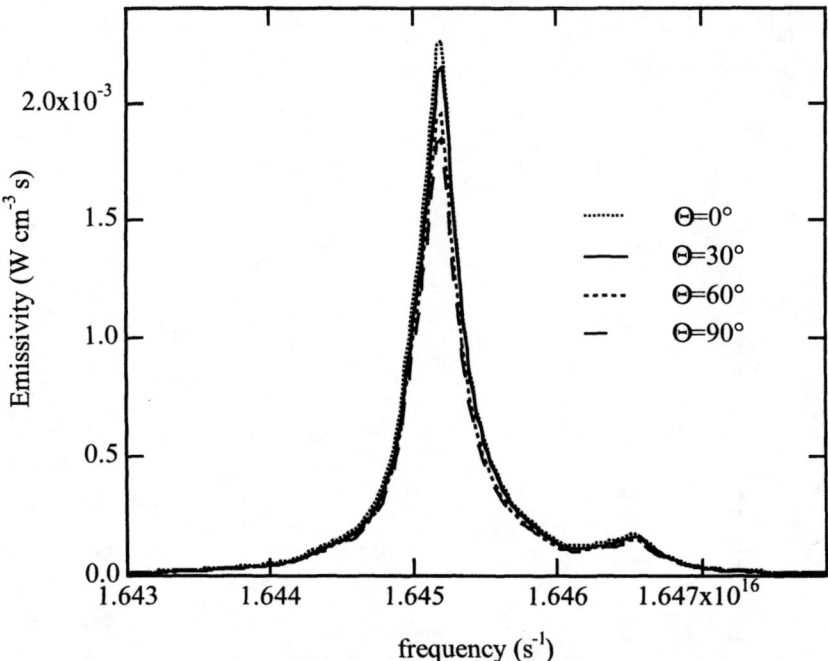

FIGURE 6. Same as Fig. 5 for the Hα line in H-like carbon. Amplification length $z = 2$cm.

The calculation of the emission and emissivity profiles requires the knowledge of the intensity of the x-ray laser. This intensity is obtained by solving the radiative transfer equation:

$$\frac{\mathbf{k}}{k} \cdot \nabla I(v,z,t) = \sigma(v,z,t) + G(v,z,t)I(v,z,t), \tag{11}$$

where **k** is the wave vector of XRL photons and G the local gain (cm^{-1}):

$$G = h\nu \frac{B_{ul}}{c} \left[N_u - \frac{g_u}{g_l} N_l \right] \Phi. \tag{12}$$

The intensity and the gain depend on the population densities which are given by the rate equations:

$$\frac{\partial}{\partial t} N_i(z,t) = -N_i(z,t)\Gamma_i(z,t) + \sum_{j \neq i} \Lambda_{ji} N_j(z,t), \tag{13}$$

where i is a label for ion levels. Γ is the total rate of population decrease and Λ a rate of population increase of the i level.

INTENSITY CALCULATIONS

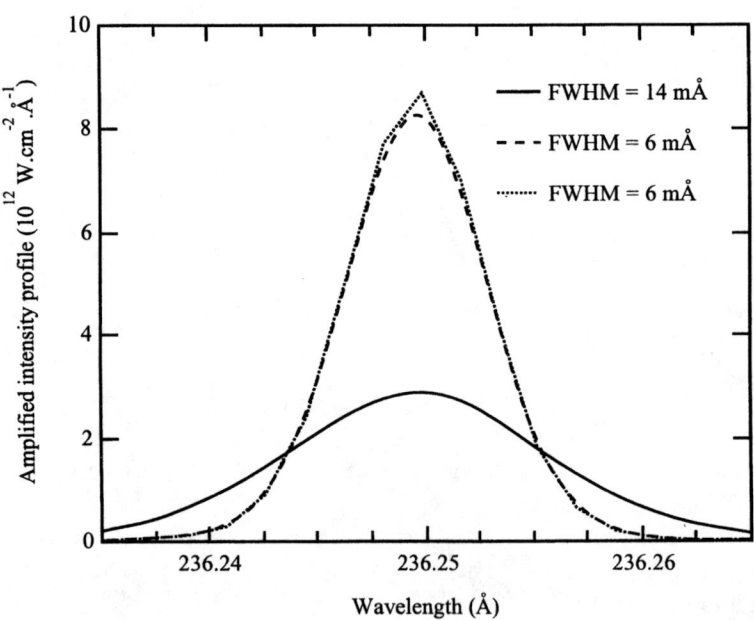

FIGURE 7. Amplified intensity profile of the 2-1 lasing transition. Full curve: line shape and frequency redistribution functions calculated in the laboratory frame; dashed curve: line shape and frequency redistribution functions calculated in the atom frame; dotted curve: lineshaphe calculated in the atom frame, without frequency redistribution. Density and temperatures, as in Fig. 1. The amplification length is fixed to 2 cm.

The coupling of the radiative transfer equation with the population equations gives the intensity as a function of z. We have applied these calculations to the 2-1 line. Figure 7 shows the intensity for large lengths, for which saturation occurs. The effect of frequency redistribution could be noticeable in this regime. The intensities, calculated in the laboratory frame and in the atom frame, with frequency redistribution taken into account in both cases, differ by an appreciable amount. Two competing effects are responsible for the difference:

(i) the inhomogeneous broadening due to frequency detuning reduces the gain by a large factor. Intensity decreases, as a result.

(ii) frequency redistribution is responsible for an increase of intensity, due to the positive sign of the correction to the emissivity.

The comparison of the two profiles relative to the atom frame shows that frequency redistribution plays a small role.

CONCLUDING REMARKS

The inhomogeneous broadening related to the Doppler shift and the redistribution function have been examined in the framework of the frequency fluctuation model. We have accounted for most broadening mechanisms, in particular for ion motion. The angular variation of the FRF is investigated with the assumption that the elastic electron-ion collisions destroying the alignment are negligible.

We have focused our study on the 23.6 nm line in neon-like germanium (for which the Doppler shift was expected to be large) and on the 18.2 nm line in hydrogen-like carbon (large Sark broadening). We have accounted for second-order processes (absorption followed by emission) in the emission, emissivity and intensity profiles. We have examined the angular variation of the FRF and its effect on the above profiles. It appears that the emission profile depends on the angle of the scattered photon and differs by a large amount from the absorption profile. Frequency redistribution affects the emissivity of the Hα line. Finally, in the conditions of the calculations (modelling of the experiment of Rus et al, Ref. 3) the output intensity of the 2-1 line is strongly affected by the inhomogeneous broadening.

REFERENCES

1. Zeitoun, P., et al, *Nucl. Instrum. Methods A* **416**, 189 (1998).
2. Albert, F., et al, *Opt. Commun.* **142**, 184 (1997).
3. Rus, B., et al, *Phys. Rev. A* **51**, 2316 (1995).
4. Larroche, O., et al, *Phys. Rev. A* **62**, 43815 (2000).
5. Omont, A., et al, *Astrophys. J* **175**, 185 (1972).
6. Talin, B., et al, *Phys. Rev. A* **51**, 1918 (1995).
7. Mossé, C., et al, *Phys. Rev. A* **60**, 1005 (1999).

8. Carillon, A., et al, *Phys. Rev. Lett.* **68,** 2917 (1992).
9. Benredjem, D., et al *Phys. Rev. A* **55,** 4576 (1997).
10. Hummer, D.G., *Mont. Not. Royal Astron. Soc.* **125,** 21 (1962).
11. Frisch, H., *Solar Phys.* **164,** 49 (1996).
12. Landi Degl'Innocenti, E., *Solar Phys.* **91,** 1 (1984).
13. Cannon, C.J., *The transfer of Spectral Line Radiation*, Cambridge University Press, 1985.
14. Zhang, J., et al, *Phys. Rev. A* **54,** R4653 (1996).

Measurement of wavelength and linewidth
of transient collisional x-ray lasers

Tetsuya Kawachi, Noboru Hasegawa, Yoshihiro Ochi, Akira Sasaki,
Takayuki Utsumi, Momoko Tanaka, Ingo Uschmann[†], Eckhart Förster[†],
Etsuya Yanase, Masayuki Suzuki, Mamiko Nishiuchi, Keisuke Nagashima
and Hiroyuki Daido

*Advanced Photon Research Center, Kansai-Research Establishment,
Japan Atomic Energy Research Institute (JAERI),
8-1, Umemidai, Kizu, Kyoto, 616-0215, Japan*

*†Institute of optics and quantum electronics, Jena University, Jena,
Germany*

Abstract.

We have conducted a preliminary measurement of the wavelength and the
intrinsic linewidh of the nickel-like silver x-ray laser at a wavelength of 13.9 nm in the
transient collisional excitaion scheme. The measured wavelength was compared with
Multi-Configuration Dirac-Fock (MCDF) code (GRASP92) and was used as a
benchmark of the accuracy of the MCDF code. The intrinsic linewidth of the x-ray
laser, which is convolution of inhomogeneous and homogeneous components, is
investigated theoretically by use of a collisional-radiative model based on the HULLAC
code.

I. Introduction

Since the demonstration of soft x-ray amplification [1,2], x-ray laser research has
been intensively studied experimentally and theoretically. Development of highly
efficient and intense output in short wavelength regions has been one of the most
important objectives. In order to optimize the plasma conditions for generating
intense x-ray lasers, detailed experimental investigations of various parameters, such as
x-ray laser output, their intensities and near- and far-field patterns under various
plasma conditions are required. Measurements of spectral lines and continua for

CP645, *Spectral Line Shapes: Volume 12, 16th ICSLS*, edited by C. A. Back
© 2002 American Institute of Physics 0-7354-0100-4/02/$19.00

determining the plasma parameters, electron density, n_e and electron temperature, T_e, lead us to better understanding of the plasma as a gain medium. And determination of the intrinsic linewidth of the x-ray laser lines may give us an important information about the mechanism of generation of the population inversion. For examples, the inhomogeneous component connects to the temperature of the lasing ions directly, and the homogeneous, or collision broadening, component corresponds to the lifetime of the lasing levels, which may determine the gain coefficient and the saturation intensity of the lasing line.

In the 1980's and early 1990's, this kind of work has been limited because the x-ray laser experiment required the large-size driver laser dedicated for the inertial confinement fusion research, and only a few groups [3-5] could conduct the measurement of the linewidth of the x-ray laser line. Koch *et al.* have measured the linewidth of the Ne-like Se laser at 20.2 nm for various gain-length product from weakly amplified region to the gain saturation region. The result showed that the intrinsic linewidth is mainly determined by the Doppler broadening and the contribution from the collision broadening is not significant. Yuan *et al.* conducted similar experiment for the Ne-like Ge ions at wavelengths of 19.6 nm and 23.4 nm, however in this experiments, they could not achieve the gain-saturation in this series of experiment.

More recently it has been demonstrated that high gain coefficients in soft x-ray amplification can be achieved using compact chirped pulse amplification (CPA) laser systems with a few 10 J energy and several ps duration in the transient collisional-excitation scheme. [6-12] Laboratory-size compact pumping laser systems dedicated to the x-ray laser research have been developed in Max Born Institute [6] and Lowrence Livermore National Laboratory [7] in late of 1990's, and quite recently JAERI (Japan Atomic Energy Research Institute) has developed compact CPA Nd:glass laser dedicated for the x-ray laser research.[13] Now we are on a stage to conduct an experimental investigations of x-ray laser in details.

In this paper, we will introduce recent our result on the strong amplification of the Ni-like silver x-ray laser at a wavelength of 13.9 nm, and a preliminary measurement of the wavelength and the intrinsic linewidth of the lasing line.

II. Demonstration of lasing of 13.9nm

Figure 1 shows the schematic energy level diagram of the nickel-like silver x-ray lasers. The nickel-like ions in the ground state in a plasma are pumped by the heating laser pulse, and substantial populations are generated in the upper levels, *e.g.*, $3d^94d$ and $3d^94p$, by electron impact excitation. Since the $3d^94p$ level has strong spontaneous transition probability to the ground state, the $3d^94p$ level population decreases rapidly, thus the population inversion is generated between the $3d^94p$ and

$3d^94d$ levels. The generated population inversion may decrease due to the collisional *l*-mixing between the lasing levels, whose time constant is typically order of 10 ps, however, if the duration of the heating pulse is a few ps, and the increase of the electron temperature is faster than the time constant of the *l*-mixing, higher population inversion density can be achieved transiently. This scheme is called as transient collisional excitation lasers, and it makes possible for us to achieve gain-saturation of the x-ray laser by use of relatively small pumping system.

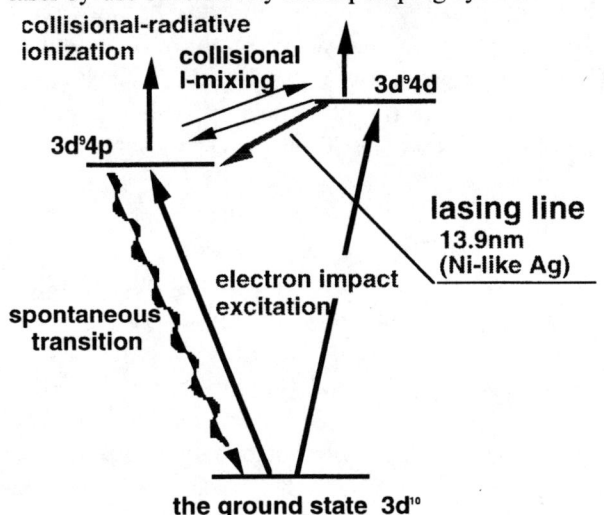

Fig.1 The Schematic energy level diagram of the nickel-like silver x-ray lasers.

The experiment was conducted by use of our compact CPA Nd:glass laser system in JAERI. Our chirped pulse amplification (CPA) Nd:glass laser system consists of a mode-locked Ti:Sapphire oscillator (Spectra Physics, Tsunami), a four-pass pulse expander, a Ti:Sapphire regenerated amplifier (Positive light, TSA-10), two amplifier chains of phosphate glass rods and two pulse compressors. The central wavelength is 1053 nm and the spectral bandwidth after the amplifier chain is 3.5 nm.

Each beam line provides a pre-pulse and a main pulse with variable pulse separation. The maximum output energy is 20 J for each beam line after the compressors, and the pulse duration is variable from 1 ps to 30 ps. The details of the pumping laser system are given in ref. 10.

The line focusing system was based on an off-axis parabolic mirror (focal length of 772 mm) and an off-axis spherical mirror (curvature of 1000 mm), which made a line-focus with 5.6 mm-length and 20 μm-width at the target position. We installed a 6 step mirror (Optical Surface Inc.) just before the off-axis parabolic mirror to generate a quasi-traveling wave. This step mirror consisted of 6 blocks and each was connected on a base plate by optical contact. Each block was parallel each other with an accuracy of 0.05 mrad in the vertical direction and of 0.1 mrad in the horizontal direction. The size of the blocks was 25 mm-width, 150 mm-height, and the difference of the thickness between the neighboring blocks was 600 μm. The angle of the incidence was 11° with respect to surface normal, which led to an optical delay of 4.07 ps between the laser beams reflected from neighboring blocks. The reflection from each block corresponded to a length of 1.2 mm in the line focus. The velocity of the traveling wave at the target position was measured to be $0.98c \pm 0.08c$ by use of a visible streak camera with a time resolution of 2 ps (HAMAMATSU-Photonics Model C4302), where c is the speed of the light. Since our streak camera was insensitive to the fundamental laser light (1.053 μm), a KDP crystal was put just before the target position, and the second harmonics was used. The speed of amplified x-ray in high-density plasmas might be different from that in vacuum, however, for the 13.9 and 12.0 nm lines, under the plasma density of 10^{21} cm^{-3}, the difference from the speed in vacuum was negligibly small. The quality of the width of the line focus was measured using a 1mm-thick titanium target, which was irradiated by a single pulse with 4 ps duration and 10 J energy. From the trace of the laser irradiation on the titanium target, the width of the line-focus was around 20 μm, and we assured that no interference pattern could be seen.

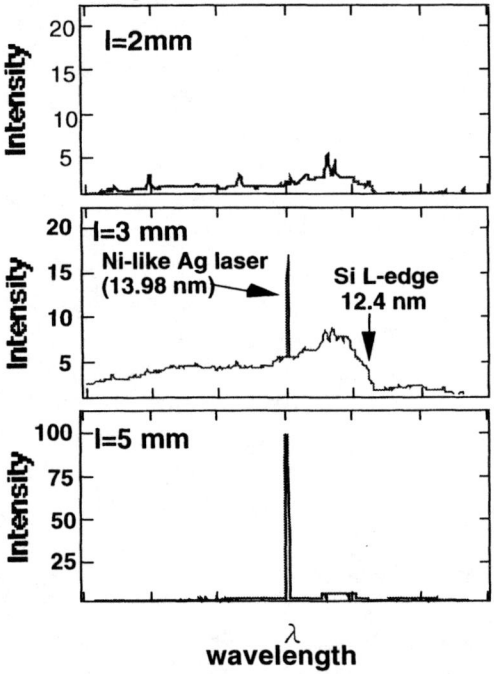

λ
wavelength

Fig.2 Typical spectra of the 13.9 nm lasing line for various plasma lengths from 2 mm through 5 mm. In order to avoid the saturation of the detector sensitivity, we used a filter with a transmittance of 9%.

A silver slab target was put at the focus position. The laser pulses consisted of 2 pulses, and the duration of each pulse was 4 ps with a pulse to pulse separation of 1.2 ns. The total pumping energy was set to 12 J. The energy ratio of the pre-pulse to that of the main pulse was 1 : 7. The irradiance of the pre-pulse and the main-pulse on target were 3.0×10^{14} and 2.3×10^{15} W/cm^2, respectively. The contrast ratio of the ASE level to the peak intensity was measured to be $3 \sim 5 \times 10^{-4}$, i.e., the pedestal level of the main pulse was around $6 \times 10^{11} \sim 10^{12}$ W/cm^2. Shot to shot fluctuation in the pumping energy was less than 7 % for the silver case. The output x-ray laser line, which corresponded to the transition of $4p$-$4d$, was observed by use of a grazing incidence spectrometer. A grazing incidence spherical mirror of 3520mm curvature was placed before the entrance slit, at a distance of 680 mm from the plasma. The acceptance angle of the spherical mirror was ± 2.5 mrad in the vertical direction and ± 10.5 mrad in the horizontal direction. The x-ray emission collected by the spherical mirror was vertically focused on a slit of 15 μm width and 20 mm length, dispersed along a vertical direction by an uneven spacing holographic grating and imaged on a focal plane. The holographic grating had a laminar-structure grooves fabricated by a collaboration with Shimadzu (average grooves was 1200 lines/mm.). [14]   Absolute diffraction efficiency of the grating was calibrated using a synchrotron radiation source of Ritsumeikan university (AURORA).   The distance from the slit to the spherical mirror and to the grating was 90 mm and 237 mm, respectively.   The detector was a back-illuminated x-ray charged-coupled-device (CCD) (Princeton Instruments, Model SX1024), whose sensitivity was proportional to the incident photon number for the photoelectron counts smaller than 60000 counts/pixel.   The resolution in wavelength of the first order light was ± 0.01 nm at a wavelength of 15 nm.

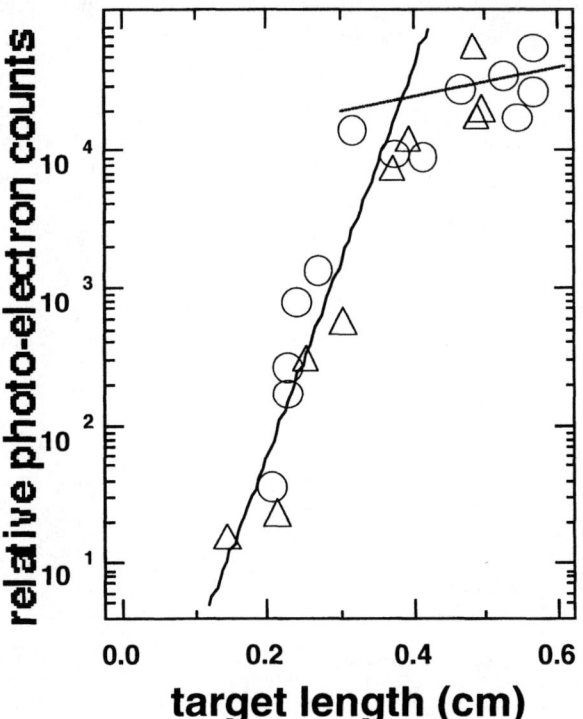

Fig.3 The output 13.9nm line for various target lengths.   The derived gain coefficient is 35 cm-1, and the achieved gain-length product is 13.6 for 3.9 mm target length.   The gain-saturation behavior can be seen for the target longer than 3.9 mm.

Figure 2 shows typical spectra of the nickel-like silver laser at 13.9 nm for various plasma lengths.   In order to avoid the saturation of the sensitivity of the CCD, the x-ray laser light was attenuated using a 0.3 μm-thick yttrium filter fabricated on a 0.2 μm-thick Si_3N_4 filter.   The transmittance of this filter was measured to be 9 % at 13.1 nm.

In Fig. 3, we plotted x-ray laser output versus target lengths.   The gain coefficient, g, derived from Linford fitting formula [15] was 35 ± 3 cm^{-1}, and the gain length product, gl, of 13.6 ± 1.2 was achieved for the 3.9 mm length target. The g and gl were improved compared with our previous result without traveling wave pumping, in which g =24 cm^{-1} and $gl \sim 10$ for 4 mm-length target.

A gain saturation-like behavior could be seen in Fig. 3 for the target length longer than 3.9 mm. We estimated output intensity of the x-ray lasers, in which we used an imaging system of Mo/Si multi-layer mirrors, whose reflectivity for the 13.9 nm was well-known. From the spot-size of the gain-region and the output energy together with the pulse duration of x-ray laser of 8 ps led to the output intensity of 4×10^{10} W/cm^2.which was almost consistent with our theoretical calculation code (Web-oriented Hierarhical Atomic Structure (WHIAM) Code) which is based on a one-dimensional hydrodynamics code copuled with a collisional-radiative model.[16]

III. Measurement of the wavelength and linewidth of x-ray lasers

In order to measure the linewidth of x-ray laser line in the saturation regime, resolution of the spectrometer required is more than $\lambda/\Delta\lambda = 15000$. However, we are now constructing this kind of high resolution spectrometer and are not on a stage to use it. Thus we conducted a preliminary measurement of the wavelength and the intrinsic linewidth of the lasing line by use of the present experimental apparatus. We set the width of the entrance slit to be 15 µm, which is assured by the Fresnel diffraction

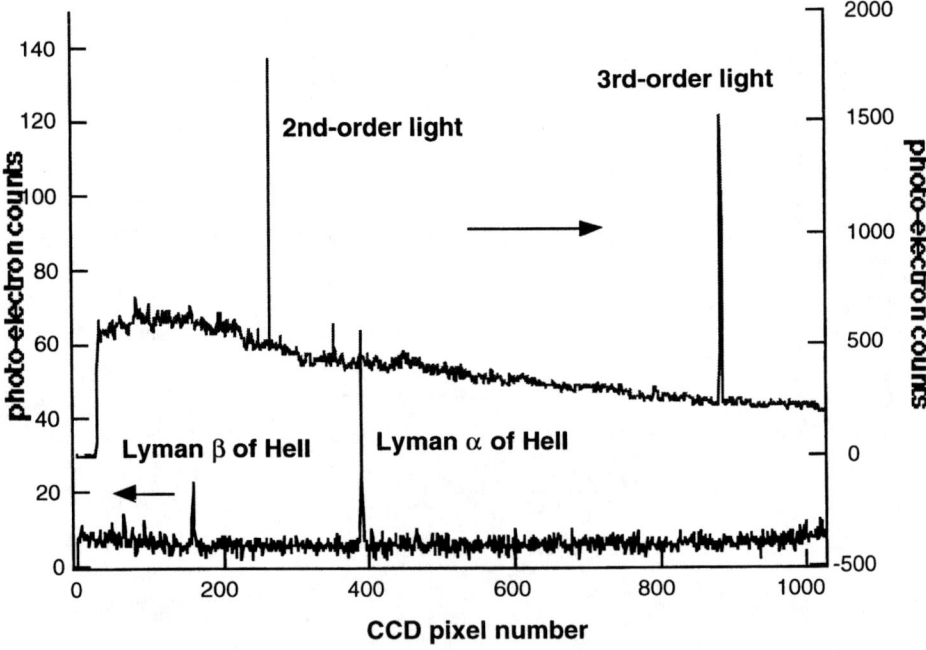

Fig.4 The typical spectrum of the second-order of the nickel-like silver x-ray laser line together with the reference HeII lines. The electron density of the helium gas was kept to be order of 10^{16} cm^{-3}, thus the Stark shift is negligible small.

irradiated by a laser pulse is used as the reference, however, since these lines are emitted from high density region, the wavelength is affected by Stark shift, and the center of the wavelength of the multiplet lines becomes diffuse due to the Stark broadening. Thus the reference lines is favorable to be emitted from a low density region and simple configuration. Standing upon these points, we use the second-order light of the x-ray laser line for the measurement the wavelength, because for this wavelength (~ 27.8 nm), we can use Lyman α (the wavelengths are 30.37804, 30.37858 nm.) and Lyman β (25.6317 nm) lines of the HeII ions as the reference lines. In the case of He plasma, the electron density can be kept low enough, the wavelength of the doublet line is well-known. Figure 4 shows the typical spectrum of the second-order of the nickel-like silver x-ray laser line together with the reference HeII lines. The electron density of the helium gas is kept to be order of 10^{16} cm^{-3}, thus the Stark shift is negligible small. From the wavelengths of the reference lines together with the wavelength calibration curve based on the angle of incidence and the focal length of the grating, the wavelength of the x-ray laser line was determined to be 13.896 ± 0.005nm.

Table 1. Comparison of measured and theoretical wavelength of the Ni-like Ag laser. "Th" represents theoretical calculations

wavelength (nm)	References
13.896 ± 0.005	Present result
13.892 ± 0.015	Li et al. [18]
13.995 ± 0.015	Ros et al. [19]
13.9 ± 0.1	Sebban et al. [20]
Th 13.906	Present result
Th 13.86	Li et al. [18]
Th 13.992	Scofield et al. [21]

The experimental result was compared with Multi-Configuration Dirac-Fock (MCDF) calculation code, GRASP92 [17], and other experimental results [18-20] as shown in Table.1. Our experimental result is almost consistent with the measurement by Li et al.[18], and also consistent with our MCDF calculation. The calculated result by Li et al. is slightly smaller than our result and that by Scofield et al.[21], is larger compared with our result. It is noted that the three calculations in table 1 are the results by use of the same atomic structure code. The differences among these results are due to the treatment of the configuration state functions in the GRASP92. Indeed we assured that for various number of the configuration state functions, the GRASP92 provided the difference results. Since the nickel-like x-ray laser line is spectral line from ions with complicated configurations and the wavelength can be determined precisely, the wavelength of the lasing line is one of the good benchmarks of treatment of the atomic structure code such as GRASP92.

The intrinsic linewidth of the nickel-like silver x-ray laser line was estimated by use of the third-order light at around 40 nm. However due to the poor resolution of our spectrometer, $\lambda/\Delta\lambda$ ~ 2800, the width of the lasing line for small gl value was ~ 0.0055 nm (55 mÅ), almost equal to the resolution of the spectrometer. For further information, high resolution spectrometer is needed.

We compared the preliminary result with our simulation code. This theoetical code consists of one-dimensional hydro-dynamics code, HYADES [22], and post-processed collisional-radiative model, in which we treated virtually all the excited levels with pricipal quantum number of n = 4 and 5 of the nickel-like ions, and the rate coefficients of all the atomic processes are calculated by use of the HULLAC code. [23] Under the present experimental conditions, the x-ray laser gain is generated around the position of 50 μm from the target surface, and the expected ion temperature is ~ 150 eV. This ion temperature corresponds to the inhomogeneous broadening with a width of 16 mÅ, which is much smaller compared with the present result. Homogeneous component, which mainly comes from the collision broadening, was estimated by use of the collisional-radiative model. Figure 5 shows the dominant depopulation flux from the laser upper levels ($3d^9 4d$ or $(3d^9{}_{5/2}, 4d_{5/2})_0$) under the quasi-steady state condition. The electron density and the temperature were assumed to be 4×10^{26} m^{-3} and 400 eV, respectively, which is typical plasma parameters of the gain region. The most dominant depopulation flux from the upper level of the lasing transition is collisional excitation to the next upper level, $3d^9 4f$ (or$(3d^9{}_{5/2}, 4f_{5/2})_1$).

The total depopulation rate from the upper level of the laser transition is 6.53×10^{12} s^{-1} or 0.153 ps, which corresponds to the Lorenzian with FWHM of 18 mÅ, which imply that in the case of the nickel-like ion laser, the contribution from the inhomogeneous and homogeneous line width is almost comparable, whereas the neon-like ion laser, contribution from the Doppler broadening is much larger than the collision broadening. [3,4] This is due to the difference of the ion temperature and is due to the difference in the atomic structure between the nickel-like and the neon-like ions. In ref. [Koch], the duration of the driver laser pulse was sub ns, and the ion is substantially heated, whereas the present case, the duration of the heating (main) pulse is a few ps, and only the plasma electrons are heated. In terms

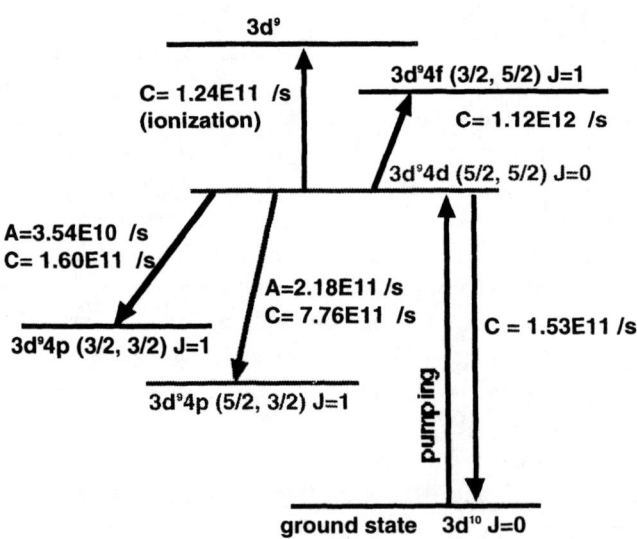

Fig.5 Dominant depopulation flux from the upper level of the lasing transition. A and C indicates radiative transition probability and collisional excitation-deexcitation rate.

of atomic structure of the nickel-like and the neon-like ions, since the neon-like ion scheme uses $n = 3 - 3$ transitions whereas the nickel-like ion use $n = 4 - 4$, the transition energy from the lasing levels to the neighboring levels of the neon-like ions is larger than the case of nickel-like ions.

The Voigt profile of the convolution of the homogeneous and the inhomogeneous contributions shows the intrinsic linewidth of 36 mÅ, which is smaller than the experimental investigation by a factor of 1.5. This implies that the resolution of our spectrometer is not sufficient for the purpose of the present experiment. We are now installing a high resolution spectrometer with a resolution of $\lambda/\Delta\lambda = 15000$, and it will be used in the next experiment.

SUMMARY

We have demonstrated substantial amplification of the nickel-like silver ion x-ray laser at a wavelength of 13.896nm. The wavelength of the lasing line has beed determined within an experimental uncertainty of 0.005nm and has been compared with MCDF code. The comparison suggested that the precise measurement of the x-ray laser was good benchmark of atomic structure code such as GRASP92. The intrinsic linewidth of the x-ray laser line was smaller than 0.005nm (or 50mÅ) which is the spectral resolution of our spectrometer. A hydrodynamics simulation coupled with a collisional-radiative model suggested that in the case of the nickel-like silvers laser, the contribution of the inhomogeneous Doppler broadening is almost comparable with homogeneous collision broadening. This implies the use of Voigt profile is indispensable to calculate the gain coefficient and the saturation intensity.

REFERENCES

1. D. L. Matthews, P. L. Hagelstein, M. D. Rosen, M. J. Eckart, N. M. Ceglio, A. U. Hazi, H. Medecki, B. J. MacGowan, J. E. Trebes, B. L. Whitten, E. M. Campbell, C. W. Hatcher, A. M. Hawryluk, R. L. Kauffman, L. D. Pleasance, G. Rambach, J. H. Scofield, G. Stone and T. A. Weave, Phys. Rev. Lett. **54**, 110 (1985).
2. S. Suckewer, C. H. Skinner, H. Milchberg, C. Keane and D. Voorhees, Phys. Rev. Lett. **55**, 1753 (1985).
3. J. A. Koch, B.J. MacGowan, L. B. Da Silva, D.L. Matthews, J. H. Underwood, P. J. Baston and S. Mrowka, Phys. Rev. Lett. **68**, 3291 (1992)
4. J. A. Koch, B.J. MacGowan, L. B. Da Silva, D.L. Matthews, J. H. Underwood, P. J. Baston, R. W. Lee, R. A. London and S. Mrowka, Phys. Rev. A. **50**, 1877 (1994)
5. G. Yuan, Y. Kato, K. Murai, H. daido and R. Kodama, J. Applied Phys. **78**, 3610 (1995)
6. M. P. Kalachnikov, P. V. Nickles, M. Schnürer, W. Sandner, V. N. Shlyatsev, C.

Danson, D. Neely, E. Wolfrum, J. Zhang, A. Behjat, A. Demir, G. J. Tallents, P. J. Warwick, C. L. S. Lewis, Phys. Rev. A **57**, 4778 (1997).

7. J. Dunn, A. L. Osterheld, R. Shepherd, W. E. White, V. N. Shlyaptsev and R. E. Stewart, Phys. Rev. Lett. **80**, 2825 (1998).

8. J. Dunn, A. L. Osterheld, J. Nilsen, J. R. Hunter and V. N. Shlyaptsev, Phys. Rev. Lett. **84**, 4834 (2000).

9. Y. Li, J. Dunn, J. Nilsen, T. W. Barbee, Jr, A. L. Osterheld, and V. N. Shlyaptsev, J. Opt. Soc. Am. B **17**, 1098 (2000).

10. A. Klisnick, P. Zeitoun, D. Ros, A. Carillon, P. Fourcade, S. Hubert, G. Jamelot, C. L. S. Lewis, A MacPhee, R. O'Rourcke, R. Keenan, P. Nickles, K. Janulewicz, M. Kalachnikov, J. Warwick, J. C. Chanteloup, A. Migus, E. Salmon, C. Sauteret, J. P. Zou, J. Opt. Soc. Am B**17**, 1093 (2000).

11. J. Kuba, A. Klisnick, D. Ros, P. Fourcade, G. Jamelot, J. L. Miquel, N. Blanchot and J. F. Wyart, Phys. Rev. A. **62**, 043808 (2000).

12. A. G. MacPhee, R. M. N. O'Rourke, C. L. S. Lewis, J. Y. Lin, A. Demir, G. J. Tallents, J. Collier, D. Neely, D. Ros, Ph. Zeitoun, S. P. McCabe, P. A. Simms, G. J. Pert, Proc. *X-ray Lasers–1998*, edited by Y. Kato, H. Takuma and H. Daido, IOP-159, 75 (1999).

13. T. Kawachi, M. Kado, M. Tanaka, N. Hasegawa, K. Sukegawa, K. Nagashima, M. Koike, A. Nagashima and Y. Kato, to be published in Applied Optics.

14. M. Koike, T. Namioka, E. Gullikson, Y. Harada, S. Ishikawa, T. Imazono, S. Mrowka, N. Miyata, M. Yanagihara, J. H. Underwood, K. Sano, N. Ogiwara, O. Yoda and S. Nagai, Proc. SPIE – Int. Soc. Opt. Eng. **4146**, 163 (2000)

15. G. J. Linford, E. P. Poressini, W. R. Sooy and M. L. Spaeth, Appl. Opt. **13**, 379 (1974).

16. A.Sasaki, T.Utsumi, K.Moribayashi, M.Kado, M.Tanaka, N.Hasegawa, T.Kawachi, H.Daido, J. Quant. Spectrosc. Radiat. Transf. **71**, 665 (2001).

17. I. P. Grant, B. J. McKenzie, P. H. Norrihngton, D. F. Mayers and N. C. Pyper, Comput. Phys. Commun. **21**, 207 (1980).

18. Y. Li, J. Nilsen, J. Dunn and A. L. Osterheld, Phys. Rev. A, **58**, R2668 (1998).

19. D. Ros, A. Klisnick, A. Carillon, P. Jaeglé, G. Jamelot, P. Zeitoun, S. Sebban, F. Albert, S. Jacquemot and J. F. Wyatt, in x-ray Lasers 1996, edited by S. Svanberg and C.G. Wahlstrom (Institute of Physics, Bristol, 1996), pp.50-52.

20.. S. Sebban, D. Ros, A. G. MacPhee, F. Albert, E. Alfred, A. Carillon, P. Jaeglé, G. Jamelot, A. Klisnick, C. L. S. Lewis, R. S. Smith, G. J. Tallents and P. Zeitoun, Proc. SPIE 3156, 11 (1997)

21. J. H. Scofield and B. J. MacGowan, Phys. Scr. 46, 361 (1992)

22. J.T.Larsen and S.M.Lane. J. Quant. Spectrosc. Radiat. Transf. **51** ,179(1994).

23. M. Klapisch, A. Bar-Shalom, J. Quant. Spectrosc. Radiat. Transf., **58**, 687 (1997).

Breakthrough to Tunable X-Ray Lasers via Dressing the Plasma by an Elliptically Polarized Radiation of an Optical Laser

V.P. Gavrilenko[a,b], E. Oks[c]

[a]*General Physics Institute, Russian Academy of Sciences, Department of Plasma Physics, Vavilov street 38, 119991, GSP-1, Moscow, Russia*
[b]*Center for Surface and Vacuum Research, Russian State Committee for Standards, Vavilov street 38, 117334 Moscow, Russia*
[c]*Physics Department, 206 Allison Lab., Auburn University, Auburn, AL 36849, USA*

Abstract. We present detailed results of calculations of modification of x-ray lasing lines of a H-like ion under a strong elliptically-polarized electric field of an optical laser. We show that using an elliptically-polarized electric field of an optical laser, it is possible to control the amplification and polarization of the x-ray radiation, as well as to vary the frequency of the x-ray lasing line.

INTRODUCTION

One of the popular schemes for soft x-ray lasing is the scheme based on recombination pumping of hydrogen-like (H-like) ions which are first stripped of electrons by optical field ionization [1-3]. The gain of an x-ray laser is determined by the product of the width of the lasing line and oscillator strength for the lasing transition. In our earlier works [4,5] we showed that by applying a high frequency electric field of electromagnetic radiation to a H-like emitter, it is possible to substantially decrease the Stark width of its spectral lines. The reason for such a decrease is that the high frequency field changes the character of interaction of a H-like emitter with plasma electric microfield. Based on this idea, it was shown in Ref. [6] that a linearly-polarized field of an optical laser can significantly narrow the profile of the absorption coefficient for some x-ray lasing spectral lines of H-like ions, thus increasing the gain of the x-ray laser. Later the effect of the narrowing of the line profile for H-like ions interacting with linearly-polarized electric field of an optical laser was also studied in Ref. [7].

This paper presents the detailed calculations of the effect of the elliptically-polarized intense field of an optical laser (EPIFOL) on the Stark broadening of some x-ray spectral lines of H-like ions. For the interaction of the EPIFOL with a H-like ion, the results obtained in our paper [8] are used. Interaction of a H-like ion with plasma microfields is considered by using the Model Microfield Method [9]. It is shown that the use of the EPIFOL can provide not only a significant enhancement of the gain on some x-ray lasing transitions, but also a possibility of tuning the x-ray laser in a wide range of frequencies.

THEORETICAL MODEL

We consider a H-like ion of a nuclear charge Z subjected simultaneously to an elliptically-polarized electric field of an optical laser

$$\vec{E}(t) = \varepsilon_0[\vec{e}_z \cos(\omega_0 t + \gamma) + \xi \vec{e}_x \sin(\omega_0 t + \gamma)] \tag{1}$$

and to the plasma microfield $\vec{F}(t)$. In Eq. (1), $\vec{e}_x$ and $\vec{e}_z$ are unit vectors along x- and z-axes of a Cartesian reference frame, ξ is an ellipticity degree. For a H-like ion under the field $\vec{E}(t)$, the Wave Functions (WFs) of quasienergy states for the energy level of the principal quantum number n=2 were found in [8] in the following form :

$$\Psi_1 \equiv \widetilde{\psi}_1(t), \quad \Psi_2 \equiv \widetilde{\psi}_2(t), \quad \Psi_3 \equiv \exp(-i\kappa t)\widetilde{\psi}_3(t), \quad \Psi_4 \equiv \exp(i\kappa t)\widetilde{\psi}_4(t),$$

$$\widetilde{\psi}_k(t) = 2^{-1}\{(-1)^{k+1}(\varphi_1 - \varphi_2) + i\varphi_3 \exp[-i\beta(t)] + i\varphi_4 \exp[i\beta(t)]\},$$

$$\widetilde{\psi}_p(t) = 2^{-1}\{i(\varphi_1 + \varphi_2) + (-1)^{p+1}\{\varphi_3 \exp[-i\beta(t)] - \varphi_4 \exp[i\beta(t)]\}\}, \tag{2}$$

$$k = 1, 2; \quad p = 3, 4,$$

where

$$\kappa = \omega_0 \xi w J_1(w), \qquad \beta = w \sin(\omega_0 t + \gamma), \qquad w = 3\varepsilon_0 /(Z\omega_0). \tag{3}$$

Here $J_1(w)$ is the Bessel function and the functions φ_s (s = 1, 2, 3, 4) are the parabolic WFs of the level n=2 of the H-like ion ($\varphi_1 \equiv |001\rangle$, $\varphi_2 \equiv |00-1\rangle$, $\varphi_3 \equiv |100\rangle$, $\varphi_4 \equiv |010\rangle$)), Oz being the quantization axis. The WFs (2) were obtained in [8] for the case of $|\xi| \le w^{-1/2}$. In Eq. (3) and below we use atomic units $\hbar = m_e = e = 1$, unless specified to the contrary.

The Schrödinder equation for the H-like ion under the field $\vec{E}(t) + \vec{F}(t)$ can be written as

$$i\frac{\partial \Phi}{\partial t} = [H_a + z\varepsilon_0 \cos(\omega_0 t + \gamma) + x\varepsilon_0 \xi \sin(\omega_0 t + \gamma) + V(t)]\Phi, \quad V(t) \equiv \vec{r}\vec{F}(t). \tag{4}$$

We seek a solution of Eq. (4) for n=2 in the form

$$\Phi_j(t) = \sum_r T_{rj}(t,0)\widetilde{\psi}_r(t) \tag{5}$$

for initial conditions

$$T_{rj}(0,0) = \delta_{rj}$$

where δ_{rj} is the Kronecker symbol. The matrix $T_{rj}(t,0)$ consists of matrix elements of the evolution operator $T(t,0)$, $T(0,0)$ being the unity operator I.

Substituting (5) in (4) we obtain

$$\frac{dT}{dt} = M(\vec{F},t)T \tag{6}$$

where elements of the matrix $M(\vec{F},t)$ are

$$M_{qr}(\vec{F},t) = -i < \widetilde{\psi}_q(t)|V(t)|\widetilde{\psi}_r(t) > \tag{7}$$

In this paper we use the Model Microfield Method (MMM) for describing the interaction of the plasma microfield with H-like ions [9]. For using the MMM, it is important to find a static evolution operator T_S, i.e., the evolution operator for the case of a time-independent field $\vec{F}$. Here we consider the situation where the optical laser frequency ω_0 is much greater than the Stark splitting of the of the level n=2 of the H-like ion in the field F

$$\omega_0 \gg 3F/Z. \tag{8}$$

Under the condition (8), we use the averaging method by Krylov-Bogoliubov-Mitropolskii [10, 11] for solving the matrix Eq. (6) for the time-independent field $\vec{F}$. In accordance to this averaging method, an approximate solution of Eq. (6) is obtained via the substitution of the matrix elements $M_{qr}(\vec{F},t)$ by their time-averaged values $\overline{M}_{qr}(\vec{F})$:

$$M_{qr}(\vec{F},t) \rightarrow \overline{M}_{qr}(\vec{F}), \qquad \overline{M}_{qr}(\vec{F}) = \frac{\omega_0}{2\pi} \int\limits_{0}^{2\pi/\omega_0} dt \; M_{qr}(\vec{F},t), \tag{9}$$

As a result we get the following equation for finding the static evolution operator $T_S(t)$

$$\frac{dT_S}{dt} = \overline{M}(\vec{F})T_S, \tag{10}$$

the matrix $\overline{M}(\vec{F})$ for the level n=2 of the H-like ion being

$$\bar{M}(\vec{F}) = \begin{pmatrix} if_y & 0 & if_x - f_z & if_x + f_z \\ 0 & -if_y & if_x - f_z & if_x + f_z \\ if_x + f_z & if_x + f_z & -i\kappa & 0 \\ if_x - f_z & if_x - f_z & 0 & i\kappa \end{pmatrix} \tag{11}$$

where

$$f_x = \frac{3}{2Z}J_0(w)F_x, \qquad f_y = \frac{3}{Z}J_0(w)F_y, \qquad f_z = \frac{3}{2Z}F_z, \tag{12}$$

Here $J_0(w)$ is the Bessel function. The principal characteristics of the model microfield $\vec{F}$, which controls the line profile in the MMM, are the distribution function $P(\vec{F})$ and the jumping frequency $v(\vec{F})$ [9]. Between two consecutive jumps, the field $\vec{F}$ is considered to be time-independent.

The MMM equation for the Fourier transform of the transition operator averaged over the realizations of the stochastic plasma microfield $\vec{F}(t)$ is as follows [9]

$$\widetilde{T}_{MMM}(\omega) = \{\widetilde{T}_S(\widetilde{\omega})\}_{av} + \{v\widetilde{T}_S(\widetilde{\omega})\}_{av}\{vI - v^2\widetilde{T}_S(\widetilde{\omega})\}^{-1}_{av}\{v\widetilde{T}_S(\widetilde{\omega})\}_{av}. \tag{13}$$

Here $v = v(\vec{F})$; $\widetilde{T}_S(\widetilde{\omega})$ is the Laplace transform of the transition operator $T_S(t)$ calculated at $\widetilde{\omega} = \omega + iv(\vec{F})$ for a static field $\vec{F}$; $\{...\}_{av}$ is the average over the probability distribution $P(\vec{F})$.

Using Eq. (13), we can represent the spectral line profile of the L_α line of the H-like ion in the form

$$I^{(\zeta)}(\omega) = \frac{1}{\pi}\sum_{\alpha',\alpha''}\mathrm{Re}[\widetilde{T}_{MMM}(\omega)]_{\alpha'',\alpha'}R^{(\zeta)}_{\alpha'\alpha''} \tag{14}$$

where ζ indicates the polarization of the emitted photons ($\zeta = x, y, z$), and

$$R^{(\zeta)}_{\alpha'\alpha''} = \left\{\langle\varphi_0|\zeta|\widetilde{\psi}^{(0)}_{\alpha''}(\tau)\rangle\langle\widetilde{\psi}^{(0)}_{\alpha'}(0)|\zeta|\varphi_0\rangle\right\}_\gamma. \tag{15}$$

In Eq. (15), $\{...\}_\gamma$ is the average over the initial phase γ of the laser field $\vec{E}(t)$. $\widetilde{\psi}^{(0)}_\alpha(\tau)$ is the zero harmonic of the periodic wave function $\widetilde{\psi}_\alpha(\tau)$, and φ_0 is the WF of the lower level $n = 1$.

CALCULATION OF THE MODIFIED PROFILE OF THE ABSORPTION COEFICIENT OF THE LASING LINE L_α

In this paper we present the corresponding calculations for the L_α line of Li III ($\lambda = 135\,\text{Å}$) subjected to an elliptically-polarized electric field $\vec{E}(t)$ of the CO_2 laser in a plasma of the electron density $N_e = 5.0 \times 10^{19}$ cm^{-3} and of the temperature $T_e = T_i = 3\,\text{eV}$. Figure 1 shows the modification of the Li III L_α line profile under a *linearly*-polarized field $\vec{E}(t)$ (i.e., for $\xi = 0$) for the following three values of the laser field amplitude ε_0: 0; 3.36×10^7 V/cm, and 4.51×10^7 V/cm. We assumed that the polarization of the L_α radiation is perpendicular to the vector $\vec{E}(t)$. It is seen that as the value ε_0 increases within the above range, the profile of the absorption coefficient of the L_α line becomes narrower – in accordance to the earlier results [4-6].

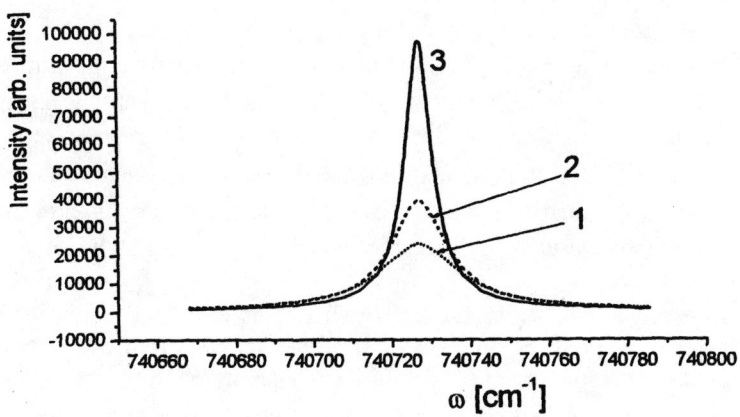

FIGURE 1. Modification of the profile of the Li III L_α line ($\lambda = 135$ Å) under a *linearly*-polarized electric field $\vec{E}(t)$ of the CO2 laser in a plasma of the electron density $N_e = 5.0 \times 10^{19}$ cm^{-3} and of the temperature $T_e = T_i = 3$ eV. Perturbing ions are Li^{3+}. Only the ionic component of the plasma microfield was taken into account. Profiles 1, 2, and 3 correspond to the dressing field amplitude 0, 3.36×10^7 V/cm, and 4.51×10^7 V/cm, respectively. The x-ray radiation is polarized perpendicularly to direction of the CO2 laser field.

471

The ratio of halfwidths (FWHM) $\Delta\omega_{1/2}^{(k)}$ for the three profiles in Fig. 1 is:

$$\Delta\omega_{1/2}^{(3)} : \Delta\omega_{1/2}^{(2)} : \Delta\omega_{1/2}^{(1)} = 0.34 : 0.70 : 1.00,$$

where $\Delta\omega_{1/2}^{(1)}$, $\Delta\omega_{1/2}^{(2)}$, and $\Delta\omega_{1/2}^{(3)}$ correspond to $\varepsilon_0 = 0$, $\varepsilon_0 = 3.36 \times 10^7$ V/cm, and $\varepsilon_0 = 4.51 \times 10^7$ V/cm, respectively.

Due to the narrowing of the profile of the absorption coefficient of the L_α line under the increase of ε_0, the intensity at the maximum of the normalized profile increases. For a lasing plasma this leads to the growth of the amplification and of the gain of the x-ray laser under the dressing by a *linearly*-polarized radiation of an optical laser. This phenomenon was studied earlier in [6].

Figure 2 shows the modification of the Li III L_α line profiles of x-, y-, and z-polarizations subjected to the CO_2 laser of $\varepsilon_0 = 3.36 \times 10^7$ V/cm as the *ellipticity* degree increases from $\xi = 0$ to $\xi = 0.3$. The most interesting is the profile of the x-polarization, i.e., the profile polarized in the direction of the minor component of the field $\vec{E}(t)$. This profile shows a splitting into three components: the central component at the unperturbed frequency $\omega_{21}^{(0)}$ of the L_α transition and two lateral components at the frequencies $\omega_{21}^{(0)} \pm \kappa$, where κ is defined in Eq. (3). For the case under the consideration, at $\xi \gtrsim 0.2$ the lateral components become more intensive in their maxima than the central component. Therefore we can expect that under certain condition it is possible to generate the x-ray radiation of the x-polarization at the shifted frequencies $\omega_{21}^{(0)} \pm \kappa$. Since the quasienergy κ is proportional to the ellipticity degree ξ (see Eq. (3)), then one can design a *tunable* x-ray laser, where the tuning would be achieved by varying ξ. Tunable x-ray lasers seem to be of a practical importance.

Another interesting result is that via the dressing by an elliptically-polarized field of an optical laser, it is possible to *control the polarization* of the generated x-ray radiation. Indeed, the employment of the elliptically-polarized dressing field produces the x-ray radiation (at the shifted frequencies $\omega_{21}^{(0)} \pm \kappa$) polarized along the minor component of the field $\vec{E}(t)$ (i.e., along the minor axis of the ellipse). As for achieving the maximum generation at the unshifted frequency $\omega_{21}^{(0)}$, one should use instead a linearly-polarized dressing field $\vec{E}(t)$. In the latter case, the x-ray radiation would be polarized perpendicular to the dressing field $\vec{E}(t)$.

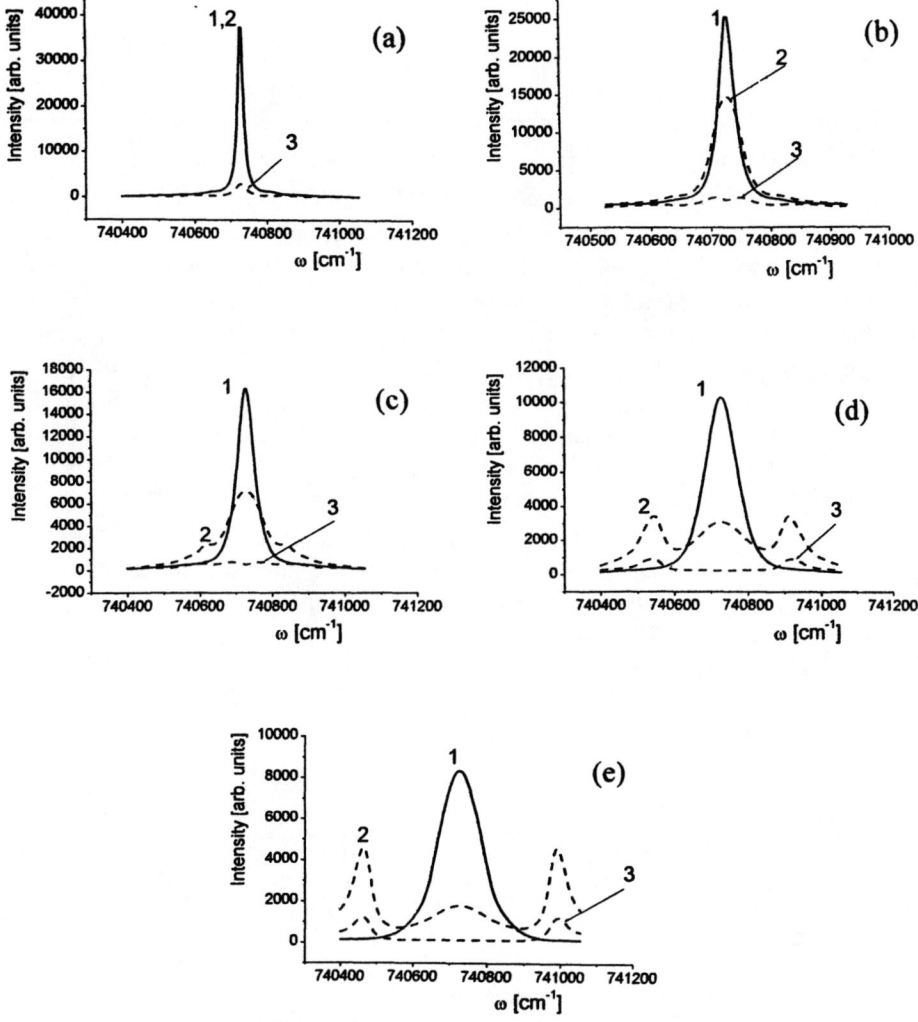

FIGURE 2. Modification of the profile of the Li III L_α line ($\lambda = 135$ Å) under an *elliptically-polarized* electric field $\vec{E}(t)$ of the CO_2 laser in a plasma of the electron density $N_e = 5.0 \times 10^{19}$ cm^{-3} and of the temperature $T_e = T_i = 3$ eV. The dressing field amplitude is 3.36×10^7 V/cm. Perturbing ions are Li^{3+}. The sub-figures 2a, 2b, 2c, 2d, and 2e correspond to the ellipticity degree 0, 0.05, 0.10, 0.20, and 0.30, respectively. In each of the sub-figures, profiles 1, 2, and 3 correspond to the x-ray radiation polarized in **y-, x-, and z-directions,** respectively, where z is the direction of the major axis of the ellipse and x is the direction of the minor axis of the ellipse (see Eq. (1)). Only the ionic component of the plasma microfield was taken into account.

DISCUSSION

In this paper we studied how the x-ray lasing at the L_α line of an H-like ion can be controlled via a dressing by an elliptically-polarized radiation of an optical laser. We showed first of all that such a dressing can *control the amplification* of the x-ray radiation at the unperturbed frequency $\omega_{21}^{(0)}$ of the L_α line. In particular, the maximum amplification can be achieved for the ellipticity degree $\xi = 0$ and it would be significantly greater than at the absence of the dressing field. Second, by using an elliptically-polarized radiation of an optical laser of $\xi > 0$, it is possible *to design a tunable x-ray laser*: the tuning would be performed by varying the ellipticity degree ξ. Third, such a dressing allows to *control the polarization* of the x-ray radiation.

Results presented in Figs. 1 and 2 were obtained by allowing only for the ionic component of the plasma microfield (via the MMM). Under the considered plasma parameters, the ionic contribution to the width of the profile of the absorption coefficient predominates over the electronic contribution. To illustrate this fact, we present in Fig. 3 two profiles of the Li III L_α line of x-polarization. Profile #1 corresponds to the allowance for only ionic contribution of the plasma microfield, profile # 2 – to the allowance for only electronic contribution of the plasma microfield. It is seen that the width of the lateral components for the profile #1 is about three times greater than for the profile #2.

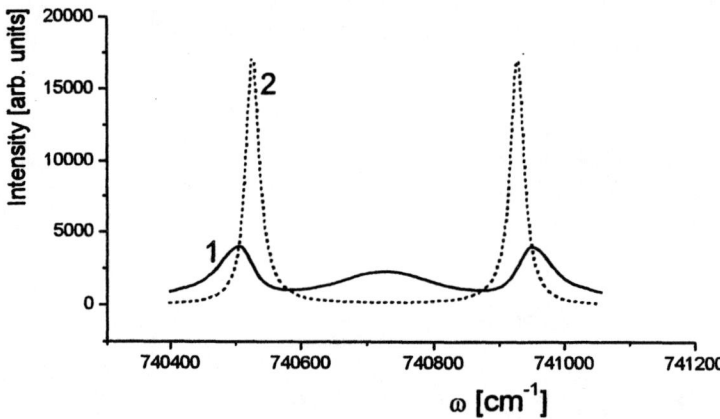

FIGURE 3. Comparison of ionic and electronic contributions to the profiles of the Li III L_α line ($\lambda = 135$ Å) under an *elliptically*-polarized electric field $\bar{E}(t)$ of the CO_2 laser in a plasma of the electron density $N_e = 5.0 \times 10^{19}$ cm^{-3} and of the temperature $T_e = T_i = 3$ eV. Profile #1 corresponds to the allowance for only ionic contribution of the plasma microfield, profile # 2 – to the allowance for only electronic contribution of the plasma microfield. It is seen that the ionic contribution to the width of the lateral components of the profile predominates over the electronic contribution. Both profiles are of x-polarization (which is the direction of the minor axis of the ellipse). The dressing field has an amplitude 3.36×10^7 V/cm and an ellipticity degree 0.25. Perturbing ions are Li^{3+}.

REFERENCES

1. Nagata, Y., Midorikawa, K., Kubodera, S., Obara, M., Tashiro, H., Toyoda, K., *Phys. Rev. Lett.* **71**, 3774-3777 (1993).
2. Donnelly, T.D., Da Silva, L., Lee, R.W., Mrowka, S., Hofer, M., Falcone, R.W., *J. Opt. Soc. Am. B* **13**, 185- 188 (1996).
3. Korobkin, D.V., Nam, C.H., Suckewer, S., Goltsov, A., *Phys. Rev. Lett.* **77**, 5206-5209 (1996).
4. Gavrilenko, V.P., Oks, E., "New Method of Local Measurements of an Amplitude- Angular Distribution of Low-Frequency Plasma Turbulence " in *Proc. 17th Int. Conf. on Phenom. in Ionized Gases* (Budapest, Hungary 1985), pp. 1081-1983.
5. Gavrilenko, V.P., Oks, E., "Theory of Stark Broadening of Quasienergy Levels and Corresponding Spectral Lines of Hydrogen Atom in Plasmas" in *Proc. 19th Int. Conf. on Phenom. in Ionized Gases* (Belgrade, Yugoslavia 1989), pp. 354-355.
6. Oks, E., *J. Phys. B: At. Mol. Opt. Phys.* **33**, L801-L805 (2000).
7. Alexiou, S., *J. Quant. Spectrosc. Radiat. Transfer* **71**, 139-146 (2001).
8. Oks, E., Gavrilenko, V.P., *Optics Communications* **46**, 205-208 (1983).
9. Brissaud, A., Frisch, U., *J. Quant. Spectrosc. Radiat. Transfer* **11**, 1767-1783 (1971).
10. Krylov, N.N., Bogoliubov, N.N., *Introduction to Non-linear Mechanics*, Princeton University Press, Princeton, 1947.
11. Bogoliubov, N.N., Mitropolskii, Yu. M., *Asymptotic Methods in the Theory of Nonlinear Oscillations*, Gordon and Breach, New York, 1961.

ULTRACOLD SYSTEMS

Formation of a molecular Bose-Einstein condensate and an entangled atomic gas by time-dependent Feshbach resonance

V. A. Yurovsky* and A. Ben-Reuven*

*School of Chemistry, Raymond and Beverly Sackler Faculty of Exact Sciences, Tel Aviv University, 69978 Tel Aviv, Israel

Abstract. Association in an atomic Bose-Einstein condensate, and the subsequent dissociation of the resulting molecular condensate, due to Feshbach resonance in a time-dependent magnetic field, are analyzed, incorporating non-mean-field quantum corrections and inelastic collisions. Calculations for the Na atomic condensate demonstrate that there exist optimal conditions under which about 80% of the atomic population can be converted to a relatively long-lived molecular condensate (with lifetimes of 10 ms and longer), if the magnetic field is temporally changed one way. Entangled atoms in two-mode squeezed states (with noise reduction of about 30 dB) are formed by molecular dissociation when the field is changed the other way.

INTRODUCTION

Similarities between the properties of Bose-Einstein condensates (BEC) – or matter waves – of dilute atomic gases, and those of coherent electromagnetic radiation led to the recent development of atom optics [1]. Quantum atom optics, like quantum photon optics, deals with non-classical properties of the fields, such as squeezing and entanglement [2]. Squeezed states of atomic waves, characterized by noise reduction, can be formed by various methods [3, 4, 5, 6, 7, 8, 9]. One of these methods is the dissociation of a molecular BEC [10, 11, 12]. The atoms can also be formed in a two-mode squeezed state, which is entangled [5, 7, 8]; i. e., it cannot be factorized into a product of one-mode states. Dissociating diatoms as a source of entangled atomic pairs have been considered only in the context of (non-condensed) individual molecules [13].

The theoretical analysis presented here is concerned with the formation of entangled and squeezed atomic fields by the dissociation of an unstable molecular BEC. The latter is temporarily created by applying a time-varying magnetic field causing a curve crossing of the atomic and molecular states through a Feshbach resonance [14]. Our analysis was carried out on the Feshbach resonances in a sodium BEC, studied experimentally at MIT [15]. Although the molecular BEC is formed in an excited molecular state, and is therefore unstable, it should still serve in this system as a good source of entangled and squeezed atomic waves.

As one of the main causes of this molecular instability is deactivation by inelastic collisions [14, 16], the lifetime of the unstable molecular condensate can be prolonged by reducing the atomic BEC density. It turns out that, under optimal conditions, lifetimes approaching the order of 1 s can be realized.

CP645, *Spectral Line Shapes: Volume 12, 16th ICSLS*, edited by C. A. Back
© 2002 American Institute of Physics 0-7354-0100-4/02/$19.00

Most impressive outcomes of the present approach are the extent of (near-total) conversion to entangled atoms or a molecular condensate, obtained under optimal conditions, as well as the extreme degree of squeezing and the relatively long molecular BEC lifetimes achievable. The theoretical analysis (see also Ref. [17]) required to obtain these results needs a solution of the quantum field equations beyond the mean-field, and even the parametric, approximations [7, 12] in a manner described below.

THE MODEL

The basic ingredients of our model are the two coupled atomic and molecular fields representing the Feshbach-resonant states (following Ref. [16]), described here by annihilation operators in the momentum representation, $\hat{\Psi}_a(\mathbf{p},t)$ and $\hat{\Psi}_m(\mathbf{p},t)$, respectively. The Feshbach coupling of the two fields (see Refs. [14, 16]) contains a product of two atomic creation operators and therefore can lead to the formation of entangled atomic pairs, in analogy with parametric down conversion in quantum optics (see Ref. [2]). Spatial inhomogeneity due to a trapping potential and elastic collisions can be neglected here (see discussion below).

The mean-field treatment of the atomic field used in Ref. [16] is insufficient here, and should be replaced by a second-quantized treatment, as in Ref. [12]. The initial state of the atomic field at $t = t_0$ (the time the magnetic ramp is switched on) is considered as a coherent state of zero kinetic energy,

$$\hat{\Psi}_a(\mathbf{p},t_0)\,|\text{in}\rangle = (2\pi)^{3/2}\,\varphi_0\delta(\mathbf{p})\,|\text{in}\rangle, \tag{1}$$

where $|\varphi_0|^2 = n_a(t_0)$ is the initial atomic density and $|\text{in}\rangle$ is the time-independent state vector in the Heisenberg representation. A pair of condensate atoms forms a molecule of zero kinetic energy. Therefore the resonant molecules can be adequately represented by a mean field $\varphi_m(t)$, such that

$$\langle\text{in}|\hat{\Psi}_m(\mathbf{p},t)|\text{in}\rangle = (2\pi)^{3/2}\,\varphi_m(t)\,\delta(\mathbf{p}), \tag{2}$$

where $|\varphi_m(t)|^2 = n_m(t)$ is the molecular condensate density. This approach (unlike Refs. [7, 12]) takes into account the time dependence of the molecular mean field. Fluctuations of the molecular field due to collisions involving non-condensate atoms are neglected.

The outcome of atom-molecule and molecule-molecule deactivating collisions is introduced, as in Ref. [16], by adding molecular "dump" states. The elimination of these states in a second-quantized description is, however, different. It is similar to the Heisenberg-Langevin formalism in quantum optics (see Ref. [2]), but takes into account the nonlinearity of the collisional dumping. In the Markovian approximation, the equation of motion for the atomic field attains the form

$$i\dot{\hat{\Psi}}_a(\mathbf{p},t) = H\hat{\Psi}_a(\mathbf{p},t) + 2g^*\varphi_m(t)\,\hat{\Psi}_a^\dagger(-\mathbf{p},t) + i\hat{F}(\mathbf{p},t), \tag{3}$$

where (using units with $\hbar = 1$)

$$H = \frac{p^2}{2m} - \mu\frac{B(t)-B_0}{2} - i\gamma|\varphi_m(t)|^2, \tag{4}$$

and m is the atomic mass. The second term in H describes the time-dependent Zeeman shift of the atom in an external magnetic field $B(t)$, relative to half the energy of the molecular state, which is chosen as the zero energy, μ is the difference in magnetic momenta of an atomic pair and a molecule, and B_0 is the resonance value of B. The coefficient g coupling the atomic and molecular fields is related to the phenomenological resonance strength Δ, as $|g|^2 = 2\pi|a_a\mu|\Delta/m$ (see Ref. [16]), where a_a is the elastic scattering length. The parameter γ describes the width of the atomic state due to deactivating collisions (see Ref. [16]). The quantum noise source $\hat{F}(\mathbf{p},t)$, which is δ-correlated in the markovian approximation, and then should obey the relation

$$\left[\hat{F}(\mathbf{p},t), \hat{F}^\dagger(\mathbf{p}_1,t_1)\right] = \gamma|\varphi_m(t)|^2\delta(t-t_1)\,\delta(\mathbf{p}-\mathbf{p}_1), \tag{5}$$

is needed in order to conserve the correct commutation relations of the field operators.

As a generalization of the parametric approximation [7, 12], let us represent the atomic field operator in the form

$$\hat{\Psi}_a(\mathbf{p},t) = \left[\hat{A}(\mathbf{p},t)\,\psi_c(p,t) + \hat{A}^\dagger(-\mathbf{p},t)\,\psi_s(p,t)\right]C(t), \tag{6}$$

including the collision damping factor

$$C(t) = \exp\left(-\int_{t_0}^{t} dt_1\,\gamma|\varphi_m(t_1)|^2\right). \tag{7}$$

The operators $\hat{A}(\mathbf{p},t)$ are expressible in terms of $\hat{\Psi}_a(\mathbf{p},t_0)$, the noise $\hat{F}(\mathbf{p},t)$, and the c-number solutions $\psi_{c,s}(p,t)$ of the two coupled equations

$$i\dot{\psi}_{c,s}(p,t) = H\psi_{c,s}(p,t) + 2g^*\varphi_m(t)\,\psi_{s,c}^*(p,t), \tag{8}$$

given the initial conditions $\psi_c(p,t_0) = 1$, $\psi_s(p,t_0) = 0$. This is achieved by substituting Eq. (6) into Eq. (3), resulting in

$$\hat{A}(\mathbf{p},t) = \hat{\Psi}_a(\mathbf{p},t_0) + \int_{t_0}^{t} dt_1\left[\psi_c^*(p,t_1)\,\hat{F}(\mathbf{p},t_1) - \psi_s(p,t_1)\,\hat{F}^\dagger(-\mathbf{p},t_1)\right]C(t_1). \tag{9}$$

The atomic density

$$n_a(t) = (2\pi)^{-3}\int d^3p_1 d^3p_2 \exp\left[i(\mathbf{p}_2-\mathbf{p}_1)\cdot\mathbf{r}\right]\langle\text{in}|\hat{\Psi}_a^\dagger(\mathbf{p}_1,t)\,\hat{\Psi}_a(\mathbf{p}_2,t)|\text{in}\rangle, \tag{10}$$

then emerges as $\mathbf{r}$-independent, and comprises the sum $n_a(t) = n_0(t) + n_s(t)$ of the densities of condensate atoms $n_0(t) = |\langle\text{in}|\hat{\Psi}_a(0,t)|\text{in}\rangle|^2$, and of non-condensate (entangled) atoms $n_s(t)$ encompassing a broad spectrum of kinetic energies E,

$$n_s(t) = \int dE\,\tilde{n}_s(E,t). \tag{11}$$

481

The equation of motion for the molecular mean field $\varphi_m(t)$ is obtained by a similar elimination of the molecular dump fields from the corresponding operator equation, followed by a mean-field averaging. We thus obtain

$$i\dot{\varphi}_m(t) = gm_a(t) - i\left(\gamma n_a(t) + \gamma_m |\varphi_m(t)|^2\right)\varphi_m(t), \qquad (12)$$

where the parameter γ_m describes molecule-molecule deactivating collisions (see Ref. [16]). Here m_a is the anomalous density

$$m_a(t) = (2\pi)^{-3}\int d^3p_1 d^3p_2 \langle\text{in}|\hat{\Psi}_a(\mathbf{p}_1,t)\,\hat{\Psi}_a(\mathbf{p}_2,t)\,|\text{in}\rangle \exp\left[i\left(\mathbf{p}_2+\mathbf{p}_1\right)\cdot\mathbf{r}\right] \qquad (13)$$

containing contributions of condensate and non-condensate atoms. The densities m_a, n_0, and $\tilde{n}_s$ are all expressible in terms of $\psi_c(p,t)$ and $\psi_s(p,t)$. A numerical solution of Eqs. (8) on a grid of values of p, combined with Eq. (12), is consistently sufficient for elucidating the dynamics of the system.

The present approach becomes mathematically equivalent to the approach of Ref. [18] if the inelastic collisions are neglected (justifiably in the case of Rb85) and the detuning is time-independent. At a low molecular density, the effect of non-condensate atoms on the molecular state is equivalent to a contribution added to its width, as introduced in Ref. [16] for the same process.

RESULTS

Molecular condensate formation

Calculations were performed for two Feshbach resonances in a condensate of Na atoms, using parameter values presented in Ref. [16]. The strong resonance, at 907 G, has the strength $\Delta = 0.98$ G, and the weak one, at 853 G, has the strength $\Delta = 9.5$ mG. The difference of magnetic momenta is $\mu = 3.65$ (in Bohr magnetons), the elastic scattering length is $a_a = 3.4$ nm, and the deactivation parameters are $\gamma = 0.8 \times 10^{-10}\text{cm}^3/\text{s}$ and $\gamma_m = 10^{-9}\text{cm}^3/\text{s}$. The neglection of elastic collisions is valid whenever $n_0(t_0) \ll 10^{15}$ cm^{-3} for the weak resonance, and $n_0(t_0) \ll 10^{17}$ cm^{-3} for the strong one. The spatial inhomogeneity can be neglected if the size of the condensate substantially exceeds $8 \times 10^{-2}\text{cm}^{-1/2} \times n_0^{-1/2}(t_0)$ and $2.5 \times 10^{-2}\text{cm}^{-1/2} \times n_0^{-1/2}(t_0)$, respectively, for the two resonances. Even when $n_0(t_0) = 10^8$ cm^{-3}, these estimates set a minimal size of 8μm for the weak resonance and 2.5μm for the strong one. The variation of the magnetic field is considered linear in time, $B(t) = B_0 + \dot{B}t$, fixing $t = 0$ as the resonance crossing time.

A relatively long-lived molecular condensate is formed more effectively in the case of a *backward sweep*, when the molecular state crosses the atomic one downwards (see Fig. 1b), as proposed in Ref. [19]. In the case of the Na resonances this can be done by starting the ramp from $B = 0$ upwards. The maximal conversion efficiency of the atomic condensate to a molecular one is $2\max(n_m)/n_0 \approx 0.8$ for the weak resonance (see Fig. 2). Increase of the atomic density, or decrease of the ramp speed,

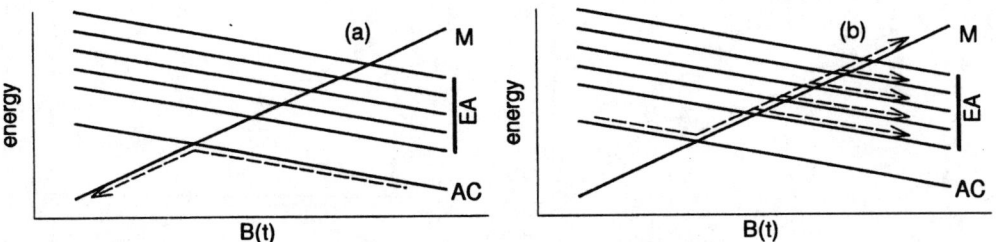

FIGURE 1. Schematic illustration of transitions between atomic (AC) and molecular (M) condensates and non-condensed atoms (EA) on backward (a) and forward (b) sweeps.

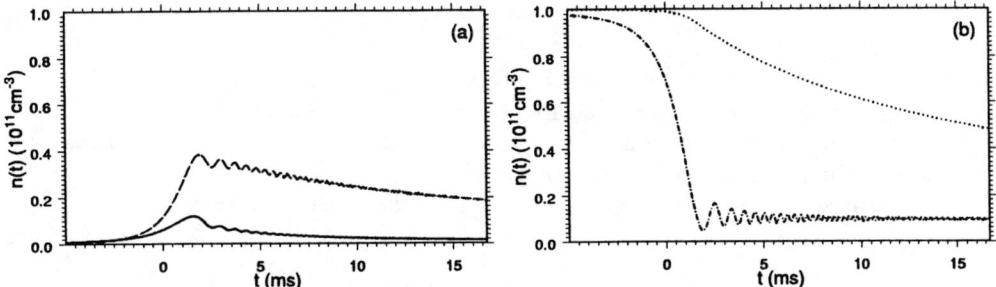

FIGURE 2. Time dependence of (a) the densities of the molecular condensate (dashed line) and the entangled atoms (solid line), and of (b) the atomic condensate density (dot-dashed line) and the total atomic density (sum of the atomic densities and twice of the molecular one) (dotted line). All plots pertain to the weak 853 G resonance in Na, with the initial atomic density $n_0 = 10^{11}$ cm^{-3}, and the ramp speed $\dot{B} = -0.1$ G/s (backward sweep). The resonance crossing time is fixed at $t = 0$.

degrade the conversion efficiency due to inelastic collisions. On increase of the ramp speed, the probability of crossing to the molecular condensate decreases, leaving more atoms in the atomic condensate (see Ref. [16]). At low atomic densities the conversion becomes less efficient due to a temporary gain of population in the non-condensate atomic states. This observation is peculiar to, and emphasizes the importance of, the simultaneous consideration of inelastic collisions and molecular dissociation within the second-quantized approach. (The approach used in [16] fails to describe the dissociation in the backward sweep.)

Figure 3 shows that a substantial conversion efficiency is retained in a wide range of the condensate density, leaving much freedom for the choice of an appropriate ramp speed. The lesser the initial density, the longer the lifetime τ_m of the molecular condensate but the higher the precision required for the control of the magnetic field. The optimal ramp speed is approximately proportional to the initial density. This dependence minimizes the effect of a variation of parameters determining the conversion of the atomic condensate to the molecular one and loss of the molecular condensate. Indeed, the conversion to the molecular condensate is (in the fast decay approximation [16]) characterized by the parameter $g^2 n_0 / \dot{B}$. Similarly, the loss is characterized by the ratio of the deactivation lifetime (which is inversely proportional to the initial density),

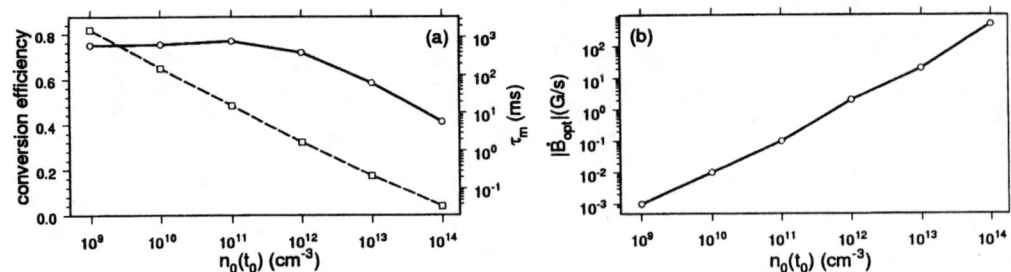

FIGURE 3. The dependence on the initial atomic density of (a) the conversion efficiency (solid line) and lifetime of the molecular condensate τ_m (dashed line), and (b) the optimal ramp speed $\dot{B}_{opt}$, all calculated for the weak resonance in Na (in a backward sweep).

and the crossing time (which is inversely proportional to the ramp speed).

The use of the strong resonance produces a lower conversion efficiency, due to a gain in the temporary formation of non-condensate atoms. (This difficulty should be exacerbated in a Rb85 condensate.) The optimal ramp speed is more than two orders of magnitude larger than in the weak resonance, given the same initial density.

Entangled gas formation

The non-condensate atoms are formed in squeezed states (as demonstrated in Ref. [12]), which now turn out to be two-mode squeezed states (as in Ref. [7]). This situation is similar to the state of electromagnetic radiation formed by parametric down conversion. As in quantum optics (see Ref. [2]), the amount of squeezing can be measured by the energy-dependent parameter $r(E,t)$ defined by

$$\exp\left[-4r(E,t)\right] = \frac{\langle \text{in}|\hat{X}\left(\mathbf{p}_1,t\right)\hat{X}\left(\mathbf{p}_2,t\right)|\text{in}\rangle_{\min}}{\langle \text{in}|\hat{X}\left(\mathbf{p}_1,t\right)\hat{X}\left(\mathbf{p}_2,t\right)|\text{in}\rangle_{\max}}, \tag{14}$$

related to the ratio of minimal and maximal uncertainties of the quadratures

$$\hat{X}\left(\mathbf{p},t\right) = \frac{1}{2}\{\left[\hat{\Psi}_a\left(\mathbf{p},t\right) \pm \hat{\Psi}_a\left(-\mathbf{p},t\right)\right]e^{i\theta} + \left[\hat{\Psi}_a^\dagger\left(\mathbf{p},t\right) \pm \hat{\Psi}_a^\dagger\left(-\mathbf{p},t\right)\right]e^{-i\theta}\} \tag{15}$$

attained at two orthogonal values of the phase angle θ. A mean squeezing parameter, weighed by the spectral density of Eq. (11),

$$\bar{r}(t) = \int dE \tilde{n}_s\left(E,t\right) r\left(E,t\right)/n_s\left(t\right), \tag{16}$$

can be introduced to describe the time variation of the squeezing.

A stable gas of entangled atoms is formed by a *forward sweep*, in which the molecular state crosses the atomic one upwards (see Fig. 1a). This process, too, is more efficient in the weak Na resonance. The molecular density is then very low and persists a shorter

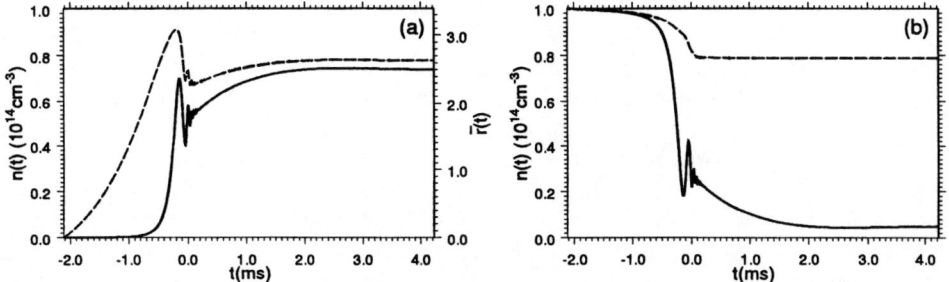

FIGURE 4. Time dependence of (a) the entangled atom density (solid line) and the the mean squeezing parameter $\bar{r}(t)$ [see Eq. (16)] (dashed line), and (b) the atomic condensate (solid line) and the total atom (dashed line) densities. All plots pertain to the weak resonance in Na with the initial atom density $n_0 = 10^{14}$ cm^{-3} and ramp speed 50 G/s in a forward sweep.

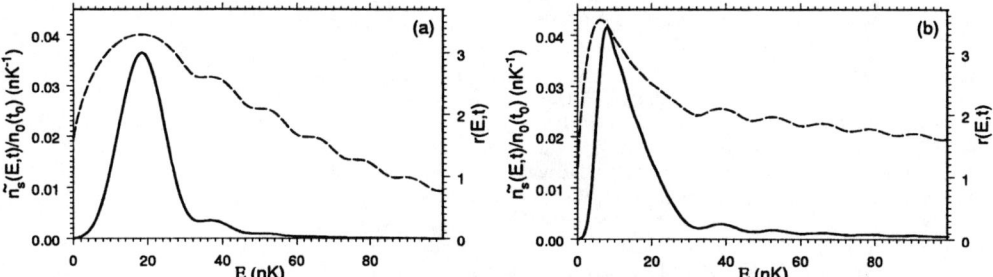

FIGURE 5. Energy spectra of the entangled-atom density $\tilde{n}_s(E,t)$ (solid line) and the squeezing parameter $r(E,t)$ (dashed line) for the weak resonance, calculated (a) at the peak, $t \approx -0.19$ ms, and (b) at the plateau, $t \approx 4$ ms. All other data are as in Fig. 4

time (compared to the one obtained in the backward sweep) due to fast dissociation. Figure 4 demonstrates that more than 70% of the atomic condensate can be transformed into a gas of atoms in two-mode squeezed states with the mean squeezing parameter $\bar{r} \approx 2.6$, corresponding to a noise reduction of about 23 dB. The time dependence of the mean squeezing parameter has a peak of $\bar{r} \approx 3.1$ at $t \approx -0.19$ ms. The state of an entangled gas can be frozen at the peak time by fast turning off of the magnetic field. The energy spectra of the entangled-atom density and the squeezing parameter are presented in Fig. 5. The density spectra are rather narrow, and the peak energy increases with time. The squeezing parameter reaches the value of $r(E,t) \approx 3.5$ (noise reduction of 30 dB) at the energy $E \approx 6$ nK and the time $t \approx 4$ ms.

CONCLUSIONS

The introduction of both quantum corrections and deactivating collisions is necessary for the proper analysis of diatom association in an atomic BEC due to Feshbach resonance in a time-dependent magnetic field. Over 80% of the atomic population can be converted to a metastable molecular condensate in a backward sweep. Under favorable conditions, such a condensate can survive for a significant fraction of a second, and thus a offer a better chance for its eventual stabilization by some means of coherent population transfer. The molecular condensate dissociates into atoms in two-mode squeezed states that are entangled. In a forward sweep, a very high degree of squeezing may be obtained, with the parameter r reaching a value of 3 and more. The availability of such an atomic system in the laboratory can provide a chance for studying a mesoscopic quantum system under conditions of unprecedented noise reduction.

REFERENCES

1. P. Meystre, *Atomic optics* (Springer, NY, 2001).
2. M. O. Scully and M. S. Zubairy, *Quantum Optics* (University Press, Cambridge, 1997).
3. L. Deng *et al.*, Nature (London), **398**, 218 (1999); M. Trippenbach, Y. B. Band, and P. S. Julienne, Phys. Rev. A **62**, 023608 (2000); J. M. Voges, K. Xu, and W. Ketterle, cond-mat/0203286.
4. C. Orzel *et al.*, Science **291**, 2386 (2001).
5. A. Sorenson, L.-M. Dian, J. I. Cirac, and P. Zoller, Nature (London), **408**, 63 (2001).
6. J. A. Dunningham, K. Burnett, and M. Edwards, Phys. Rev. A **64**, 015601 (2001).
7. V. A. Yurovsky, Phys. Rev. A **65**, 033605 (2002).
8. D. C. Roberts, T. Gasenzer, and K. Burnett, cond-mat/0106060.
9. J. Rogel-Salazar, S. Choi, G. H. C. New, and K. Burnett, quant-ph/0110076.
10. U. V. Poulsen and K. Molmer, Phys. Rev A **63**, 023604 (2001).
11. A. Vardi, V. A. Yurovsky, and J. R. Anglin, Phys. Rev. A **64**, 063611 (2001).
12. V. A. Yurovsky, A. Ben-Reuven, and P. S. Julienne, Phys. Rev. A **65**, 043607 (2002).
13. T. Opatrny and G. Kurizki, Phys. Rev. Lett. **86**, 3180 (2001).
14. E. Timmermans, P. Tommasini, M. Hussein, and A. Kerman, Phys. Rep., **315**, 199 (1999).
15. S. Inouye *et al.*, Nature **392**, 151 (1998); J. Stenger *et al.*, Phys. Rev. Lett. **82**, 4569 (1999).
16. V. A. Yurovsky, A. Ben-Reuven, P. S. Julienne and C. J. Williams, Phys. Rev. A **60**, R765 (1999); **62**, 043605 (2000).
17. V. A. Yurovsky and A. Ben-Reuven, cond-mat/0205267.
18. M. Holland, J. Park, and R. Walser, Phys. Rev. Lett. **86** 1915 (2001).
19. F. H. Mies, E. Tiesinga, and P. S. Julienne, Phys. Rev. A **61**, 022721 (2000).

Formation and Detection of Ultracold Molecules

Goran Pichler

Institute of Physics, P. O. Box 304, Bijenicka cesta 46, HR-10000 Zagreb, Croatia
e-mail: pichler@ifs.hr

Abstract. We shall describe a few new possibilities of making ultracold molecules using the recently acquired knowledge from the alkali vapor spectroscopy experiments at thermal energies. In addition, we shall discuss detection schemes of ultracold molecules by means of cesium and rubidium vapor at thermal energies will be described to illuminate possible connection with high and ultra-low temperature experiments.

INTRODUCTION

Making ultracold molecules recently became a very interesting and challenging task of modern atomic and molecular physics. There are several already well-developed methods of making ultracold molecules in their lowest triplet or singlet ground states. Photoassociation as proposed in ref. [1] offered the first successful pathway to make ultracold molecules in their excited state, which were actually experimentally detected by means of the ion detection spectrum [2] and the trap loss spectrum [3]. However, ultracold molecules connected with the electronic states emerging from the $s_{1/2}+s_{1/2}$ ground asymptote, were formed and detected using an additional feature. Namely, the photoassociation as a first step, is not sufficient for making stable ground state molecules. It is necessary that from the excited molecular state the spontaneous or the driven transition goes into the bound ground state. To make it more probable there should exist a relatively close lying turning point in the long-range region [4] or a near lying perturbing molecular state [5], as shown in the case of Cs_2. Photoassociation into the pure long-range rubidium molecule, was also used to make ultracold rubidium dimer in the lowest triplet state [6]. Ultracold alkali dimers in the $X\,{}^1\Sigma_g^+$ state have been also observed [7] and even the lowest vibrational level of the $X\,{}^1\Sigma_g^+$ state was populated [8]. In ref. [9] it was shown that Cs_2 molecules could be trapped in the light trap (focus of the infrared laser beam). It is interesting that all this development relies on the experimental [10] and theoretical [11] studies of self-broadening in alkali vapor, which led to the proposition of existence of pure long-range alkali molecules [12]. They became the useful tool in making ultracold molecules, and will probably serve in many future experiments.

CP645, *Spectral Line Shapes: Volume 12, 16th ICSLS*, edited by C. A. Back
© 2002 American Institute of Physics 0-7354-0100-4/02/$19.00

FORMATION

Double minimum potential

Above the first excited asymptote it is very difficult to excite molecules at long ranges. However, there are many states with double minimum or any other peculiar shape, which may facilitate the formation of the ultracold molecules in the ground states. In the case of alkali dimer molecules the double minimum state is well known due to the beautiful interference continuum phenomenon in LIF experiments.

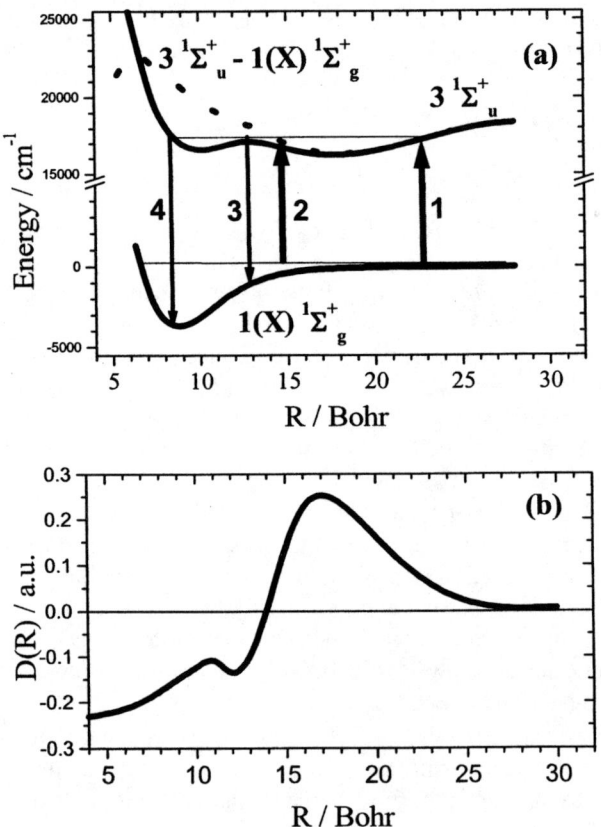

Figure 1. Double minimum and ground state potentials. Corresponding difference potential is shown by dotted line. Lower panel shows transition dipole moment adapted here from lithium dimer.

In Fig.1 we may observe that from the ultracold collision state the photoassociation into the vibrational level just above the barrier can take place with large Franck-Condon factor at locations 1 and 2 (bold arrows). At location 1 the transition dipole moment is much smaller than at position 2. However, the number of collisions, which enable photoassociation at location 2, is small.

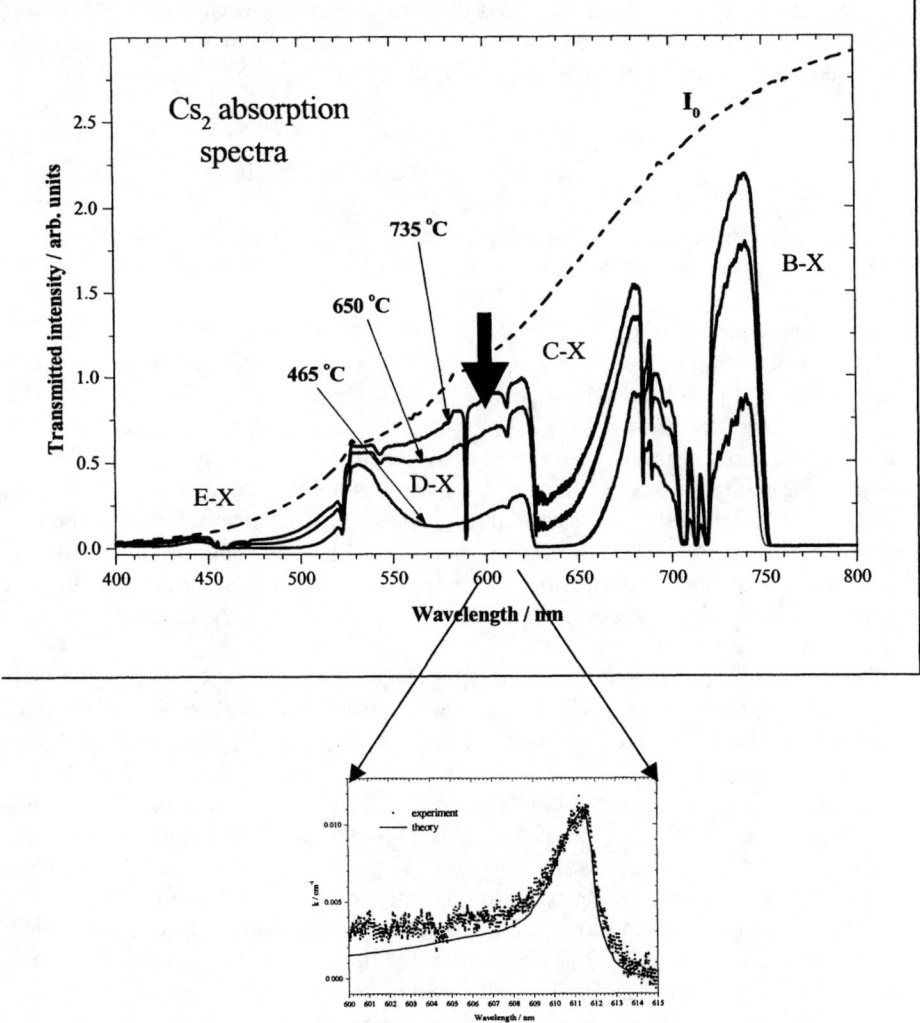

Figure 2. Transmission through the superheated cesium vapor at different temperatures, and the absorption coefficient at 613 nm (lower panel). There the absorption from the free ground state excites the outer well of the double minimum potential.

Spontaneous transition from the location 3 (top of the barrier) will end up in the excited bound levels of the X $^1\Sigma_g^+$ state. The inner turning point (location 4), will enable transition, fortuitously, into the lowest vibrational level of the ground molecular state. Similar situation prevails in almost all alkali dimers, with exception of K_2 and Rb_2, where the height of the barrier in the relevant double-minimum state is very small. Theoretical calculations and the detailed discussion of the photoassociation and spontaneous emission processes involving cesium dimer double minimum state will be reported in ref. [18].

The possibility of making ultracold molecules in the lowest vibrational state from the atomic Bose-Einstein condensate is very attractive, but in the present case the difference of photon momenta would heat up BEC.

Ion-pair long-range molecules

Even more appealing case of a highly excited molecular state is the pure ion-pair long-range molecule, which can be formed at ultracold conditions in almost all alkali molecules.

In ion-pair molecule (like $Cs^+ + Cs^-$) attractive Coulomb potentials, of $^1\Sigma_g^+ (0_g^+)$ and $^1\Sigma_u^+ (0_u^+)$ designation, cross many asymptotes at large interatomic distances. When they cross the potential curves of the same symmetry, then, by means of Wigner-Neumann theorem the avoided crossings (or pseudocrossing) appear. This provides minima and maxima in many potential curves, which may be observed in the quasistatic wings of the atomic spectral lines as satellite bands or shoulders (provided the ground state is relatively flat at large interatomic distances). Some of the potential curves possess minimum and maximum at large distances, and the minimum provides bound states of ion-pair molecule. We believe that at ultracold conditions these bound states could serve as an energy reservoir for some interesting physical and chemical collision processes.

In Fig. 3 we present such ion-pair long-range molecule on the left panel [19]. The relevant difference potentials are actually almost identical with the excited state potentials, since the ground state potentials are entirely flat at these long-range distances. The minima and maxima (or inflections) in the difference potentials lead to the formation of the satellite bands (dash arrows point to these satellite bands). The absorption line profile, on the right panel, shows that the minimum at about 36 bohr corresponds to the strong satellite band. Thus, we confirmed experimentally, that there is nonzero absorption probability directly into this ion-pair long range molecule. Laser excitation at 454.4 nm at ultracold conditions could make ultracold molecule of this ion-pair character. Similar satellite bands have been observed in rubidium, potassium and sodium, and will be reported in details soon.

We note that at these large distances excited molecules of both $^1\Sigma_g^+ (0_g^+)$ and $^1\Sigma_u^+ (0_u^+)$ types can be formed. They may be used as narrow Franck-Condon windows to reach even higher states within the small interval of interatomic distances. Chemical activity of such molecules will be of considerable interest for studies in the near future.

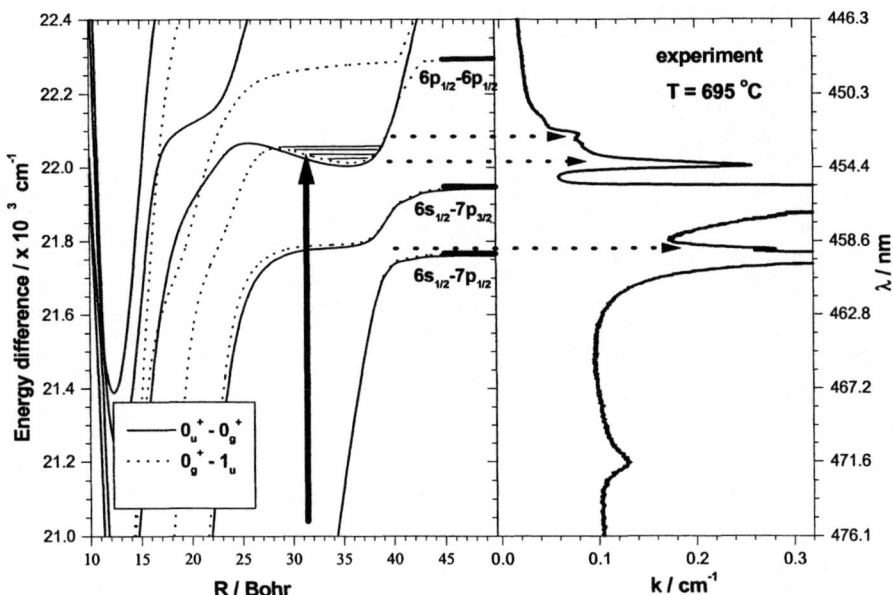

Figure 3. Left panel shows a possible excitation of the pure ion-pair long-range molecule (bold arrow). Absorption measurements at thermal energies revealed satellite bands corresponding to extrema and inflections in the difference potential curves (horizontal dash arrows towards the left panel).

Blue detuned intermediate long-range molecules

Another case of making ultracold molecules is connected with blue detuned potential curves that exhibit double minimum shape. Again, we shall deal with the potential curve, of 0_g^+ symmetry, that possesses two minima. The outer minimum is located in the intermediate long-range region, at about 20 bohr. Direct excitation from the ultracold collision pair into the outer minimum is possible, and it can serve as a Franck-Condon window for the additional excitations to even higher potential curves, of 0_u^+ or 1_u symmetries, at intermediate long-range distances. In our experiments at thermal energies the double minimum potential was observed in absorption, for K_2 [20], Rb_2 [21] and KRb [22] cases. Blue detuned formation of the intermediate long-range molecule at ultracold conditions is possible, although with smaller chances than pure long-range molecules [21,22]. Absorption and emission measurements at thermal energies confirm the existence of such peculiar molecular states, and it would be very interesting to make them at ultracold conditions.

As an example, in Fig. 4 we show KRb blue detuned potential curves with corresponding satellite bands (right panel).

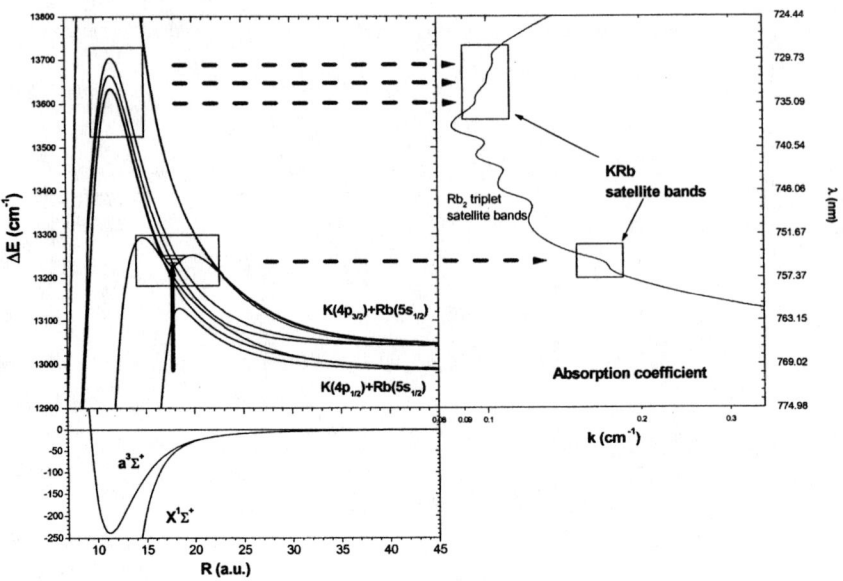

Figure 4. Blue detuned excitation into the intermediate long-range molecule of KRb. Right panel shows the corresponding KRb satellite bands and lower left panel shows the triplet and a portion of the singlet ground state of KRb.

DETECTION

Cesium dimer detection

A very efficient detection method of ultracold molecules was employed in the first experiment [4]. Researchers searched almost the whole visible spectrum for the REMPI signal. In their paper Fioretti et al. [4] reported two spectral regions with high detection sensitivity, one being in the spectral region around 716 nm, and the other at about 540 nm. It is well known that in the region between 705 and 720 there are three distinct diffuse absorption bands, peaking at 707, 713 and 718 nm, and the obvious explanation of this successful new detection technique was the resonance enhancement in the first photon step of excitation. Diffuse bands of all alkali dimers [23] are ubiquitous in absorption and emission spectra, due to the large transition dipole moment connected with (a $^3\Sigma_u^+ \rightarrow 2\ ^3\Pi_g$) transition. Diffuse appearance is connected with a shallow lowest triplet state, and therefore many transitions in experiments at thermal energies are of bound-free nature, which produce continuum spectrum.

The REMPI process in the first step may be described by the following equation:

$$Cs_2(a\ ^3\Sigma_u^+) + h\nu_{diff.band} \rightarrow Cs_2^{**}(2\ ^3\Pi_g),$$

where $2\ ^3\Pi_g$ state in Cs_2 is actually split into four different states, 2_g, 1_g, and almost degenerate 0^+_g and 0^-_g states, due to relatively large spin-orbit interaction in cesium. The second photon directly produces cesium dimer ion in certain rovibrational levels:

$$Cs_2^{**}(2\ ^3\Pi_g) + h\nu_{diff.band} \rightarrow Cs_2^+(X\ ^2\Sigma_g^+) + e^-(E_{kin} \approx 0),$$

and we assume that the ejected electrons have kinetic energy close to zero. We may conclude that in order to use efficient REMPI process the following equation should hold:

$$2 \times h\nu_{diff.band} < I.P.\ (atom),$$

where $\nu_{diff.band}$ is the frequency of the diffuse band transition and I.P. is the ionization energy of the atom. We assume a small binding energy in the $a\ ^3\Sigma_u^+$ state.

If we plot the photon energies for diffuse bands of all alkali dimers and compare them with their corresponding half value of the ionization potential, then we may readily conclude that cesium dimer detection is best suited for this REMPI process. In the case of rubidium dimer Gabbanini et al. [6] used the same diffuse bands in order to detect lowest triplet state molecules. Beside this, they also found that $2\ ^3\Sigma_g^+$ state could be also used for the detection purposes [24], which was confirmed even for the case of cesium dimer [25].

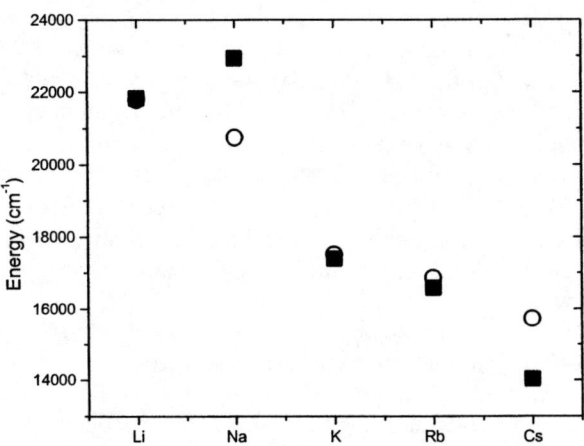

Figure 5. Photon energies of the diffuse bands (squares) compared with half value of the ionization potentials (open circles) for all alkali cases.

Even if the 2 x $h\nu_{diff.band}$ is larger than the ionization energy of the alkali atom, like in the case of sodium (2 x $h\nu_{diff.band}$ > I.P.), it is still possible to form a dimer. However, in that case the ejected electron takes a certain amount of energy in order to leave behind the dimer ion in the bound state. This was actually demonstrated in the first successful detection of ultracold cesium molecules [4]:

first step:

$$Cs_2(a\,^3\Sigma_u^+) + h\nu_{diff.band} \rightarrow Cs_2^{**}(3\,^3\Pi_g)$$

second step:

$$Cs_2^{**}(3\,^3\Pi_g) + h\nu_{diff.band} \rightarrow Cs_2^+(X\,^2\Sigma_g^+) + e^-(E_{kin}>0)$$

or

$$Cs_2^{**}(3\,^3\Pi_g) + h\nu_{diff.band} \rightarrow Cs^+ + Cs + e^-(E_{kin}=0).$$

We observed diffuse bands of cesium dimer at 543.5 nm and 557 nm [26]. It is interesting to note that Pillet's group actually detected cesium dimer ions at about 554 nm, although the efficiency of that channel was smaller than of 716 nm channel [4]. This certainly leads to the conclusion that the diffuse bands should be located midway to the dimer ion bound states, if we want the lowest triplet state molecules to be efficiently detected.

If the REMPI process with two-photons cannot be conveniently applied it is possible to find REMPI process with three photons as was shown, quit recently, in the case of rubidium dimers in the lowest triplet state [27]. Three-photon ionization was performed with an 805 nm wavelength pulsed laser. Researchers could estimate more than 10^4 of these molecules in magnetic trap at temperature of about 200 μK.

In Fig. 5 we present the absorption spectrum of superheated cesium vapor in the vicinity of cesium dimer diffuse bands. At shorter wavelengths we may see quadrupole transitions with underlying molecular contribution. This spectral region was, in fact, used for the detection of singlet ground state cesium dimer formed in the high vibrational levels [5].

In Fig. 6 we also present the absorption spectrum of dense rubidium vapor in the visible spectral region, where electric quadrupole spectral lines and Rb_2 diffuse bands can be readily observed. Spectral region in-between Rb forbidden lines and (inclusive) Rb_2 diffuse bands are of great interest for the detection of lowest triplet and singlet molecules. The excitation around 580 nm within the blue shaded wing of rubidium diffuse bands, should photoassociate a pair of ultracold rubidium atoms into the double minimum state. Spontaneous emission from such levels could form stable ultracold rubidium dimer in the lowest vibrational levels.

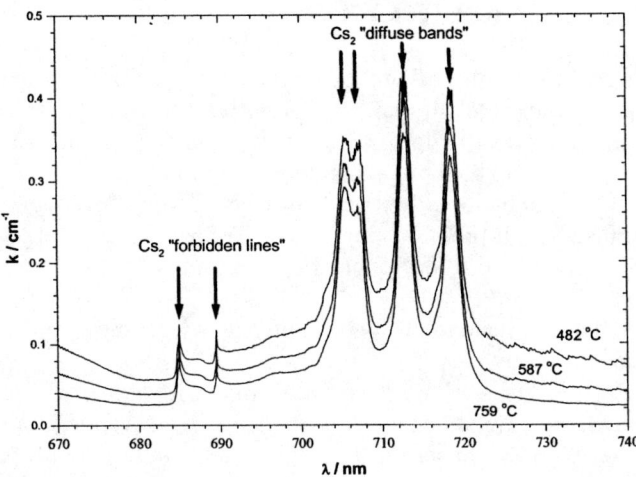

Figure 6. Absorption coefficient profiles of superheated cesium vapor in the region of cesium diffuse bands and forbidden lines.

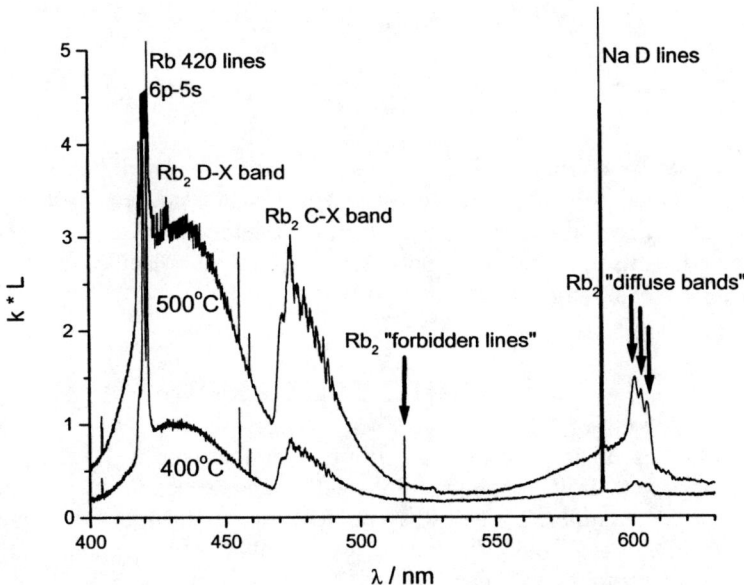

Figure 7. Absorption coefficient profiles of rubidium vapor in the region of diffuse bands and forbidden lines.

APPLICATIONS

In order to better apply the results of our experiments at thermal energies it is necessary to make pure atomic alkali vapor. Currently, we investigate several schemes for the alkali dimer destruction. We use superheating method [28,29], resonance line destruction [30] and the optical pumping method [31].

Out of many interesting applications that rely on spectroscopic investigations at thermal energies are pulsed high-pressure pure alkali lamps. A very large color-rendering index can be achieved, but the efficiency is not quit satisfactory yet [32, 33].

Brown dwarfs atmospheres offer another opportunity to widen the application of the basic alkali vapor spectroscopy [34].

There are many interesting tasks in the near future that will ultimately need the basic spectroscopic knowledge on the behavior of the dense alkali vapor. We usually obtained this knowledge from the experiments at high temperatures and densities. However, the findings can be, without any doubt, applied to ultracold conditions on one hand and very high temperatures on the other hand. It seems that ultracold alkali molecules provide additional mean to study the structure, physical and chemical behavior, which can be further applied at elevated temperatures. The distinction between triplet and singlet manifold can be best investigated at ultra low temperatures, as well as tiny perturbations between singlet and triplet states.

Outlook

We may expect in the near future, that an efficient method will be found by means of which the cloud of molecular Bose-Einstein condensate can be achieved. This ultimate goal will certainly bring about a new quantum mechanical object able to perform wonder in some futuristic applications, coherent molecular optics and many others [35,36,37,38,39,40].

ACKNOWLEDGMENTS

This review relies on the recent work from several groups in Zagreb, which was enabled through very fruitful collaborations with many groups in Europe and U.S.A. I am grateful to D. Aumiler, T. Ban, R. Beuc, I. Labazan, S. Milosevic, M. Movre, H. Skenderovic, D. Veza and V. Zivcec for invaluable help and many fruitful discussions. We all thank to Orsay group: M. –L. Almazor, D. Comparat, C. Dion, O. Dulieu, C. Drag, F. Masnou-Seews, P. Pillet, Storrs group: H. Wang, M. Pichler, W. C. Stwalley, Lyon group: M. Aubert-Frecon, A. R. Alouche, S. Rousseau, Ashtarak group: S. Ter-Avetisyan, D. Sarkisyan, and W. Meyer from Kaiserslautern for experimental and theoretical collaboration.

REFERENCES

1. Thorsheim, H. R., Weiner, J., and Julienne, P. *Phys. Rev. Lett.* **58**, 2420-2423 (1987).
2. Lett, P. D., Helmerson, K., Phillips, W. D., Ratliff, L. P., Rolston, S. L., and Wagshul, M. E., *Phys. Rev. Lett.* **71**, 2200-2203 (1993).
3. Miller, J. D., Cline, R. A., and Heinzen, D. J., *Phys. Rev. Lett.* **71**, 2204-2207 (1993).
4. A. Fioretti, A., Comparat, D., Crubellier, A., Dulieu, O., Masnou-Seeuws, F., and P. Pillet, *Phys. Rev. Lett.* **80**, 4402-4405 (1998).
5. Dion, C. M., Drag, C., Dulieu, O., Laburthe Tolra, B., Masnou-Seeuws, F., and P. Pillet, *Phys. Rev. Lett.* **86**, 2253-2256 (2001).
6. Gabanini, C., Fioretti, A., Lucchesini, A., Gozzini, S., and Mazzoni, M., *Phys. Rev. Lett.* **84**, 2814-2817 (2000).
7. Nikolov, A. N., Eyler, E. E., Wang, X., Li, J., Wang, H., Stwalley, W. C., and Gould, P., *Phys. Rev. Lett.* **82**, 703-706 (1999).
8. Nikolov, A. N., Ensher, J. R., Eyler, E. E., Wang, H., Stwalley, W. C., and Gould, P. *Phys. Rev. Lett.* **84**, 246-249 (2000).
9. Takekoshi, T., Patterson, B. M., and Knize, R. J., *Phys. Rev. Lett.* **81**, 5105-5108 (1999) and *Phys. Rev. A* **59**, R5-7 (1999).
10. Niemax, K., and Pichler, G., *J.Phys.B:Atom.Mol.Phys.* **8**, 179-184 (1975).
11. Movre, M., and Pichler, G., *J.Phys.B:Atom.Mol.Phys.* **10**, 2631-2638 (1977).
12. Stwalley, W. C., Uang, Y., and Pichler, G., *Phys. Rev. Lett.* **41**, 1164-1167 (1978).
13. Tellinghuisen, J., Pichler, G., Snow, W. L., Hillard, M. E., and Exton, R. J., *Chem.Phys.* **50**, 313-330 (1980).
14. G. Pichler, G., Milosevic, S., Veza D., and Vukicevic, D., *J.Phys.B:At.Mol.Phys.* **16** (1983) 4633-4642.
15. Pichler, G., Bahns, J. T., Sando, K. M., Stwalley, W. C., Konowalow, D. D., Li, L., Field, R. W., and Müller, W., *Chem.Phys.Lett.* **129** 425-428 (1986).
16. Luh, W. T., Bahns, J. T., Sando, K. M., Stwalley, W. C., Henneghan, S. P., Chakravorty, K. P., Pichler, G., and Konowalow, D. D. *Chem.Phys.Lett.* **131** (1986) 335-338, and Milosevic, S., Kowalczyk, P., and Pichler, G. *J.Phys.B:At.Mol.Phys.* **16** (1983) 4633-4642.
17. Gondal, M. A., Shahdin, S., Shaban, A. A.,Omar, M. S., *Opt. Commun.* **86**, 128 (1991).
18. Pichler, M., Stwalley, W. C., Beuc, R., and Pichler, G. to be published.
19. Ban, T., Skenderović, H., Beuc, R., Krajcar Bronić, I., Rousseau, S., Allouche, A. R., Aubert-Frécon, M., and Pichler, G., *Chemical Physics Letters*, **345**, 423-428 (2001).
20. Beuc, R., Milosevic, S., Movre, M., Pichler, G., and Veza, D., *FIZIKA (YU)* **14** 345-349 (1982).

21. Almazor, M. –L., Masnou-Seeuws, F., Dulieu, O., Beuc R., and Pichler, G., *Eur. Phys. J. D*, **15**, 355-363 (2001).
22. Skenderovic, H., Beuc, R., Ban, T., and Pichler, G., *Eur. Phys. J. D*, **19**, 49-56 (2002).
23. Pichler, G., Milosevic, S., Veza, D., and Beuc, R., *J.Phys.B:At.Mol.Phys.* **16** 4619-4631 (1983).
24. Fioretti, A., Amiot, C., Dion, C. M., Dulieu, O., Mazzoni, M., Smirne, G., and Gabbanini, C., *Eur. Phys. J. D*, **15**, 189-198 (2001).
25. Dion, C. M., Dulieu, O., Comparat, D., W. de Souza Melo, W., N. Vanhaecke, N., Pillet, P., Beuc, R., Milošević, S.,and Pichler, G., *Eur. Phys. J. D*, **18**, 365-370 (2002).
26. Ban, T., Skenderović, H., Beuc, R., and Pichler, G., *Europhys.Lett.* **48**, 378-384 (1999).
27. Ryu, C., Freeland, R. S., Shim, W., and Heinzen, D. J., DAMOP, G5 10 (2002).
28. Sarkisyan, D. H., Sarkisyan, A. S., Yalanusyan, A. K., *Appl. Phys. B* **66**, 241-244 (1998).
29. Ban, T., Skenderović, H., Ter-Avetisyan, S., and Pichler, G., *Applied Physics B*, **72** (2001) 337-341
30. Lintz, M., and Bouchiat, M. A., *Phys. Rev. Lett.* **80**, 2570-2573 (1998).
31. Berheim, R. A., and Xu, J. H., *Phys. Rev. Lett.* **83**, 3394-3397 (1998).
32. Liu, J. Guenther, K., Kaase, H., and Serick, F., *Proc. 8ᵗʰ Int. Symp. On the Science and Technology of Light Sources*, Greifswald, Germany, paper CO2 (1998).
33. Gu, H. Muzeroll, M. E. Chamberlain, J. C., and Maya, J., *Plasma Sources Sci. Technol.* **10**, 1-9 (2001).
34. Burrows, A., Hubbard, W. B., Lunine, J. I., and Liebert, J., *Rev. Mod. Phys.* **73**, 719-765 (2001).
35. Bahns, J. T., Gould, P. L., and Stwalley, W. C., *Adv. Atomic Mol. Opt. Phys.***42**, 171-224 (2000).
36. Stwalley, W. C., and Wang, H., *J. Mol. Spectrosc.* **195**, 194-228 (1999).
37. Weiner, J., Bagnato, V. S., Zilio, S. C., and Julienne, P.S., *Rev. Mod. Phys.* **71,** 1-86 (1999).
38. Wynar, R. H., Freeland, R. S., Han, D. J., Ryu, C., and Heinzen, D. J., *Science*, **287**, 1016-1019 (2000).
39. Gerton, J. M., Strekalov, D., Prodan, I., and Hulet, R. G., *Nature*, **408**, 692-695 (2000).
40. Donley, E. A., Claussen, N. R., Thompson, S. T., Wieman, C. E., *Nature*, **417**, 529-532 (2002).

Ultracold Molecules: Formation, Detection

Olivier Dulieu

Laboratoire Aimé Cotton, CNRS, Campus d'Orsay, 91405 Orsay cedex, FRANCE

Abstract. Since the first observation of ultracold molecules in Orsay, investigation of new ways of creating ultracold molecules inside cold atom traps is a challenging task for both theoretical and experimental groups. In this paper, we review some of our recent results on formation processes using cw laser photoassociation of cold atoms, involving population of long-range molecular potential wells, or resonant coupling. We implemented accurate numerical methods based on a mapped Fourier grid approach, to solve time-independent Schrödinger equation, in order to obtain quantitative predictions for ultracold molecule formation rates and detection efficiency.

INTRODUCTION

During the last decade, laser cooling and trapping of atoms has allowed researchers to obtain gaseous samples of atoms at ultralow temperatures, i.e. well below 1 milliKelvin. Among the most spectacular results, are undoubtely the achievement of quantum degenerate gases of ultracold alkali atoms [1-4]. In contrast, direct laser cooling of molecules, which may appear as a simple extension of laser cooling of atoms, encounters major difficulties [5]. Indeed, the much more complicated energy level structure of molecules is not suitable for a large number of optical pumping cycles within a closed level scheme.

Since their first observation in 1997 [6], cold (T=0.01-1K) and ultracold molecules (T<0.001K) have encountered a growing interest from many research groups around the world, investigating several kinds of formation processes. Up to now, the photoassociation (PA) of a pair of ultracold atoms [7] is the only way to produce molecular samples with temperatures lying in the microKelvin range (demonstrated for Cs_2 [6,8], K_2 [9,10], and Rb_2 [11]), as it overcomes the inherent difficulties of direct laser cooling of molecules. Use of a number of non-optical methods to cool molecules to the 0.1 - 1 K temperature range has also been reported, such as buffer gas cooling techniques [12] or cooling with electrostatic forces [13]. Other alternative methods have also been proposed [14-16].

The aim of the present paper is to report on the formation of ultracold molecules via photoassociation. We first recall the basics of cold atom photoassociation, and present the various mechanisms for ultracold molecule formation which have been reported. The theoretical methods required to treat such systems are briefly described. We finally emphasize on the formation process based on molecular resonant coupling, which allowed a precise comparison of theory with an experiment showing a spectrum with a specific line shape.

CP645, *Spectral Line Shapes: Volume 12, 16th ICSLS*, edited by C. A. Back
© 2002 American Institute of Physics 0-7354-0100-4/02/$19.00

PHOTOASSOCIATION OF COLD ATOMS

Up to now, the coolest molecules have been formed through photoassociation (PA) reaction in a cold alkali atom trap, where a pair of cold atoms A in their ground state, separated by a large distance R, absorbs a photon with a frequency $v_L = v_0 - \delta_{red}$ red-detuned by $\delta_{red} > 0$ from an atomic transition v_0, to populate a rovibrational level (v,J) of a long-range excited electronic molecular state Ω:

$$A + A + h(v_0 - \delta_{red}) \rightarrow A_2^* (\Omega; v,J). \qquad (1)$$

In this reaction, the kinetic energy distribution of the atoms is so sharp ($E/h \approx 5$ MHz at 100 μK), that it is smaller than any other characteristic energy of the system. The PA process acts now as a resonant free-bound transition (figure 1): the radial motion of the colliding atoms is represented by a stationary continuum wave function with a large amplitude at large interatomic distances, determined by the R^{-6} variation of the long-range potential correlated to $Cs(6s) + Cs(6s)$. In contrast with the initial state, the R^{-3} long-range behaviour of the electronic excited state correlated to $Cs(6s) + Cs(6p_{1/2,3/2})$ implies the existence of bound levels with a vibrational motion reaching distances of $100a_0$ (where a_0 is the Bohr radii), or more.

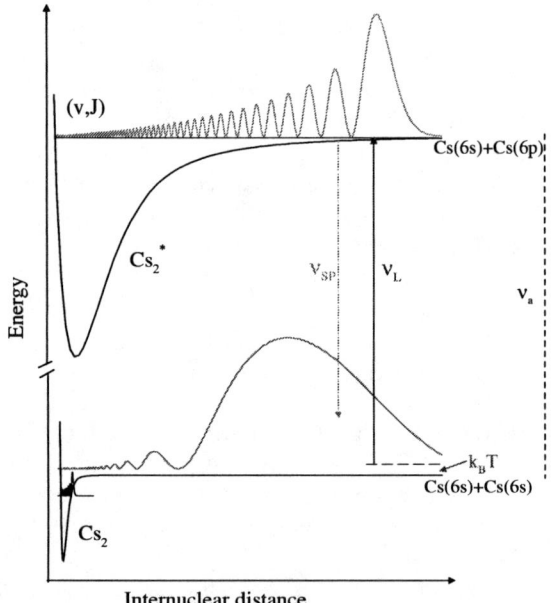

Internuclear distance

FIGURE 1. Schematic representation of the photoassociation process between cold cesium atoms. The PA photon frequency v_L is detuned to the red of the $6s \rightarrow 6p_j$ $(j=1/2,3/2)$ transition frequency.

Such excited molecules decay back into a pair of hot free atoms, which escape from the cold atom trap. This loss process is actually the simplest way to observe the PA process. Several review papers have been devoted to the PA process and its various applications [17-20], in which the interested reader can find extensive lists of references. The resonant character of the PA process has resulted into the development of the so-called photoassociative spectroscopy, which gives access to the determination of the long-range part of the electronic potential curves. Among many remarkable achievements, the so-called pure long-range molecules [21], i.e. molecules with excited electronic states exhibiting a potential well at interatomic distances well beyond the standard chemical bond length, have been observed for all alkali dimers except Li$_2$. The long-range coefficients C_3 governing the R^{-3} variation of the excited electronic states have been extracted, providing new determinations of the radiative lifetime of the first excited $P_{1/2,3/2}$ levels of alkali atoms with a better accuracy [22-25] than standard atomic spectroscopy experiments. Determination of the scattering length for elastic collisions, which a key parameter for BEC studies, also benefited from PA studies in several cases.

FORMATION OF ULTRACOLD MOLECULES

The PA process creates only short-lived molecules, decaying by spontaneous emission of a photon ν_{sp} back to a pair of free atoms, as shown in figure 1 and 2a. This is due to the poor overlap between the vibrational wave function of the PA level, and the vibrational wave functions of the lowest electronic states correlated to $6s+6s$. The way to stabilize the photoassociated molecules into long-lived ultracold molecules has been first demonstrated with cesium dimer in our group in 1997 [6], and is illustrated in figure 2b.

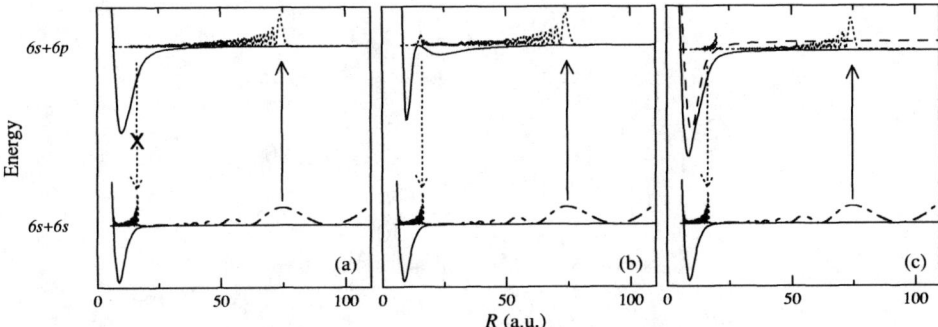

FIGURE 2. Cold molecule formation schemes discussed in the text, through cold atom photoassociation (arrow 1). (a) usual case: radial wave functions in the upper and lower potentials overlap weakly, yielding mainly free atoms (arrow ii); (b) the excited double-well potential ensures an efficient transfer of the probability density towards short distances; (c) the resonant coupling mechanism, results from the pseudo-crossing of two coupled potential curves.

The existence of a long-range potential well (see the pure long-range molecules quoted above) in an excited electronic state of 0_g^- symmetry (in Hund's case c notation), correlated to the $(6s+6p_{3/2})$ dissociation limit, makes the radiative decay towards bound vibrational levels very efficient, through the presence of a turning point in the 15-20a_0 range for the PA vibrational level. Such molecules are actually created in vibrational levels of the lowest triplet state in Cs$_2$, and are metastable. They are not trapped by the standard magneto-optical trap used for the cesium atoms in the experiment, so they escape from the trap due to gravity within 10 ms. The balistic expansion of the molecular cloud provides an estimate for their temperature, found in the 20-100 µK range, i.e. similar to the atom trap temperature. The same 0_g^- double-well state is also found to be responsible of ultracold Rb$_2$ molecule formation [11], while in Na$_2$ and K$_2$, the 0_g^- external well is located at too large distance to yield ultracold molecules.

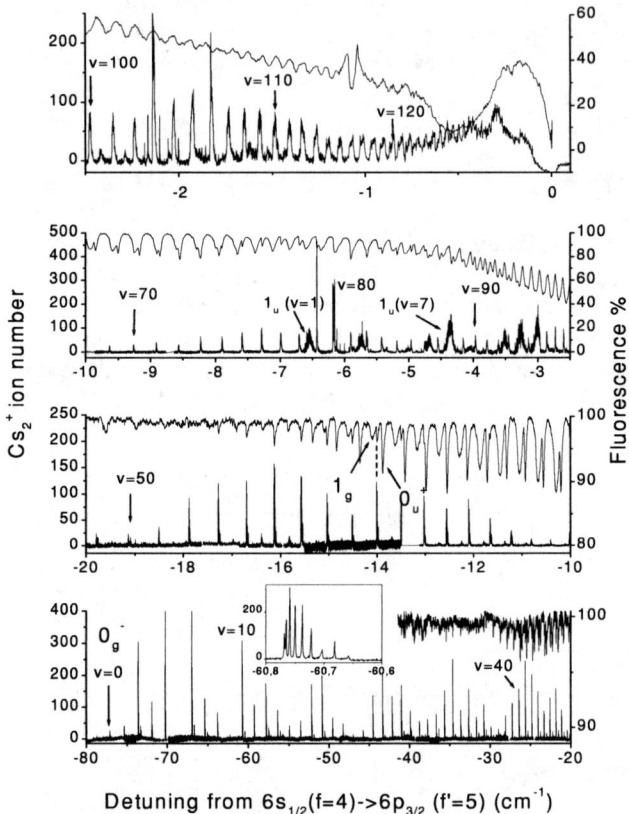

FIGURE 3. Experimental Cs$_2^+$ spectrum recorded after photoassociation of rovibrational levels of the external well of the 0_g^- long-range state, as a function of the PA laser detuning [26]. The assignement of selected levels is reported. The broad structures are produced by photoassociation of another long-range state in Cs$_2$, labelled by the 1_u symmetry [27] . Stars indicates lines which are due to the coupling between vibrational levels lying in the external well and in the internal well of the 0_g^- state, due to tunneling through the potential barrier (see figure 2b).

The ultracold molecules are detected using an ionizing probe laser located below the atom trap, by counting the Cs_2^+ molecular ions produced by a resonant two-step ionization process. In a typical experiment, the atom trap is illuminated continuously by the PA laser, while an intense laser pulse of ≈ 1 ns duration is applied at a 10 Hz repetition rate. All ultracold molecules accumulated within the trap between two pulses are then ionized in a single shot. A typical Cs_2^+ ion spectrum is shown in figure 3, as a function of the PA laser detuning δ_{red}. In this example, the lines are assigned to the PA of rovibrational levels lying in the external well of the 0_g^- long-range state.

THEORETICAL METHODS

As illustrated by figures 1 and 2, ultracold molecule formation involves both the large distance region, governed by weak interactions and low kinetic energies, and the short distance region, which is dominated by strong chemical interactions, and large kinetic energies. The theoretical description of these processes should treat accurately such a situation where the local de Broglie wavelength associated to the radial motion changes dramatically from large R towards small R. For this purpose, we developed in our group the Mapped Fourier Grid Hamiltonian (MFGH) method, described in detail in refs. [28,29].

In the standard Fourier Grid Hamiltonian method, wave functions are represented on a grid of N equally-spaced points, equivalent to an expansion on a basis of N plane waves [30-32]. The Hamiltonian for an atom pair in a given electronic state is expressed as a $N{\times}N$ matrix, whose diagonalization provides N eigenvalues and eigenvectors.

The MFGH method relies on the definition of an *adaptive coordinate*, which is controlled by the local de Broglie wavelength of the system. This leads to a drastic reduction of the required number of grid points, particularly suitable when large distances are studied. The MFGH method is easily generalized to the calculation of eigenvalues of a system with p coupled electronic states (figure 2c), obtained after diagonalization of the corresponding pN×pN Hamiltonian matrix. The resonant coupling mechanism described below has been treated with the MFGH method.

RESONANT COUPLING MECHANISM

The ultracold molecule formation process described above relies on a particular feature of the molecular potential excited by photoassociation, i.e. its double-well shape present in Rb_2 and Cs_2. Recently, we proposed [34] a formation scheme for ground state ultracold molecules, represented in figure 2c. Due to spin-orbit coupling, two excited electronic states of 0_u^+ symmetry, correlated to $(6s+6p_{1/2})$ and $(6s+6p_{3/2})$, interact through an avoided crossing at short distance (around $10a_0$ in Cs_2. The vibrational series of each electronic state are resonantly coupled each other (figure 4), inducing mixing of vibrational wavefunctions, according to the example shown in figure 2c. As in the double-well case, a turning point in the $15-20a_0$ range for the PA vibrational level enhances the radiative decay towards bound levels of the ground state (of $^1\Sigma_g^+$ symmetry) of the molecule. This process is expected to be quite general, as avoided crossings is a common feature in diatomic molecules.

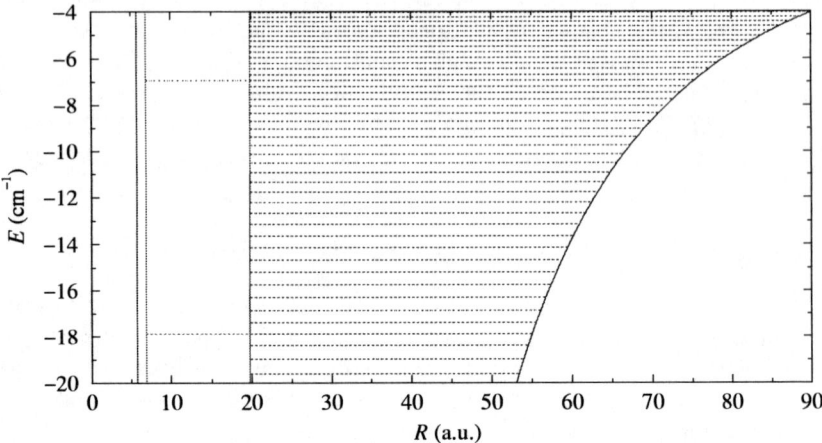

FIGURE 4. Blow-up of the 0_u^+ coupled potentials schematized in figure 2c, showing the resonant coupling occuring when a vibrational level of the sharpest potential (correlated to $6s+6p_{3/2}$) is embedded in a dense vibrational series of the broadest potential (correlated to $6s+6p1_{/2}$). The avoided crossing between the two electronic states is out of the range of the figure.

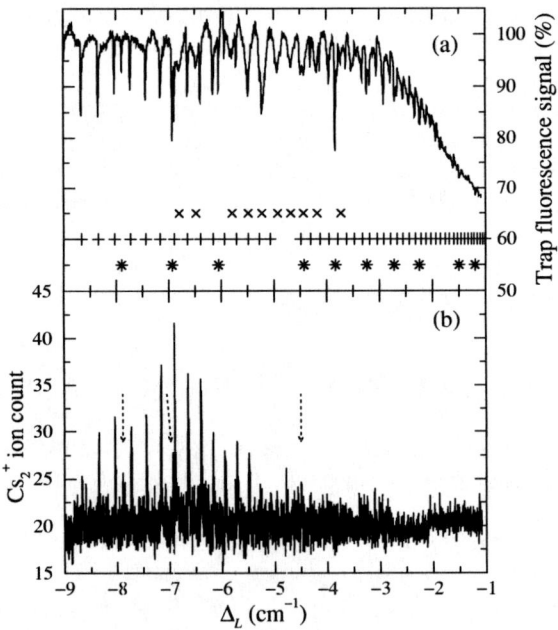

FIGURE 5. Experimental spectrum obtained after photoassociation of vibrational levels of the 0_u^+ coupled potentials. The detunings are taken from the $6s+6p_{1/2}$ limit.

In figure 5, we reproduce a set of isolated lines, recorded after PA with a laser detuned by a few cm^{-1} from the $6s$-$6p_{1/2}$ atomic transition [34]. None of the potentials correlated to the ($6s$+$6p_{1/2}$) dissociation limit exhibits a double-well structure. Following recent work on channel coupling between the 0_u^+ (ns+np_j) states of the heavy alkali dimers [29,33], we used the MFGH method to model this Cs_2^+ spectrum (figure 6), assuming that resonantly coupled levels of the 0_u^+ potentials interacting through spin-orbit interaction are populated by PA.

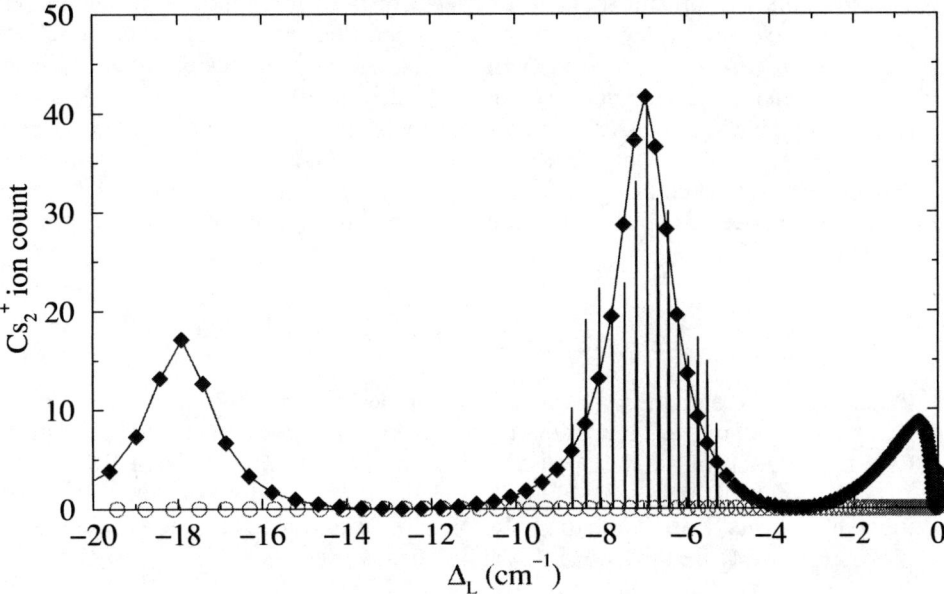

FIGURE 6. Cs_2^+ ion count for molecules formed through the 0_u^+ coupled channels in Cs_2. Solid line: representation of the experimental signal; circles: calculated signal (absolute value arbitrarily scaled).

Details on the cold molecule and ion formation rate calculation are available in ref. [34]. These rates are governed mainly by the overlap between radial wave functions: (i) the initial continuum wave function and the vibrational wave function of the 0_u^+ levels populated by PA, (ii) the wave function of the photoassociated level with the ground state vibrational wave functions, (iii) the latter ones with the vibrational wavefunctions of the intermediate state of the ionization process. We see an oscillation in the rate value attributable to the oscillations of the continuum wave function. The position and the enveloppe of the experimental signal is well reproduced by our model. The width of this enveloppe, which can be qualified as a line shape, illustrates the efficiency of the channel coupling, as about 14 levels of the broader well are affected by the presence of the isolated level in the sharp well (figure 4).

CONCLUSION

Many issues remain open in the research field of ultracold molecule formation. Once dense and cold molecular samples are produced, the next step is to trap them in particular configurations of electromagnetic fields. This has been demonstrated with a small number of ultracold cesium molecules in ref. [35], and with much more efficiency in Orsay [36]. A coherent decay process (like stimulated Raman emission) can be used to control the final state of the formed ultracold molecule [37]. Such achievements will open a way for a better control of these molecules, including the possibility to increase their number density and temperature to reach the quantum degenerate regime. Further works will be motivated by various possible applications such as precision measurements (accurate spectroscopy, determination of data such as atomic radiative lifetimes [38], measurement of dipole moment of the electron as a critical test of elementary particle physics beyond the standard model [39]), condensation of complex systems (achievement of a molecular condensate or of an assembly of oriented dipolar molecules), coherent control of molecule formation, ultracold chemistry.

ACKNOWLEDGMENTS

The work presented in this article has been performed within the "Cold Molecule" group at Laboratoire Aimé Cotton, involving both the theoretical team (C. Amiot, A. Crubellier, O. D., and F. Masnou-Seeuws) and the experimental team (D. Comparat, S. Guibal, and P. Pillet). Several Ph.D. students and post-docs have also been strongly involved in this work, including C. Dion, C. Drag, A. Fioretti, R. Gutterres, V. Kokoouline, B. Laburthe-Tolra, N. Vanhaecke, M. Vatasescu.

REFERENCES

1. M. H. Anderson, J. R. Ensher, M. R. Matthews, C. E. Wieman, and E. A. Cornell, *Science*, **269**, 198(1995).
2. K. B. Davis, M.-O. Mewes, M. R. Andrews, N. J. van Druten, D. S. Durfee, D. M. Kurn, and W. Ketterle, *Phys. Rev. Lett.*, **75**, 3969(1995).
3. C. C. Bradley, C. A. Sackett, J. J. Tollett, and R. G. Hulet, *Phys. Rev. Lett.*, **75**, 1687 (1995).
4. B. DeMarco, D. S. Jin, *Science, 285,* 1703 (1999).
5. J.~T. Bahns, P.~L. Gould, and W.C. Stwalley, *J. Chem. Phys.,* **104**, 9689 (1996).
6. A.~Fioretti, D.~Comparat, A.~Crubellier, O.~Dulieu, F.~Masnou-Seeuws, and P.~Pillet, *Phys. Rev. Lett.,* **80**, 4402 (1998).
7. H.~R. Thorsheim, J.~Weiner, and P.~S. Julienne, *Phys. Rev. Lett.,* **58**, 2420 (1987).
8. T.~Takekoshi, B.~M. Patterson, and R.~J. Knize, *Phys. Rev. A,* **59**, R5 (1999).
9. A.~N. Nikolov, E.~E. Eyler, X.~T. Wang, J.~Li, H.~Wang, W.~C. Stwalley, and P.~L. Gould, *Phys. Rev. Lett.,* **82**, 703 (1999).
10. A.~N. Nikolov, J.~R. Enscher, E.~E. Eyler, H.~Wang, W.~C. Stwalley, and P.~L. Gould, *Phys. Rev. Lett.,* **84**, 246 (2000).
11. C.~Gabbanini, A.~Fioretti, A.~Lucchesini, S.~Gozzini, and M.~Mazzoni, *Phys. Rev. Lett.,* **84**, 2814 (1996).
12. R.~deCarvalho, J.~M. Doyle, B.~Friedrich, T.~Guillet, J.~Kim, D.~Patterson, and J.~D. Weinstein, *Eur. Phys. J. D, 7,* 289 (1999).

13.H.~L. Bethlem, G.~Berden, and G.~Meijer, *Phys. Rev. Lett.,* **83**, 1558 (1999).

14.J.~A. Maddi, T.~P. Dineen, and H.~Gould, *Phys. Rev. A,* **60**, 3882 (1999).

15.M.~Gupta and D.~Herschbach, *J. Phys. Chem. A,* **103**, 10670 (1999).

16.V.~Vuletic and S.~Chu, *Phys. Rev. Lett.,* **84**, 17 (2000).

17.P.~D. Lett, P.~S. Julienne, and W.~D. Philips, *Annu. Rev. Phys. Chem.,* **46**, 423 (1995).

18.J.Weiner, V.~S. Bagnato, S.~C. Zilio, and P.~S. Julienne, *Rev. Mod. Phys.,* **71**, 1 (1999).

19.W.C. Stwalley and H.~Wang, *J. Molec. Spect.,* **195**, 194 (1999).

20.F.~Masnou-Seeuws and P.~Pillet, *Adv. At. Mol. Opt. Phys.,* **47**, 53 (2001).

21.W.~C. Stwalley, Y.~H. Uang, and G.~Pichler, *Phys. Rev. Lett.,* **41**, 1164 (1978).

22.W.~I. McAlexander, E.~R.~I. Abraham, and R.~G. Hulet, *Phys. Rev. A,* **54**, R5 (1996).

23.K.~M. Jones, P.~S. Julienne, P.~D. Lett, W.~D. Phillips, E.~Tiesinga, and C.~J. Williams, *Europhys. Lett.,* **35**, 85 (1996).

24.H.~Wang, J.~Li, X.T. Wang, C.J. Williams, P.~L. Gould, and W.C. Stwalley, *Phys. Rev. A,* **55**, R1569 (1997).

25.R.~Gutterres, C.~Amiot, A.~Fioretti, C.~Gabbanini, M.~Mazzoni, and O.~Dulieu, *Phys. Rev. A,* **66**, 024502 (2002).

26.A.~Fioretti, D.~Comparat, C.~Drag, C.~Amiot, O.~Dulieu, F.~Masnou-Seeuws, and P.~Pillet, *Eur. Phys. J. D,* **5**, 389 (1999).

27.D.~Comparat, C.~Drag, B.~Laburthe Tolra, A.~Fioretti, P.~Pillet, A.~Crubellier, O.~Dulieu, and F.~Masnou-Seeuws, *Eur. Phys. J. D,* **11**, 59 (2000).

28.V.~Kokoouline, O.~Dulieu, R.~Kosloff, and F.~Masnou-Seeuws, *J. Chem. Phys.,* **110**, 9865 (1999).

29.V.~Kokoouline, O.~Dulieu, and F.~Masnou-Seeuws, *Phys. Rev. A,* **62**, 022504 (2000).

30.R. Kosloff, *J. Phys. Chem.,* **92**, 2087 (1988).

31.C.~C. Martson and G.~G. Balint-Kurti, *J. Chem. Phys.,* **91**, 3571 (1989).

32.O.~Dulieu and P.~S. Julienne, *J. Chem. Phys.,* **103**, 60 (1995).

33.V.~Kokoouline, O.~Dulieu, R. Kosloff and F.~Masnou-Seeuws, *Phys. Rev. A,* **62**, 032716 (2000).

34.C. M. Dion, C.~Drag, O.~Dulieu, B.~Laburthe Tolra, F.~Masnou-Seeuws, and P.~Pillet, *Phys. Rev. Lett..,* **86**, 2253 (2001).

35.T. Takekoshi, B. M. Patterson and R. J. Knize,*Phys. Rev. Lett.,* **81**, 5105 (1998).

36.Nicolas Vanhaecke, Wilson de Souza Melo, Bruno Laburthe Tolra, Daniel Comparat, and Pierre Pillet,*Phys. Rev. Lett.,* **89**, 063001 (2002).

37.B. Laburthe Tolra, C. Drag and P. Pillet,*Phys. Rev. A,* **64**, 061401(R) (2001).

38.C. Amiot, O. Dulieu, R. Gutterres, and F. Masnou-Seeuws, *Phys. Rev. A,* in press (2002)

39.J. J. Hudson, B. E. Sauer, M. R. Tarbutt, and E. A. Hinds, *Phys. Rev. Lett.,* **89**, 023003 (2002).

APPENDICES

Minutes of the ICSLS International Committee Meeting

The ICSLS business meeting took place on June 3, 2002 and reconvened on June 5. Items that came up for discussion were:

1. participation in the conference
2. finances
3. tradition of the chair
4. choice of site for 2006
5. change in Committee membership

Most of the discussion centered around the lower number of participation at the meeting. The registration fee was comparable to other years. Low cost dormitory housing was also offered, but apparently it did not sufficiently offset the travel costs for those from Europe.

The chair this year will pass on electronic copies of the following:
- mailing list (more inclusive than the participant list)
- participant list for the current year
- budget
- the meeting announcement mailer
- program
and
- a reserve of money

The chair hopes that these will updated and continue to be passed along without interruption to the future chairs. This would serve as a guide to the next chairperson and ensure that no more time is wasted creating and recreating the information. More importantly, it will lead to more time that could be spent advertising the conference and broadening participation.

A reserve of money was successfully established this year. It is slightly larger than expected, partially due the addition of money allocated to financially-supported people who were unable to come this year due to visa problems. A similar sum of money should be passed on to future chairs to help secure the conference center, which often requires a deposit before registration money is collected. It also provides a cushion for emergencies.

At the last meeting it was adopted, but not recorded that the chair should receive full registration and lodging at the next ICSLS conference in recognition of service to organize the conference. At this meeting the Committee agreed to continue to waive the registration fee as a "thank you". However, covering lodging fees is no longer automatic and is dependent on whether the chair successfully passed a reasonable amount of money to seed the next conference.

The next ICSLS conference will be held in Paris, France, hosted by Elisabeth Leboucher-Dalimier in June of 2004. Discussions for the 18th ICSLS conference were split between the generous offers by Eugene Oks and John Lewis. After reflection and some investigation of travel costs to each of the two host institutions, Auburn was selected as the next site for 2006. The Committee thanks John Lewis and hopes he will consider the next round.

The final item covered during the business meeting was the change in the membership of the international organizing committee. Roger Herman has stepped down and John Lewis was welcomed as his replacement.

Adams, Mark	Lawrence Livermore National Lab P.O. Box 808, L-18 Livermore, CA 94550	adams@mit.edu
Alexiou, Sprios	20 Chiou St., Kato Pefki 15121 Athens, Greece	moka@hol.gr
Andiel, Ulrich	Max Planck Institut für Quantenoptik Hans-Kopfermann Str. 1 Garching 85748 Germany	ura@mpq.mpg.de
Back, Christina	Lawrence Livermore National Lab P.O. Box 808, L-021 Livermore, CA 94550	tinaback@llnl.gov
Behar, Ehud	Columbia University 550 W. 120th St. MC 5427 Columbia University New York, NY 10027	behar@astro.columbia.edu
Benredjem, Djamel	LIXAM, Université Paris-Sud Centre d'Orsay Orsay 91405 France	djamel.benredjem@lixam.u-psud.fr

Boulet, Christian	Université Paris-Sud Laboratoire P.P.M. Bat 350 Campus d'Orsay Orsay 91405 France	christian.boulet@ppm.u-psud.fr
Brandt, William	Pennsylvania State University Pennsylvania State Astronomy 525 Davey Lab University Park, PA 16802	niel@astro.psu.edu
Brodbeck, Claude	CNAS-LPM Bâtiment 350, Campus d'Orsay Orsay Cedex 91405 France	claude.brodbeck@ppm.u-psud.fr
Calisti, Annette	CNRS Université de Provence Centre St. Jerome, Case 232 Marseille Cedex 70, 13397 France	annette.calisti@piimdgp.univ-mrs.fr
Chrysos, Michel	Laboratoire POMA UMR CNRS 6136 Université d'Angers 2 Boulevard Lavoisier 49045 Angers, Cedex France	michel.chrysos@univ-angers.fr
Chung, Hyunkyung	Lawrence Livermore National Lab P.O. Box 808, L-411 Livermore, CA 94550	hchung@llnl.gov

Ciurylo, Roman	Institute of Physics Nicholas Copernicus University Grudziadzka 5/7 Torun 87-100 Poland	rciurylo@phys.uni.torun.pl
Cormier, John	National Institute Of Standards & Technology Chemical Science & Technology Lab Gaithersburg, MD 20899	jcormier@mail.nist.gov
Dufour, Emmanuelle	Lab de Physique des Interactions Toriques et Moleculaires, Université de Provence, Centre de St. Jérome Case232 Marseille 93397 cedex 20 France	edufour@piima1.univ-mrs.fr
Dulieu, Olivier	Laboratoire Almé Cotton, CNRS Bat. 505, Université Paris-Sud Orsay 91405 France	olivier.dulieu@lac.u-psud.fr
Fournier, Kevin	Lawrence Livermore National Lab P.O. Box 808, L-41 Livermore, CA 94550	fournier2@llnl.gov
Gavrilenko, Valeri	General Physics Institute Russian Academy of Sciences Plasma Physics Department Vavilov Street, 38 Moscow 119991, GSP-1, Russia	gavrilen@fpl.gpi.ru
Ghatass, Zekry	Institute of Graduate Studies & Res. Alexandria University Alexandria, Egypt	z_ghatass@yahoo.com

Godbert-Mouret, Laurence	Université de Provence PIIM, case 232, Centre de St. Jérome Marseille 13397, France	laurence.mouret@piimdgp.univ-mrs.fr
Graf, Alexander	Lawrence Livermore National Laboratory P.O. Box 808, L-260 Livermore, CA 94550	graf2@llnl.gov
Gregori, Gianluca	Lawrence Livermore National Laboratory P.O. Box 808, L-399 Livermore, CA 94551	gregori1@llnl.gov
Gunderson, Mark	Los Alamos National Lab P. O. Box 1663 Los Alamos, NM 87545	magx@lanl.gov
Gustafsson, Magnus	University of Texas at Austin Dept. of Physics Austin, TX 78712	magnus@physics.utexas.edu
Hammer, Dominik	Dept. of Physics University of Texas at Austin Austin, TX 78712	dhammer@mail.utexas.edu
Helbig, Volkmar	Institut für Experimentalle und Augewandte Physik Universitat Kiel, Leibuiz str 19 D24098 Kiel Germany	helbig@physik.uni-kiel.de
Herman, Roger M.	Pennsylvania State University Dept. of Physics, 104 Davey Lab University Park, PA 16802	rmh@phys.psu.edu

Hey, John D.	School of Pure & Applied Physics University of Natal Durban, KZN 4041 South Africa	hey@nu.ac.za
Hoover, Todd	Lawrence Livermore National Laboratory P.O. Box 808, L-015 Livermore, CA 94550	hoover4@llnl.gov
Iglesias, Carlos	Lawrence Livermore National Lab P.O. Box 808, L-041 Livermore, CA 94550	iglesias1@llnl.gov
Isler, Ralph	Oak Ridge National Laboratory Fusion Energy Division, Box 2009 Oak Ridge, TN 37830-8072	isler@fed.ornl.gov
Kallman, Timothy	Lab of High Energy Astrophysics NASA/Goddard Space Flight Center Code 662, NASA/GSFC Greenbelt, MD 20771	tim@xstar.gsfc.nasa.gov
Kawachi, Tetsuya	Advanced Photon Research Center Japan Atomic Energy Res. Institute 8-1 Umemidai, Kizu Kyoto 619-0215, Japan	kawachi@aprms.apr.jaeri.go.jp
Kielkopf, John	University of Louisville Physics Department Louisville, KY 40292	kielkopf@louisville.edu
Kreye, Warren	Wright State University 140 Library Annex Dayton, OH 45435	wkreye@desire.wright.edu

Lewis, John C.	Dept. of Physics & Phys. Ocean Memorial University of NFLD. St. John's NF A1B3X7, Canada	court@physics.mun.ca
Leboucher-Dalimier, Elisabeth	Physique Atomique dans les Plasmas Denses LULI-Université Paris VI Case 128, Tour 12-22 4 Place Jussieu 75252 Paris cedex 05, France	lebda@ccr.jussieu.fr
Mancini, Roberto C.	Physics Department University of Nevada, Reno Reno, NV 89523	rcman@physics.unr.edu
Marandet, Yannick	Université de Provence PiiM, case 232 Centre de St. Jérome Marseille 13397, France	marandet@piimdgp.uni-mrs.fr
Martel, Pablo	Dept. of Physics, Universidad de Las Palmas de Gran Canaria Madrid, Spain	minguez@denim.upm.es
McCrorey, Diana L.	Physics Department University of Nevada, Reno Reno, NV 89523	mccrorey@physics.unr.edu
Miller, Michael C.	Lawrence Livermore National Laboratory P.O. Box 808, L-183 Livermore, CA 94550	miller124@llnl.gov
Minguez, Emilio	Instituto de Fusion-Denim Universidad Politecnica de Madrid Madrid, Spain	minguez@denim.upm.es

Nash, Jeff	Lawrence Livermore National Lab P.O. Box 808, L-015 Livermore, CA 94551	jknash@llnl.gov
Oks, Eugene	Auburn University Physics Department 206 Allison Lab Auburn, AL 36849	goks@physics.auburn.edu
Pichler, Goran	Institute of Physics Bijenicka Cesta 46 Zagreb, Croatia	pichler@ifs.hr
Rachet, Florent	Laboratoire POMA UMR CNRS 6136 Université d'Angers 2 Boulevard Lavoisier 49045 Angers, Cedex France	florent.rachet@univ-angers.fr
Rebentrost, Frank	Max-Planck-Institut für Quantenoptik 85748 Garching Germany	far@mpq.mpg.de
Roston, Gamal	Faculty of Science Alexandria University Alexandria, Egypt	gamal_daniel@yahoo.com
Rozsnyai, Balazs	Lawrence Livermore National Lab P.O. Box 808, L-023 Livermore, CA 94550	rozsnyai1@llnl.gov
Sauvan, Patrick	Instituto de Fusion-Denim Universidad Politecnica de Madrid Madrid, Spain	patrick@denim.upm.es

Schott, Romain	Physique Atomique dans les Plasmas Denses LULI-Universite Paris VI Tour 12-22 5ème 4, place Jussieu 75252 PARIS Cedex 05 France	schott@ccr.jussieu.fr
Scott, Howard	Lawrence Livermore National Laboratory P.O. Box 808, L-018 Livermore, CA 94550	scott6@llnl.gov
Stamm, Roland	Université de Provence PIIM, case 232, Centre de St. Jérome Marseille 13397, France	roland.stamm@piimdgp.univ-mrs.fr
Stehlé, Chantal	Laboratoire de l,Univers et de ses Théories FRE 2462 du CNRS Observatoire de Paris 5 Place J. Janssen, 92195 Meudon, France	chantal.stehle@obspm.fr
Steiger, Andreas	Physikalisch-Technische Bundesanstalt Abbestr, 2-12, D-10587 Berlin, Germany	andreas.steiger@ptb.de
Szudy, Jozef	Instutute of Physics, Nicholas Copernicus University, ul. Grudziadzka 5, 87-100 Torun, Poland	szudy@phys.uni.torun.pl

Tabisz, George	Department of Physics and Astronomy University of Manitoba 301 Allen Building Winnipeg, Manitoba Canada R3T 2N2	tabisz@cc.umanitoba.ca
Talin, Bernard	CNRS France PIIM, Université de Provence Carte de St. Jérome, case 232 Marseille, 13397, France	btalin@piima1.univ-mrs.fr
Tipping, Richard	University of Alabama Box 870324 Tuscaloosa, AL 35487	rtipping@bama.ua.edu
Vitcu, Adrian	University of Toronto 60 St. George St. Toronto, Ontario M5S1A7 Canada	vitcu@atmosp.physics.utoronto.ca
Vitel-Lepinay, Yves	Laboratoire des Plasmas Denses Université Pierre et Marie Curie T12-E5 casier 90 4 Place Jussieu 75252 Paris cedex 05 France	yves.vitel@wanadoo.fr
Voslamber, Dietrich	Physiklaisch-Technische Bundesanstaef Abbestrasse 2-12 D10587 Berlin Germany	dietrich.voslamber@t-online.de
Wan, Alan	Lawrence Livermore National Lab P.O. Box 808, L-015 Livermore, Ca 94550	wan1@llnl.gov

Wehr, Rick	Univ. of Toronto, Dept. of Physics 60 St. George Street Toronto, Ontario, Canada	rick@atmosp.physics.utoronto.ca
Welser, Leslie	Physics Department University of Nevada, Reno Reno, NV 89523	lwelser@physics.unr.edu
Wiese, Wolfgang	National Institute of Standards & Technology 100 Bureau Drive, Mailstop 8420 Gaithersburg, MD 20899-8420	wiese@nist.gov
Yurovsky, Vladimir	Tel Aviv University School of Chemistry Tel Aviv, 69978, Israel	volodia@post.tau.ac.il

AUTHOR INDEX

A

Adams, M. L., 40
Alexiou, S., 302
Allard, N. F., 310
Altug, Z., 79
Angelo, P., 332, 340, 352

B

Babb, J. F., 211
Baily, D., 181
Bastea, M., 359
Beiersdorfer, P., 74
Benredjem, D., 445
Ben-Reuven, A., 479
Bielski, A., 151, 185, 188
Borysow, A., 181
Bouanich, J.-P., 181
Brandt, W. N., 119
Brodbeck, C., 181
Brooks, N. H., 3
Bureyeva, L. A., 378

C

Calisti, A., 247, 252, 259, 268, 352
Capes, H., 15, 67
Celliers, P., 359
Cherkasova, E. K., 413
Chrysos, M., 174
Chu, C. C., 26
Chung, H.-K., 211
Ciuryło, R., 151, 161, 166
Collins, G. W., 359
Cormier, J. G., 401

D

Daido, H., 457
Dalhed, H. E., 294
de la Rosa, M. I., 86
Delettrez, J. A., 294
Demura, A., 79, 318

Drummond, J. R., 151, 161, 166, 401
Dufour, E., 252, 268, 332, 340
Dufty, J. W., 252, 268
Dulieu, O., 499

E

Eggert, J., 359
Escarguel, A., 15
Esch, D., 79

F

Faenov, A. Y., 259
Ferri, S., 247, 259, 445
Flih, S. A., 99
Förster, E., 457
Frank, A. G., 421
Freese, W., 94
Frommhold, L., 216, 237
Fuhr, J. R., 99

G

Gavrilenko, V. P., 94, 413, 421, 467
Geissel, M., 259
Ghatass, Z. F., 200, 426, 434
Gigosos, M. A., 268
Gil, J. M., 340, 352
Glenzer, S. H., 359
Godbert-Mouret, L., 15, 60, 67
Golovkin, I. E., 294
González, M. A., 268
Graf, A., 74
Gregori, G., 359
Grosser, F., 206
Grützmacher, K., 86
Guirlet, R., 15, 60
Gunderson, M. A., 276
Gustafsson, M., 216

523

H

I

K

L

M

N

O

P

R